D1485390

An Approach to Chemistry

An Approach to Chemistry

F. D. de Körösy

Former Associate Professor of Applied Physical Chemistry
at the University of Budapest
and now Director of the Chemical Laboratory,
Negev Institute for Arid Zone Research,
Beer Sheva, Israel

London: Sir Isaac Pitman & Sons Ltd

First published 1969

SIR ISAAC PITMAN AND SONS LTD.
Pitman House, Parker Street, Kingsway, London, W.C.2
Pitman House, Bouverie Street, Carlton, Victoria 3053, Australia
P.O. Box 7721, Johannesburg, Transvaal, S. Africa
P.O. Box 6038, Portal Street, Nairobi, Kenya

PITMAN PUBLISHING CORPORATION
20 East 46th Street, New York, N.Y. 10017

SIR ISAAC PITMAN (CANADA) LTD.
Pitman House, 381-383 Church Street, Toronto

SBN: 273 42253 7

MADE AND PRINTED AT THE PITMAN PRESS, BATH
(T. 1135)

Dedication

To the group of Teenage Zionists in Budapest who in the fearful circumstances of 1942 found time to force me to build the foundations of chemistry for them, without conventional circumventing of difficulties or compromise.

NESHER ADLER
YOCHEVED BARMAT
ROBERT BEINHAUER
SHMUEL BERGER
MOSHE EHRENSTEIN
YESHAYAHU FETTMAN
CHANNA FUCHS
NAOMI GLUCK
URI HERMANN
YEHUDIT SHLOMI-KEMENY
MOSHE KLEIN
AVRI LISSAUER
YITZCHAK MENDELSSOHN
EZRA REICHMAN
MIRYAM RENYI
DIZZI WEINSTOCK
ANTI WEISS

Preface

This book is meant for the first-year degree-level science student and particularly for those studying chemistry. It is not a textbook and does not replace one, but should be helpful in understanding the way of thinking of chemists and the logical approach to chemistry. It is also meant for intelligent people who are not chemists but would like to know what chemistry is about and for the bright high-school student who wants to get closer to the way of thinking in chemistry. It shows how the concepts of chemistry, which today pervade everyday life but are seldom understood (atoms, electrons, compounds, structural formulae, nuclei and nuclear reactions), have arisen from observation and experiment by logical thinking.

Great care has been taken to build up the book step by step, beginning from primitive observations and taking nothing for granted. Whenever the discussion leads beyond the grasp of the author, as in quantum mechanics, this is clearly stated and the reader is advised to consult more competent sources. In these few areas and throughout the book, intellectual honesty has been one of the foremost aims. Very little formal mathematics has been used and the few derivations are mostly simple arithmetic. The Maxwell–Boltzmann formula, however, has been introduced without its derivation.

Properties of elements and of compounds are described only when they help to elucidate a general argument. The book is the skeleton rather than the flesh and blood of chemistry. Naturally it will be read more easily by those already familiar with the factual data of chemistry, but those who are not will find sufficient examples to get along with. The main emphasis is on the existence, size, movement and bonding of atoms, with the last part dealing with nuclei. Sections on physical chemistry are included or excluded from the discussion according to their degree of importance from this point of view.

In general the historical sequence of evolution of chemical thought has been followed. However, blind alleys such as the phlogiston hypothesis have been excluded and in some cases more recent approaches have been given priority over the historical ones, as for instance when discussing the actual size of atoms.

I am extremely indebted to Dr. A. G. Briggs for having gone through the whole manuscript and having improved and corrected my English. Being an excellent chemist himself, it must have been difficult for him to remain loyal to my original conception of the subject. Several of his valuable suggestions have been incorporated into the text and his critical comments have always been very helpful. I shall always gratefully remember his co-operation.

Half of the original drawings have been executed by Mr. J. Vágó. Personal reasons prevented him from finishing the work, which was then taken over by Mr. I. D. Korngold. I wish to express my thanks to both of them—their job was certainly not an easy one.

THE NEGEV INSTITUTE FOR
ARID ZONE RESEARCH,
BEER SHEVA, ISRAEL   FRANCIS D. DE KÖRÖSY
November, 1967

Contents

PART VII. ATOMIC NUCLEI

Part 1

The Existence of Atoms

1. ABRUPT AND GRADUAL CHANGES IN NATURE

A candle burns, its wax disappears in the luminous flame, while black soot and drops of water appear on a cold surface held in the flame. The air above the flame has changed too. It has become heavier and will extinguish another candle which is set alight below the former one within a common, small enclosure. A part of this air (carbon dioxide) will dissolve to a moderate extent in water and even more readily in caustic alkali, whereas normal air will not do so. There is no gradual change from wax to the final products: any small piece of wax is either unchanged or completely transformed.

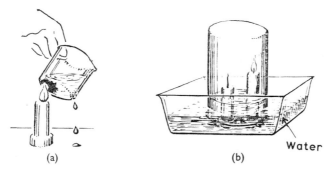

(a)   (b)

Fig. 1. **Burning candles**

(a) Soot and water appear on cooling the flame of the candle.
(b) A heavy gas is formed when the candles burn and puts out the lower candle first. The gas dissolves in water; the water rises within.

A piece of iron lies about in a damp place for some time. Its surface gradually becomes covered with rust, and after a while it may even change completely into rust. If it were enclosed in a small compartment we could easily observe that the pressure of air above it would decrease possibly down to 80 per cent of its original value if there were enough iron present, but not below 80 per cent. If the pressure becomes as low as this, the remaining gas will be somewhat lighter than air and nothing will burn in it. Even the remaining iron will not continue to rust. There is no gradual change from iron to rust: any particle, originally of iron, is either unchanged iron at any moment of observation or has become rust; it is never anything in between the two.

We immerse two electrodes in a glass of water and pass a direct current between them. Gas bubbles appear immediately at each electrode. We may collect the gases and readily ascertain that exactly twice as much has been

1

evolved at the negative pole (the so-called cathode) as at the anode (the positive pole). Twice as much by volume, but not by weight! It just happens that the one volume of anodic gas weighs eight times as much as the double volume of gas at the cathode. The latter (hydrogen) burns if it is led into air and ignited; it even explodes if it has been mixed with air before ignition.

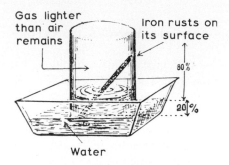

Fig. 2. **The rusting of iron**

Iron rusts in damp air until 20 per cent of the volume of air has disappeared; the process then ceases.

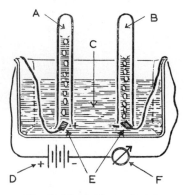

Fig. 3. **Electrolysis of water**

The volume of hydrogen evolved at the negative electrode is twice that of oxygen at the positive electrode.

A	= oxygen	D	= battery
B	= hydrogen	E	= electrodes
C	= water with traces of sul- phuric acid	F	= ammeter

The former gas (oxygen), on the other hand, makes everything burn much more vigorously in it than in air, and if it is used to burn the gas from the cathode, both gases disappear and water is reformed. Exactly the same amount of water is produced as was consumed during electrolysis. There is no gradual change from water to the gases nor from the gases to water on their reunion. Any drop of water is absolutely water as long as it is not gas and conversely; it is never anything in between.

We could go on to discuss thousands of similar processes. A metal is dissolved in acid: it is either metal or has become dissolved salt, nothing in between. The oxide of mercury is heated until the same gas is evolved as we have already encountered around the anode on electrolysis: oxygen. But at the same time drops of metallic mercury are formed, quite distinct from the original oxide. The characteristic feature of all these transformations is their discontinuity, their abrupt nature. Matter exists in certain forms and if it changes chemically it does so into different, quite distinct forms without any gradual transition.

Changes in nature are by no means confined to abrupt (chemical) transitions of this sort. On the contrary, we are familiar with a lot of changes which are gradual in their nature. Think of thermal expansion: in the process of becoming warmer a body expands little by little. By making the rise of temperature very small any measure of expansion, even the smallest, can be arrived at. Think of the velocity of a moving body: it can be changed by any small amount provided the force acting upon it is small enough. Consider an electric current which increases or decreases in a continuous manner with increasing or decreasing voltage. All changes which we assign to the realm of classical physics are continuous. One or more properties of a body or of a system are changed by a small increment if one or more of its circumstances are changed by small increments. In the language of mathematics these changes may be described by differential (another word for "incremental") equations.

The abrupt changes mentioned at the beginning of this chapter belong to another realm of nature—to that of chemistry. We may very well define the difference between chemical happenings (chemical reactions) and physical changes as one between abrupt and gradual transitions. Only very late in the development of physics have transitions been observed, and have even achieved fundamental importance, which are also discontinuous in their nature. The light emitted by isolated atoms or molecules, for instance, may correspond to sets of well-defined, discontinuous wavelengths. The position of elementary particles relative to a magnetic field cannot vary in a continuous manner but is confined to a small set of permitted directions. The specific heat of matter at extremely low temperatures shows characteristic deviations from its behaviour at higher temperatures and this can be accounted for by the assumption that the movement of small particles at the lower temperatures cannot change in a continuous manner. The absorption and emission of light as an elementary process is also governed by laws of discontinuity.

Thus, the deeper we probe into the laws of physics, the more often we arrive at fundamental discontinuities. The impression that all physical changes are gradual was partly caused in classical physics by the fact that we always observed a great multitude of elementary particles simultaneously. Indeed for a long time we did not even know with certainty that elementary particles such as atoms really existed. The more we know about them, the more we see that the laws of classical physics are only statistical averages over a multitude of discontinuously changing states of elementary particles and are only applicable to these as limiting laws for very large momenta—the momenta of high velocities or of macroscopic bodies.

The division of Nature into chemistry, physics and so on is of course man-made, a purely human invention of convenience. Therefore we must never wonder that the borderlines of these divisions are unclear and faint and only enforced by a certain amount of dogmatism. Although the division we have proposed above seems to be quite adequate on the whole, in some cases it runs contrary to generally accepted definitions. We are now going to discuss two such cases.

When a solid body melts or conversely when a liquid solidifies such particle of substance is either a liquid or a solid. Similarly during the evaporation of a liquid or a solid or the condensation of a vapour a particle is either a liquid or a solid or in the gaseous state. The transitions are abrupt. Thus according to our proposed definition these so-called "changes of state" would come under the heading of chemical processes, whereas they are generally assigned to the realm of physics. As a matter of fact it could have as easily happened that they were assigned to chemistry. Different compounds bear different names. The three states of the very common compound water, have three different words to denote them in ordinary language also: ice, water and steam. The primitive mind dealt with these three as if they had been three separate chemical entities and gave them names of their own. Nothing need have prevented us from denoting the three different states of all substances (or, whenever a compound has more than one solid modification, each one of these too) by different names. Instead of saying for instance sulphur vapour, liquid sulphur and sulphur (or more specifically rhombic or monoclinic sulphur because it crystallizes in one of these modifications), we might as well invent three different names. As a matter of fact mineralogists have chosen the former solution and, for example, call the two solid modifications of calcium carbonate "calcite" and "aragonite" respectively. Nothing would change at all, but we should have to learn three times as many names as we have to do now. Also we should need to memorize which trios of names belong together in the sense that one body changes over into the other two by heating or cooling without giving rise to or absorbing any other body during this transformation (e.g. ice–water–steam). It is only a convention that we call these transformations physics instead of chemistry, a convention which saves us the trouble of having to learn sets of new names. No metaphysical or physical meaning should be attached to our having adopted this convention.

The second case where our convention of drawing the borderline between classical physics and chemistry in terms of abrupt and continuous changes leads us somewhat astray, is the process of dissolution. A piece of sugar or salt dissolves in water and at a given moment any fragment of the original solid either remains solid or has already dissolved. Are we dealing with a physical or with a chemical process? According to our original convention concerning abruptness we ought to decide that it is a chemical one. But if we fix our attention on the solution instead of on the dissolving solute the situation is reversed. The taste, the density, the refractive index and often the colour of the solvent change gradually as more and more of the solute is dissolved in it. From this point of view the process of dissolution as well as the mixing of solutions or gases with each other should come under the heading of physics. As far as energy is concerned, heats of solution or of mixing are often not less than those of chemical reactions. Think of the heat evolved when water and sulphuric acid are mixed! Solvent and solute thus often attract each other with "affinities" comparable to those we encounter in chemistry and we are not simply dealing with the inclination of all matter to pervade space uniformly, as occurs when gases diffuse into each other. It seems to be wisest not to stress definitions beyond their limits. Let us drop dogmatism and conclude that dissolution stands midway between chemistry and physics, having some aspects of the former and some of the latter.

2. MIXTURES AND COMPOUNDS

Suppose we mix sand with salt, sugar with coal-dust and oil with ultramarine. We know then, that we have made the mixture ourselves and we have only to look at it through a magnifying lense or a microscope in order to be convinced that the originally distinct materials still remain distinct. The only

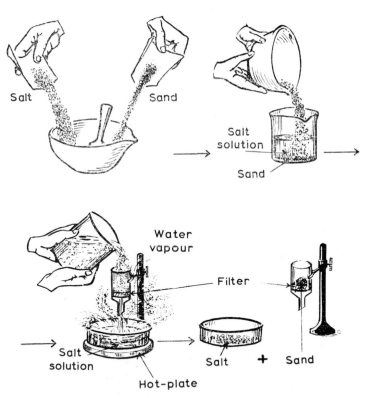

Fig. 4. **Separation of sand–salt mixture by dissolution of the salt, filtration and evaporation of the solution to dryness**

change is that their minute particles have come very close together and are irregularly intermixed one beside the other. We can pour some water on to the mixture of sugar and coal dust, whereupon the sugar dissolves and may be separated from the coal-dust by filtration. It is only necessary to evaporate the water from the sugar solution in order to regain the sugar once more. Salt and sand may be separated in the same manner (*see* Fig. 4), and by using benzene as a solvent, oil and ultramarine may also be separated.

However, it is a question of good luck whether we are able to separate a mixture into its components as easily as described above. What about the separation of salt from sugar or of sand from coal dust? It is no use trying to dissolve them in water this time. We may however try to invoke their different densities: that of sand is 2·65 whereas that of coal is 2·17. If we manage to find a liquid with a density in between those of sand and coal, then coal will float and sand will sink to the bottom if we pour the mixed powder into the liquid.   Methyl iodide has a density of 2·28 g/ml and if we

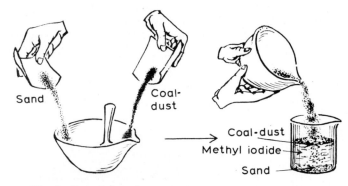

Fig. 5. **Separation of sand from coal-dust by preferential flotation in methyl iodide**

use it as separating liquid we shall achieve our aim in this case. Of course it is not easy to find appropriate liquids to separate mixtures of compounds, all of high density, but salt and sugar may be easily separated, e.g. in carbon tetrachloride. The method is often used, after pulverization, for separating ores from the rocks in which they were enclosed.

Separation is often simpler if we use the process of crystallization. An aqueous solution which contains about equal amounts of salt and sugar yields salt at first on gradual evaporation of the water because the solubility of sugar in water exceeds that of salt. When we have concentrated the solution to such an extent that sugar begins to crystallize out too, both compounds crystallize *separately, one beside the other* in well-defined crystals. With some luck we shall be able to separate them under a magnifying glass with a pair of tweezers, because they look different.

We turn to another method if we intend to separate two liquids, e.g. water and alcohol. On cooling a dilute solution of alcohol in water below 0°C ice crystals appear and may be filtered off; or we can boil the mixture and condense the vapours which are richer in alcohol than is the original solution. If we repeat the distillation a great many times, almost pure water remains behind while nearly pure alcohol condenses as the ultimate distillate. The whole repetitive process is carried out in a so-called "fractionating column"

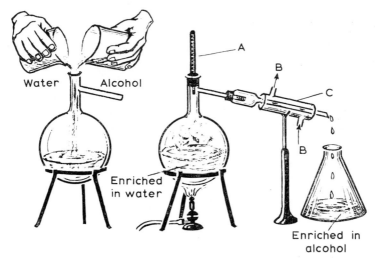

Water   Alcohol

Enriched
in water

Enriched in
alcohol

Fig. 6. **Partial separation of water from alcohol by distillation**

A = thermometer
B = cooling water
C = Liebig condenser

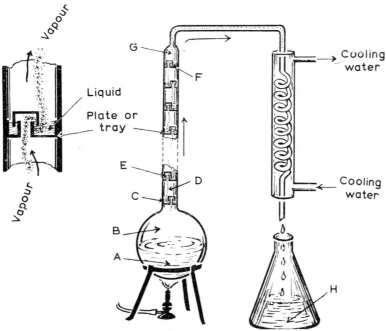

Vapour

Liquid

Plate or
tray

Vapour

Cooling
water

Cooling
water

Fig. 7. **More efficient separation of alcohol from water by
multiple-stage distillation in a fractionating column**

A = water + alcohol
B = original mixed vapour
C = first condensate, enriched in alcohol, boils at lower
    temperature
D = second vapour further enriched in alcohol
E = second condensate, still further enriched in alcohol
F = last condensate, lowest boiling point
G = last vapour, richest in alcohol
H = condensate of final vapour, maximally enriched in
    alcohol (96 per cent): lowest-boiling mixture

where the flask at the bottom is heated. Mixed vapour from this flask condenses and re-evaporates repeatedly in a set of "plates" or "trays" within the column. Condensed liquid in each tray is heated by the vapour rising from the tray below and alcohol is driven towards the top. The boiling-point of the liquid in the trays decreases step by step as we ascend the column and eventually the lowest possible boiling-point of the mixed liquid is reached. In the case of alcohol–water, this lowest-boiling mixture contains 96 per cent of alcohol. Similar columns are used to separate liquids in industry.

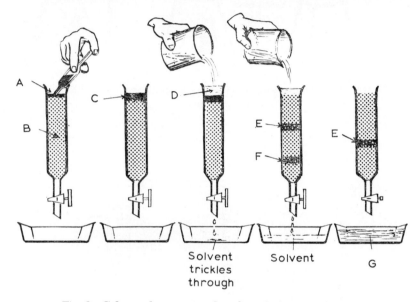

Solvent
trickles
through

Solvent

G

Fig. 8. **Column chromatography of a solution containing two solutes**

A = solution of mixture   E = more strongly adsorbed
B = adsorbent powder   component
C = adsorbed mixture   F = less absorbed component
D = eluting solvent   G = solution of less absorbed
                                     component

Another very ingenious method consists in passing solutions (or gases) through a column of powder or a strip of filter paper from one end to the other by allowing a solvent (or an indifferent gas) to wash the solutes gradually through the system. Adsorption of a solute on to the surface (or solubility of the solute in another immiscible liquid which wets the surface) varies from compound to compound. Those which are absorbed more strongly remain behind the others and if the path of separation is long enough even very similar compounds may be separated along it. If the compounds are coloured we can easily distinguish between them at different places in the column or on the paper and this process is therefore called chromatography. Colourless compounds have to be located by using colour reactions or other means of characterization. The method is specially suitable for micro-scale separation and has achieved immense importance, e.g. in the field of biochemistry.

However, it is never enough to use a single method for separating a material that is suspected of being a mixture. We could not get more alcohol than 96 per cent into our final distillate no matter how elaborate the column was, simply because liquid of this composition happens to have the lowest boiling-point at atmospheric pressure of any water–alcohol mixture. At another pressure the composition of the "head" (lowest-boiling) fraction would have been different: with a different boiling-point, refractive index, specific gravity, etc. Evidently a liquid which changes its properties when distilled at different pressures cannot be homogeneous. We find that the last traces of water can be removed from these mixtures by treating them with some material that absorbs water eagerly, e.g. calcined limestone (quick-lime) or anhydrous copper sulphate. After this treatment all properties of the alcohol remain unchanged, no matter how often and at what pressures we distil it. The first drop of its distillate has the same properties as the last. We cannot find any means, save chemical destruction, which would enable us to divide it into parts with different properties. This is what we mean when we say that this purified alcohol is a chemical entity, a pure compound.

A similar situation where we might not be able to separate a mixture by a given method might occur during the freezing of some specific solutions. Consider salt in water. Below 23 per cent salt in the solution ice begins to freeze out if we cool sufficiently. This ice can be separated and the solution becomes more and more concentrated. Above 23 per cent salt concentration salt separates on cooling the solution because its solubility decreases with temperature. This salt can be separated in its turn and the solution becomes more dilute. But *at* a concentration of 23 per cent salt we are at the boundary between the two cases. The temperature of freezing will by now have reached its minimum at $-22\cdot4°C$, and salt and ice will crystallize out simultaneously. The concentration of the solution does not change any further and the solution has the same composition as the frozen mixture. The freezing-point now remains constant because the composition of the solution is not changing. It will be seen that the whole behaviour of this particular solution suggests the freezing of a pure compound from its first drop to its last. But further inspection of the frozen solid, say under a magnifying glass, reveals that cubic crystals of salt are intermingled with hexagonal ice crystals. It was just bad luck that this time they happened to crystallize together at this specific composition of the solution. There is a special name for such solid mixtures, which melt at constant temperature without change of composition: the are called "eutectics". It becomes evident that we are dealing with such a mixture if, when it is warmed water evaporates from it leaving pure salt behind. Once more a single method of separation—freezing—was misleading and a second method has to be invoked.

We are therefore only entitled to call something a pure substance if it cannot be divided into unlike fractions from the first until the last drop during a distillation, no matter what the pressure may be, and if it yields the same material on freezing from the first drop to the last or on melting from the first part to the last. Nor should it be possible to divide it into substances of different properties by crystallization from any solvent whatsoever as shown by a comparision of the properties of the first crystal fraction with the last one; nor should it show more than one zone in any chromatographic process.

As we have seen we have had to speak about melting, boiling or dissolving our material in order to decide whether it was a mixture or a pure compound. But what about something which happens to be insoluble in all solvents, which on melting and chilling does not yield the original material again and which does not distil without decomposition? There do exist such substances: diamond, carborundum, Bakelite and cellulose are among them. Truly, all our definitions forsake us now and the chemist is often at a loss to find an unambiguous answer. In some cases, as we shall see later on, the internal structure of the crystals may be of help to us, but in the case of thermosetting plastics (for instance, like Bakelite) even this criterion is no use, because they are not crystalline. We have reached the limit of our definitions.

Let us now envisage the process of mixing or dissolving as compared with a chemical reaction. On the progressive admixture of coal-dust, white sugar will change its colour gradually through shades of grey towards black. On dissolving pure sugar in water the latter becomes gradually sweeter and sweeter; when the solute is potassium permanganate, the solution is at first light pink and changes gradually to deep violet. As the concentration of solute increases, the freezing-points of these solutions are gradually lowered and their boiling-points gradually increase. This gradual change of properties on bringing different substances into contact with each other is characteristic of the mixing process. In contradistinction, a burning piece of coal is either still coal or has become an oxide of carbon; a drop of water during electrolysis either remains water or becomes a mixture of two gases, oxygen and hydrogen. These are the chemical reactions which were dealt with in the preceding chapter.

Now we are beginning to understand what a chemist has to do if he is given an unknown material. First of all he applies all methods at his disposal to separate pure components from the material which, of course, might be a mixture for all he knows. The separated components of the mixture must be pure in the sense given above, that it they must retain all their properties through further attempts at separation. Only after he is convinced that the components are pure, does the chemist proceed to their chemical analysis.

3. FROM WHAT IS MATTER BUILT?

To analyse means to ascertain which components are present in a compound or mixture and in what proportions. But what are these components? We know they are called the atoms of the elements or the molecules of compounds, but it happens only too often that the exact meaning of these popular words remains empty and one is ashamed to admit that one does not even understand such a "simple" concept. This is not something of which to be ashamed. The concepts are not quite so simple as some books on chemistry profess, but it is not very difficult to explain them in their logical sequence and thus establish a solid basis for the monumental edifice of chemistry. The higher parts of this edifice are often easier to understand than the apparently trivial fundamentals—what is an element, what do we mean by speaking of atomic weights and by writing chemical formulae? We shall now discuss these problems and, whenever necessary, discuss them in context with relevant events in the history of chemistry.

It is one of the most imposing achievements of science that we have learned to divide matter into its building stones. We speak about molecules, atoms, nuclei, electrons and some more esoteric particles but many people do not know why we assert that this complicated world of our senses is really built from them. They *believe* the men of science who tell them so but the proofs of this assertion remain unknown to them. However, science is not merely something to believe in; it is the outcome of many an observation and much reasoning.

Knowledge of nature is conveyed to us by our senses. Scientific observation, however, is generally not fortuitous. We ourselves produce the circumstances of our observation—we experiment. All books on science describe observations and experiments as a basis for further reasoning about them. The reader must believe of the authors that their descriptions are true. However, this is not blind faith in the authority before us, because the proper scientific description of observations must be so exact that anybody who is willing to take the pains and repeat them himself must be able to do so, even if this may involve great expense and serious work on his part.

We can then proceed to reason about our observations in order to sum them up into a coherent and clear system. As the number of observations multiplies it may of course become necessary to expand our set of reasonings accordingly and this process of expansion sometimes involves very great mental effort indeed. Our generation lives under the strain of such an expansion but it would be a grievous fault to speak about a crisis in science when we are only forced to expand the realm of ideas to which we have become accustomed.

Many people look towards science as if it were able to answer the last questions of being and to fill human life as entirely as religion did fill it some time ago. Such people will necessarily be disappointed. Science never reaches the limits of time, space, and chains of causal connexions which are probably infinitely remote. The ultimate cause of everything is certainly beyond the reach of science as also it most probably exceeds the reach of human comprehension.

Science comes to grips with the world from an in-between position. Starting from present experience it tries to draw inferences relating to the misty past and to the distant future. It examines at first the bodies which are visible to the unaided eye and slowly extends its domain towards both the minute electron and the inconceivably great universe around us. Its results are difficult to arrange into a deductive system because they are all derived from induction. Yet general laws which emerge from the primordial chaos of observations are the uniting principles of our contemplation.

This painstaking way of observation, induction and reasoning has led in the last centuries to identification of the chemical elements, ninety natural and about a dozen or so artificially-prepared building stones of the material universe. But they are not the ultimate building stones; we know as much already. They themselves are composed of even more primitive particles: electrons, protons, neutrons and a small host of newly discovered, elusive particles which do not yet fit into a unified system. The desire of the human soul to unify the world drives us to seek out their interconnexions and as far as possible decrease the number of primary particles until perhaps one day they will give place to some single principle. Maybe, but maybe not. We are in the midst of becoming acquainted with the world and do not yet see an end.

4. LOOKING INTO THE PAST

Primitive man performed chemical reactions when he smelted metals from their ores and when he made wine from grapes. But it took a long time for man to look for something more than practical results by so doing and for ages he did not seek connexions between the manifold and seemingly independent experiences he encountered in Nature. The question as to what forces moved the world around him and what were the common building stones of the material universe arose only late in history.

Among the people whose history is known to us, the Greeks were the first to have asked these questions unmingled with religious ideas. It was in the nature of classical Greek thought that they tried to deduce everything from contemplation, by reasoning from a set of given axioms and for a long time they were conspiciously indifferent towards facts, towards reality. Thus they developed mathematics, a science of axioms and reasoning, to an incredible degree but even their greatest early philosophers were remarkably uninformed as far as natural science was concerned. *Plato* (about 427–347 B.C.), for instance, wrote that "we should not be concerned about the heavenly bodies if we desire to understand the real essence of astronomy." No wonder that they failed to get any nearer to the heavenly bodies and at the same time made no advance towards that "real essence of astronomy", whatever it may have been.

However, the Greek philosophers were the first to formulate a fundamental question of chemistry and physics in such an exact way that it has not changed since: *Can matter be divided infinitely or is it composed of small, finite particles?* The question evidently allows two answers and accordingly these became the basis of two Greek schools of philosophy. *Anaxagoras* (about 500 B.C.) taught that matter may be divided without limit and that its properties remain unchanged after any extent of subdivision. It fills space completely and therefore it is inconceivable to speak about free space in itself. *Demokritos* (about 470 B.C.) on the other hand, maintained that all matter is composed of minute particles and the properties of tangible matter are caused and determined by properties of these ultimate building stones.

Water was supposed to be fluid because its particles were smooth and rolled on each other, iron strong because its particles had a raw surface and stuck into each other. From the greek verb τέμνω (cut up) he derived the word ἄτομος (atomos), indivisible and named the ultimate particles with it. He thought of his atoms as being in perpetual movement, attracting and repelling each other, giving rise to dense or light matter according to their degree of proximity. It was impossible to create or to destroy atoms according to Demokritos. This was the first time that anyone had stated the law of conservation of matter. We cannot but bow our heads with admiration before this hypothesis born exclusively from the creative mind of man without consulting the facts of nature and still vindicated in many of its details by inductive science. It is one of the most glorious triumphs of human intuition. Demokritos is said to have followed in his ideas his teacher *Leukippos* whose works have been lost to us. Demokritos himself denied the existence of the gods and belonged to the moderate stoic philosophers of high moral principles. His ideas concerning atoms were taken over by *Epikuros* (342–270 B.C.) and later by *Lucretius*, the Roman (98–55 B.C.) who fortunately incorporated them into his grand opus, *De rerum natura* and thus preserved them for later ages.

Notwithstanding his enormous merits it would be wrong to overestimate the hypothesis of Demokritos from the point of view of natural science. It is not sufficient to choose one of two possible answers to a well-conceived question. The answer must be proven, it has to be shown that matter really does behave as if it were composed of atoms and if natural observations do not show this, experiments have to be devised and performed to prove it or disprove it. Such an idea seems to have been utterly beyond the grasp of Greek philosophy and therefore, with all our admiration for the great intuitive philosopher, we must exclude Demokritos from among the protagonists of natural science.

Aristoteles (384–322 B.C.) who impressed his system of thinking on the Middle Ages and partially on the beginning of Modern Times is very much to blame for the fact that human knowledge of the chemical elements remained so primitive for nearly two millennia. He proclaimed that the material world is built from four elements. Following *Empedokles* (about 440 B.C.) he taught that the cold and dry earth, the cold and wet water, the hot and wet air, and the hot and dry fire were the ultimate elements, with a fifth, the "essence" (ὑλή), joining them to yield all matter as we know it. It is difficult to reconstruct today what the ancients meant when they were speaking of elements. It seems probable that it must have been much nearer to what we now call states of matter than to the chemical elements. Anyway the postulate of the four first elements did not do much to enlarge our knowledge, whereas the fifth, the "essence" was the ultimate goal of alchemists throughout the centuries to come. It was the marvellous *quinta essentia*! Nobody had any idea what it was but the most wonderful effects were ascribed to it. It was to conquer death, bring happiness and make gold. Aristotle told that it existed and he must have known; he must have been right, so it was up to the alchemists to find it. This is what happens if authority takes the place of independent critical thinking and unbiased observation.

Throughout the Middle Ages we see a parallel evolution of quite practical chemistry producing useful recipes and processes mostly due to the Arabs and a mystical, sometimes more sometimes less honest, pseudo-science, searching for the *quinta essentia*, the philosophers' stone and all that was held to be connected with it. Alchemists thought that they would manufacture gold from less noble metals such as lead, retain youth and achieve great power by finding that which they sought. They conceived magic relations between the planets and some metals and denoted both a planet and its metal with the same Chaldean symbol. Chemistry and astrology became mixed up to a hopeless extent. Some new elements were gradually added to the four of Aristotle, namely, mercury as the most remarkable of metals, sulphur as a representative of the combustible, and salt as representative of the soluble forms of matter. If we realize that prehistoric man had managed to smelt beautiful, shining metals from very plain-looking ores, we should understand that the idea of producing gold from some lesser metal was not absurd in itself. Only the method of the alchemists was not what one would today consider to be the right one. They tried to force Nature into fulfilling their wishes. They tried without coordination, unsystematically now here, now there and hoped for good luck to help them. Such was their approach, instead of trying to enter with humble respect into the enormous kingdom of Nature and progressing stepwise slowly and reasonably into the unknown. The Arab alchemists conformed somewhat more to what we should call the inductive method of science. Much of their interesting material is ascribed to their most renowned member, *Geber* (900?–1200?), but the Arabic manuscripts in our hands are so full of forgery that we may not even be certain that a single person ever existed bearing this name. Nevertheless the discoveries must have been made by someone or possibly by more than one person.

The German *Paracelsus* (1490–1541) was already attacking the alchemists and saw that the main task of chemistry lay no longer in the effort to make gold but to make medicine for the people. Many of his recipes seem fantastic today but he must have been a very keen observer because recent research has frequently corroborated his findings. He was not much concerned about the building stones of matter. The school of chemistry which developed around and after him was named iatrochemistry ($\iota\alpha\tau\rho\acute{o}\varsigma$ = physician).

It is certainly the merit of the God-seeking English physicist *Robert Boyle* (1627–1691) that he grasped with determination at the venerable but empty concepts of two millennia. He upset this house of cards and laid the foundations of that other, solid building which has grown to become our modern chemistry. He really only inverted the classical question. Instead of asking from what elements can we build up matter, he asked into which components are we able to decompose it. The right question invoked the right solution. In his fundamental work, *The Sceptical Chymist* (1661) he even described some of his experiments which served to refute absurd statements of alchemists. Among others he showed that the statement that living organisms contain a lot of mercury was utterly false. After having demonstrated the absurdity of a series of superstitions he set out new, rational problems to be solved and thus opened the road towards systematic analysis of matter.

5. CHEMICAL ANALYSIS

Half a century after Boyle, a few years after the great Russian scientist, *M. J. Lomonosoff* (1711–1765), had reaffirmed Demokritos' statement as to the conservation of matter, the scales entered the laboratories of experimental chemistry, never to leave them again. *A. L. Lavoisier* (1743–1794), who was sent to the guillotine during the French Revolution with the remark that "the Republic does not need scientists," was the first to measure the weights of compounds participating in chemical reactions. Boyle had asked: into what can materials be decomposed? And the scales made it possible for chemists to answer his question.

Whenever a substance gave rise to light products during a reaction it became certain that something must have "left" and that the process was a decomposition. The products were taken one by one and chemists did their best to decompose them further. Often they succeeded in doing so and they went on preparing simpler and simpler materials with different properties. The sum of the weights of all decomposition products of a given material was always exactly (as far as exactitude went in those days) equal to the weight of the decomposed material itself. The law of conservation of matter was experimentally confirmed and Demokritos was vindicated. But at the end of all these series of decompositions there remained a set of (not very many) substances which underwent no further decomposition, no matter what methods were employed. *All their reactions consisted in combination with other substances, the products of the reactions were always heavier than the original substance and never weighed less than it had done. These substances which did not undergo further decomposition are now named, indeed defined, as the chemical elements.*

Let us look at an example. If we take 1·00 gramme (g) of mercury, dissolve it in excess nitric acid and evaporate the excess acid by heating, a new compound is formed, since heavy, white, water-soluble crystals remain. Provided we have taken care to work without loss (as chemists say, working "quantitatively"), we ascertain that after energetic desiccation at low temperature the new compound weighs 1·62 g. During the reaction brown, disagreeable-smelling gases are evolved, showing that the reaction has more than one product. In which product is our mercury to be found? This question can be answered only by trying to find the mercury again. Experience shows us that mercury can be recovered only from the white crystals and not from the gases or vapours. Like so many other compounds, the white crystals decompose on further heating. The same brown gas is evolved that we have already met in the first reaction and a red powder remains. This turns black reversibly on heating. However, its weight remains unchanged during this change of colour between red and black. If we work carefully, decomposing the white crystals until the last trace of brown gas has gone without, however, exposing the new material to higher temperatures than necessary, we observe that it now weighs 1·08 g. We have performed a decomposition, because without having given access to any new material we have obtained a lighter, red powder from our white crystals and have also obtained a brown gas. If we now heat this red powder more strongly in a test-tube it begins to decompose further and drops of pure mercury will settle out on the cooler upper part of the tube. At the same time it may be observed that a gas is evolved from the tube. It has no colour, but a glowing match will burn with a bright flame if it is inserted into the test-tube. We have encountered this gas already in our electrolysis of water, at the anode. It is oxygen. Having performed this last decomposition quantitatively, we may weigh the mercury regenerated in the test-tube and will find that the weight is exactly 1·00 g, the amount we used at the beginning. *J. Priestley* (1733–1804) was the first to examine this thermal decomposition of red mercuric oxide, as we call it, and Lavoisier found that the amount of oxygen evolved during the decomposition is exactly equal to the amount consumed by simply heating in it 1·00 g of mercury, to a temperature which is below the temperature of decomposition of the oxide.

A circle of reactions has been completed. We started with 1·0 g of mercury and subsequently regained exactly the same amount. It was established that neither the white crystals (mercuric nitrate) nor the red mercuric oxide could be elements, because they were subject to decomposition. They yielded products which weighed less than they themselves weighed. But what about the mercury? In the course of our experiments we found it only in compounds which were heavier than itself. Therefore it seems advisable to perform with it new sets of experiments in order to see whether it can be decomposed into other substances which weigh less. However, it is never possible to find a reaction as a result of which something lighter is formed from mercury. Its weight either increases or, if no reaction occurs, remains unchanged. *All these unsuccessful attempts at preparing something from mercury which weighs less than the mercury*, all these negative results are summed up in the short statement that *mercury is a chemical element*.

In the hands of Lavoisier and his great contemporary colleagues the chemical balance rapidly made us acquainted with most of the simple elements of chemistry. *H. Cavendish* (1731–1810) demonstrated that hydrogen, known already to Paracelsus, forms water when it combines with oxygen. He showed too that air is a mixture containing 80 per cent nitrogen and 20 per cent oxygen (compare Fig. 2, p. 2). Thus water and air from among the four classical "elements" lost their elementary character while nobody really believed any longer that earth was a chemical element. However, Lavoisier still included fire in his list of elements, notwithstanding the fact that it has no weight. Such is the inertia of human advance! It was left to the great Swedish analytical chemist, *J. J. Berzelius* (1779–1822) to eliminate fire from the table of chemical elements, and so the last classical element disappeared. Berzelius was the first to introduce the custom of symbolizing elements with their initials, adding a second small letter if necessary for distinction. This convention is valid up to the present day and greatly helps us to manipulate formulae and visualize chemical reactions and structures. *J. Black* (1728–1799) was the first to prepare carbon dioxide from calcium carbonate (limestone) while Lavoisier demonstrated that this gas was the same as originates from carbon on its complete combustion in oxygen.

Carbon itself withstood all efforts to decompose it further: it is a chemical element.

Let us now deal with a few cases where the advance of science brought about the decomposition of substances which for a while were held to be elements. Lavoisier thought that the oxides of the alkali and alkaline earth metals (sodium oxide, burnt limestone and burnt magnesia) were elements because nobody was able to decompose them at that time. He acted according to the definition of page 15 in good faith. A few years later, however, *H. Davy* (1779–1829) applied the newly discovered electricity to the decomposition of chemical compounds and on leading an electric current through molten sodium oxide or the molten chloride of calcium or magnesium found that it was possible to decompose these compounds into new, light metals. These metals readily combined with oxygen, increasing their weight thereby and yielding the oxides again which thus had to be struck off the list of elements. Later on, aluminium, the metal so important for our industry, was prepared in the same manner from its oxide.

Thus we are only allowed to call a substance a chemical element as long as we remain unable to decompose it. The question emerges: is our table of elements always to be a provisional one? After what we have said up to now this should be the case. The probability that an element will be decomposed in the future is very slight because of the fact that a very long time has elapsed since such an event occurred in the history of chemistry. More than a hundred years of continuous effort plead for the correctness of our table of elements. But this, in itself would be no definite proof. However, much more recently we have learned more about atoms and thus about elements and we therefore feel quite certain that it will not be possible to decompose our known elements by any chemical means. For instance it was found that all elements, whether free or combined in compounds, emit a series of X-rays of characteristic wavelengths on the impact of high-energy electrons. Any compound emits all the X-rays of its constituent *elements*, but none of its own, and so the compound or elementary nature of a substance may be ascertained. Furthermore no substance emits X-rays which cannot be assigned to known elements.

But what about radioactivity, what about atomic explosions? Well yes, this is really a destruction, a decomposition or sometimes a synthesis of chemical elements. If we are prepared to stretch definitions and logic beyond their limits we may say that our elements are in reality not elements, or atoms are not "indivisibles" any more. They are composed of protons, neutrons and whatever may be beyond them. We shall become acquainted with these particles later on.

A comparison may serve to clarify the situation. A house is built of bricks, mortar, concrete, wood, nails and so on. However, the bricks, mortar, etc., all consist of chemical compounds. The smallest particles of the compounds are the molecules and these molecules are built from atoms. Does it then make sense to say that the house is built from atoms or from electrons, neutrons and protons? The chemist's point of view is that it makes more sense to say that houses are built from bricks; it is better to break down an entity step by step than to skip the steps. However, it is important that the steps should be real and well defined. Thus the difference between the bricks and the atoms is as well established as that between the atom and its nuclear components. The energy (heat) which is liberated in the course of a typical chemical reaction lies within a certain range of values. The energies of nuclear reactions have their typical ranges as well but the two ranges are far from overlapping. The smallest energy of a nuclear reaction is about a million times as great as the energy of the most intense chemical explosion. There is nothing in between. A process belongs either to the one class or to the other. Therefore we may safely define our elements in chemistry as being those ultimate substances into which we are able to decompose all matter using energy of an intensity typical of chemical reactions or for that matter a hundred or a thousand times greater. We shall learn later that not only are nuclear energies millions of times larger than chemical energies, but that nuclear dimensions are many orders of magnitude smaller than the dimensions of atoms.

We now understand what it means to analyse a material. First of all it has to be ascertained that the substance is pure in the sense that it is not a mixture. If it is not pure, the mixture has to be separated into its constituents using the techniques we have learned to know as "physical" methods. After this, each pure compound has first to be examined by qualitative analysis, to establish which elements (and sometimes also which groups of elements) it contains. Any characteristic property of an element or of its well-known compounds is suitable for its qualitative identification, e.g. colour, smell, melting- and boiling-points, spectrum, etc. Qualitative analysis as a science is nothing else than a systematic collection of these properties and of methods which enable us to transform unknown compounds into known and characteristic compounds of the elements. When this has been achieved quantitative analysis begins. Its aim is to ascertain the percentage composition of a compound with respect to its constituent elements. It is seldom performed by actual decomposition to the elements; generally well-known compounds of the elements are preferable as the ultimate stage of analysis.

6. CONSTANT AND MULTIPLE PROPORTIONS

At this point we must call a halt. How is it possible not to decompose a material into its elements and still be able to tell how much of its constituent elements it contains? Why are its well-known compounds sufficient as end-products to be weighed in quantitative analysis? The reason is that the percentage composition of any given compound is constant and invariable.

This emerged as an experimental fact after about half a century of analytical work (the second half of the eighteenth century). In our imaginary experiment with mercury and nitric acid the ratio $1\cdot00:1\cdot62$ of mercury metal which has reacted, to mercuric nitrate which has been formed is always constant, no matter how much excess nitric acid remains after the reaction.

The same argument applies to the thermal decomposition of the nitrate. We need not decompose all of it; that which remains can be extracted with water and thus separated from the insoluble oxide. The ratio of decomposed nitrate to newly-formed oxide will always be 1·62:1·08. Again, during the ultimate decomposition of mercuric oxide, the weight ratio 1·08:1·00 between oxide decomposed and mercury formed is invariably obtained.

Take another example. "Pure" coke will burn in an oven practically without residual ash provided that it obtains sufficient air to keep it burning until it disappears. Chemically pure carbon would burn away without any residue at all. The coke disappeared in some form up the chimney and we know from Lavoisier's work that it combined with the oxygen of the air. Lavoisier ascertained experimentally that 1·00 g of carbon consumes 2·67 g of oxygen and yields 3·67 g of carbon dioxide. The law of conservation of matter was thus obeyed. It should be realized that gases are weighed as simply as any other kind of matter by first weighing an evacuated container and filling it afterwards through a stop-cock with the gas that is to be weighed.

Now let us return to the coke in the oven. If we completely shut off the air, the fire will go out. But if we let a limited amount of air pass through, burning continues but a gas emerges which is different from carbon dioxide. This becomes evident if we open an upper door of the stove and peer in. As fresh air enters through the door the new gas which has already passed through the burning coke itself begins to burn with a blue flame. Carbon dioxide would not do so since it cannot be burnt. On collecting the new gas before it burns we find that it is indeed very different from carbon dioxide. It is very poisonous and scarcely dissolves in water or alkali whereas the dioxide did so fairly readily. The two gases may be separated by passing the mixture through a solution of caustic alkali. The second gas goes through while the first is absorbed and may subsequently be liberated from its alkaline solution by acidification: bubbles emerge.

By weighing the amount of oxygen consumed by 1 g of carbon under such circumstances that only the new gas evolves (that is by passing a small amount of oxygen or air through a long column of red-hot carbon), we find that 1·33 g of oxygen is consumed and 2·33 g of gas is formed. Furthermore, this gas needs exactly a further 1·33 g of oxygen to burn and be totally converted into 3·67 g of carbon dioxide, the same quantity as was formed on direct combustion in the first experiment. Thus carbon has reached the same end state through a well-defined intermediate compound. It is worth noting that it consumed exactly the same amount of oxygen in each half of its two-stage combustion. The "new" gas is called carbon monoxide.

We should have obtained an analogous result if we had not given the mercury of our previous experiment sufficient nitric acid in which to dissolve. The unreacted mercury could have been separated by filtration and on evaporation of the solution to complete dryness another white crystalline compound would have been found. However 1 g of mercury only yields 1·31 g of these new crystals, not 1·62 g as was found for the mercuric nitrate. Heating the new crystals—mercurous nitrate—with excess nitric acid changes them to mercuric nitrate and their weight increases from 1·31 to 1·62 g.

Once more it is noteworthy that the weight increase in both stages of the reaction is the same, being 0·31 g per gramme of mercury.

The reaction of chlorine (a corrosive, nasty-smelling yellow-green gas) with iron powder has points in common with the other two reactions though it is not quite the same. 1 g of iron yields 2·92 g of a dark readily subliming crystalline solid if it is allowed to combine at an elevated temperature with 1·92 g of chlorine. But a different compound still containing nothing except iron and chlorine, may be obtained by dissolving iron in hydrochloric acid and evaporating the resulting solution to dryness. This white crystalline mass melts at red heat without sublimation and contains 1·28 g of chlorine for every 1 g of iron. When heated in chlorine gas it takes up 0·64 g and is converted to the same product as we have encountered in the direct reaction. Now, 1·28 and 1·92 are not in the ratio of 1 : 2, but they are in another simple relation: division by 0·64 reveals it to be 2 : 3.

Let us take another example. The lightest compound of natural gas found trapped over oil deposits is called methane. This gas reacts with chlorine gas, especially if illuminated and according to the relative amounts of the reactants we get a mixture of four distinct new compounds which may be separated by fractional distillation or gas-phase chromatography. At room temperature one of them is a gas, the others are liquids. Some of their properties, together with those of methane are given in Table I. Quantitative

TABLE I

Name	Boiling-point	Carbon %	Hydrogen %	Chlorine %	1 g carbon to	
					g H	g Cl
Methane . . .	−161°C	75·0	25·0	—	0·333	—
Methyl chloride .	+22	23·8	5·9	70·3	0·250	2·96
Methylene chloride .	+42	14·1	2·4	83·6	0·167	5·92
Chloroform . .	+61	10·0	0·84	89·1	0·083	8·88
Carbon tetrachloride .	+77	7·8	—	92·3	—	11·84

analysis gives us the percentage compositions. Looking at these we can merely observe that going downwards the carbon : hydrogen ratio decreases whereas the carbon : chlorine ratio increases. The numbers become more significant if we use them to calculate the amounts of hydrogen or of chlorine which combine with 1 g of carbon. We see that methylene chloride contains just twice as much hydrogen per gramme of carbon as does chloroform, methyl chloride three times as much and methane four times as much. On the other hand 1 g of carbon combines with twice as much chlorine in methylene chloride as it does in methyl chloride, with three times as much in chloroform and four times as much in carbon tetrachloride.

One last example: Normal salt contains a metal called sodium and chlorine in the ratio 1·00 : 1·54. However, we know a series of other compounds with just the same sodium : chlorine ratio but with increasing amounts of oxygen in them. The bleaching compound sodium hypochlorite contains 0·696 g of oxygen for 1 g of sodium, the new bleaching compound sodium chlorite 1·392 g, sodium chlorate (which yields explosive mixtures) 2·088 g and sodium perchlorate 2·784 g. We see that the ratio of oxygen combining with the same amount of sodium (or chlorine) is 1:2:3:4 in these four compounds.

It is time to sum up all that we have discussed in this section. We have seen that *in every compound the constituent elements are present in constant proportions*. The percentage composition of a compound is just as characteristic as its melting-point, colour or any other of its properties. This is the law of constant proportions (*J. B. Richter*, 1792). Further, we have seen that *whenever we compare the weight ratios of the same elements in different compounds they are either constant*, as in the case of the sodium:chlorine ratio in the sodium chloride–sodium perchlorate series, *or they are in the ratio of small integral numbers*, as the sodium:oxygen ratio in the same series or the C:H:Cl ratios in the series of chlorination products of methane. This second law is called the law of multiple proportions (*J. L. Proust*, 1808).

Now we can understand why it is not necessary to decompose all compounds into their elements in the course of quantitative analysis. It is sufficient to transform a compound into one of its characteristic products under circumstances known to guarantee a complete transformation, because such a product will have a well-established percentage composition accurately known from previous analyses or syntheses. Thus it is very convenient, for example, to determine the amount of chloride ions in a solution by the addition of silver nitrate solution to it. It is known that silver chloride is very sparingly soluble (virtually insoluble for purposes of analysis) and its exact composition has been determined with great accuracy by synthesizing it from elementary silver and chlorine. It is known, that every gramme of silver chloride contains 0·247 g of chlorine. Hence the amount of silver chloride obtained after careful, quantitative precipitation, filtration, washing and drying must be multiplied by this "factor" to yield the amount of chloride ions in the original solution. Often it is not necessary to resort to weighing the product. In the present case, for example, it is sufficient to possess a silver nitrate solution the exact concentration of which is known to us. Now silver chloride is formed as a white curdy precipitate. It is much less soluble than silver chromate which is reddish brown. Therefore suppose we add a small amount of sodium chromate to the chloride solution to be analysed and then begin to add the known silver nitrate solution dropwise. The resulting precipitate will remain white as long as there is any chloride to be precipitated but turns brown at the first drop of excess silver nitrate when all chloride has been consumed and the precipitation of silver chromate begins. From the number of millilitres of silver nitrate *solution* added we can calculate how many *milligrammes* of silver nitrate were used and it is known from previous experiments that each gramme of silver nitrate will combine with 0·209 g of chloride. Thus multiplication by this factor converts the milligrammes of silver nitrate into milligrammes of chloride. Naturally it is more convenient to jump over the calculation of milligrammes of silver nitrate and establish once for all the weight of chloride equivalent to one millilitre of our standard solution. The two methods above are examples of a gravimetric and a volumetric (= titrimetric) determination, the two classical methods used in quantitative analysis.

7. THE ATOMIC HYPOTHESIS

This was as far as the chapter of chemistry dealing with the relative weight composition of compounds (so-called stoichiometry from στοιχεῖα = elements) had advanced when an English chemist, the father of modern atomic hypothesis, *J. Dalton* (1766–1844), began to meditate over the possible cause of these remarkable laws. In science it is very often most important to be able to be astonished at facts which have either seemed natural to other people or which have become accepted by them after long familiarity without thinking about them any further. Dalton could not and would not believe that such simple laws of Nature were just isolated facts but was convinced that they must be the outcome of some fundamental property of matter. He published his ideas on this question during the years 1807–1808 and these publications mark the beginning of what we may now call the scientific hypothesis of the existence of atoms.

He went back to ask the old Greek question once more: *Is it possible to go on dividing matter indefinitely or is matter composed of finite, ultimate small particles?* Dalton had the impression that the laws of stoichiometry would help in answering this question. What should we expect to occur if it were possible to divide matter *ad infinitum*? Evidently something along the lines that all combinations of different materials should be continuous, the materials being able to permeate both each other and space with infinitesimal "fineness", in a way similar to the process of dissolution, of absolutely intimate mixing. Just as water and alcohol can be mixed in any ratios to give mixtures whose properties vary in a continuous manner (as far as macroscopic observation is concerned), so should we expect that all kinds of matter would mix and permeate.

However, this is not the way things happen in Nature. The mercury we considered previously did not gradually loose its metallic properties to exchange them for those of the white mercuric nitrate crystals: any speck of the original mercury was either still mercury at any moment or had been converted to mercuric nitrate. Similarly the piece of black carbon did not gradually become transparent and of greatly increased volume in changing over into its gaseous oxides during oxidation but, at any given moment, was either a solid speck of carbon or an invisible gas. There is no gradual change observable in chemical transformations.

Dalton cut through this Gordian knot at one stroke. He said that chemical reactions behaved as they did because matter was not infinitely divisible but was built from fundamental particles, atoms, which had a characteristic mass for each element. Thus all atoms of oxygen are identical and all atoms of mercury are identical but the masses of an atom of mercury and an atom of oxygen are different.* As a matter of historical fact Dalton used the term "atom" for those smallest particles into which a pure substance could be divided without changing its properties. For example he referred not only to atoms of mercury but also to atoms of mercuric oxide. It was the famous Italian physicist *A. Avogadro* (1776–1856) about whom we shall hear much subsequently, who made the necessary finer distinction. He differentiated between the smallest particles of compounds, which could be further decomposed into still smaller elementary particles, and the latter which were the smallest particles of the elements. He named the latter "elementary molecules" whereas for the former he reserved the term "molecules". *Laurent* in 1842 re-established the term "atoms" for Avogadro's "elementary molecules".

What, then, is supposed to happen on chemical combination of elements A and B according to the atomic hypothesis? The simplest possible case will occur when each atom A catches for itself an atom B, the resulting pair hence forth going together as AB molecules until some force separates them. Since each atom has a characteristic mass of its own, the compound AB will contain the two elements in a weight ratio corresponding to the atomic weight ratio A:B. (It should be remembered that for all quantitative considerations it is customary to denote by the symbol of an atomic species a quantity of the element equal to its atomic weight in grammes.)

If our molecule AB combines with other atoms, say C and D, to form a more complex molecule, the original ratio A:B remains unchanged because the molecule has neither lost nor received either A or B. Here we encounter again the law of constant proportions: the fixed proportions of the elements in their compounds are a consequence of the fixed weights of the individual atoms, constant for each element.

What happens, however, if a molecule contains more than one atom of an element? Suppose that an atom of A is bound up with two atoms of B in one compound, with three in another and so on. Atomic weights being constant, the relative amount of B to A in compound AB_2 will be twice the relative amount in AB and that in AB_3 will be three times as large. In just the same way we may imagine that a compound is formed from two atoms of A and three atoms of B. The relative amount of B to A in this new compound, symbolized by the formula A_2B_3 will be 3/2 times as high as in compound AB. The above argument may be extended along the same lines. The more complicated a molecule, the more atoms it may contain of a given sort and so the weight ratios of its constituent elements will be given by larger integral numbers. Larger, that is, when compared with the ratios in their other, more simple compounds, for example the Fe:Cl weight

---

* We shall see later (p. 412) that this assumption was not absolutely correct. Atoms of the same element may differ with respect to mass, but the number of different masses for atoms of a particular element is small and well-defined. Atoms of the same element but with different masses are called isotopes. The properties of isotopic atoms are very similar, so that for a long time it was impossible to separate the isotopic mixtures which occur in Nature into their components. Therefore what Dalton, and after him generations of chemists, observed as atomic weights were actually the mean atomic weights of the isotopes.

ratios in ferric chloride (*see* p. 20) and in ferrous chloride. In the larger molecules the integral ratios will not be, say 3:2, but perhaps 37:19. However, the ratios will always be found to be between integral numbers because by definition only integral atoms can exist. Putting it another way only integral atoms, with integral multiples of their weight may combine into molecules. This is how the law of multiple proportions is quite naturally explained by the atomic hypothesis.

Let us go through the argument once more. Suppose that elements are nothing but a lot of identical atoms with characteristic weights and that compounds are conglomerations of a mass of molecules built from these individual atoms, in the way that buildings may be constructed from sets of different bricks. Then, as we have seen, it follows that the elements must combine in the weight ratios of their atomic weights or in ratios which are derived therefrom by multiplication by a ratio of integral numbers—the number of each kind of atom in the molecule. Thus the atomic hypothesis is sufficient to explain the laws of stoichiometry. But is it necessarily the only explanation? There have been great scientists right up to the beginning of the present century who viewed the atomic hypothesis with great scepticism. They only began to believe that atoms really do exist when quite a number of other proofs of atomic existence besides the laws of stoichiometry had been found.

Dalton's hypothesis only demonstrates that in the course of subdividing matter we somewhere reach a point where particles cannot be divided any further. Nothing is said about the actual magnitude of the weight of the ultimate particles, and for all that it matters, they could be of any degree of smallness as long as they remain distinct in the end. Mathematicians call this concept a discontinuum in contradistinction to continua such as time and space.

Now let us turn the argument round. So far we have deduced the laws of constant and multiple proportions from the hypothesis that elements were composed of atoms with a characteristic weight for each element. Let us depart from the weight ratios we have found by experiments on compounds chosen at random and try to determine *the weight ratios of the elementary atoms themselves.* It must be already clear to us that we shall not be able to determine the absolute weight of the atoms, but only the ratios of their weights. The determination of these relative atomic weights seems to be a more modest task but unfortunately even this is not so.

For example let us consider further the two oxides of carbon. We have seen that 1·00 g of carbon combines with 1·33 g of oxygen to form carbon monoxide and with 2·66 g to form carbon dioxide. Let us resort to a graphical representation (Fig. 9). Our analysis showed us only that the molecule of carbon monoxide contains 1·33 times the weight of oxygen as of carbon. This observation is accounted for in Fig. 9 by drawing the rectangle which represents oxygen 1·33 times as broad as that for carbon. But how do we know that the carbon monoxide molecule really contains one atom of carbon and one atom of oxygen only? We do not know this at all. Subsequent lines of the pictorial representation show a few from among the many possible combinations which would all yield the same analytical result. There is only one condition which has to be fulfilled in every case: the sum of the weights of all the carbon atoms in the molecule and the sum of weights of all the oxygen atoms in it must give the ratio 1·00:1·33. The right-hand side of the graph shows the same set of possibilities for the carbon dioxide molecule for which the condition is that the sums of the weights of carbon and oxygen atoms respectively should yield the ratio 1·00:2·66. This, and only this, is the fact given by quantitative analysis.

Analysis revealed the following weight ratios of carbon to oxygen in carbon monoxide and carbon dioxide.

These ratios can be accounted for by an infinite set of assumptions regarding relative atomic weights, each with the corresponding formulae, as for instance:

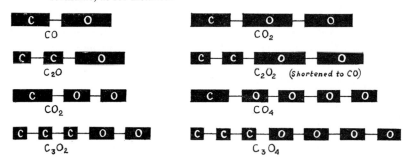

As long as the sum of the weights (represented here as areas) of C and O remain constant and in the ratio given by analysis, all assumptions are equally justified.

Fig. 9

We assume in the first row in Fig. 9 that the oxygen atom is 1·33 times as heavy as the carbon atom. By this assumption the analytical data are accounted for if one carbon atom combines with one oxygen atom to give carbon monoxide and with two to give carbon dioxide. In the second row we have assumed that the weight of the carbon atom is only half of that which was taken in the first row, but that we are dealing with twice as many atoms of carbon. Thus the analytical results will not change. The formulae of the first row would be represented by the symbols CO and CO_2 and those of the second row by C_2O and C_2O_2. Remember however, that the quantity represented by a C symbol has been halved in the meantime. The third row is based on the assumption that carbon has the atomic weight assigned to it in the first row but that the atomic weight of oxygen is only half as great. Of course, we have to double the number of these halved oxygen atoms to preserve the analytical ratios. The formulae would change to CO_2 and CO_4, the sign O representing half the amount which it did in row one. It is already evident that there will be no end to the possible assumptions concerning the relative atomic weights, but let us try just one more example. Suppose we subdivide the original carbon atom into three equal parts and the original oxygen atom into two parts (fourth row in Fig. 9). The ratio of atomic weights of carbon and oxygen would thus change to $0\cdot33:0\cdot67$ while the respective formulae would become C_3O_2 and C_3O_4. Once more the analytical results will be fulfilled.

There is, therefore, no reason whatever to be complacent about the situation. The analytical data, the only experimental data we possess, may be explained by an infinite set of assumptions concerning the relative atomic weights, as long as each set is combined with the appropriate chemical formulae of the compounds. It is only necessary that the assumed atomic weight multiplied by the assumed number of atoms in the molecule shall have a constant value. In the graphical representation only the sum of areas which represent an element in the compound has to be kept constant. It is unimportant into how many equal parts we subdivide the areas. The analytical result only yields the ratios of areas belonging to the individual elements. An unlimited number of possible atomic weights leads to an unlimited number of formulae which we have to assume accordingly: that is all.

Half a century elapsed (1811 to 1860) during which this unhappy situation prevailed. Chemists had virtually accepted the view that all sets of assumptions regarding atomic weights were logically equally consistent, though some did yield sets of simpler formulae than did others. It was a matter of convenience rather than of intrinsic importance which set was agreed upon and open to debate whether it would ever be sensible to speak about real atomic weights if it was impossible to determine them from analytical evidence.

It seemed quite logical therefore when sets of *equivalent* weights were established, based on the convention that the equivalent weight of hydrogen was unity and the equivalent weight of another element or group was defined as the quantity which would combine with one unit of hydrogen or would displace it from one of its compounds. Hydrogen was chosen because it was the lightest of all matter and because its weight percentage in all of its compounds was the smallest of any element. Thus there was good—though not cogent—reason to believe that the atom of hydrogen was the lightest atom of all.

For example, analysis revealed that 1 g of hydrogen combines with 8 g of oxygen to form water. Thus 8 was the equivalent weight of oxygen. According to the symbolism of Berzelius the signs H and O now represented an equivalent weight of an element and so the formula of water was written HO, meaning quantitatively 9 g of water. In the same way the analysis of hydrogen chloride gas gave the formula HCl with $35\cdot5$ as the equivalent weight of chlorine. Metals did not easily combine with hydrogen, therefore they were either used to displace it from acids in order to define their equivalent weights or the analysis of an oxide or chloride was used for calculation on the assumption that the amount of substance that combined, e.g. with $35\cdot5$ g of chlorine, was equivalent to 1 g of hydrogen. Sodium thus received the equivalent weight of 23, calcium of 20.

So far, so good. But what about, for instance, carbon and oxygen? Starting from carbon monoxide the analytical ratio $C:O = 1\cdot00:1\cdot33$ would mean that 6 g of carbon was equivalent to the standard amount of 8 g of oxygen. However, starting from the analysis of carbon dioxide for which the experimental ratio is $1\cdot00:2\cdot67$ the amount of carbon equivalent to 8 g of oxygen turns out to be only 3 g. Which should we decide upon? A glance at Fig. 9 reveals that both assumptions are absolutely consistent with the analytical results if we assign the formulae CO and CO_2 or C_2O and C_2O_2 respectively, to carbon monoxide and carbon dioxide. We have already discussed this situation and even more complicated versions of it in discussing Fig. 9. There was then, and there is now, no hope of deciding logically which was the correct one of an unlimited number of possibilities.

However, this time we would be content with a convention instead of a logical decision because we only want to obtain a set of consistent equivalent weights without arguing that they really represent the relative weights of the real atoms. To this end, of course, we could decide upon any of the many possibilities, choose the formulae accordingly (Fig. 9) and stick to the decision once for all. Dalton's advice for such cases was to choose those equivalent weights which yielded the simplest formulae. But in our case even this principle does not work. Remember that we have up to now no means of distinguishing a formula CO from C_2O_2 or for that matter from C_nO_n. All of them stand for the same ratio! So instead of C_2O_2 in the second row of Fig. 9 we may as well write CO and then the alternative propositions become CO and CO_2 (with $C = 6$) or C_2O and CO (with $C = 3$). Those alternatives lead to sets of formulae which are equally simple. The same dilemma arises every time we deal with elements which combine with each other in more than one proportion: in each and every case that was considered under the heading of multiple proportions!

In practice all this would not have mattered very much if the chemists had come to an understanding as to which set of conventional equivalent weights to use. The corresponding set of formulae would then have represented the compounds and everybody would have known what a formula meant and what percentage composition it stood for. But they could not agree, because in the depth of their hearts they never gave up the hope (or illusion?) that atomic weights were not merely conventional units to express analytical results but that they really existed. They believed in real atoms and their real grouping around each other and wanted to express chemical properties of compounds by their formulae. Looking at it from this angle it does matter whether we write HO or H_2O for water and CO_2 or CO for carbon monoxide. A lot of chemical facts had accumulated during these fifty years and according to the group of facts that seemed more fundamental to him, each chemist supported this or that list of equivalent weights.

What do we mean by this sort of chemical interpretation? In the case of water, for example, it is observed that half of its combined hydrogen behaves chemically in one way and the other half in another way. If we throw a piece of metallic sodium into water it melts from the heat of reaction to a light white sphere of metal which runs hither and thither on the surface of the water until it disappears, evolving an equivalent quantity (1 : 23) of hydrogen gas. On evaporation of the now alkaline aqueous solution to dryness a solid, sodium hydroxide, remains and after analysis it is found that this compound still contains hydrogen. It contains just as much hydrogen as was evolved during reaction and so half of the hydrogen in water behaves differently from the other half. This evidence is in favour of a formula with at least two H atoms in the molecule of water or at any rate for a formula with an even number of hydrogen atoms, since this could be halved. The simplest assumptions are therefore, H_2O for the formula of water and 16 for the atomic weight of oxygen. We may adhere to this hypothesis so long as different experiments do not involve us in further complications. However, the

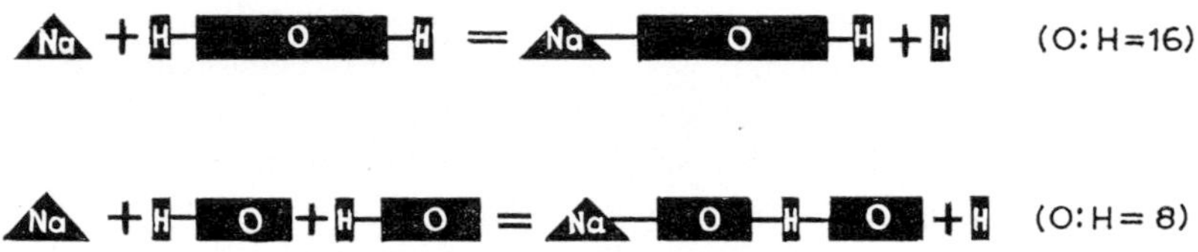

Fig. 10. **Two tentative explanations of the analytical results of the reaction of sodium metal with water**

The results are explained by either assumption.

inference drawn was not necessarily completely valid. Another chemist could have argued, that water was HO and that after Na had displaced the single atom of hydrogen to yield NaO (with O = 8 this time!) there was immediate combination with another molecule of water to form $NaHO_2$ which would accordingly stand for sodium hydroxide in this system. The explanation is hardly more complicated than the alternative one.

Thus every bit of chemical evidence was weighed and discussed but without definite result. The rival schools understood each other of course because each knew about the conventions of the other, but formulae had to be literally "translated" between the systems. It was an exasperating situation. At last the German chemist *Weltziehn* called a congress of chemists to Karlsruhe in 1860 to try to arrive at a unified system of atomic (or equivalent) weights. It must be clear to us by now that purely stoichiometric reasoning was not going to be of any help in this decision, something entirely new had to be added in order to settle the dispute. We are going to deal with the fundamentally new set of observations in the following chapters.

However, we should not take leave of stoichiometry before having a glance at a fundamental question concerning the law of multiple proportions. Scientists have been able to discover this law because the combination ratios of elements in different compounds were *small* multiples of each other or at least stood in the ratio of *small* integral numbers. In our examples we never went beyond the number 4, though it would have been possible to do so. It was good luck, or rather the good intuition of our chemist "fathers" which made them analyse relatively simple compounds. Otherwise the law of multiple proportions would never have occurred to them at all!

Suppose for instance that they had isolated and analysed compounds which are common in organic chemistry, compounds which we now know to be built of very many atoms. Take for argument's sake a high-molecular-weight paraffin, $C_{60}H_{122}$. In this compound 1 g of carbon is combined with 0·171 g of hydrogen. Among the many compounds of organic chemistry there exists one (today we know it is an "unsaturated" derivative of the former; *see* p. 82), which has the composition $C_{60}H_{120}$. (These formulae assume we adhere to the set of atomic weights which has by now been definitely accepted.) In this second compound the amount of hydrogen which combines with 1 g of carbon is 0·168 g. Suppose that the techniques of analysis had been sufficiently refined in those days to establish such a small difference beyond the limits of experimental error. Certainly nobody would have exclaimed that the ratio of 0·171 to 0·168 is exactly 120:122! But even if this fact had occurred to anybody, he would have dismissed it with a shrug of his shoulders. The much more probable conclusion from these analytical results would have been that evidently carbon and hydrogen are apt to combine in a quasi-continuous series of weight ratios.

The main difference between the compounds of carbon and those of other elements is just the fact that the former often contain very many atoms within a molecule. Therefore the number of possible combinations, that is the number of possible carbon compounds is enormous also and even today we describe them under the omnibus heading of organic chemistry. For a long time it was wrongly supposed that only living organisms were able to synthesize organic compounds. We owe the discovery of the law of multiple proportions and thus the progress of chemistry to the fact that simple inorganic compounds were analysed before the complicated organic ones. Dalton was able to build his atomic hypothesis on this foundation, and after this was accomplished it became possible to fit the complicated organic compounds into the established system.

8. THE LAWS OF GASES

The missing link in the logical chain which eventually led to the definition of real relative atomic weights came from experiments on gases. Gases resemble one another in their physical properties to a much greater extent than do the condensed systems of solids and liquids. Coefficients of thermal expansion of gases and their coefficients of elasticity are the same no matter what the chemical composition of the gas. Gases enter into and are evolved from chemical combinations in such a manner that the volumes of the reacting gases and the volumes of the product gases are all in ratios of small whole numbers. (*See* Section 9.)

Consider the process of thermal expansion. Every solid and every liquid has its own characteristic coefficient of thermal expansion and these are in general different from each other. *J. A. C. Charles* (1746–1823) announced, however, as early as 1787 that all gases, when their temperature is increased by 1 deg C, expand by 1/273 of the volume they have at the melting-point of ice.

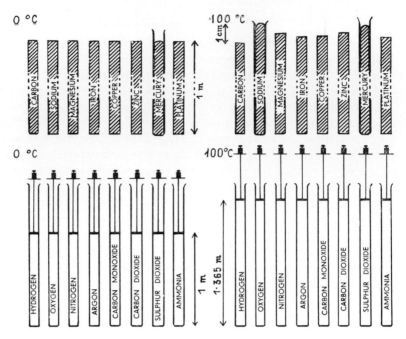

Fig. 11. Thermal expansion of various substances

(*upper*) Thermal expansion of solids and liquids: expansion always small and different for different materials.

(*lower*) Thermal expansion of gases: expansion large and the same for all gases.

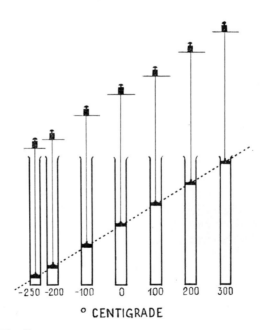

° CENTIGRADE

Fig. 12. Gases at constant pressure and various temperatures
The law of Charles and Gay-Lussac.

J. L. Gay-Lussac (1778–1850), the famous French chemist, confirmed these results and by some historic misunderstanding the law is generally associated with his name instead of with that of *Charles*.*

The relation between the pressure and the volume of a gas is also very simple and the same for all gases, whereas solids and liquids behave individually. The volume of a gas is inversely proportional to its pressure: increasing the pressure *n*-fold decreases the volume to $1/n$ of its initial value, and

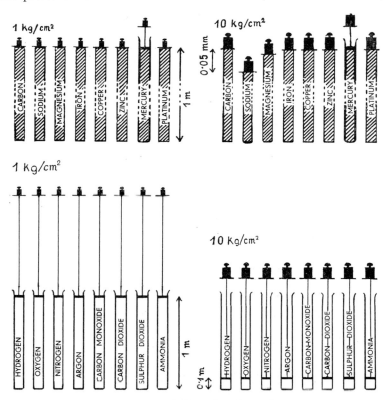

Fig. 13. **Compressibility of various substances**

(*upper*) Compressibility of solids and liquids: compressibility is always small and different for different materials.

(*lower*) Compressibility of gases: compressibility is large and the same for all gases.

conversely the volume increases with decreasing pressure. Once more the result would be quite different with solids or liquids where each material has its own characteristic coefficient of compression or elastic expansion. The

---

* As a matter of fact the expansion of gases seemed to be a more adequate phenomenon upon which to base the definition of temperature differences, than the expansion of mercury or alcohol, just because all gases behaved in the same way. Temperature as defined by the gas thermometer was superseded by the so-called thermodynamic definition only when it became clear that the relative amount of heat which can be gained as work, when heat is transferred from a warmer body to a colder body under ideal circumstances is an even more appropriate basis of a definition.

law which governs the pressure–volume relation of gases is due to *Robert Boyle* (1660), the author of *The Sceptical Chymist*, and to the French abbé *E. Mariotte* (1620–1684) who discovered it independently but somewhat later (1676). It is known as the Boyle–Mariotte law.

The combination of these two laws gives of course the dependence of gas pressure on temperature. Because the thermal expansion per degree is 1/273 of the volume the gas possesses at 0°C ($=v_0$) and because pressure is inversely proportional to volume, it follows that when a gas is recompressed to its

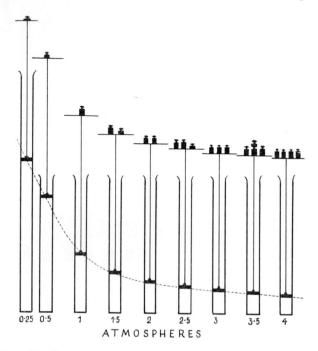

0·25 0·5   1   1·5   2   2·5   3   3·5   4

ATMOSPHERES

Fig. 14. **Gases at constant temperature and various pressures**
The law of Boyle and Mariotte.

original volume after thermal expansion, its pressure necessarily increases by 1/273 of the value at 0°C for each degree centigrade it has been heated. In the first case it is the pressure, in the second case the volume, that has been kept constant.

It is evident that the temperature of −273°C should represent a limit. If it did not do so, then either the volume of a gas would necessarily become negative if the gas were further cooled at constant pressure or its pressure would become negative at constant volume. These results would, of course be nonsensical. As a matter of fact, experimentally we are only able to *approach* a zero of temperature but never to reach it. The closest approach at the present time is to within one millionth of a degree centigrade. Many properties of matter undergo a drastic change at these extremely low temperatures. The theoretical limiting temperature has the value −273·16°C and is called absolute zero.

Temperatures which are reckoned from this absolute zero are recorded in degrees absolute (°abs.), nowadays often referred to as degrees Kelvin (°K), after *Lord Kelvin* (Sir W. Thomson, 1824–1907), the physicist. Degrees Kelvin, being measured from a zero point, are necessarily always positive. One degree centigrade or Celsius corresponds to the same temperature interval as one degree Kelvin, and any centigrade temperature is convertible to an absolute temperature by the addition of 273·16. Centigrade temperatures may be positive or negative.

If we write v_0 and p_0 respectively, for the volume and the pressure of a given mass of gas at 0°C, and v_t and p_t respectively for the same quantities at any temperature $t°C$, we have

$$v_t = v_0 \left(1 + \frac{t}{273}\right) \qquad . \qquad . \qquad . \qquad . \quad (8.1)$$

and

$$p_t = p_0 \left(1 + \frac{t}{273}\right) \qquad . \qquad . \qquad . \qquad . \quad (8.2)$$

which are of course valid for negative as well as positive values of t. The equations have identically the same meaning but look somewhat simpler if we change over from the centigrade (t) to the absolute temperature scale (T). In the latter case we simply put

$$T = t + 273 \qquad . \qquad . \qquad . \qquad . \quad (8.3)$$

and by substitution obtain respectively

$$v_t = v_0 \times T/273 \qquad . \qquad . \qquad . \qquad . \quad (8.4)$$

and

$$p_t = p_0 \times T/273 \qquad . \qquad . \qquad . \qquad . \quad (8.5)$$

Thus the volume (at constant pressure) and the pressure (at constant volume) of a given mass of gas are proportional to the absolute temperature. This is the law of Charles and Gay-Lussac.

The simple formula for the Boyle–Mariotte law is on the other hand

$$v_1/v_2 = p_2/p_1 \qquad . \qquad . \qquad . \qquad . \quad (8.6)$$

if we write v_1 for the volume of the gas at pressure p_1 and v_2 for its volume at p_2. It must be emphasized that both laws represent only first approximations when applied to real gases and are the more valid the lower the pressure and the higher the temperature of the gas. However, they are simple and are quite good approximations so that we use them often. A gas which would behave exactly as these laws prescribe is called an ideal gas.

It is convenient to decide upon a standard temperature and a standard pressure for comparing gases. In spite of the fact that neither the freezing-point of water, nor the pressure of the atmosphere above us at sea-level have any special physical significance they have been agreed upon as reference points because they are well defined and *some* standards had to be chosen anyway. Thus whenever we speak about a gas at "normal temperature and pressure" (N.T.P.) we mean that it is at 0°C and 1 atmosphere.

Let us now proceed to obtain an expression for the general case when *both* the temperature *and* the pressure of a gas are changed at the same time and see how its volume is affected by these external changes. According to our above-mentioned convention, T_0 now stands for 0°C, that is 273·16° abs. and p_0 for 1 atm pressure, which is the pressure exerted by a column of mercury 760 mm high (it is a more reproducible standard than the actual pressure of the air which of course varies from moment to moment).

Suppose that the volume of our gas is v_0 at N.T.P. and we first change its temperature to $T°$abs. Its volume changes accordingly (Charles, Gay-Lussac law) to

$$v_T = v_0 \frac{T}{T_0} \qquad . \qquad . \qquad . \qquad . \qquad (8.7)$$

We now go on to change the pressure also, say to p. The Boyle–Mariotte law governs this part of the process and determines the eventual volume of the gas at T and p, written as $v_{T,p}$

$$v_{T,p}/v_T = p_0/p \qquad . \qquad . \qquad . \qquad . \qquad (8.8)$$

Substituting for v_T in this expression we have

$$v_{T,p}/v_0 \frac{T}{T_0} = p_0/p \qquad . \qquad . \qquad . \qquad . \qquad (8.9)$$

or rearranging

$$\frac{v_{T,p} p}{T} = \frac{v_0 p_0}{T_0} \qquad . \qquad . \qquad . \qquad . \qquad (8.10)$$

This is the general gas law for "ideal" gases.

9. VOLUME OF REACTING GASES

What we really owe to Gay-Lussac concerning the uniform behaviour of gases is not the law of their thermal expansion, which as we have said was discovered by Charles, but the law governing the volume ratios which appertain during the course of chemical reactions. He discovered this law in collaboration with the renowned German philosopher *A. v. Humboldt* (1769–1859) who made a stop in Paris during his world-wide journeys and worked on this topic with Gay-Lussac in the years 1805–1809. They found that *whenever gases disappear in the course of chemical reactions, their volumes are in the ratio of small integers.* These volumes must of course be compared at the same pressure and temperature.

Let us consider some examples. If we consume just one litre of oxygen to convert some carbon into carbon monoxide we obtain just two litres of the latter. Further, in order to burn this gas until it is completely converted to carbon dioxide one more litre of oxygen is consumed whereas the volume of carbon dioxide formed is exactly two litres.

A mixture of chlorine and hydrogen gases reacts explosively when exposed to sunlight and experiments show that one litre of hydrogen combines with exactly one litre of chlorine, the product being two litres of hydrogen chloride, all measurements being made at N.T.P.

If we explode hydrogen with oxygen or combine them catalytically on the surface of hot platinum, we find that two volumes of hydrogen are consumed for every volume of oxygen: the reverse of what we found in Chapter 1, on the electrolytic decomposition of water. The product water, as we know, is a liquid at ordinary temperatures. But we can measure the volumes of the substances at a temperature above 100°C and in this case our product is water vapour. Its volume may be compared with those of the hydrogen and oxygen at this same temperature and will be found to be exactly equal to that of the hydrogen or twice that of oxygen.

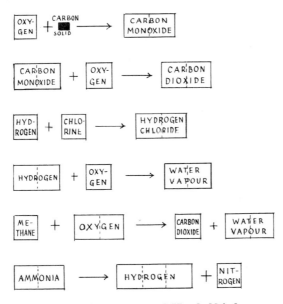

Fig. 15. **Gay-Lussac and Humboldt's law**

Total combustion of methane gas (CH_4) consumes two volumes of oxygen for each volume of methane yielding one volume of carbon dioxide and two volumes of water vapour.

One last example. Ammonia gas may be decomposed by the prolonged action of an electric spark. We find that two volumes of ammonia gas yield one colume of nitrogen and three volumes of hydrogen.

The whole process reminds us very much of the rules of stoichiometry but it is much simpler. The volumes in themselves show ratios of small integers whereas in the weight rules of stoichiometry small numbers only appear when two elements combine in more than one proportion. For example 1 g of iron combines either with 1·28 or with 1·92 g of chlorine: neither of these is a small integral number. But the ratio of 1·28 to 1·92 is 2:3!

Of course we must be aware that the weight laws of stoichiometry apply to gases as well as to matter in the condensed state. The fact that a rule concerning combining volumes exists for gases is additional. It has no counterpart for condensed systems. We would try in vain to arrive at any simple volume ratios between mercury and mercury oxide or the mercury nitrates. Thus something very fundamentally simple must be sought for in the structure of gases, since as we have seen, they all react quantitatively alike with respect to volume changes caused by pressure and/or temperature changes and in showing simple volume ratios in the course of chemical reactions.

10. AVOGADRO'S HYPOTHESIS

It was the Italian physicist, *A. Avogadro* (1776–1856), who found the common structural feature of gases when he proposed in 1811 (the same year that Dalton put forward the atomic hypothesis!) that probably *equal volumes of different gases, at the same pressure and the same temperature, contained identical numbers of molecules.* He proved that this hypothesis was sufficient to explain why gases combine in simple volume ratios and immediately saw that it was to yield us the coherent system of relative atomic weights which were not given by the laws of stoichiometry (*see* Fig. 9). Later on we shall see that the Gay-Lussac–Charles law and the Boyle–Mariotte law are also a consequence of this assumption and are explained by means of it.

Let us look at the chemical argument. Every molecule naturally contains an integral number of atoms because no fractions of atoms exist separately. Whenever molecules of a gas decompose, the total number of molecules increases, according to the number of new molecules that arise during the decomposition. This must of course again be an integral number. Thus the total number of initial molecules has to be multiplied by an integer to arrive at the number of molecules existing after the decomposition. On the hypothesis that equal volumes of gases at the same temperature and pressure must always contain the same number of molecules, the final gas mixture has to expand until the original number of molecules per unit volume is arrived at. This will be an integral multiple of the original volume because the number of molecules finally obtained will be an integral multiple of the number decomposed. Owing to the fact that a molecule generally does not decompose into very many others, this multiplication factor of volumes is generally a *small* integral number. The same is true when molecules unite. They generally unite in small numbers for a given reaction and therefore their number and consequently their volume decreases by a small integral ratio. This is the Gay-Lussac–Humboldt law.

Let us now discuss some of these gas reactions in detail. Hydrogen and chlorine react in a volume ratio of 1 : 1 and the volume of the resulting hydrogen chloride is 2. The simplest assumption, that one molecule of hydrogen combines with one molecule of chlorine to give one molecule of hydrogen chloride, leads us now to a difficulty because if this were true we should obtain 1 volume of hydrogen chloride from 1 volume of hydrogen and 1 volume of chlorine. We must look for some other assumption. One which might help us is the following. Let us assume that each molecule of hydrogen contains two hydrogen atoms and each chlorine molecule two atoms of chlorine. The two "twin atoms" may simply change their partners according to the scheme

$$H_2 + Cl_2 = 2 HCl \qquad . \qquad . \qquad . \qquad (10.1)$$

Thus the number of molecules has not changed during the reaction and according to Avogadro the volume of the reactants must be equal in this case to that of the products. This is found by experiment to be so.

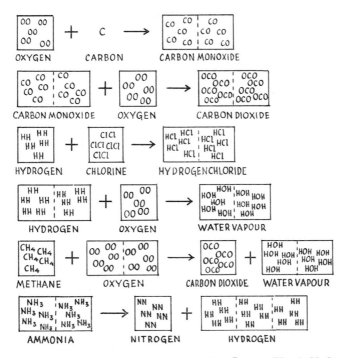

Fig. 16. **Avogadro's hypothesis and the Gay-Lussac–Humboldt law**
"All gases at the same temperature and pressure contain the same number of molecules in unit volume."

Figure 16 illustrates the above process and also the other gas reactions dealt with. Let us see which is the simplest set of assumptions which explains our experimental facts if we only take care that equal volumes of gases should always contain an identical number of molecules (e.g. 5 in each square of Fig. 16). Changing over to chemical symbols from pictures we have

$$O_2 + 2 C \text{ (solid)} = 2 CO \qquad . \qquad . \qquad . \qquad (10.2)$$

$$2 CO + O_2 = 2 CO_2 \qquad . \qquad . \qquad . \qquad (10.3)$$

$$2H_2 + O_2 = 2H_2O \qquad . \qquad . \qquad . \qquad (10.4)$$

$$CH_4 + 2O_2 = CO_2 + 2H_2O \qquad . \qquad . \qquad . \qquad (10.5)$$

$$2NH_3 = N_2 + 3H_2 \qquad . \qquad . \qquad . \qquad (10.6)$$

A molecule of oxygen must contain at least two oxygen atoms because its volume expands, as a result of reaction, into two volumes of carbon monoxide. Thus one molecule of oxygen is capable of subdivision, which would be impossible for one atom. If we thus agree that hydrogen gas as well as oxygen gas contains two atoms per molecule the equation for their combustion into water is written quite naturally according to equation (10.4) above. The volume ratio of $3:2$ between the reacting gases and the product (water vapour) which was a result of experiments, follows naturally from this equation and Avogadro's hypothesis.

Our examples have indicated to us that the molecules of gasous hydrogen, oxygen, nitrogen and chlorine each contain two atoms of the element. More exactly, they have shown that we can understand the volume ratios found experimentally by *assuming* that the molecules each contain two atoms. But can we not explain the same results by different assumptions? We certainly can. For instance we might assume that the molecule of any of these gases is built from four, six or for that matter from any even number of atoms. After all, we only know that their number within the molecule is capable of being halved, and this condition is satisfied by any even number of atoms. Thus we could write for the hydrogen–chlorine reaction

$$H_4 + Cl_2 = 2H_2Cl \quad . \qquad . \qquad . \qquad . \quad (10.7)$$

or
$$H_6 + Cl_2 = 2H_3Cl \quad . \qquad . \qquad . \qquad . \quad (10.8)$$

or in general
$$H_{2n} + Cl_2 = 2H_nCl \quad . \qquad . \qquad . \qquad . \quad (10.9)$$

Considering that the chlorine molecule too may be imagined to be built from any even number of atoms we may further write

$$H_{2n} + Cl_{2m} = 2H_nCl_m \qquad . \qquad . \qquad . \quad (10.10)$$

without contravening experimental facts and Avogadro's hypothesis.

In just the same manner, instead of H_2O for water, we should legitimately be able to write H_4O or H_2O_2 according to whether we double the assumed number of atoms in the hydrogen or in the oxygen molecule. The only thing that is certain is that neither of the said gases may contain an odd number of atoms because we are aware of reactions in which their molecules gave rise to two identical product molecules. Thus Avogadro's hypothesis allows us to exclude a number of formulae from among those we were unable to decide upon after a discussion of the relevant stoichiometry (p. 28). A set of several possible formulae still survives, however. Water cannot be represented by HO, H_3O, HO_3, etc., but may well correspond to H_2O, H_4O, H_2O_2, H_2O_4, etc.

A full discussion of the experimental facts does however permit the resolution of this ambiguity. Gay-Lussac's and Humboldt's results show that all the gaseous compounds of the elements may be segregated into classes according to how much of the particular element is contained in a given volume of gas (at some fixed pressure and temperature, of course). A few examples will make this point clear. Thus we have learned that a certain volume of water vapour contains the same amount (same weight) of hydrogen as does an identical volume of hydrogen gas. Other experiments show that the gas methylene chloride (CH_2Cl_2) we referred to earlier (p. 20) also contains the same amount of hydrogen in the same volume. The bad smell of rotten eggs is due to a gas called hydrogen sulphide (H_2S) and this gas too contains just as much hydrogen per volume. In contradistinction to the above, hydrogen chloride gas (HCl) contains exactly half as much hydrogen per volume as those mentioned previously and its hydrogen content is equal to those of chloroform ($CHCl_3$) and hydrogen bromide (HBr) for instance. Each group of gases forms a class in itself. *Nobody however has ever discovered a compound in all the million or so that have been investigated which contains less hydrogen in unit volume than do the compounds of the "hydrogen chloride class."*

This is the experimental fact that we use to bring order into the system of chemical formulae. Owing to the fact that no molecule can contain less than one atom of hydrogen we are going to assume that these compounds of hydrogen which belong to the class of lowest hydrogen per volume (in the gaseous state) contain one H atom per molecule. Thus hydrogen chloride, hydrogen bromide and chloroform must contain one H atom per molecule and we have therefore decided upon their formula so far as hydrogen is concerned.

This problem is sufficiently important to merit further discussion. Let us consider further classes of the hydrogen-containing compounds. In ammonia (NH_3) and in methyl chloride (CH_3Cl) we find three times as much hydrogen (in terms of weight) per unit volume as we do in compounds of the first, the hydrogen chloride, class. In methane (CH_4), ethylene (C_2H_4), methyl alcohol (CH_3OH) vapour and acetic acid (CH_3COOH) vapour we find four times as much hydrogen. Naturally, then, we always encounter small integral numbers for the volume ratios occurring in gas reactions, because the gas molecules can only contain the appropriate elements in small multiples of an "elemental" quantity per molecule: i.e. the atom. Or, using the language of Avogadro, these small integral numbers occur because equal volumes of all gases contain equal numbers of molecules and every molecule can only contain an integral number of atoms.

Now at last we are able to compare the atomic weights of the elements with that of hydrogen and so establish a series of relative atomic weights. Let us arbitrarily assume that the atomic weight of hydrogen is 1 and let us denote by the name "atomic weight" of an element the factor by which an atom of the element is heavier than an atom of hydrogen.* Suppose we determine by experiment that volume of the "class one" hydrogen compounds which contains just one gramme of hydrogen. We shall find that this volume is 22·4 litres at normal pressure and temperature. Next we have to arrange the gaseous compounds of each of the other elements in classes, in

---

* This quite arbitrary definition was subsequently changed for reasons of convenience which we shall discuss later, to an alternative arbitrary definition in which the atomic weight of the most abundant isotope of oxygen was taken as 16·000. The difference between the two scales is less than 1 per cent (*see* p. 412). Today the most abundant isotope of carbon is accepted as the basis of the system with $^{12}C \equiv 12·0000$.

the same way as we did with hydrogen. If this is done with the oxygen compounds, for example, we find that the least amount of oxygen per volume is contained in water vapour, carbon monoxide and nitric oxide; twice as much in oxygen gas, in carbon dioxide and sulphur dioxide; three times as much in ozone and sulphur trioxide vapour; four times as much in the vapours of oxalic ester, malonic ester, etc. The compounds of carbon too may be fitted into classes in the same manner, as shown in Table II.

At this stage we may proclaim, for example, that substances which belong to the first class of oxygen compounds contain one oxygen atom per molecule, while those which belong to the first class of carbon compounds contain one carbon atom per molecule and so forth. Suppose we have examined a compound which contains say hydrogen, oxygen, carbon and nitrogen and have assigned it to its respective classes in regard to these four elements. We are now in a position to say how many atoms of each element are present in the molecule since the number will be that of the respective class of the element to which it was found to belong. And thus we are at last able to write the so-called empirical formulae. For instance it happens that hydrogen chloride not only belongs to the first class of hydrogen compounds but also to the first class of the chlorine series and therefore its formula is HCl. Water belongs to the second hydrogen and the first oxygen class and so its formula is H_2O.

What would happen if we found a gaseous compound of an element which contained less of the element in unit volume than did the lowest class previously known to us? Suppose this hypothetical compound was one of hydrogen and contained half as much hydrogen per unit volume as did hydrogen chloride. We should then have to revise our system of formulae and our system of relative atomic weights as well, because we should have proved that what we had formerly assumed to be an indivisible hydrogen atom was really an "H_2" group which we were unable, up to now, to split into two. The atomic weight of hydrogen would become 0·5. Water would have the formula H_4O and so on. Everything would remain perfectly logical and we should accept the new fact as an improvement on our previous knowledge of chemistry. It would be just the same situation as we were in when we defined a chemical element. The fact that hitherto it had never undergone a reaction which made it lighter, was the reason for its being called an element. Here the meaning of writing a formula with one atom in it is that the gas belongs to a class containing the minimal amount of the element per volume. The definition is precise and purely experimental. However, it was never necessary to revise the system after Avogadro discovered it, the reason for this being quite simple. The compounds which contain the smallest number of atoms are the simplest compounds and thus they were among the first to have been analysed by chemists. At the time of Avogadro they had already been analysed and so no difficulties were caused subsequently by the finding of simpler compounds, belonging to a lower class.

The system of relative atomic weights now follows without difficulty. We know that one gramme of hydrogen combines with, and is equivalent to, 35·5 grammes of chlorine and we also know that one atom of hydrogen

TABLE II. Classification of some Gaseous Compounds based on the Amount of Elements found in Equal Volumes of them

A. Classes based on hydrogen content

Class I. Found: 0·0446 g hydrogen per litre.
(No compound is known which contains less.)

Hydrogen chloride (HCl)
Hydrogen bromide (HBr)
Chloroform* (CHCl₃)

Class II. Found: 0·0892 g hydrogen per litre.
(No compound is known which lies between I and II.)

Water vapour* (H₂O)
Hydrogen (H₂)
Hydrogen sulphide (H₂S)
Methylene chloride* (CH₂Cl₂)

Class III. Found: 0·1338 g hydrogen per litre.
(No compound is known which lies between II and III.)

Ammonia (NH₃)
Methyl chloride (CH₃Cl)

Class IV. Found: 0·1784 g hydrogen per litre.
(No compound is known which lies between III and IV.)

Methane (CH₄)
Ethylene (C₂H₄)
Methyl alcohol* (CH₄O)
Acetic acid* (C₂H₄O₂)

B. Classes based on oxygen content

Class I. Found: 0·714 g oxygen per litre.
(Never less.)

Water vapour* (H₂O)
Carbon monoxide (CO)
Methyl alcohol (CH₄O)
Nitric oxide (NO)

Class II. Found: 1·428 g oxygen per litre.
(Nothing between I and II.)

Oxygen (O₂)
Carbon dioxide (CO₂)
Sulphide dioxide (SO₂)
Acetic acid* (C₂H₄O₂)

Class III. Found: 2·142 g oxygen per litre.
(Nothing between II and III.)

Ozone (O₃)
Sulphur trioxide** (SO₃)

Class IV. Found: 2·856 g oxygen per litre.
(Nothing between III and IV.)

Dinitrogen tetroxide (N₂O₄)
Dimethyl oxalate* (C₄H₆O₄)
Diethyl malonate* (C₇H₁₂O₄)

C. Classes based on carbon content

Class I. Found 0·536 g carbon per litre.
(Never less.)

Carbon dioxide (CO₂)
Carbon monoxide (CO)
Methane (CH₄)
Methyl alcohol* (CH₄O)
Chloroform* (CHCl₃)
Methylene chloride* (CH₂Cl₂)
Methyl chloride (CH₃Cl)
Carbon tetrachloride* (CCl₄)

Class II. Found: 1·072 g carbon per litre.
(Nothing between I and II.)

Ethylene (C₂H₄)
Ethane (C₂H₆)
Ethyl alcohol* (C₂H₆O)
Acetic acid* (C₂H₄O₂)

Class III. Found: 1·608 g carbon per litre.
(Nothing between II and III.)

Methyl acetate* (C₃H₆O₂)
Acetone* (C₃H₆O)

Class IV. Found: 2·144 g carbon per litre.
(Nothing between III and IV.)

Dimethyl oxalate* (C₄H₆O₄)
Ether* (C₄H₁₀O)

* All compounds marked * are liquids at room temperature. The carbon, hydrogen and oxygen contents are based on measurements made on the vapour and are recalculated according to the law of Boyle and Mariotte so as to apply to 1 litre of gas at atmospheric pressure at 0°C.
** Solid at room temperature.

combines with one atom of chlorine; thus the atomic weight of chlorine is
35·5. Furthermore one gramme of hydrogen is combined in water with eight
grammes of oxygen and we know that two atoms of hydrogen and one atom
of oxygen are present in each molecule of water (H_2O). Thus the atomic
weight of oxygen must be 16 and not 8.

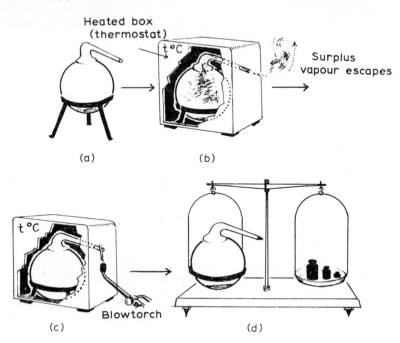

Fig. 17. **Determination of the vapour density of a liquid by
Dumas' method**

(a) Vessel of known volume (V) and known weight when evacuated
(W_0), contains liquid at the bottom.

(b) It is heated to known temperature above the boiling point of the
liquid.

(c) Tip is sealed off when all liquid has evaporated and the vessel
is full of vapour at $t°C$ and 1 atm pressure.

(d) The sealed vessel filled with vapour is weighed. Weight $= W$.

$$\text{Vapour density at } t°C = \frac{W - W_0}{V}$$

$$\text{Vapour density at } 0°C = \frac{W - W_0}{V} \times \frac{t + 273}{273}$$

(Compare p. 33).

Another line of reasoning leads to the same conclusion. It is known that
22·4 litres of such hydrogen compounds as contain the least amount of hydrogen per unit volume contain just one gramme of hydrogen. Let us take the
lowest class of compounds of some other element and determine by analysis
how much of this element is to be found in 22·4 litres of each of these compounds. This amount in grammes is the relative atomic weight of the element.
Now the number of molecules of the hydrogen compound in 22·4 litres equals
the number of molecules of the compound of the other element in 22·4

litres (Avogadro!). The fact that the two compounds belong to the "lowest class" means that each of their molecules contains one atom of hydrogen or one atom of the element in which we are interested. Relative atomic weight is by definition the weight referred to H = 1 and this means that the atomic weight of the element is that weight found in 22·4 litres of the compound. The volume 22·4 litres is therefore called the gramme-molecular volume (at N.P.T.) since the weight of the gaseous compound in grammes contained in this volume, simply tells us how many times the molecule is heavier than an atom of hydrogen. (*See* however, footnote on p. 39.)

As a matter of fact we can find the molecular weight of gases in this manner. We determine the gas density, calculate how much 22·4 litres of gas would weigh (at N.T.P.) and this weight is the molecular weight. The reason for this conclusion is that 22·4 litres of those gases which contain the least amount of hydrogen contain just one gramme of hydrogen, and "molecular weight" means the number of times a molecule is heavier than an atom of hydrogen. (*See* however, footnote on p. 39.)

This reasoning of Avogadro was much beyond the intellectual capacity of his fellow scientists. Decades after he had published it, chemists were still fighting each other on the problem of HO versus H_2O. This was the reason, as we mentioned before, why Weltziehn organized the international congress of chemists in Karlsruhe in 1860. The chemists convened, discussed the matter and disagreed. Avogadro was dead by this time, but a pupil of his, the well-known organic chemist *S. Cannizzaro* (1826–1910) took part and gave a lecture along the lines of Avogadro. He made little impression. His lecture was then printed and distributed among the members of the congress. They dispersed and many of them took to reading the lecture over again; after a while they found out that it really contained the solution of the problem. Thus, though the congress ended in dissent, the young Italian's pamphlet did not miss its aim and the very annoying question of relative atomic weights and empirical formulae was at last settled for ever. The reasoning of Avogadro was not, perhaps, the simplest and this may have been the reason why it took so long for his contemporaries to grasp it. The reader should not be dismayed if he too has some difficulty in following it and it is for this reason that we have discussed the theory in such detail. Avogadro's hypothesis has set the stage for well-defined molecular formulae and for the whole edifice of logical chemistry.

11. MOLECULAR WEIGHTS OF NON-GASEOUS COMPOUNDS

From all that has been discussed it is evident that Avogadro's hypothesis allows us to find the molecular weights of gases or vapours only. We determine the gas or vapour density at any appropriate temperature and pressure and then proceed to calculate how great it would be at N.P.T. (p. 33) and how much 22·4 litres of the gas or vapour would weigh under these "normal" conditions. Apparatus made of platinum has enabled chemists to work with materials that evaporate only above 1000°C. However, many compounds decompose before they are hot enough to evaporate sufficiently for measurements to be made and other means must be found to enable us to determine the molecular weights of such substances. It is therefore important to become acquainted with other methods which give information on molecular weights, it being understood that their validity must have been tested on compounds whose molecular weights have been determined in the gas phase by Avogadro's method.

In 1819, not long after Avogadro had published his ideas, two French physicists, *P. L. Dulong* and *A. Th. Petit* discovered an empirical rule concerning the specific heats of elements. The specific heat of a substance is the quantity of heat necessary to raise the temperature of one gramme of the substance by one degree centigrade. The atomic heat of an element is its specific heat multiplied by its atomic weight and is the heat necessary to raise the temperature of one gramme-atom of the element by one degree centigrade. It is evidently the quantity which should be compared when considering equal numbers of atoms of different elements. The empirical rule was that with few exceptions (mostly among the elements of lowest atomic weight) this atomic heat was in the neighbourhood of 6 calories per gramme-atom per degree. The average value was 6·2.

For instance, 0·18 calorie is necessary to raise the temperature of 1 g of sulphur by 1 deg C and hence 5·6 calories ($= 32 \times 0·18$) are required to raise the temperature of one gramme-atom (32 g) by 1 deg C. Similarly the specific heat of phosphorus (atomic weight $= 31$) was found to be 0·20, giving 6·2 for its atomic heat. The atomic heat of sodium is 6·7 that of aluminium 5·8, and so on.

Naturally such a method yields no accurate atomic weights but may be used to decide which multiple of the chemically determined equivalent weight is the true atomic weight of an element. Thus gold can form two chlorides, one of them containing 65·73 g of gold for each 35·5 g of chlorine (35·5 is the atomic weight of chlorine). Hence 65·73 is the "equivalent weight" of gold in this particular compound. But what is its atomic weight? It is out of the question to measure the vapour density of gold chloride because it rapidly decomposes on heating. Atomic heat measurements enable us to solve our problem, however. Experiments show that 0·0316 calorie is necessary to raise the temperature of 1 g of gold by 1 deg C (this quantity of heat is the "specific heat" of gold). We must find out what weight of gold would require 6·2 calories to produce a temperature rise of 1 deg C; the answer is about 196 g. The next step is to ascertain which integral multiple of 65·73 is nearest to this number and we find that $65·73 \times 3 = 197·2$ g. This is therefore the atomic weight of gold and the chloride we have been dealing with is $AuCl_3$. Our result is corroborated by the purely chemical fact that the other chloride of gold is found to contain exactly 197·2 g of gold for each 35·5 g of chlorine and thus has the simple formula $AuCl$.

The physical meaning of the Dulong–Petit rule is itself connected with the physical meaning of temperature. It will be shown in Section 30 (p. 150) that if two bodies are at the same temperature each of them will possess the same average kinetic energy per particle. Hence the increment of this average energy for equal rises in temperature must be constant when a constant number of particles is considered, for instance one gramme-atom of an element. The exceptions to this rule, however, can only be understood on the basis of quantum mechanics (p. 207).

Another quasi-empirical rule which may often serve to distinguish between possible atomic weights was discovered by the Hungarian physicist *L. Eötvös*. His rule concerns the surface tension of liquids. As we know, liquids tend to contract their surface to a minimum area so long as other forces do not intervene, and so energy (force) is necessary to increase the surface area of a liquid. This is the force that makes liquids form into drops. Eötvös found in 1886 that the energy which is necessary to form one face of a cube which contains one gramme-molecule of any liquid decreases by 2·12 ergs for each degree centigrade. Thus it was only necessary to determine the temperature dependence of the surface energy of a fluid to be able to estimate its molecular weight. This could then be compared with possible values obtained from stoichiometric determinations and so a choice among the latter became possible. The rule does not hold for all fluid compounds, however.

It was fortunate indeed for chemistry, when *F. M. Raoult* (1830–1901) discovered relationships which permitted the calculation of the molecular weights of dissolved compounds (1886). Experiments show that the freezing-point of a liquid is lowered when a substance is dissolved in the liquid. For example we know that a salt–ice mixture is colder than pure ice since salt solution freezes at a lower temperature than does pure water. The boiling-point of a liquid will also change if the liquid contains a solute. When the solute itself is not volatile the boiling-point always rises, e.g. salt water boils a few degrees above 100°C. These freezing and boiling phenomena are interrelated through the fact that the vapour tension above a liquid decreases if anything is dissolved in the liquid (Fig. 18). If the vapour

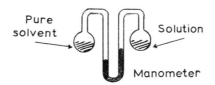

Pure solvent   Solution

Manometer

Fig. 18. **Vapour pressure depression of solutions**

tension (vapour pressure) of salt water has not reached a value of one atmosphere at 100°C, the solution simply cannot boil at this temperature and has to be heated further. On the other hand, if the vapour pressure of salt water is less than that of pure ice at 0°C, ice crystals will evaporate and condense on the surface of the solution where the vapour tension is lower. Alternatively, leaving out the process of sublimation they may simply dissolve in the solution. (The route selected to attain an equilibrium in a particular system is itself never decisive and all routes must lead to the same equilibrium of the system; the phase with the lowest vapour pressure is always the stable one to which the others will eventually revert.)

All three phenomena may be visualized, albeit crudely, by imagining that the solute molecules, which are themselves unable to leave the solution, obstruct the solvent molecules, e.g. when they are about to leave the liquid surface and enter the vapour phase or leave the surface of their own solid crystals. Hence the vapour pressure of the solvent above the solution will be lowered. In case of the freezing-point depression, the number of water molecules passing from the ice surface into the solution remains unchanged, whereas the number of water molecules reaching the same surface from the solution has decreased because of the solute molecules impeding their passage. The ice crystals will therefore dissolve. A fourth phenomenon related to the depression of vapour pressure and freezing-point, and the elevation of the boiling-point, of a solution is called the "osmotic pressure" and all four

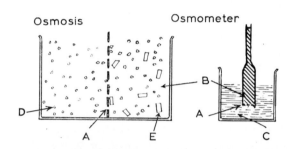

Osmosis   Osmometer

Fig. 19. **Osmosis and osmotic pressure**

(*left*)   Solvent molecules (D) are able to traverse the semi-permeable membrane (A), solute molecules (E) are not. Their passage is blocked. Result: net flow of solvent (D) from left to right.

(*right*)   Influx of solvent (C) raises level of solution (B) until the hydrostatic "head" becomes equal to the osmotic pressure.

are often called "osmotic" phenomena. Osmotic pressure can be detected across a membrane which is permeable to the solvent molecules but impermeable to the solute molecules. Collodion membranes, for example, can be used for aqueous solutions. If such a membrane is interposed between a pure solvent and its solution, once again the solute molecules can be regarded as blocking the passage of solvent molecules through the membrane from the solution. Solvent molecules are free to enter from the pure solvent side, however, and thus solvent keeps streaming across the membrance into the solution until the hydrostatic pressure of the solution is sufficiently large to stop it. The counter-pressure necessary to stop this solvent flow is called the osmotic pressure and it is evident on the basis of the elementary theory how it is related physically to the other three phenomena.

Raoult found, that all four effects were proportional to the amount of solute in a solution. A much more important experimental result was that if two or more solutions were made, consisting of the same number of moles (= gramme-molecular weights) of different solutes dissolved in the same weight of a given solvent (i.e., with identical molal concentrations), then the changes in freezing-point, boiling-point, etc., were the same for each solution. Thus, once a solvent was "calibrated" by solutes of known molecular weight, that is once it had been ascertained by just how much the freezing-point, etc., changed when one mole of a solute was dissolved in a kilogram of solvent, the molecular weight of unknown compounds could be determined. It was only necessary to determine what weight of the unknown material per kilogram of solvent would yield the same freezing-point depression or boiling-point elevation, etc., as one mole of a calibrating compound. This amount of the unknown compound, in grammes, was a gramme-molecule or a "mole" of it. The method based on freezing-points is called cryoscopy, that based on boiling-points, ebullioscopy.

One important and extensive list of exceptions must be noted. All solutions which conduct electricity in a non-metallic manner, as, e.g. aqueous solutions of salts, strong acids and strong bases, show osmotic effects which are considerably greater than expected on the basis of the molecular weights of the solutes. Solutions of non-electrolytes such as sugar, urea or glycerine or of weak electrolytes such as boric acid or weak organic bases behave normally. However, it will become evident later on (p. 63) that the apparently anomalous behaviour of these strong electrolytes can be fitted logically into the picture.

12. MOLECULAR WEIGHTS AND FORMULAE

Let us now be content that we have managed to determine the molecular weight of quite a range of compounds by one method or another. Our first interesting observation in this field is the fact that the molecular weight of a compound is just as characteristic of it as its quantitative composition. This means, that two or more "unknown" substances will represent individual compounds if their molecular weights differ, notwithstanding the fact that their percentage compositions might be the same. Perhaps the simplest example to consider is that of oxygen and ozone. Each is 100 per cent oxygen as far as elementary composition is concerned but their formulae are O_2 and O_3 respectively, corresponding to the molecular weights 32 and 48. Ozone is formed from oxygen by passing the latter through a silent electric discharge or by irradiation with ultraviolet light. It is a very corrosive, poisonous oxidizing agent, has a boiling-point and condensation point of its own, distinct from those of oxygen and it dissolves in solvents differently from oxygen. By all chemical and physical criteria we are dealing with two distinct compounds although they only differ in molecular weight and not in elementary composition.

A similar situation exists with respect to acetylene, a gas used for lighting and welding, and benzene, the well-known solvent. Each compound has the elementary composition 7·7 per cent H and 92·3 per cent C. It is hardly necessary to point out in detail their different properties. It will suffice to realize that benzene is a liquid at ordinary temperatures whereas acetylene is a gas. If we determine experimentally the density of the gas and of benzene vapour and calculate the weight of each in 22·4 litres at N.T.P., (that is, if we apply Avogadro's hypothesis) we arrive at the molecular weights 26 and 78 for acetylene and benzene respectively. What are their chemical formulae? In the case of acetylene, 7·7 per cent of 26 is 2 and 92·3 per cent is 24. Bearing in mind that the atomic weight of hydrogen is 1 and of carbon is 12 we see that a mole of acetylene contains two gramme-atoms of hydrogen and two gramme-atoms of carbon. In our "class system" acetylene belongs to the second H-class and the second C-class. Its formula must therefore be C_2H_2. In the same manner it can be shown that benzene has the formula C_6H_6. These two compounds are by no means the only ones with the same quantitative composition: 7·7 per cent H and 92·3 per cent C. The simplest member of the series however (CH) does not exist as a stable compound, although the reason for this fact can be understood only after a discussion of the principles of chemical bonding and so-called "valency".

A last short example involves rather more complex molecules. For-maldehyde gas, acetic acid and grape-sugar (glucose) differ widely in their chemical and physical properties but all three substances have the elementary composition 6·7 per cent H, 40·0 per cent C and 53·3 per cent O. Determination of the freezing-points of their solutions in water yields the molecular weights 30, 60 and 180 and hence their formulae are CH_2O, $C_2H_4O_2$ and $C_6H_{12}O_6$ respectively, if we remember that the atomic weights of H, C and O are 1, 12 and 16.

13. ISOMERISM

The foregoing is not all of the story that chemistry has to tell us. Up to now we have dealt only with examples of compounds with the same elementary composition but different molecular weights. We saw that such compounds were quite distinct from each other. Next we shall become acquainted with compounds which differ in spite of the fact that not only their composition but even their molecular weights are the same! Their physical constants and their chemical reactions are entirely different so there cannot be the least doubt as to their being different chemical entities.

Most of us are familiar with the volatile liquid called ether (ethyl ether) which is used as an anaesthetic. This compound has a "junior sister" called methyl ether, an inflammable gas under normal conditions. It formula is C_2H_6O. Ordinary alcohol, the intoxicating component of hard drinks, happens to have the same formula, meaning that its percentage composition *and* its molecular weight are identical with those of methyl ether. Another example of the above type concerns hydroquinone, a compound used in photographic developers, the white crystals of which melt at 172°C and another developer, pyrocatechol which melts at 105°C and is more soluble in water than hydroquinone. A third compound, resorcinol, is of no use for developing but is employed in pharmacy. It melts at 111°C and its solubility in water is intermediate between that of hydroquinone and that of pyrocatechol. All three substances have the same formula, $C_6H_6O_2$.

Do not think that the above can happen only with carbon compounds.

that is, in the vast domain of what we call organic chemistry. In the great majority of cases it does happen there, but it also occurs among inorganic compounds. For example, there are two different compounds of platinum, chlorine and ammonia having the same empirical formulae $PtCl_2N_2H_6$. One is sulphur-yellow, the other greenish-yellow and all their properties are different. They are certainly separate compounds.

Such sets of compounds which are built from the same number of the same sort of atoms but still are distinct from each other in their properties and reactions are called isomeric compounds or in short isomers (from the Greek ἴσος equal, μέρος part). What is the cause of this interesting phenomenon?

If we decide to take the atomic hypothesis seriously, that is if we really believe that bulk matter is built from molecules and every molecule from a specified number and kind of atoms, we should not be surprised to learn that it is possible to arrange these atoms in more than one way within the molecule. It is analogous to erecting different buildings from the same number and types of bricks. The significant question, one which we could not have answered previously, is whether the atoms within a molecule are free to move around each other like balls in a bag or are confined to definite relative positions as are the bricks of a house. The fact that isomeric compounds do exist enables us to decide in favour of the latter assumption.

It seems therefore, that the formulae we have used up to now, the so-called empirical formulae, cannot give us all the information that there is concerning the constitution of molecules. They inform us about percentage composition and molecular weight, in other words about the kind and number of atoms within a molecule, but say nothing about molecular internal structure. It has therefore been necessary to evolve another set of formulae which do show the relative position of the atoms in a given molecule. These are the structural formulae of the chemist.

The first pair of isomeric compounds was prepared by the young *J. v. Liebig* (1803–1873) in 1824–1825. He was a quite exceptional boy who began to play around with chemistry at the age of 15. In the market place of his native town he learned from a juggler how to prepare a detonating material from mercury, alcohol and nitric acid. He loved to fiddle around with this compound, which, incidentally happens to be one of the most sensitive and dangerous detonating compounds known (mercury fulminate). He was sent away from school for playing jokes with the substance and later on he lost his job in a pharmacy because he caused an explosion with it. However, he soon managed to join the famous Gay-Lussac in Paris and worked in his laboratory for a couple of years. It was there that he prepared the parent compound of his mercury fulminate, as we now call it. It proved to be an acid with the formula CHON. The Academy of France published the experimental results of this young man of twenty. One year later however, *F. Wöhler* (1800–1882) using red potassium ferricyanide prepared a compound of the same formula which gave a series of salts (cyanates) in its turn. None of them proved to be explosive, though, and they were absolutely distinct from Liebig's compounds. At first each of the experimenters thought that the other was wrong, and they were quite offended by each other. However, they soon met and in a common series of experiments proved that both sets of analysis were correct. They remained good friends ever after. It was Gay-Lussac, who finally drew the conclusion that the arrangement of the atoms must be different in the two sets of compounds, while it remained for Berzelius to define exactly the concept of isomerism.

14. RECURRING GROUPS: RADICALS

It is now up to us to determine by some means the relative position of atoms within a given molecule, knowing from our experience of isomerism that such relative positions really do exist. Evidently we ought to follow chemical changes in the compounds subjected to a series of reactions and see which atoms usually remain together as a group and which are more easily split off. Logically we must assume that those atoms which do remain as a group through many reactions are close together and more strongly bound to each other. It was again Liebig and Wöhler who made a historic set of experiments on the aromatic oil of bitter almonds (benzaldehyde) in 1832, along the following lines. The formula of the oil itself was C_7H_6O whereas a set of derivatives had the formulae C_7H_5OCl, $C_7H_5O_2$, $C_{14}H_{10}O_3$, etc., the group C_7H_5O being common to all of them. It became customary to call such groups which seemed to form the root, or radix, of a series of compounds "radicals", and the name Liebig and Wöhler gave to the radical C_7H_5O was benzoyl. Using this concept the formulae of the four compounds mentioned above could be written: $(C_7H_5O)H$, $(C_7H_5O)Cl$, $(C_7H_5O)OH$ and $(C_7H_5O)_2O$. The brackets simply mean that the atoms within them remain together throughout a set of reactions; it would be logical to name the four compounds benzoyl hydride, benzoyl chloride, benzoyl hydroxide and benzoyl oxide.

E. Frankland (1825–1899), an English chemist who studied in Germany, and *A. W. H. Kolbe* (1818–1884), a German chemist, found an even simpler group which tended to remain intact. This had the formula (C_2H_5) and they named it ethyl. Its hydride is the gas ethane, C_2H_6, whereas its hydroxide is common alcohol, $(C_2H_5)OH$, and its chloride, $(C_2H_5)Cl$, is the sweet-smelling gas used for light anaesthesia. Our old acquaintance for deep anaesthesia, ethyl ether, belongs to this series, too, being its oxide, $(C_2H_5)_2O$. It is worth while to have a look into the reactions which connect these compounds. The action of not too much chlorine on ethane in the presence of light yields ethyl chloride:

$$C_2H_6 + Cl_2 = C_2H_5Cl + HCl \qquad . \qquad . \qquad . \quad (14.1)$$

which in turn yields alcohol on reaction with caustic soda:

$$C_2H_5Cl + NaOH = C_2H_5OH + NaCl \qquad . \qquad . \quad (14.2)$$

On removing water from two alcohol molecules with hygroscopic concentrated sulphuric acid we obtain ethyl ether:

$$2C_2H_5OH = (C_2H_5)_2O + H_2O \qquad . \qquad . \qquad . \quad (14.3)$$

Now ether can be split into two molecules of ethyl iodide by concentrated hydrogen iodide:

$$(C_2H_5)_2O + 2HI = 2(C_2H_5)I + H_2O \qquad . \qquad . \quad (14.4)$$

and ethyl iodide in its turn yields alcohol again on treatment with caustic soda:

$$(C_2H_5)I + NaOH = (C_2H_5)OH + NaI \qquad . \qquad . \quad (14.5)$$

There is no need to illustrate further reactions, although it would of course be possible.

The main point is that in one case the benzoyl, in the other the ethyl group of atoms remain together throughout a long series of chemical reactions. We seem to be justified in our initial assumption that the atoms within brackets must be more strongly bound together than those atoms of the molecule which change more readily. However, it is possible to proceed a step further. If we react metallic sodium with alcohol it dissolves, evolving hydrogen gas at the same time. The product, which is a white solid recoverable by driving off excess alcohol, is called sodium ethoxide:

$$2(C_2H_5)OH + 2Na = 2(C_2H_5)O \cdot Na + H_2 \qquad . \quad (14.6)$$

(It is necessary to show two molecules of alcohol reacting with two atoms of sodium in order to obtain the two atoms of hydrogen necessary to form a hydrogen molecule.) During this reaction we find that not only does the ethyl group hold together, but even the oxygen atom of the original alcohol remains in the new product. We could thus become acquainted with a whole set of compounds derived from ethyl alcohol by replacement of only the hydrogen atom of its hydroxyl (OH) group with something else, leaving the (C_2H_5O) group unchanged. This so-called ethoxy radical may be regarded as a composite radical $[(C_2H_5)O]$ which includes the simple ethyl radical as a component. The situation is as follows: the ethyl group is a part of the ethoxy group because it is present each time we find the latter. It should be realized that the converse is not true—we do not find an ethoxy group each time we encounter an ethyl radical. The oxygen atom can be separated from the ethyl group, e.g. in the reaction with hydrogen chloride:

$$[(C_2H_5)O]H + HCl = (C_2H_5)Cl + H_2O \qquad . \qquad . \quad (14.7)$$

This is the behaviour of a composite radical. The round brackets in our formula correspond to a stronger binding than the straight ones. However, they do still represent a binding force and therefore we are justified in saying that the oxygen atom in an alcohol molecule is bound to the ethyl group, while the sixth hydrogen atom of the molecule is only bound to the ethyl *through* this oxygen atom, since it is always lost from the molecule whenever the oxygen atom is lost. If we decide to formulate the position of binding forces by drawing lines between the different parts of a molecule we arrive at the partial structural formula for ethyl alcohol C_2H_5—O—H, and not, for example, C_2H_5—H—O because it *is* possible to remove the lone H atom while leaving the O atom connected to the C_2H_5 radical. This structural formula is nothing else than the sum of all our chemical experience which has been described above. It is essentially in this laborious way that chemists go on building their structural formulae, step by step by a study of chemical reactions.

15. VALENCY

It would be interesting to know the detailed internal structure of the ethyl radical itself, because until now we know only that it tends to remain intact so long as it is not exposed to violent reactions. (Such a violent reaction would be, for instance, the complete burning of an ethyl compound with oxygen, which would yield carbon dioxide and water.) By becoming more intimately acquainted with the structure of the ethyl radical we are able to introduce a new concept of chemistry, that of valency.

As was mentioned above, the formula of the gas ethane is C_2H_6 and on chlorination with a small quantity of chlorine it yields C_2H_5Cl. However, we could use much more chlorine in this reaction and in such a case we should obtain a long series of new chlorinated derivatives. These are liquids, the boiling points of which increase with increasing chlorine content. The last member of the series, which no longer contains hydrogen and has the formula C_2Cl_6 is a solid. The formulae are: C_2H_5Cl (one isomer), $C_2H_4Cl_2$ (two isomers), $C_2H_3Cl_3$ (two isomers), $C_2H_2Cl_4$ (two isomers), C_2HCl_5 (one isomer) and C_2Cl_6 (one isomer). What is noteworthy in this series? Just the same thing that would have caught our attention among the chlorinated derivatives of methane in Table I (p. 20). If we had written their formulae instead of their percentage compositions we should have obtained CH_4, CH_3Cl, CH_2Cl_2, $CHCl_3$ and CCl_4 (only one isomer in each case). Chlorine is substituted for hydrogen, atom for atom, in this series ("Substitution theory" of *J. B. A. Dumas* 1800–1884, *A. Laurent* 1807–1853, *Ch. F. Gerhardt* 1816–1856). There are always four atoms associated with a carbon atom in the methane series. Whenever the number of hydrogen atoms decreases by one, one more atom of chlorine enters in its place. In the same way there are always six atoms around the two carbon atoms in the ethane series and here also hydrogen atoms are substituted one by one.

We thus gain the impression that a carbon atom always binds four hydrogen atoms, or four atoms or atomic groups equivalent to a hydrogen atom, in these compounds. We regard as "equivalent to hydrogen" any atom or group which is able to combine with or to replace one atom of hydrogen. It is the same definition as we have already used when introducing the concept of equivalent weights in a previous chapter (p. 26). Thus a chlorine atom is equivalent to one hydrogen atom because we know they are combined in HCl. An oxygen atom is equivalent to two hydrogen atoms because water has the formula H_2O. Similarly a nitrogen atom is equivalent to three hydrogen atoms because ammonia is NH_3. However, the hydroxyl radical, OH, is equivalent to only one hydrogen atom because it combines with one to form water, H_2O. *We refer to the number of atoms of hydrogen (or atoms or radicals equivalent to an atom of hydrogen) which another atom is able to combine with or displace, as the valency of this atom* (*Frankland*, 1852).

As we are acquainted with the idea that a pair of elements may combine with each other in more than one weight ratio (law of multiple proportions) it is clear that the valency of an element cannot always be represented by a single number. However, this simple truth did not become acknowledged for a long time because, as we have seen, the concepts of atomic, equivalent and molecular weights were not sufficiently clarified until Cannizzaro helped Avogadro's hypothesis to victory. It is obvious to us today that we are unable to formulate many of even the simplest compounds using constant valency values.

Consider for instance, carbon dioxide and monoxide (CO_2 and CO). We know of carbon that it combines with four hydrogen atoms in its simplest hydrogen compound, methane (CH_4). Oxygen, on the other hand, is divalent in its simplest hydrogen compound, water (H_2O). Thus we are able to formulate with consistency CO_2 as $O{=}C{=}O$ with tetravalent carbon and divalent oxygen provided we denote each bond formed by one valency each of both bonded atoms with a straight line. But we have difficulties with CO. Either we must assume that carbon is divalent, to satisfy oxygen, or that oxygen is tetravalent to satisfy carbon. On the other hand they could each have a valency of three or any other number, for all we know from the previous discussion. We have to invoke detailed chemical arguments to reach a decision in this matter. Experiments show us that CO is capable of combining with different atoms or groups and every time such an addition happens, the added parts are bound in the new molecule directly to carbon and not to oxygen, e.g. CO unites with Cl_2 and the resulting $COCl_2$ (the poisonous gas phosgene) may be converted by reaction with alkali into carbonic acid $CO(OH)_2$ in which the two OH groups are bound to the carbon atom. This fact is established by the existence of a whole series of salts (carbonates) which are derived from carbonic acid by the substitution of metals for the H atoms, while the O atoms always remain attached to carbon:

$$
\begin{matrix}
Cl \\
| \\
Cl
\end{matrix}
+ CO =
\begin{matrix}
Cl \\
\diagdown \\
\diagup \\
Cl
\end{matrix}
CO
\qquad . \qquad . \qquad . \qquad . \quad (15.1)
$$

$$
\begin{matrix}
NaOH + Cl \\
\diagdown \\
\diagup \\
NaOH + Cl
\end{matrix}
CO =
\begin{matrix}
HO \\
\diagdown \\
\diagup \\
HO
\end{matrix}
CO + 2NaCl
\qquad . \qquad . \quad (15.2)
$$

$$
\begin{matrix}
NaOH + HO \\
\diagdown \\
\diagup \\
NaOH + HO
\end{matrix}
CO =
\begin{matrix}
NaO \\
\diagdown \\
\diagup \\
NaO
\end{matrix}
CO + 2H_2O
\qquad . \qquad . \quad (15.3)
$$

the same compound as from the reaction

$$2NaOH + O{=}C{=}O = \quad \begin{matrix} NaO \\ \diagdown \\ \diagup \\ NaO \end{matrix} C{=}O + H_2O \quad . \qquad . \quad (15.4)$$

One could of course say that the two new oxygen atoms became attached to the original O of CO, which superficially could also be true. But a vast body of experience shows that two O atoms bonded to each other are always energetic oxidants. This we cannot say of the carbonates, so it is fairly certain that the oxygen atoms of the OH groups which have displaced the chlorine atoms are attached to carbon. We refer to the carbon atom in carbon monoxide as being "unsaturated". The isomeric fulminates and cyanates found by Liebig and Wöhler and referred to earlier also differ in the fact that the carbon in fulminic acid is unsaturated (C$=$N—O—H) whereas that in cyanic acid is in its usual state of tetravalency (N$\equiv$C—O—H).

We have already encountered multiple valency in compounds of carbon but it must be added immediately that carbon is nearly always tetravalent whereas other elements such as chlorine, sulphur, iron, etc., are frequently found in different valency states. *Kolbe* was the first to insist on the general tetravalency of carbon while *F. A. Kekulé* (1829–1896) who postulated the ring structure of benzene, pronounced it as a rule (1858).

Hydrogen is always monovalent, the alkali metals are always monovalent, the alkaline earth metals divalent and aluminium always trivalent. However, as we have seen earlier, there are two chlorides of iron, corresponding to its di- and trivalent states, (i.e. $FeCl_2$ and $FeCl_3$). We have also dealt with the nitrates of mono- and divalent mercury, $Hg(NO_3)$ and $Hg(NO_3)_2$. Another example is vanadium which is a metal often alloyed with steel; it may occur in any valency state between one and five.

16. STRUCTURAL FORMULAE I: INORGANIC COMPOUNDS

In the chapter on multiple weight ratios we have mentioned the series of compounds from sodium hypochlorite to sodium perchlorate (p. 21). Now at last we are able to convert the percentage composition of these compounds into the empirical formulae: $NaOCl$, NaO_2Cl, NaO_3Cl, NaO_4Cl. By considerations similar to those of the preceding chapter we are able to show that the Na atom is always bound to Cl via an O atom, but this only settles the total structure for the first compound: Na—O—Cl. In the case of the other compounds we only know that O atoms tend to remain combined with Cl, but we do not know whether they are each completely bonded to it or are partly bonded to each other, say in a chain or ring. For example, the structure of sodium perchlorate might conceivably be

$$Na{-}O{-}O{-}O{-}O{-}Cl \quad \text{or} \quad Na{-}O{-}Cl \begin{matrix} \diagup O \diagdown \\ \diagdown O \diagup \end{matrix} O$$

However, the O—O bond is improbable in this instance because it exhibits a marked tendency to oxidize other substances or even to evolve molecular oxygen, as for example in the case of the decomposition of hydrogen peroxide under the influence of powdered platinum metal or a drop of blood. (This "influence" in helping a reaction is called catalysis). Therefore we write H—O—O—H for hydrogen peroxide but hesitate very much in the case of perchlorates. As a matter of fact perchlorates do evolve oxygen after they have been heated to some hundreds of degrees, but they evolve it even less easily than e.g. NaO_3Cl (sodium chlorate) which contains one oxygen atom *less*. In the whole series it is remarkable that the compound with the lowest oxygen content, sodium hypochlorite, is the strongest oxidizing agent and this we have already agreed to formulate as Na—O—Cl. Sodium perchlorate is formulated as

$$Na-O-\overset{\displaystyle O}{\underset{\displaystyle O}{\overset{\|}{\underset{\|}{Cl}}}}=O \quad \text{or rather as} \quad Na\begin{bmatrix} O & O \\ & Cl & \\ O & O \end{bmatrix}$$

In order to reach this conclusion it is instructive to glance at a series of similar compounds which include not only the salts of oxygenated acids but other salts too, the acid radicals of which contain a plurality of atoms of a sort different from the "central" atom. Such are, for example, sodium fluoroborate, $NaBF_4$; sodium sulphate, Na_2SO_4; potassium nitrate, KNO_3; potassium hexachloroplatinate, K_2PtCl_6. Their aqueous solutions conduct electricity and the sodium or potassium ions (cations) migrate towards the cathode whereas the "acid radicals" (anions) migrate towards the anode. The characteristic atom of the complex anion and its cluster of allied atoms all remain together; they are strongly bonded to each other whereas the cationic metal (Na, K) is readily exchangeable for another and seems to live a life of its own in solution, as judged by its opposite direction of migration.

It seemed probable a long time ago, that in the complex anions the allied atoms were bonded in a symmetrical manner around the central atom. This symmetry has now been definitely established but only by physical methods which we shall consider later on. Chemical proof came when isotopic labelling became possible. We have mentioned already (p. 23) that the atomic weight of individual atoms of one element is not necessarily always the same but may extend over a range of values which are not very different from each other and as we shall see (p. 415) are nearly integral numbers. In the case of hydrogen, for instance, we have besides the common isotope (H) the heavy hydrogen isotope, deuterium (often designated by D) with atomic weight 2 and also radioactive tritium (T) with atomic weight 3. Oxygen has isotopes of mass 15, 17 and 18 besides the common one of mass 16. The isotopically pure samples of elements behave from the point of view of chemistry in the same way, but their atomic weights and accordingly their specific gravities are different. Thus they can be differentiated from each other by density measurements or by more intricate physical methods. Now, if we want to know whether atoms of the same element occupy identical positions in a molecule or not, we may make use of these isotopic atoms. They are in effect "marked" by their characteristic masses as though a ribbon had been tied to them and they allow us to "trace" their fate during the course of reactions.

All this sounds a bit esoteric at first but an example will show what is meant. If we burn elementary sulphur and oxygen to SO_2 and proceed to unite the latter with further oxygen over a "catalytic" platinum gauze we obtain sulphur trioxide, SO_3. This compound reacts very energetically with water, H_2O, to yield sulphuric acid, H_2SO_4. As a matter of fact sulphuric acid can be made this way. We want to know whether, in the H_2SO_4 molecule, the oxygen atom which originated from the water is bound differently from those which were already beside the S atom in SO_3; i.e. whether the O atoms in the sulphate group (SO_4) can "indicate," can "remember," their origin. To decide the question we can use water containing a heavy isotope of oxygen for our reaction; let us write it $\underline{O}$ in our formulae. We then have a sulphuric acid $H_2SO_3\underline{O}$. If it were possible to regenerate SO_3 from sulphuric acid we could see, by suitable means, where the tracer $\underline{O}$ atom went to—whether the three O atoms originally attached to S remained there or whether $\underline{O}$ had intruded. It is possible to regenerate the SO_3. For instance by mixing the sulphuric acid with a salt of lead (Pb) we can precipitate lead sulphate, $PbSO_4$ (lead is divalent and replaces two hydrogen atoms). This in its turn decomposes on heating to give PbO (lead oxide) and SO_3. Experiments have proved that our tracer $\underline{O}$ atom is distributed between PbO and SO_3 exactly according to the law of chance, being found three times as frequently among the three O atoms of SO_3 as in the single O of PbO. This means that the O atom from water "forgot" its origin in the SO_4 radical and was indistinguishable from its sister atoms. Similar tracer reactions may be carried out with $K\underline{Cl}$ and $PtCl_4$ to yield information on K_2PtCl_6 and in other similar cases, always with the same result. This means that these allied atoms are distributed with equal probabilities, in symmetrical positions around the central atom of their anion. When writing formulae we represent this state of affairs by, for example

$$\begin{bmatrix} & O & \\ O & S & O \\ & O & \end{bmatrix} \text{ for the sulphate radical}$$

$$\begin{bmatrix} & F & \\ F & B & F \\ & F & \end{bmatrix} \text{ for the fluoroborate}$$

and so on.

This equivalence of atoms from different origins within a complex group of atoms (complex, because originating through the combination of different molecules) is not limited to anions. Ammonia gas and hydrogen chloride combine to "sal ammoniac", ammonium chloride, H_4NCl. As it happens, Cl is found to leave the molecule with ease: it can be precipitated with a soluble silver salt to form insoluble AgCl. Furthermore it migrates to the anode under the influence of an electric field, whereas the new complex "ammonium" group, H_4N, remains together throughout a long series of reactions and on electrolysis migrates to the cathode: it is a complex cation. Again we may ask ourselves whether the H atom from the original HCl is bound in a different way from those which had been in the NH_3 already, or not. All we have to do is to repeat the reaction with the chloride of heavy hydrogen, DCl. The labelled ammonium chloride is now H_3DNCl. It may be decomposed to yield ammonia again by reaction with caustic soda:

$$H_4NCl + NaOH = H_3N + NaCl + H_2O \qquad . \qquad . \quad (16.1)$$

If the D atom had "remembered" its origin we should have NH_3 and HOD. But this happens only with one-quarter of our molecules, three out of four yield ordinary H_2O and H_2DN. The D atom is distributed between the two products according to pure chance. Every fourth hydrogen atom in our labelled ammonium chloride is a heavy one and hence every fourth one of the resulting water molecules will have "drawn" the heavy atom from the bag which contained the four hydrogen atoms in equivalent positions. Once more we shall write

$$\begin{bmatrix} & H & \\ H & N & H \\ & H & \end{bmatrix}$$

for the ammonium group.

This concept of complex groups around a central atom, with other atoms further away, "in the second sphere" is due to *A. Werner* (1866–1919), the eminent Swiss inorganic chemist at the beginning of our century. He was the first to disapprove definitely of writing, for example, ammonium salts with pentavalent nitrogen,

$$\begin{array}{ccc} H & & H \\ & \diagdown \quad \diagup & \\ H\!\!-\!\!N\!\!-\!\!\!-Cl & \\ & H & \end{array}$$

and introduced the symbol

$$\begin{bmatrix} H & & H \\ & N & \\ H & & H \end{bmatrix} Cl$$

instead, indicating the complex nature of the ammonium group, the equivalence of its hydrogen atoms and the independence of the chloride ion. Isotopic tracers were not available in his time; unfortunately he died at an early age and did not live to see them. It needed an immense effort on the part of Werner to systematize the vast amount of chemical information which had accumulated in inorganic chemistry, in order to make plausible on chemical grounds views which today can readily be proved with tracer experiments. But he did succeed in convincing his contemporaries that he was right and that inorganic compounds fit much more easily into a unified system if they are represented in this manner, than if one tries to enforce a rigid system of valence bonds between pairs of atoms, representing them with straight lines in the formulae.

This system is not yet so forgotten that we should be allowed to by-pass it. It is applicable in some cases but not in others, having given much trouble to the generation of chemists before Werner. Let us look at some examples.

We can prepare SO_3 from S and O_2 and owing to the fact that it exerts no exceptional oxidative properties there is no sense in arranging its formula to contain O—O bonds. Thus the simplest formula results,

$$O=S=O$$
$$\|$$
$$O$$

with hexavalent sulphur at the centre. A hexafluoride, SF_6 is known, so there was really no reason to doubt the hexavalent S and this arrangement. Simple addition of water should not change the valency of an element (because this changes only during oxidations or reductions), so sulphuric acid can be represented by the formula

conserving the hexavalency of sulphur. Let us note, by the way, that the four oxygen atoms in this formula are by no means equivalent. There are two within OH groups and two doubly bonded to sulphur. Our heavy O would have had a chance of 1/2 instead of only 1/4 to remain with the Pb atom in $PbSO_4$ if the above formula really represented sulphuric acid.

But F is monovalent both in NaF and in BF_3, and thus evidently in their addition compound $NaBF_4$. The same applies to H in HCl and NH_3 and so in NH_4Cl. Therefore the trick we were able to employ with divalent oxygen in the sulphate ion whereby we used it to interconnect the H (or metal) atom to S by a valency bond, will not work here. A monovalent atom cannot interconnect two others in this manner, having but one bond available. Therefore these compounds were given either a definitely false representation as ammonium compounds with pentavalent N and equally bonded H and Cl (*see* above) or were not seriously represented at all until Werner brought order and system into their chemistry. They were considered to be somewhat beyond the bounds of regular, well-behaved compounds. Werner made it clear that the "coordinately" bound atoms around the central atom of a complex structure are fundamentally equal to each other. There is for instance no difference between the O atoms expected to be singly or doubly bonded within a complex ion, and therefore it seems of no importance what valency they possess. Whatever the force which binds them to the central atom may be, it is distributed equally among all of them. As far as the valencies are concerned it is only necessary that the sum total of the valency bonds of the coordinative atoms should be equal to the sum total of the valencies of the central atom plus those of the atoms bound more loosely, "in the second sphere". Thus in H_2SO_4, the four O atoms together have a valency of 8; that of the two H's is 2 and that of S (in SO_3) is 6, making a total of 8. The same consideration applies to $NaBF_4$ with the 4 valencies of the F atoms balancing the $(1 + 3)$ of Na and B. Physics has only comparatively recently given us more insight into the nature of these unlocalized valencies in certain compounds and the phenomenon is now given the name mesomerism (or, much less fittingly, resonance). We shall deal with it later on when the nature of the chemical bond comes under discussion (p. 317).

Let us return to the simpler structural formulae in inorganic chemistry. We have mentioned earlier that the presence of O—O bonds would bring about easy decomposition of a compound into elementary oxygen, the O_2 molecule being preformed to some extent in these compounds. The simplest member of the series is hydrogen peroxide, H_2O_2, i.e. H—O—O—H. The principle holds for other elements as well. Compounds containing chains formed of atoms of only the one element, readily decompose to yield this element in its pure state. Hydrazoic acid and its salts for example contain the N_3 group and are liable to explode violently with evolution of elementary nitrogen, N_2. In this case we formulate the acid HN_3 as HNNN, with three N atoms bonded together. The polysulphides, e.g. sodium pentasulphide Na_2S_5, easily decompose yielding elementary sulphur. We formulate them with S—S bonds. Similarly we depict sodium triiodide, (NaI_3) with I—I bonds because it splits up with ease into simple sodium iodide (NaI) and elementary iodine (I_2). Exact geometrical structures of such compounds, or let us say of compounds in general, are determined by physical methods, e.g. X-ray analysis (p. 125), but it is very remarkable how far purely chemical arguments have led us.

If there is no special reason to suppose that atoms of the same element are connected to each other within a molecule, we generally assume that a chemical reaction between elements brings the atoms of these elements in chemical contact with each other:

$$
\begin{array}{ccc}
 & & \text{H} \\
 & & | \\
\text{Cl—Fe—Cl} \quad \text{O=S=O} \quad & \text{H—C—H} \\
 & || & | \\
 & \text{O} & \text{H}
\end{array}
$$

This simple formulation is possible as long as the valency of one element is an integral multiple of the valency of the other as in the above examples. If this is not the case, no molecule can be assembled on this basis, but must needs contain more than one of each kind of atom. Thus P is tri- or pentavalent, allowing for two simple phosphorus chlorides because in chlorides Cl is monovalent:

$$
\begin{array}{ccc}
 & & \text{Cl} \quad \text{Cl} \\
 & & \diagdown \quad \diagup \\
\text{Cl—P—Cl} \quad \text{and} & & \text{P} \\
| & & \diagup | \diagdown \\
\text{Cl} & & \text{Cl} \; | \; \text{Cl} \\
 & & \text{Cl}
\end{array}
$$

However, in divalent oxygen it burns (according to whether P or O_2 is present in excess) to the trioxide or pentoxide, respectively, each with two P atoms in a molecule. This arrangement brings the total valency of the phosphorus and the oxygen atoms to their smallest common multiple, 6 in the trioxide and 10 in the pentoxide.

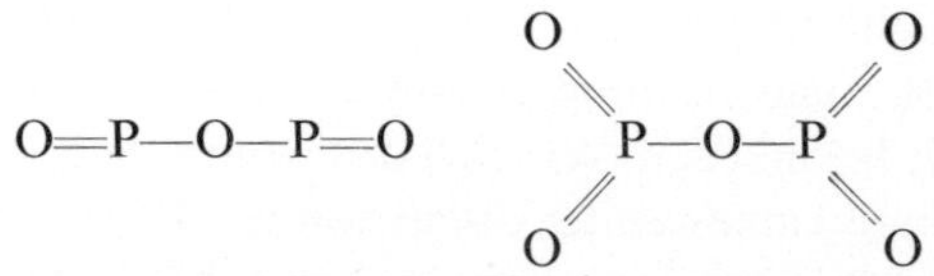

$$O=P-O-P=O \qquad \begin{array}{ccc} O & & O \\ \diagdown & & \diagup \\ P-O-P \\ \diagup & & \diagdown \\ O & & O \end{array}$$

An oxide molecule formed, for example, from trivalent phosphorus cannot contain a single P atom, since the latter's valency of three could only be satisfied by one and a half atoms of divalent O, a state of affairs manifestly impossible. Actually their structure is more complicated; their formulae are P_4O_6 and P_4O_{10}, respectively.

The two chlorides and the two oxides of phosphorus all decompose on contact with water (are "hydrolysed"), yielding phosphorous acid from the trivalent series and phosphoric acid from the pentavalent series:

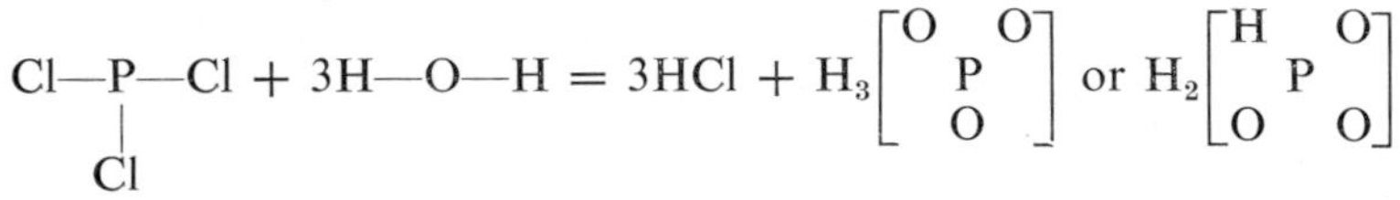

$$Cl-P-Cl + 3H-O-H = 3HCl + H_3\begin{bmatrix} O & O \\ & P \\ & O \end{bmatrix} \text{ or } H_2\begin{bmatrix} H & O \\ & P \\ O & O \end{bmatrix}$$

(with Cl below P)

$$(16.2)$$

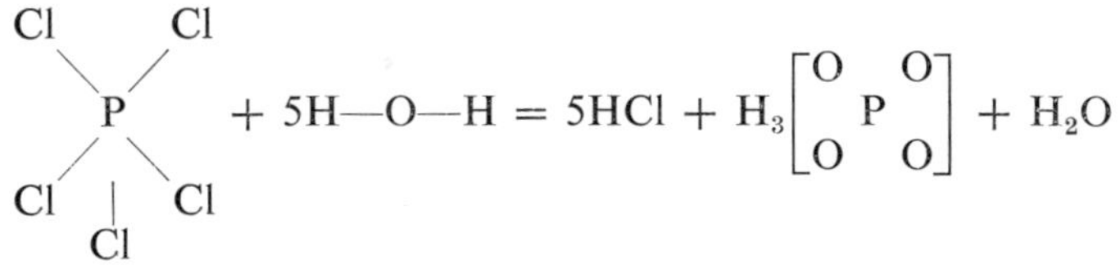

$$\begin{array}{cc} Cl & Cl \\ \diagdown & \diagup \\ & P \\ \diagup & \diagdown \\ Cl & Cl \\ & | \\ & Cl \end{array} + 5H-O-H = 5HCl + H_3\begin{bmatrix} O & O \\ & P \\ O & O \end{bmatrix} + H_2O$$

$$(16.3)$$

and

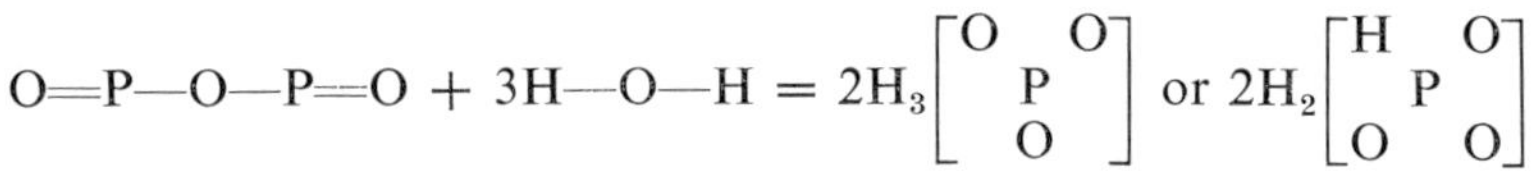

$$O=P-O-P=O + 3H-O-H = 2H_3\begin{bmatrix} O & O \\ & P \\ & O \end{bmatrix} \text{ or } 2H_2\begin{bmatrix} H & O \\ & P \\ O & O \end{bmatrix}$$

$$(16.4)$$

$$\begin{array}{ccc} O & & O \\ \diagdown & & \diagup \\ P-O-P \\ \diagup & & \diagdown \\ O & & O \end{array} + 5H-O-H = 2H_3\begin{bmatrix} O & O \\ & P \\ O & O \end{bmatrix} + 2H_2O$$

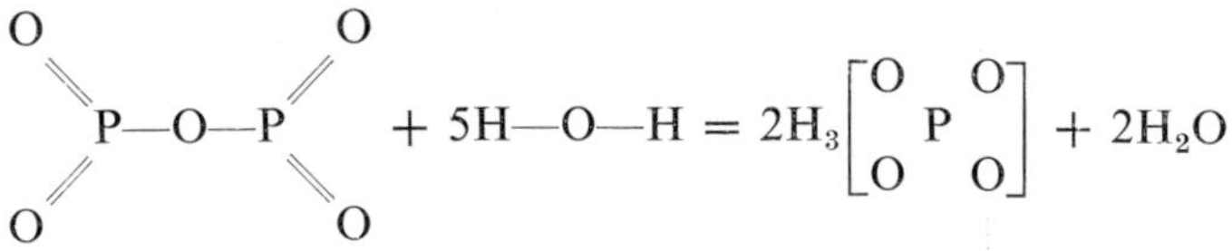

It would have been formally simpler and more in accord with general usage to write respectively only 4 HOH and 3 HOH reacting with the two pentavalent compounds and to omit H_2O molecules from the products of the reaction. We have only used the above representation to show that we do not always arrive at the formally simplest compounds, e.g. $P(OH)_5$ for phosphoric acid which would be the analogue of PCl_5. The number of O atoms around a

P atom in practice never exceeds 4, so water must be split off from our theoretical $P(OH)_5$. It is also only fair to state, that the chemical equivalence of the three H atoms in phosphorous acid is not complete and the formula

$$H_2 \begin{bmatrix} O & H \\ & P & \\ O & O \end{bmatrix}$$

must be given consideration rather than the one written in the equations:

$$H_3 \begin{bmatrix} O & O \\ & P & \\ & O & \end{bmatrix}.$$

None of its metal (Me^+) salts has the formula: Me_3PO_3 but only Me_2HPO_3.

Let us now return to perchloric acid, which we began to discuss on p. 54 before we learned about complex anions. Perchloric acid is obtained by reacting the highest oxide of chlorine, Cl_2O_7, with water in perfect analogy to eqn. 16.5:

$$O\!\!=\!\!\overset{\displaystyle O}{\underset{\displaystyle O}{Cl}}\!\!-\!\!O\!\!-\!\!\overset{\displaystyle O}{\underset{\displaystyle O}{Cl}}\!\!=\!\!O + HOH = 2H \begin{bmatrix} O & O \\ & Cl & \\ O & O \end{bmatrix} \qquad (16.6)$$

The formula is analogous to that of other oxy-acids; the valency of chlorine in this compound is seven.

17. ACIDS, BASES, SALTS: ELECTROLYTES

We have already used these concepts but did not define them as precisely as we should have done. It is very difficult to build up chemistry in a linear deductive fashion, always using concepts which have been already defined. In this book it is our special aim to do this as far as possible, but having to deal with a science based on innumerable facts which are multiply interconnected, we sometimes do not succeed. Let us next try to rectify this matter.

Chemists have used the words acid, base and salt since the centuries of alchemy and they worked out pretty close practical definitions so that there could be hardly any doubt at all in particular cases. But the generalization of these concepts proved to be a very hard task and in some fine details definitions are apt to differ even in the present day.

Practically speaking acids are supposed to taste sour and they mostly do so. They are also said to be capable of changing the colours of a series of special compounds (indicators) which can occur in two different colours. Most acids make litmus and methyl orange turn red, and phenolphthalein turn colourless. Not all of them do so. They are supposed to dissolve the less noble metals with the evolution of hydrogen gas and many of them do this. Their aqueous solutions should conduct electricity and hydrogen gas should be evolved at the cathode during this electrolysis. In so far as they are soluble in water acids do behave in this manner.

Electrolytically mobile hydrogen, the so-called positive hydrogen ion (from ἰών meaning "going" in Greek) is in the end found to be the characteristic common constituent of acids and for our purposes it is best to agree upon this as a definition, i.e. an acid is a substance whose aqueous solution contains hydrogen ions. The strength of the force which binds the hydrogen ion to the remainder of the molecule may vary within extremely wide limits. If the hydrogen ion is bound very tightly it will be more difficult to detect it— it may not make the acid taste sour, it may not make it able to change the colour of all our indicators or to dissolve metals which tend to be less reactive, or more "noble". If a hydrogen ion is bound tightly we are dealing with a very weak acid and if it is bound loosely, we have a strong acid. But the hydrogen ion is always there and it is always able to break away in solution ("dissociate") to a greater or lesser extent. We will deal subsequently with Lewis's extension to the concept of acids (p. 333).

Looking through our repertoire we find HCl among the strong acids, H_2SO_4 somewhat weaker, and H_3PO_4 weaker still. But all three remain among the strong acids. HF and H_2CO_3 are substantially weaker and H_3BO_3 is very weak indeed.

Bases change the colours of indicator compounds in the opposite sense to acids, e.g. they turn litmus blue, methyl orange yellow and phenolphthalein purple-red. Their aqueous solutions conduct electricity more or less and at the anode oxygen gas is evolved, arising from OH groups which have migrated there. These hydroxyl groups are either already present in the compound as in the case of caustic soda (NaOH) or are generated from the compound and water by combination of a molecule of the compound with a hydrogen ion from a molecule of water, leaving the hydroxyl ion behind. Ammonia gas can be supposed to react with water as follows:

$$NH_3 + HOH = NH_4^+ + OH^- \qquad . \qquad . \qquad (17.1)$$

(It has become customary to denote electrically charged groups, that is ions, with + or − signs as right-hand superscripts to the chemical symbol). Positively charged ions are attracted during electrolysis of a solution by the negative pole, the cathode, and are therefore called cations, whereas negatively charged ions go to the anode and are called anions.

The definition of a base which is most suitable for our use is any compound or ion which can add a hydrogen ion. Thus NH_3 is a base because it forms NH_4^+, and OH^- is a base because it forms HOH with H^+. The latter process between acids and bases is called neutralization, because the acidic H^+ and the basic species disappear simultaneously. The more avidly a species attracts H^+, the stronger base it is.

The neutralization of a strong alkali and a strong acid could be written as a reaction between the molecules, e.g.

$$NaOH + HCl = NaCl + HOH \qquad . \qquad . \qquad (17.2)$$

Common salt, NaCl, may be recovered as a matter of fact from the neutralized solution. But it is important to remember that sodium ions (Na^+) migrate towards the cathode during the electrolysis of caustic soda solution in the same way as they do when a solution of common salt is used. Likewise the chlorine ions (Cl^-) in HCl and in salt solution migrate towards the anode. Thus nothing seems to have happened to these ions during neutralization. All their characteristic chemical reactions and their lack of colour have remained unchanged too. Therefore it is much more appropriate to write the two ions of electrolytically conducting compounds ("electrolytes"), separately and so the above reaction becomes

$$Na^+ + OH^- + H^+ + Cl^- = Na^+ + Cl^- + HOH \qquad . \quad (17.3)$$

or, simplifying by using only the changing species,

$$H^+ + OH^- = HOH \qquad . \qquad . \qquad . \quad (17.4)$$

The particular metal cation and acid anion present during the neutralization do not matter since they undergo no change.

The assumption, that the ions in solutions of electrolytes are separate from each other in space is made more plausible by a quite different experimental fact. On dealing with Raoult's laws concerning freezing-point depression and boiling-point elevation, etc., of solutions (p. 47) we have remarked that these methods are consistently at fault for solutions which conduct the electric current. In dilute solutions we find depressions and elevations which are twice as big for NaCl, HCl or NaOH as their molecular weights would warrant according to the quoted formula. It seems as if the number of molecules present in the dilute solution is double the number we actually dissolved. This conclusion is in excellent agreement with our finding that the two ions of such compounds seem to lead separate lives as far as mobility in an electric field is concerned. It was *S. A. Arrhenius* (1859–1927), the great Swedish physicochemist, who in 1883 assumed that these compounds did not dissolve as molecules but as "dissociated" ions and thus imparted electrical conductivity to their solutions. At the same time they acted as if as many separate entities were present to depress the freezing point, etc., as there were ions, and not undissociated molecules in the solution.

The compounds which are built from positive and negative ions which are neither H^+ nor OH^- are classed as salts. Electrolytes are by no means confined to those which contain only one cation and one anion; the number of ions is determined by valency considerations. Thus

$$H^+Cl^-,\ Na^+Cl^-,\ Ca^{++}(SO_4)^{--},\ H^+(NO_3)^-,\ Na^+(NO_3)^-$$

each contain one positive and one negative ion;

$$H_2^{++}(SO_4)^{--},\ Na_2^{++}(SO_4)^{--}$$

each contain two positive and one negative ion;

$$Ca^{++}Cl_2^{--},\ Ca^{++}(NO_3)_2^{--},\ Ca^{++}(OH)_2^{--}$$

each contain one positive and two negative ions;

$$Fe^{+++}Cl_3^{---}, \quad Fe^{+++}(NO_3)_3^{---}$$

each contain one positive and three negative ions;

$$Fe_2^{+++}(SO_4)_3^{---}$$

contains two positive and three negative ions. In every case very dilute solutions of such electrolytes behave, as far as their freezing points, vapour pressures, etc., are concerned as though they contained as many times more particles than dissolved molecules as there are ions in the molecules. That is, the expected effect on freezing point, etc. has to be multiplied by a factor in the range two to five in the series of examples given above. It is important to note, that this holds exactly only for extremely dilute solutions. This is because the electrostatic interactions of the dissociated ions do not allow them absolute independence in their movements if their solutions are not very dilute. Hence they cannot quite act as entirely independent entities.

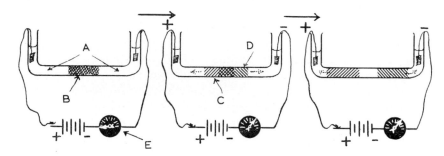

Fig. 20. **Electrolysis of copper chromate solution**

On completing the circuit blue copper ions migrate towards the cathode and yellow chromate ions towards the anode, from the originally green copper chromate solution.

    A = sodium chloride solution
    B = cupric chromate solution (green)
    C = chromate ions (yellow)
    D = copper ions (blue)
    E = ammeter

Having referred to the migration of ions in an electric field it is instructive to consider a simple experiment which reveals this movement. The simplest observations are performed using ions of different colours because we can then follow them by eye without having to make chemical analyses. A suitable experiment can be carried out by interposing, for example, a green solution of cupric chromate between two common salt solutions containing immersed electrodes. Salts of the cupric ion are blue, like copper sulphate, those of the chromate ion, yellow like potassium chromate or the pigment called chrome yellow. We should already suspect that the green colour of cupric chromate is a simple mixture of the colours of its individual ions, but on switching on the current this state of affairs becomes manifest: the green section in the middle of the tube develops a blue margin at its end nearest the cathode and a yellow one on the anodic side. If we use a tube which is long enough and have time to wait, some twenty minutes, say, the two regions will eventually separate and there will then be no green region at all, in between. Although the experiment can be performed using aqueous solutions it becomes entirely insensitive to mechanical disturbances if we add a few per cent of agar-agar which causes the formation of a jelly. The ions can still move through the jelly as though through the mesh of a microsieve.

18. OXIDATION–REDUCTION

In the original narrow sense, combination of a substance with elementary oxygen was called oxidation, and loss of oxygen was reduction. However, in the broader sense, combination with elementary chlorine is also called oxidation whereas combination with elementary hydrogen is a reduction process. What do we mean by these broad definitions?

Consider for example sodium metal. It is able to combine simply with oxygen to form the oxide. This dissolves in water and reacts with it, evolving appreciable heat and yielding sodium hydroxide (caustic soda):

$$4Na + O_2 = 2Na_2O \qquad . \qquad . \qquad . \quad (18.1)$$

$$Na_2O + H_2O = 2NaOH \qquad . \qquad . \qquad . \quad (18.2)$$

Evidently oxidation has taken place in the first step, while the second step is a reaction between two oxides. The sodium and the hydrogen atoms have simply changed places but still remain bonded to an oxygen atom as they were initially (though obviously not to the same oxygen atom). But the same result can be achieved by dropping the sodium metal into water:

$$2Na + 2H_2O = 2NaOH + H_2 \quad . \qquad . \qquad . \quad (18.3)$$

This reaction, having produced sodium hydroxide from sodium metal, must also be an oxidation. We have oxidized sodium with water, at the same time reducing water to hydrogen.

Sodium also reacts with chlorine:

$$2Na + Cl_2 = 2NaCl \qquad . \qquad . \qquad . \quad (18.4)$$

A second reaction, that of shaking a solution of NaCl with brown, sparingly soluble silver hydroxide, yields a white precipitate of silver chloride and once again a solution of sodium hydroxide:

$$NaCl + AgOH = AgCl + NaOH \qquad . \qquad . \quad (18.5)$$

Here we must ask ourselves at which step the sodium was oxidized. The sensible answer is: during the course of the first reaction. Sodium loses its metallic character not only by reaction with oxygen but also in reaction with chlorine. In the resulting compounds it is now present as the colourless, cationic sodium ion. The oxide, the hydroxide and the chloride, either molten or in solution, conduct an electric current, the Na^+ migrating towards the cathode. From the melts it is being reduced back to elementary sodium.

The silver, on the other hand, evidently underwent oxidation when its hydroxide was formed from the metal. During the reaction of NaCl with AgOH two already oxidized metals simply exchanged anions. Let us note by the way that the silver ion is a cation, Ag^+, both in the hydroxide and in the chloride.

Iron in its divalent state, forms ferrous oxide, FeO. On heating this in oxygen it is converted to ferric oxide, the oxide of trivalent iron:

$$4FeO + O_2 = 2Fe_2O_3 \qquad . \qquad . \qquad . \quad (18.6)$$

By definition this is an oxidation. The same result could, however, have been brought about by adding chlorine to ferrous chloride, and "hydrolysing" the resultant ferric chloride with steam to ferric hydroxide. This on heating again forms ferric oxide:

$$2FeCl_2 + Cl_2 = 2FeCl_3 \qquad . \qquad . \qquad . \quad (18.7)$$

$$FeCl_3 + 3HOH = Fe(OH)_3 + 3HCl \qquad . \qquad . \quad (18.8)$$

$$2Fe(OH)_3 = Fe_2O_3 + 3HOH \qquad . \qquad . \qquad . \quad (18.9)$$

Chlorine is present in ferric chloride $(FeCl_3)$ as a negative ion (anion) and exists in the same form in HCl. On the other hand during the reaction between iron and chlorine the iron is converted to a trivalent cation and the chlorine from the element to a monovalent anion.

These reactions which go hand in hand with changes in valency and ionic nature are termed oxidation or reduction according to the direction of change. Iron was oxidized by chlorine from the ferrous state to the ferric state, an increase in valency from two to three. In the same manner it can be oxidized from the metal to the ferrous state by chlorine

$$Fe + Cl_2 = FeCl_2 \qquad . \qquad . \qquad . \qquad . \quad (18.10)$$

Iron has increased its positive valency while chlorine has become negative in both cases. Electricity cannot be created, and so the sum of the charges on all the ions must be zero, i.e. the chemical system is electrically neutral. However, the distribution of electric charges among the ions must have changed during the reaction, corresponding to simultaneous oxidation and reduction. These processes are inseparable from each other. One cannot oxidize something without reducing something else at the same time because the balance of electric charges must be preserved. Whatever species has become more positive has been oxidized, whatever more negative has been reduced.

Instead of chlorine we could have discussed similar examples, using fluorine, bromine or iodine. For that matter we might have employed sulphur because on melting sulphur with iron powder a violent, incandescent reaction unites them forming ferrous sulphide:

$$Fe + S = FeS$$
$$\qquad . \qquad . \qquad . \qquad . \quad (18.11)$$

This again yields ferrous chloride and hydrogen disulphide on reaction with HCl:

$$FeS + 2HCl = FeCl_2 + H_2S \qquad . \qquad . \qquad . \text{ (18.12)}$$

The zero-valent elementary sulphur has been reduced to a negative ion in its hydrogen compound whereas the zero-valent iron metal has been oxidized to its bivalent positive ion.

Reduction of a substance may be brought about, for example, by reaction with sodium metal:

$$FeCl_2 + 2Na = Fe + 2NaCl \qquad . \qquad . \qquad . \text{ (18.13)}$$

Here iron has lost its cationic state and is reduced to zero-valent metal while metallic sodium has been oxidized to its cation.

The products of reaction are not always ionic compounds, but having considered matters so far, it is evident by analogy, that carbon, for example, is not only oxidized when it reacts with oxygen to form carbon dioxide, but also during its reaction with chlorine to form carbon tetrachloride:

$$C + O_2 = CO_2 \qquad . \qquad . \qquad . \qquad . \text{ (18.14)}$$

$$C + 2Cl_2 = CCl_4 \qquad . \qquad . \qquad . \qquad . \text{ (18.15)}$$

On reacting carbon tetrachloride with alcoholic alkali it is possible to hydrolyse away its chlorine atoms to sodium chloride, while the carbon goes over into the carbonate ion:

$$CCl_4 + 4NaOH = 4NaCl + H_2CO_3 + H_2O \qquad . \text{ (18.16)}$$

Evidentally the carbon was already in the oxidized state in carbon tetrachloride.

19. STRUCTURAL FORMULAE, II: ORGANIC COMPOUNDS

We have gained some insight into general structural problems of chemistry, but have scarcely dealt with molecules which contain more than some eight to ten atoms at the most. Certainly there do exist more complicated molecules among inorganic compounds but they are rather rare and their discussion generally does not involve more fundamental problems than we have at least glanced at in the preceding chapters. We shall see later that ionic crystals may be considered as immense molecules.

The situation among the compounds of carbon, in the domain of what we call organic chemistry, is quite different. The compounds of carbon differ from most other compounds in that the number of atoms in their molecules can easily exceed a hundred or even many more. This peculiarity of carbon compounds is due to the fact that carbon atoms are easily and indeed often, bonded to each other, forming chains, rings and even more intricate structures. In inorganic compounds bonding between atoms of the same kind usually occurs as an exception, whereas among organic compounds it becomes the rule at least with respect to carbon atoms. It is evident that the number of isomeric compounds which may be built from a given number of atoms increases immensely with the number of atoms involved. A tetra-valent carbon atom needs only two of its valencies to enable it to fit into a chain or a ring system. The remaining two valencies are free to combine with other atoms or to partake in the formation of other chains or rings. From this viewpoint organic chemistry is much more complicated than inorganic. However it is simpler in so far as carbon is nearly always tetra-valent and its so-called maximal coordination number, i.e. the largest num-ber of atoms it is able to accommodate around itself, is four also. This explains why it was not necessary to resort to the ideas of complex chemistry in order to describe the structure of organic compounds. The old-fashioned way of drawing lines for valency bonds between atoms was sufficient to explain most of their properties for a long time.

Let us take up matters where we left them in Section 14, dealing with the ethyl radical. We learned to know that this group, C_2H_5, recurs unchanged throughout a whole series of reactions. But of necessity we had to side-track to learn something about chemical valency before we could start to discuss the internal structure of this radical, which we are now in a position to do. Let us take the simplest approach and build up the ethane molecule from parts which contain only one carbon atom. A classical method of doing this involves the use of the "Wurtz synthesis" using two molecules of methyl chloride (which we mentioned in Table I (p. 20) among the chlorides of methane) and metallic sodium:

$$H_3C—Cl + Na + Na + Cl—CH_3 = H_3C—CH_3 + 2NaCl \quad (19.1)$$

The result of this reaction is common salt and a gas, ethane. Knowing how it was synthesized and knowing that carbon is tetravalent, we conclude that it must have the structure

$$
\begin{array}{c}
\ \ \ \ H \ \ H \\
\ \ \ \ | \ \ \ | \\
H—C—C—H \\
\ \ \ \ | \ \ \ | \\
\ \ \ \ H \ \ H
\end{array}
$$

The formula of the ethyl radical would thus be that of ethane after one H atom has been removed from it. But which one? Is there any difference between them? Experiments have proved that whenever we derive a new compound from ethane by substituting anything else in place of one of its H atoms, we always get only one derivative (one isomer) of the new compound. We may chlorinate ethane singly with all the methods of chlorination at our disposal but the resulting compound will always be one and the same ethyl chloride and no other isomer is ever to be found. This should mean, then, that the six hydrogen atoms in the formula of ethane are equivalent, because it is extremely improbable that among all experimental conditions and methods there would be none to substitute each separate and different hydrogen atom in its turn. Is this in accord with our formula? Not quite. If we imagine that the formula of ethane really represents the spatial arrangement of the atoms in the plane of the paper, there is a difference between the situation of the two terminal hydrogen atoms and those which are above or below the chain. The two terminal atoms are positioned identically as are the other four. As a consequence of this, the molecule as written looks exactly the same from behind as it does from the front, even though in this process what was originally the top left-hand atom has become the top right-hand one. Turning the formula upside-down in the plane of the paper changes the top left-hand atom into a bottom right-hand atom and so forth. However, the difference between the terminal and the top or bottom hydrogen atoms remains, no matter how we turn our model. Translated back into the concept of chemistry that means that we should expect two isomeric ethyl chlorides, according to whether the terminal or the top or bottom hydrogen atom was substituted by chlorine:

$$
\begin{array}{ccc}
\text{H} \quad \text{H} & & \text{H} \quad \text{Cl} \\
\mid \quad \mid & & \mid \quad \mid \\
\text{H—C—C—Cl} & \text{or} & \text{H—C—C—H} \\
\mid \quad \mid & & \mid \quad \mid \\
\text{H} \quad \text{H} & & \text{H} \quad \text{H}
\end{array}
$$

But something is wrong, because only one isomer has been found.

Now why did we insist on keeping our atoms within the plane of the drawing while the three dimensions of space seem to have *a priori* equal rights to accommodate them? Really there was no other reason for doing this than

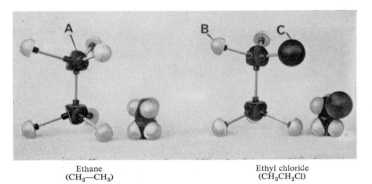

Ethane
(CH$_3$—CH$_3$)

Ethyl chloride
(CH$_3$CH$_2$Cl)

Fig. 21. **Conformations of ethane and ethyl chloride**

All positions are equivalent around carbon atoms. In the model on the *left* in each pair, the atoms are only set far apart to make the structure visible. The more realistic model is that on the *right*, in each case.

A = carbon; B = hydrogen; C = chlorine

the fact that we normally draw on a flat surface. This in itself is not really a sufficient reason for restricting the atoms to a given plane. The moment we give up this restriction and go over to a three-dimensional spatial arrangement as seen in Fig. 21 there is no longer any contradiction with experiments. The six hydrogen atom positions are now absolutely equivalent and can be transformed into each other simply by turning the molecule in space. We can now grasp the fact that an ethane molecule exists in space and not in a plane and can see that the figure in perspective accounts perfectly for the fact that ethyl chloride has only one isomer. As all the hydrogen atoms of ethane are equivalent, it makes no difference which one we substitute. However, we still go on drawing formulae on paper but it is tacitly agreed that this drawing of four atoms around a carbon atom should always be taken to mean that they are equivalent in space, notwithstanding their apparent differences in the plane of the paper.

What happens if we go on to substitute further hydrogen atoms of ethane by chlorine? Let us proceed another step and substitute a second atom. It is evident that this second atom may either be attached to the same carbon atom as was the first chlorine atom or to the other carbon:

$$
\begin{array}{cc}
\underset{|}{\overset{|}{\text{Cl}}}\ \underset{|}{\overset{|}{\text{H}}} & \underset{|}{\overset{|}{\text{H}}}\ \underset{|}{\overset{|}{\text{H}}} \\
\text{Cl—C—C—H} \quad \text{or} \quad \text{Cl—C—C—Cl} \\
\underset{}{\overset{|}{\text{H—H}}} & \underset{}{\overset{|}{\text{H}}}\ \underset{}{\overset{|}{\text{H}}}
\end{array}
$$

We cannot expect that more than these two isomers exist as long as the three valencies left on each carbon atom are symmetrically distributed in space, as we have assumed them to be, provided the two parts of the molecule are free to rotate around the C—C axis. If so, the different "conformations"

Fig. 22. **Conformations of 1,2-dichloroethane**

All conformations change into each other if relative rotation of top and bottom parts is allowed.

of Fig. 22 can be shown to mean the same thing by simply turning the two parts around the axis. Now, this question can once more only be answered by experiments and these show that no more than two isomers of dichloroethane ($C_2H_4Cl_2$) exist at ordinary temperatures. At a very low temperature, which as we shall see corresponds to less energy within the molecule and in its movement, it has been shown that species do exist according to the different conformations of Fig. 22 but the energy of transposition of one conformation into the other, i.e. the energy necessary to make free rotation around the C—C axis possible, is very small indeed. Therefore such differences in conformation around single bonds are only to be observed under special circumstances in organic chemistry; for our purposes we may safely disregard them.

Thus in chemistry at normal temperatures we only have to reckon with two isomeric dichloroethanes, those we have depicted in the above formulae.

Which is which, however? We must perform reactions on each of them and draw our conclusions from the results. For instance, let us exchange the two chlorine atoms for hydroxyl groups by reacting the chlorides with lead hydroxide:

$$\begin{array}{c} \text{H} \quad \text{H} \\ | \quad | \\ \text{Cl}-\text{C}-\text{C}-\text{H} + \text{Pb(OH)}_2 \\ | \quad | \\ \text{Cl} \quad \text{H} \end{array}$$

$$\begin{array}{c} \text{H} \quad \text{H} \qquad\qquad\qquad \text{H} \quad \text{H} \\ | \quad | \qquad\qquad\qquad | \quad | \\ = \text{PbCl}_2 + \text{HO}-\text{C}-\text{C}-\text{H} = \text{PbCl}_2 + \text{O}{=}\text{C}-\text{C}-\text{H} + \text{H}_2\text{O} \\ | \quad | \qquad\qquad\qquad\qquad\quad | \\ \text{HO} \quad \text{H} \qquad\qquad\qquad\qquad\quad \text{H} \qquad (19.2) \end{array}$$

Acetaldehyde

$$\begin{array}{c} \text{H} \quad \text{H} \qquad\qquad\qquad\qquad \text{H} \quad \text{H} \\ | \quad | \qquad\qquad\qquad\qquad | \quad | \\ \text{Cl}-\text{C}-\text{C}-\text{Cl} + \text{Pb(OH)}_2 = \text{PbCl}_2 + \text{HO}-\text{C}-\text{C}-\text{OH} \quad (19.3) \\ | \quad | \qquad\qquad\qquad\qquad | \quad | \\ \text{H} \quad \text{H} \qquad\qquad\qquad\qquad \text{H} \quad \text{H} \end{array}$$

Ethylene glycol

As shown in the upper equation, the dihydroxy compound which might be supposed to form, immediately loses the elements of water and yields the compound called acetaldehyde. The end product of the second reaction is a stable dihydroxy compound, ethylene glycol (used in antifreeze solutions). The normal assumption is that water is more easily split off from two OH groups on the same carbon atom than from those on two adjacent ones and this allows us to allocate the two formulae to the isomeric dichloro-ethanes. The one which yields acetaldehyde is named conventionally 1,1-dichloroethane, or ethylidene dichloride, while the other is 1,2-dichloro-ethane or ethylene dichloride. (The numbers signify the sequence of the carbon atoms in the chain.)

Without going into all the details, it should be clear that we can arrange three chlorine atoms in two different ways on the ethane chain, if free rotation is allowed for:

$$\begin{array}{cc} \text{Cl} \quad \text{H} & \text{Cl} \quad \text{H} \\ | \quad | & | \quad | \\ \text{Cl}-\text{C}-\text{C}-\text{H} \quad \text{and} & \text{Cl}-\text{C}-\text{C}-\text{Cl} \\ | \quad | & | \quad | \\ \text{Cl} \quad \text{H} & \text{H} \quad \text{H} \end{array}$$

Substitution with four chlorine atoms also yields two isomers, a fact evident if we consider that it is the same symmetry problem as the arrange-ment of four hydrogen atoms in the dichloroethanes, except that the hydrogen and chlorine atoms have now been exchanged:

$$\begin{array}{cc} \text{Cl} \quad \text{Cl} & \text{Cl} \quad \text{Cl} \\ | \quad | & | \quad | \\ \text{Cl}-\text{C}-\text{C}-\text{H} \quad \text{and} & \text{H}-\text{C}-\text{C}-\text{H} \\ | \quad | & | \quad | \\ \text{Cl} \quad \text{H} & \text{Cl} \quad \text{Cl} \end{array}$$

To end the list we have again only one way of substituting either five or all six hydrogen atoms by chlorine atoms:

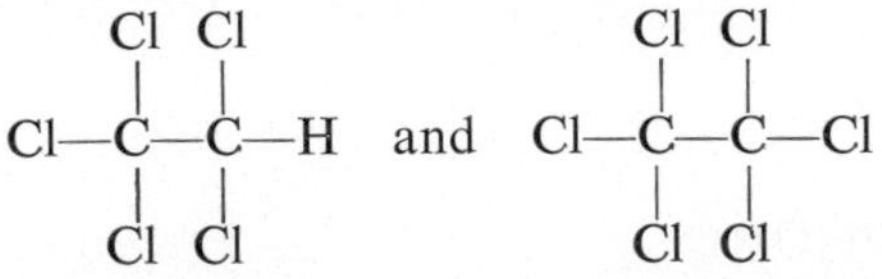

$$
\begin{array}{ccc}
& \underset{\overset{|}{Cl}\ \ \underset{|}{Cl}}{\overset{Cl\ \ Cl}{Cl-C-C-H}} & \text{and} & \underset{\overset{|}{Cl}\ \ \underset{|}{Cl}}{\overset{Cl\ \ Cl}{Cl-C-C-Cl}}
\end{array}
$$

The interesting point in all this is the fact that chemists have found *as many, and only as many*, isomers of each set of compounds as are predicted by the above theory of geometrical arrangements. We could proceed to show by a series of appropriate reactions which formulae must be ascribed to the different isomers that can be isolated from the crude mixture obtained on chlorination. But this would lead us into detailed organic chemistry, without adding to our knowledge more than we have learned already from the above example of acetaldehyde and ethylene glycol.

It is easy to see that the number of possible isomers increases rapidly if we use two or three different atoms instead of only chlorine to substitute the hydrogen atoms in ethane. It would be quite a good exercise for the reader to write down all the isomers of the compound

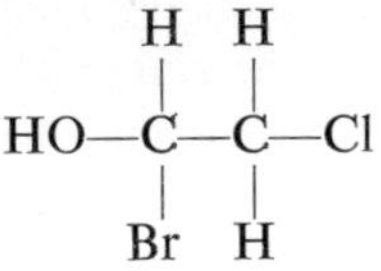

$$
\underset{\overset{|}{Br}\ \ \underset{|}{H}}{\overset{H\ \ H}{HO-C-C-Cl}}
$$

(There are four of them.)

Instead, we may proceed to lengthen the carbon chain itself using the Wurtz synthesis or other suitable reactions. Reaction of two ethyl chloride molecules with sodium yields butane gas, which is a well-known by-product of the "thermal cracking" of oil and is sold in steel containers as a fuel:

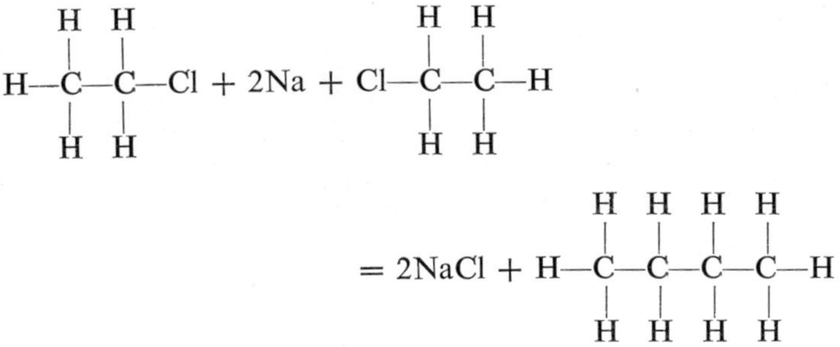

$$
\underset{\overset{|}{H}\ \ \underset{|}{H}}{\overset{H\ \ H}{H-C-C-Cl}} + 2Na + \underset{\overset{|}{H}\ \ \underset{|}{H}}{\overset{H\ \ H}{Cl-C-C-H}}
$$

$$
= 2NaCl + \underset{\overset{|}{H}\ \ \underset{|}{H}\ \ \underset{|}{H}\ \ \underset{|}{H}}{\overset{H\ \ H\ \ H\ \ H}{H-C-C-C-C-H}} \tag{19.4}
$$

n-Butane

At this stage we have arrived at a compound containing only carbon and hydrogen but which has already a second isomer because the carbon chain may be built in two ways. Performing the Wurtz synthesis with a compound called symmetrical propyl chloride, or 2-chloropropane, and methyl chloride we arrive at another hydrocarbon with the same empirical formula C_4H_{10} as the above "normal" butane:

$$
\begin{array}{c}
\text{H H H} \\
| \ | \ | \\
\text{H—C—C—C—H} \\
| \ | \ | \\
\text{H Cl H} \\
+ \\
2\text{Na} \\
+ \\
\text{Cl} \\
| \\
\text{H—C—H} \\
| \\
\text{H}
\end{array}
\quad = 2\text{NaCl} +
\begin{array}{c}
\text{H H H} \\
| \ | \ | \\
\text{H—C—C—C—H} \\
| \ | \ | \\
\text{H} \quad \text{H} \\
\\
\text{H—C—H} \\
| \\
\text{H} \\
\text{Isobutane}
\end{array}
\quad =
$$

(19.5)

The chemical and physical properties of normal and isobutane are different, e.g. the former condenses at normal pressure to a liquid if cooled to 1°C whereas the latter has to be cooled to as low as −17°C. We are obviously dealing with two distinct compounds.

It is not necessary to describe all the possible isomers of other members of the hydrocarbon series. Those of the series with not too many carbon atoms in the molecule have all been prepared and their number is always exactly that predicted by the structural theory. The number of possible isomers of hydrocarbons containing many carbon atoms soon reaches astronomical proportions and, for this reason only, they have not all been examined. Suffice it to say that in each case where the structural theory, due to *Kekulé* and clarified by the Russian chemist *A. M. Butlerow* (1818–1886) has been put to a test, the number of isomers found and those prescribed by the theory always agree. For more than a hundred years the models with tetravalent carbon have remained the foundation upon which the vast edifice of organic chemistry has been built.

20. STEREOCHEMISTRY, I: OPTICAL ISOMERS

We have agreed upon the four valencies of carbon being distributed in space, in order to account for the observation that the three H atoms in the CH_3 group of ethane were equivalent. As a matter of fact we could have stated the case in a more general manner by affirming that no isomers exist whenever a carbon atom is only substituted by three different atoms: CX_2YZ. If the valencies were all in a given plane we could arrange the substituents of such a molecule in two ways, corresponding to two isomers:

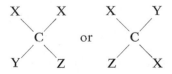

Never throughout the vast domain of organic chemistry have such isomers been found. This experimental fact was explained by the assumption that the four valencies of a carbon atom extend towards four different directions of space.

However if we go on to draw conclusions from this assumption of a regular or irregular tetrahedral arrangement of valencies around each carbon atom, we arrive at very striking results indeed. As a matter of fact, experiments preceded theory in this field.

L. Pasteur (1822–1895), the famous French scientist, worked with tartaric acid as far back as 1860 and by crystallization from water he isolated two kinds of crystals. These were alike in all physical and chemical properties except that those of one kind had the appearance of being the reflected image of those of the other kind, like right and left hands. It was impossible to

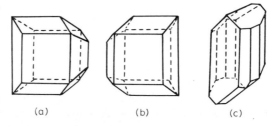

(a)   (b)   (c)

Fig. 23. **Crystal forms of isomeric tartaric acids**
(*a*) *d*-form; (*b*) *l*-form; (*c*) racemic form.

turn them around in space in such a way as to bring them into the same position, just as one cannot turn a left glove to make it look like a right one. Solutions of these crystals or of their derivatives retained a part of this asymmetry in so far as the angle of polarization of polarized light was rotated in one direction by one of these compounds and to the same extent in the other direction, by the other.

Polarized light arises, for example, when a beam of light is reflected from a glass mirror at an angle of 57 degrees to the normal. It does not appear different from normal, unpolarized light to the human eye, though insects are able to discriminate between the two. In the case of a second reflection at the same angle from another mirror, however, it does make a difference whether the plane of reflection coincides with that of the first reflection or is perpendicular to it. The polarized beam is reflected if it remains in the same plane, but it is not reflected if it has to turn through a right angle (Fig. 24). It has acquired a special asymmetry about its line of propagation. Knowing today that light is electromagnetic radiation, we can say that polarization is related to the direction of the oscillating electric field around the line of propagation. (Radiation emitted from a linear antenna is polarized this way, the electric field being concentrated in planes which include the antenna itself.) Normal light is a mixture of light polarized in random directions. It can be polarized in certain directions only, without needing to be reflected, by letting it pass through certain crystals of "lower symmetry" in which case two beams may emerge having planes of polarization perpendicular to each other. "Polaroid" light filter sheets allow only one of the beams to pass.

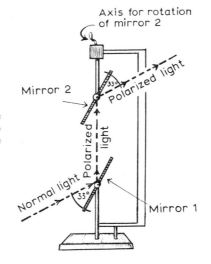

Fig. 24. **Polarization of light
by reflection**

By rotating mirror 2 around the
axis we obtain no light in the two
directions perpendicular to the
plane of the drawing.

An oversimplified picture (Fig. 25) may assist our imagination. Suppose we
attach a rope to a hook and move its other end with our hand until waves are formed
along it. So long as our hand movement is not confined to a straight line the waves
in the rope are not confined to a plane. But let the rope pass through a fence of
parallel planks and only waves in one direction, in that of the planks, will remain.
This polarized rope-wave will readily pass through another fence if the planks of

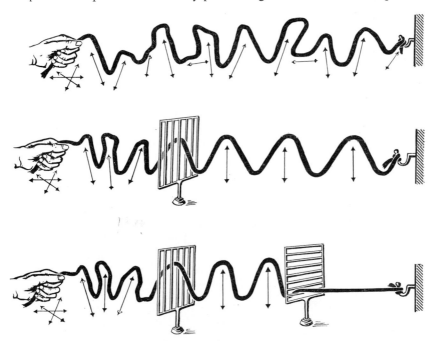

Fig. 25. **Mechanical model of wave polarization**

If there is no obstacle a rope can be made to swing in all planes.
A fence only allows movements to pass it which are parallel to its
openings. Two perpendicular fences prohibit the passage of all
motion.

the latter are parallel to those of the first one, but they will be totally extinguished if the planks run in a perpendicular direction.

If we speak about a medium rotating the plane of polarized light, we simply mean that on passing through this medium the plane necessary to extinguish the beam, or the plane necessary to let it through unhindered, has changed from its original orientation. We observe this phenomenon *only* if the internal structure of the material is asymmetric in the sense that it does not coincide with its own mirror image.

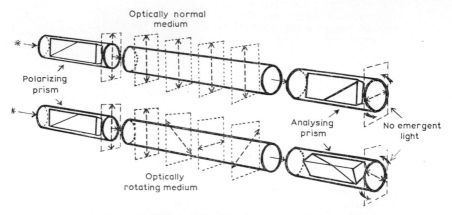

Fig. 26. **Rotation of the plane of polarization of light**

The analyser is set perpendicular to the plane of polarization of the light that enters in order to give total extinction. In the presence of a medium which shows optical rotation, the analyser setting has to be altered in order to maintain the condition of total extinction.

Pasteur was able to separate the two sorts of crystals simply by picking them out with a pair of tweezers. He also prepared a third isomer of tartaric acid, but one which crystallized in a different form from the other two, always gave only a single type of crystals, and did not act upon polarized light.

J. Wislicenus (1835–1903) in Germany, prepared isomers of a relatively simple compound, lactic acid, which behaved in the same manner as the two "mirror" isomers of Pasteur. This acid occurs in muscle and also results from certain bacterial fermentations. According to its origin one or other of its "mirror" isomers is obtained. No third lactic acid was found, which would have corresponded to the optically inactive tartaric acid. The formula of lactic acid is

$$H_3C-\overset{\displaystyle\overset{H}{|}}{C^*}-\overset{\displaystyle\overset{H}{|}\;\;\overset{O}{|}}{C}=O$$

whereas the formula of tartaric acid is

$$
\begin{array}{cccc}
\text{H} & \text{H} & \text{H} & \text{H} \\
| & | & | & | \\
\text{O} & \text{O} & \text{O} & \text{O} \\
& | & | & \\
\text{O}{=}\text{C}{-}\text{C*}{-}\text{C*}{-}\text{C}{=}\text{O} \\
& | & | & \\
& \text{H} & \text{H} &
\end{array}
$$

In each case we are dealing with carbon atoms (those marked by the asterisk) which have four different substituents attached to them.

One of the founders of modern physical chemistry, the Dutchman *J. H. van't Hoff* (1852–1911) and independently the French chemist *J. A. Le Bel* (1877–1930) found an explanation for the results of Pasteur and Wislicenus (1874). This explanation soon expanded into an important branch of chemistry, the chemistry of space relation or stereochemistry (this latter word is derived from the Greek στερεός = solid).

They argued as follows. If the four valencies of carbon (visualized as short rods) are really pointing towards four non-coplanar directions in space, as Kekulé and Butlerow assumed, their ends lie at the apices of a regular (or deformed) tetrahedron (Fig. 27). Now imagine four different substituents to be attached to the carbon atoms, and designated by *a, b, c, d*. If we attach only *three* different substituents it will always be possible to turn the model around until it comes into a desired position (Fig. 27). The moment, however, that we go over to *four* different groups this is no longer possible, as anyone may easily convince himself by making a model of four differently coloured matches stuck tetrahedrally into a central piece of Plasticine. We can stick the four matches in many positions but after turning them around for a while we shall soon discover that all the arrangements of matches belong to one of two groups. Any member of a group can be brought into exact coincidence with any other member of the same group but it is impossible to turn a member of either group so as to make it coincide with any member of the other. Molecular structures which can be brought into the same position by simply turning them around are evidently identical; it is not important from which side we look at them. But what about the other group, that is the other non-superimposable molecular structure? We see from Fig. 28 that this is always the mirror image of the first one, differing only in this single structural feature. For example let us build two models, each having the substituent *a* on top and *b* pointing towards us at the bottom. We may now place *c* or *d* either in the left-hand or right-hand bottom position, and so arrive at two models which are mirror images of each other. They cannot be brought into coincidence.

Van't Hoff and Le Bel's fundamental assumption was that all molecules with formulae where object and mirror image do not coincide according to the tetrahedral model, exist in two isomeric forms. These are identical in all their physical and chemical properties except crystal form and behaviour towards polarized light. With respect to these two properties they behave like object and mirror image; their crystal forms can only be brought to

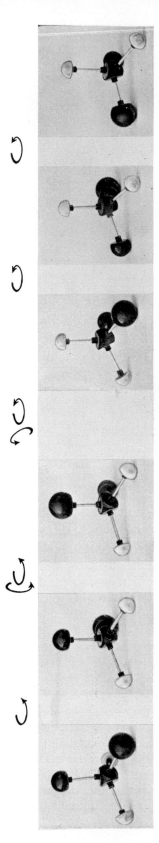

Fig. 27. **Three different substituents on a carbon atom**

Appropriate rotations of the model show that with three different substituents all positions of the molecule are equivalent.

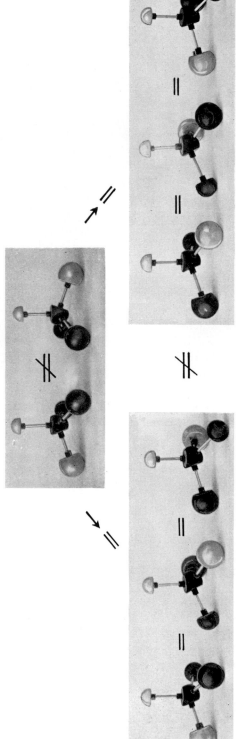

Fig. 28. **Four different substituents on a carbon atom**

Two sets of configurations exist which are mirror images of one another. A configuration belonging to one set can never be brought into coincidence

coincidence if one is seen in a mirror, and they rotate the plane of polarized light to the same extent in two opposite directions. This hypothesis was very plausible: a mirror asymmetry of the "building stones" of a compound would express itself in a mirror asymmetry of its physical properties, so far as such can be found.

The most convincing vindication of their hypothesis is the observation that each time the tetrahedral space model of a molecule predicts two distinct structures of object and mirror-image symmetry, i.e., in all cases when a carbon atom is substituted by four different groups or atoms, it is always possible to find experimentally such mirror-image isomers of the compound, called optical isomers. Conversely, never has an optically stereoisomeric pair of compounds been found without the tetrahedral formula allowing a

 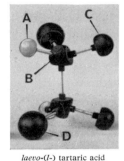 

| *dextro-(d)* tartaric acid | *laevo-(l-)* tartaric acid | *meso-(m-)* tartaric acid |
| (enantiomorphic) | | (internally symmetrical) |

Fig. 29. **Stereo-isomeric tartaric acids**

mirror-image difference in their structures. Stereochemistry is one of the most glorious proofs of the atomic hypothesis. It shows us that molecules are really built from definite spatial arrangements of atoms and that all demands of symmetry which follow from such models are satisfied by experiment.

The situation becomes of course more involved if more than one centre of asymmetry, e.g. asymmetric carbon atom,* is built into the molecule. However, it is always possible to build a space model of a molecule from its atoms and we may always determine what logical implications follow from the given structure. The number of isomers according to the theory and the number actually found are always the same.

Tartaric acid with two asymmetric carbon atoms bound to each other is a classical example. Fig. 29 shows us that beside the first two asymmetrical structures which are mirror images of each other, it is possible to build a third structure, where the symmetry of one half of the molecule has been changed into that of the opposite structure. This molecule can be turned around so as to coincide with its own mirror image. It is not asymmetric in its overall structure, the asymmetries of its two parts just compensate each other. Hence it should not have two asymmetric crystal structures and should

---

* Although this nomenclature is always used, it should be noted that it is the arrangement of the four groups about the carbon atom which is asymmetric, not the carbon atom itself.

not rotate the plane of polarized light, because being symmetric itself, it would not be able to cause rotation one way or another. Indeed the third acid found by Pasteur with the same structural formula as the other two tartaric acids, did not rotate the plane of polarized light. It was not "optically active" as we would now say, and has been named mesotartaric acid. Whenever two asymmetric halves of a molecule are mirror images of each other, thus compensating their optical activity, we arrive at optically inactive, meso-compounds.

It is worth mentioning that two mirror-image compounds of the non-meso type sometimes crystallize together in such a manner that the two antisymmetric parts are placed in mirror image positions within the crystal. In these cases the *crystal* is optically inactive and has a plane of symmetry. The two antisymmetric tartaric acids can crystallize in this manner within a certain range of temperature and such crystals are called racemic acid. From this name all similar compound crystals and mixed solutions of left- and right-handed forms are called racemates. They have to be separated into their respective components in order to arrive at the two active modifications. The fundamentally easy way to do this is by Pasteur's first method of crystallizing them under circumstances in which they prefer to grow separately and separating them with a pair of tweezers. Another method of separation is based on the fact that all living organisms are built from one set of the possible optically isomeric compounds they contain, and thus ferment, assimilate, or metabolize the two mirror-image isomers differently, very ofter being able to use only one of the two for their metabolic reactions. Either one of the two isomers is formed by the organisms or one is left unattacked.

Pasteur ingeniously found a third, chemical method of separating optically stereoisomeric compounds. Let us designate the compounds which rotate the plane of light towards the right by the letter d* (from the Latin *dexter*, right) and the other by l (for *laevus*, left). Now suppose we want to separate the two optical isomers of a racemic acid which is the mixture of d- and l-forms. Let us assume that we already possess an optically active, pure d' isomer of a base, say an alkaloid occurring in this form in nature or separated by the "tweezers" method. We shall be able to form a salt mixture from the acids and the base which contains the two salts of the $d–d'$ and the $l–d'$ configuration. These salts are not mirror images any more, because their basic component is the same. (The types $l–l'$ and $d–l'$ respectively, would be the mirror images of the preceding two, and we should have obtained them if we had used the laevorotatory isomer base, l'.) Since they are not mirror images, the salts have different physical characteristics, solubilities, crystal forms, melting-points, etc., and we are able to separate them by conventional methods. After separation the two acids can be regenerated from their salts. The above is the method generally in use for separating optical stereoisomers.

---

* In current nomenclature the letters d or l before the name or formula of a compound no longer stands for the sign of optical rotation but for the structural relation with dextrorotatory d-glucose. Sign of rotation is designated by $+$ or $−$, so that a compound with the prefix d- may actually rotate the plane of polarized light to the left while being chemically derived from the glucose stereoisomer which rotates to the right.

Stereoisomerism is by no means restricted to organic compounds, though it was first discovered in organic chemistry and because of the tetravalency of the carbon atom it is very common in that field. Natural quartz (SiO_2) crystals also exhibit the phenomenon of optical activity. From X-ray analysis of the crystal structure we know that the internal asymmetry is due to the OSiO atoms being wound around left-handed and right-handed endless screw axes within the crystal lattice.

Fig. 30. **Segregation of racemic compounds**

A racemic mixture of *d*- and *l*-forms of a racemic compound (e.g. an acid) in the lower part of the picture, is reacted with either the *d*- or the *l*-form of another asymmetric compound (e.g. base) represented in the upper part. The result is two compounds which though asymmetric, are not mirror images of each other and may therefore be separated by the conventional methods used in separating mixtures. Regeneration of the acids from the separate salts yields the *d*- and *l*-forms of the original racemic mixture.

The first true molecular optical isomerism in inorganic chemistry was discovered by A. Werner on complex salts of cobalt. He found that the trivalent cobalt ion is in most cases surrounded by six substituents (ions, molecules) in the "first sphere", (p. 57) that is firmly attached to the central ion. The substituents may be NH_3, Cl^-, SO_4^{--}, etc., or even water. A Cl^- ion in this first, coordinative sphere does not react with silver ions, its favourite partners, so firm is the bond which holds it to the complex. Now it could be shown that these coordinated groups are distributed in space around the central cobalt ion to form a more or less regular octahedron around it. The argument is exactly as for the carbon atom, with the difference that we now have an octahedron instead of a tetrahedron. A complex with two substituents of one sort and four of another, that is [(Co)A$_2$B$_4$] occurs in two isomeric experimentally observable forms (not optical isomers) which agrees with the octahedral arrangements

Three isomers would have been expected if the substituents were arranged in a plane:

$$
\begin{bmatrix}
 & A & A & \\
B & & Co & & B \\
 & B & B &
\end{bmatrix}
\qquad
\begin{bmatrix}
 & A & B & \\
B & & Co & & A \\
 & B & B &
\end{bmatrix}
\qquad
\begin{bmatrix}
 & A & B & \\
B & & Co & & B \\
 & B & A &
\end{bmatrix}
$$

We can go on to compare the number of isomers actually found and the number predicted by the geometry of the octahedron or the hexagon for the more complicated cases of substitution.

At the stage where at least five of the six coordinated groups are different, the geometrical model demands the existence of two mirror-image molecules:

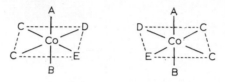

Werner actually succeeded in resolving similar isomers by combining them with optically active acids. This achievement brought him the Nobel prize.

21. STEREOCHEMISTRY, II: MULTIPLE BONDS

Let us go one step further in our study of the valency of carbon, bearing in mind that all we are going to learn is valid for other elements also. It was well established, right at the start of modern chemistry, that the number of atoms or groups bound to the carbon skeleton of a molecule was often less than it should be according to the principle of the tetravalency of carbon. But the number of unused carbon valencies was always even: two, four, six, etc. atoms or groups were "missing". We may easily explain this state of affairs by the assumption that the two valencies which have become free through two carbon atoms each loosing a substituent are used in bonding these two carbon atoms together with a new, extra, bond. This implies either ring formation, if the two carbon atoms are not adjacent, or the formation of a double bond between them, if they are. For instance, ethyl chloride loses a molecule of hydrogen chloride on being heated to 500–600°C and a C=C double bond is formed, analogous to the double bonds we have encountered, for example, in CO_2: $O{=}C{=}O$

$$
\underset{\underset{\displaystyle \text{H} \quad \text{Cl}}{|\quad\;|}}{\overset{\overset{\displaystyle \text{H} \quad \text{H}}{|\quad\;|}}{\text{H—C—C—H}}} = \underset{}{\overset{\overset{\displaystyle \text{H} \quad \text{H}}{|\quad\;|}}{\text{H—C=C—H}}} + \text{HCl}
$$

(21.1)

The new hydrocarbon, C_2H_4 is called ethylene. A hydrocarbon containing even fewer hydrogen atoms per two carbon atoms can be prepared by reacting calcium carbide with water:

$$
CaC_2 + 2HOH = H{-}C{\equiv}C{-}H + Ca(OH)_2 \qquad . \quad (21.2)
$$

We can justify our writing the triple bond between the two carbon atoms by the fact that our new gas, the very explosive acetylene, is capable of adding two successive molecules of chlorine gas, yielding eventually symmetrical tetrachloroethane:

$$H—C{\equiv}C—H + Cl_2 = \underset{\underset{Cl}{|}}{H—C}{=}\underset{\underset{Cl}{|}}{C—H} \qquad (21.3)$$

$$H—C{=}C—H + Cl_2 = \overset{\overset{Cl}{|}}{H—C}—\overset{\overset{Cl}{|}}{\underset{\underset{Cl}{|}}{C}}—H \qquad (21.4)$$

It is always necessary to take away two substituents at a time in order to establish a new double bond—our model has been justified once more.*

If we take the tetrahedral model seriously, then to represent a double bond, we have to attach two carbon tetrahedra along a common edge (Fig. 31). The remaining two pairs of substituents on the doubly bound carbon

cis-form   *trans*-form

Fig. 31. **A double bond between two carbon atoms**

---

* In some relatively uncommon cases a single bond is sometimes broken without elimination of a second atom or group from a molecule. In inorganic chemistry we have, for instance, the reaction

$$\underset{\underset{O}{\diagup}}{\overset{\overset{O}{\diagdown}}{N}}—\underset{\underset{O}{\diagdown}}{\overset{\overset{O}{\diagup}}{N}} = 2 \; \underset{\underset{O}{\diagup}}{\overset{\overset{O}{\diagdown}}{N}}— \qquad (21.5)$$

with a "free valency" on the nitrogen atoms. Among carbon compounds such a bond-cleavage only occurs if adjoining carbon atoms are very heavily substituted with groups which occupy much space and belong to a special bond-weakening type. Three benzene rings (p. 86) on the same carbon atom constitute one example of this sort. The compound hexaphenylethane: $(C_6H_5)_3C—C(C_6H_5)_3$, splits thus into two $(C_6H_5)_3C—$ groups, called triphenylmethyl, where the heavily substituted carbon becomes trivalent. It is customary to call such compounds free radicals because of their analogy to radicals such as e.g. methyl ($—CH_3$) within a molecule. In all such cases the "radical" compound is very reactive with respect to its "free valency". Furthermore it is always paramagnetic (p. 284), that is it behaves as an elementary magnet, a state not generally occurring among organic compounds.

atoms thus lie within one plane, the plane perpendicular to that containing the double bonds. It is extremely interesting that this simple geometrical assumption has been fully borne out by the modern, electronic theory of valency bonds, as we shall see in a later chapter (p. 327). If this be so, however, its consequence is that it gives rise to a new type of isomerism. Let us look, for example, at the compound which is derived from ethylene by substituting the characteristic "carboxyl" group of organic acids:

$$-C\underset{\textstyle OH}{\overset{\textstyle O}{\big<}}$$

in place of a hydrogen atom on each carbon atom. The C=C double bond makes the molecule rigid, so that the two halves cannot rotate relative to each other about the C=C axis as they could around a single bond at normal temperature. In fact, two isomeric acids are known to us, and have the required formulae. The first is called fumaric acid and has its carboxyl groups on opposite sides of the double bond whereas the second, maleic acid, has them on the same side. This latter assumption is based on the fact that maleic acid readily loses a molecule of water to form an intramolecular anhydride, whereas fumaric acid is unable to do this:

Fumaric acid   Maleic acid   Maleic anhydride

(Conventionally we write on paper the double bond *and* the four substituents in the same plane, though according to Fig. 31 they are perpendicular to each other.)

We always encounter this type of isomerism whenever we have to deal with a double bond and different substituents on the atoms joined by the double bond. It is called geometric isomerism. The isomer of the type of fumaric acid is called the *trans* isomer, the other one with similar groups on the same side of the double bond the *cis* isomer. Which compound is *trans* and which *cis* must be decided after weighing all the chemical evidence available, as we have shown in the simple case of maleic anhydride. The more double bonds in a molecule, the greater the number of possible isomers. As far as it was possible to examine them, their number was always in accord with that which followed from the considerations we have just learned. The whole problem was clarified by van't Hoff and Wislicenus as a consequence of the tetrahedral carbon model.

The triple bond in acetylene and its homologues can be readily imagined as arising from two carbon tetrahedra being joined through one of their planes. The two remaining substituents thus lie on the same straight line as the carbon atoms and no isomers should exist. None has ever been found in experiments.

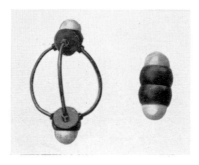

Fig. 32. **A triple bond between two carbon atoms**

22. THE BENZENE RING

Let us recall the benzoyl radical of Liebig and Wöhler: its formula was C_7H_5O (p. 50). Compounds containing this radical are derivatives of benzene, a well-known, inflammable organic liquid. The formula of benzene is C_6H_6 and its six carbon atoms remain together as a group throughout such a very large series of chemical reactions, that this arrangement was a long time ago considered to be one of the fundamental radicals of organic chemistry.

The six hydrogen atoms of benzene may be substituted one after the other, e.g. by chlorine, and in the course of such substitutions we arrive at a series of interesting isomeric compounds. Only one monosubstituted compound, chlorobenzene exists; there are three dichlorobenzenes, three trichlorobenzenes, three tetrachlorobenzenes but only one penta- and one hexachlorobenzene. In the same way as we deduced the structure of ethane from the number of its chlorine-substituted derivatives, we can deduce the internal structure of benzene from the number and relation of the above isomers. Of course, instead of chlorine we could have chosen any other substituent or even more than one kind of substituent. It is to Kekulé's everlasting credit that he found the solution to this important problem in 1858.

He argued that the C_6 group must necessarily have all six atoms in equivalent positions because the monosubstituted derivatives have no isomers. One of the possible ways to achieve this is to arrange the six carbon into a plane hexagon, or ring as we are accustomed to call it. We shall see that this assumption demands just the number of isomeric derivatives which have been shown to exist by experiment. But the valencies of the six carbon atoms must be disposed in some way or other. The arrangement which is in best agreement with chemical facts was proposed by *H. E. Armstrong* (1878–1937) who drew one bond between each pair of adjacent carbon atoms of the ring, one bond between each carbon atom and its associated hydrogen atom and made the fourth valencies of the carbon atoms point symmetrically towards the middle of the hexagon, as a sign that in one way or another they counteracted each other there:

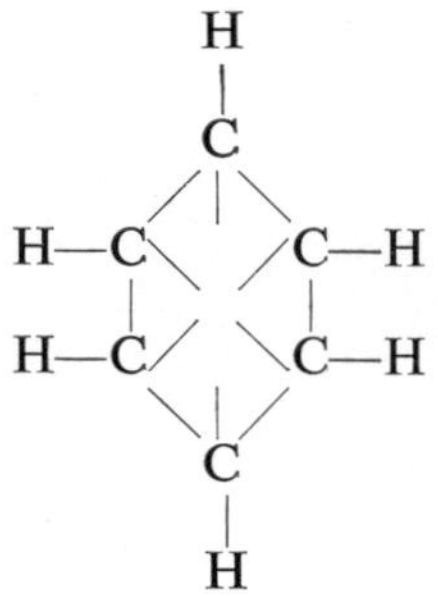

Kekulé himself preferred another solution, doubling every second bond around the ring. His formula has become accepted in spite of the fact that it demands the existence of two distinct isomers when neighbouring hydrogen atoms are substituted—there should be a difference according to whether there is a double bond between them or not. As it was not found to be possible to demonstrate such isomers experimentally, there seems to be no real difference between the double and single bonds in Kekulé's formula. On the other hand it does account for the fact that each of the three double bonds is able to add two hydrogen atoms.

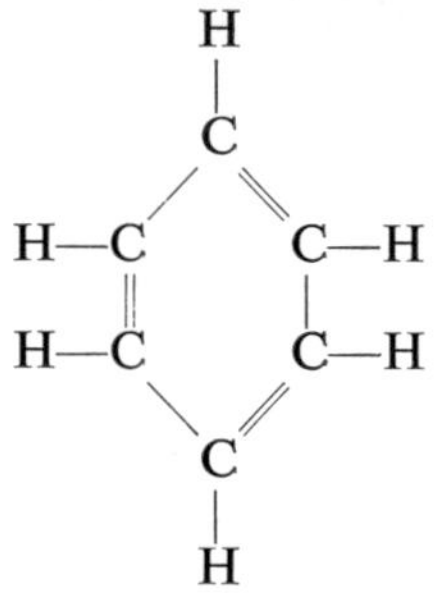

Kekule's formula 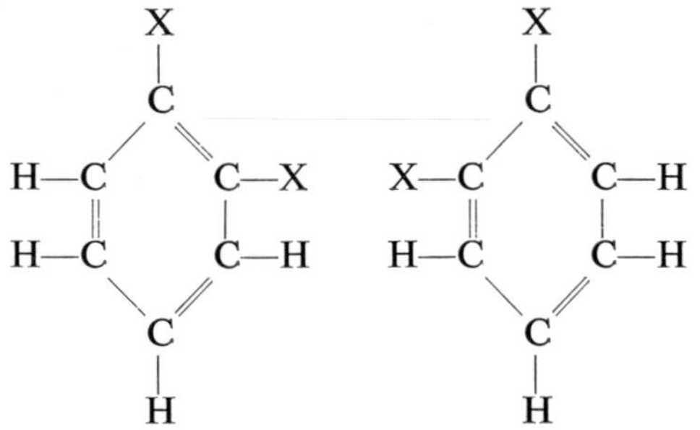   Isomers which should but do not exist

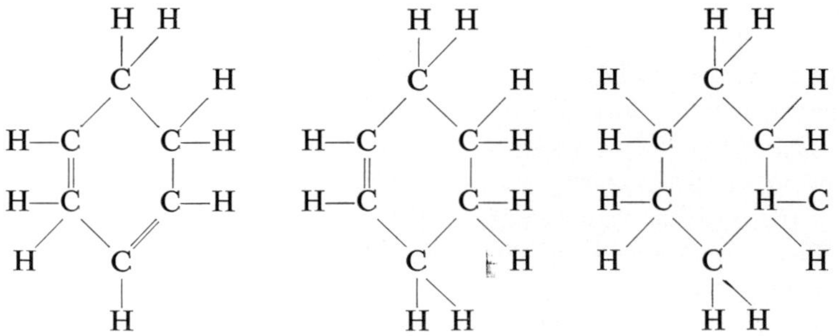

Successive hydrogenation of double bonds

It should be noted that the Armstrong theory is not in discord with this hydrogenation process.

It is a pity, but the Armstrong formula is not in use today, the simplified Kekulé formula being used. The six carbon atoms are often omitted and

only a hexagon with three double bonds is drawn:

Often even the double bonds are omitted and only a hexagon remains, but this can give rise to confusion because it is also the symbol of the fully hydrogenated benzene, cyclohexane, which is shown on the extreme right-hand side of the formulae above. Our modern concepts about the benzene ring agree better with the Armstrong formula. According to these concepts, whatever the cause of the double bonds may be (we now know it to be an arrangement of electrons; *see* p. 328), they rotate with enormous speed within the ring and thus cannot be localized between any pair of carbon atoms. All six carbon atoms and all six bonds between them are strictly equivalent—so all our experiments tell us. Therefore nowadays the symbol

is sometimes used.

Now let us see how the isomerism of the whole series of chlorine-substituted benzenes is explained by the hexagonal arrangement, writing for the sake of convenience the hexagons with the circle inside them (as shown on p. 88).

There are just as many isomers predicted as have been found experimentally. One may be inclined to think that the above list is incomplete, but it is always possible to turn any other configuration in or through the plane of the paper until it coincides with one of the above formulae. Thus for instance the structure

is identical with the second formula of the fourth row on p. 88. It is only necessary to turn it 120° counter-clockwise to see that this is so.

It is not necessary here to speak about the names we assign to such isomeric configurations. If the number of different substituents increases, these names are of no use anyway, and it is much better to assign a number to each carbon atom of the hexagon, beginning at the top carbon atom. Thus 1-chloro-3-bromo-4-nitro-benzene is represented by the formula:

Carbon atoms not enumerated in the name are assumed to have hydrogen atoms attached to them.

We have the compounds and we have the same number of formulae, but how are we to decide which is the appropriate formula for a given compound? Once more we cannot do better than follow sets of reactions logically and draw our conclusions therefrom. This process is sometimes very complicated indeed and may involve a lot of hard thinking and laborious experimentation. Just to show one relatively not-so-complicated line of argument, let us see how the structures of the three dichloro- and the three trichlorobenzenes are ascribed to the respective compounds. Let us take each of the three distinct dichlorobenzene isomers and react them with more chlorine until they take up a third chlorine atom. We find that there is one dichlorobenzene which yields only a single trichloro-benzene on chlorination.

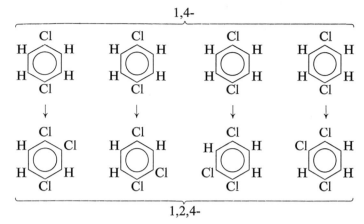

1,4-dichlorobenzene can yield only 1,2,4-trichlorobenzene,

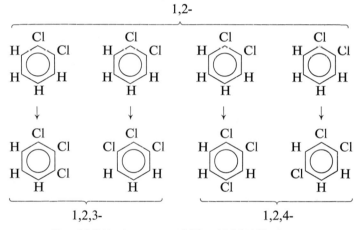

From 1,2-dichlorobenzene we get 1,2,3- and 1-2,4 trichlorobenzene.

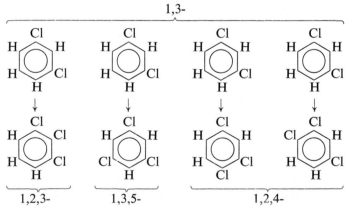

From 1,3-dichlorobenzene we get all three possible trichlorobenzenes.

We have to look carefully at the three dichlorobenzene formulae to see which of them gives the above result. We find that the required isomer is 1,4-dichlorobenzene, with the two chlorine atoms opposite each other. Thus, as we know that the compound giving a single trichlorobenzene must have the 1,4-structure, then the trichloro-compound derived from it has the 1,2,4-structure. Thus two of our six compounds have already been identified.

One of the two remaining dichlorobenzenes gives two trichlorobenzenes on further chlorination, the other gives three. One of the two trichloro-compounds generated from our second dichlorobenzene is identical with the trichlorobenzene obtained from our first series of reactions, where it was the sole product. We know then that this is the 1,2,4-isomer. Looking at our table of reactions we see that it is 1,2-dichlorobenzene which yields two isomers on further chlorination and we see that indeed one of these must of necessity be the same compound as the one which was obtained as a sole product before. Thus the other trichlorobenzene we obtained from 1,2-dichlorobenzene must have the 1,2,3- formula. By the principle of exclusion we might already assign the remaining two formulae to the remaining two compounds, but we are in the agreeable situation that we can use the chlorination of the remaining dichlorobenzene isomer as a control of our logic and our experiments. If our reasoning is correct, this final dichloro-compound has to yield all three possible trichloroisomers. As we have seen, it really does so and the problem has been solved.

The logic of the whole operation is only complicated at first sight; one grows accustomed to it. This, and similar sets of operations, are used to decide the formulae of isomeric benzene derivatives. The fact that all isomers which follow from the Armstrong formula are always found and that there have never been found more isomers than this formula predicts is the experimental proof of its correctness. Now that chemists have solved these riddles it has become possible to examine the physical properties of certain types of isomers, and today we are often able to decide between possible isomers simply by measuring some of their appropriate physical constants. But the logical foundation of these physical methods is based on the tiresome procedures of the above "chemical group theory".

It is customary to discuss the organic compounds which only contain open-chain systems separately from those which contain closed rings. This distinction is even greater when we consider the so-called "aromatic" ring systems, with formally alternating single and double bonds, as in benzene. Compounds which contain other atoms apart from carbon in the ring, the so called "heterocyclic" compounds are normally discussed separately too. However, we do not wish to enter the realm of special organic chemistry. We only require to show that it is permissible to think of atoms as the real building stones of compounds, having definite places within the molecules.

It may be as well to mention, that there are much more complicated ring systems among organic compounds than the benzene ring, but their formulae are just as soundly based on logical sets of experiments. The next member to benzene is naphthalene which has the formula

$$
\begin{array}{ccc}
 & \underset{C}{H} & \underset{C}{H} \\
HC & C & CH \\
HC & C & CH \\
 & \underset{H}{C} & \underset{H}{C}
\end{array}
\quad \text{or}
$$

$$
\begin{array}{ccc}
 & \underset{C}{H} & \underset{C}{H} \\
HC & C & CH \\
HC & C & CH \\
 & \underset{H}{C} & \underset{H}{C}
\end{array}
\quad \text{or} \quad \text{⬡⬡} \quad \text{or} \quad \text{◯◯}
$$

The two rings are equivalent, thus the Armstrong formula only makes sense if we imagine that the six and four central valencies alternate very quickly between the rings. We have reason to suppose that this is the case because of the special diamagnetic behaviour of the compound (p. 281).

To end this section let us reproduce the formula of the coloured part of the red haemoglobin molecule of blood, (haemin) which carries oxygen to our cells throughout the body. It is the result of the classical research done by *H. Fischer* in Munich, who spent many years in its elucidation:

$$
HC \text{---} C \text{---} CH \text{=} CH_2
$$

(haemin structure with central Fe, four N, Cl, and substituent groups CH_3, HC, CH, CH_3, $C\text{---}C$, $O\!\!=\!\!C\text{---}CH_2\text{---}CH_2\text{---}C\text{---}C$ / HO, $C\text{---}CH_2\text{---}CH_2\text{---}C\text{==}C\text{---}CH_3$ / HO, CH, CH_2, etc.)

23. CHEMISTRY AND THE ATOMIC HYPOTHESIS

Let us cast a glance back upon all we have learned up to now. Dalton explained the laws of constant and multiple combining proportions by assuming that a given number of elementary atoms are built together to form a molecule of a compound. Avogadro assumed that gases behave so much like one another because every unit volume of a gas contains (at a given temperature and pressure) the same number of molecules. Thus it became possible to arrange gaseous compounds into series and to tell which of them contained one, two, three, etc., atoms of a given element. Taking the atomic weight of hydrogen as the unit (more precisely that of $C = 12 \cdot 000$) it became possible to arrange a table of relative atomic weights and to ascertain molecular weights by more than one method.

It soon became apparent that it was not correct to assume the identity of two compounds if the percentage compositions were the same. They would really be *different* if their molecular weights differed, but even the same percentage composition and the same molecular weight did not guarantee that two substances were identical. It was still possible that they could be different compounds. Thus we had to assume that the relative positions of the atoms within a molecule are also important and that different positions of the same set of atoms corresponded to different compounds, i.e. it was possible to build a series of different compounds from the same set of atoms. We learned that this phenomenon, which we call isomerism, was frequently encountered among the compounds of carbon in particular, because carbon atoms very readily formed complicated chain and ring systems. Carbon generally has a valency of four and the number of isomers demanded by the "geometrical" theory only coincided with the number actually found in experiments, if we assumed that the four valencies acted along the four directions from the centre towards the corners of a tetrahedron. We also learned that four different substituents at the four corners of such a tetrahedron can always be arranged in two ways, one the mirror image of the other. As a matter of fact every time this occurred with the formulae, it was possible to separate two isomers of the compound in actual practice—two isomers which were identical in everything except the properties related to symmetry in space. So far as these properties are concerned, optically isomeric compounds behaved like mirror images of each other. We have seen the consequences of double bonds upon the number of possible isomers of a compound and have learned a little about aromatic ring structures in connexion with benzene.

From beginning to end we have drawn all logical consequences from some basic assumption, e.g. the tetrahedral arrangement of the four valencies of carbon, or the hexa-coordination of groups around some inorganic ion, and have always compared the number of isomers actually found with that demanded by theory. These numbers always agreed exactly. The features of proposed formulae were such that they well represented the chemical behaviour of the corresponding compounds. Atoms near to each other in the formula often tended to remain together in fact, while those further apart were easier to separate, etc. After all this we may rightly ask the question: Have we definitely proved the existence of atoms and of molecules built from them, or have we not? One may well feel that the far-reaching agreement between experimental results and theoretical demands has made the assumption that atoms really exist more than extremely probable. Indeed it is hardly possible to doubt their existence. One could only make the challenge: "It is true that everything gives the impression that compounds could be composed of molecules which are built to specific configurations from their atoms, but surely all this is only fortuitous and there is some other reason which explains the whole set of experimental results we have discussed in so much detail? Another explanation might exist which explains all the facts as well as the atomic hypothesis does, but which had not been discovered as yet". It is impossible to refute such an argument. One can never argue with so much certainty from effect to cause as to say "No" if someone says: "But could not all these facts also be explained by an entirely different mechanism?" One generally admits after a certain amount of positive evidence that a hypothesis is probably valid. While every consequence that can be experimentally checked is confirmed, then there is more than a fair chance that this will continue to be so in the future. So long as, but only so long as, we find nothing that contradicts a hypothesis we shall suppose it to be true.

However, it seems to be a matter of individual taste just how much evidence is sufficient for somebody to argue on the above lines. No lesser a chemist than *W. Ostwald* (1853–1932), the father of physical chemistry, was not convinced by all the enormous amount of evidence of structural chemistry and for a long time treated the atomic hypothesis only as an assumption and said that we have no immediate proofs of the existence of the atoms. The author personally does not think that the evidence for the earth being a sphere is any more convincing. We shall see, however, that it was possible to gather an overwhelming amount of evidence from an entirely different realm of science which affirmed the existence of atoms. As a matter of fact we shall see that atomic dimensions could be calculated, and that quite different methods gave identical numerical results. Only after this had happened did Ostwald feel convinced that atoms "really existed" and he stated as much in the preface to one of his books in 1903. He was the last great scientist to be sceptical about this assumption.

After all, what is the guarantee that the table I write upon "really exists" and is not merely a result of delusion, of a conspiracy of my senses? We are once more up against an ultimate and hopeless question of philosophy to which there is no answer. One culminates a long series of sensory perceptions by saying "Here is a table." After another long set of sensory perceptions and very hard thinking one says that matter is composed of molecules and atoms. Both assertions are true or not true to the same degree.

Part 2

The Size of Atoms

24. HOW BIG IS AN ATOM?

Now that we have come so far we cannot help asking how big are these atoms, if we are so certain that they do exist. Can their size be measured, or are they so small that they keep their secret in spite of the best instruments invented by us? And if we are able to measure them, do we always arrive at the same size for a given atom irrespective of the methods we apply?

We have indeed managed to measure atoms and always obtain the same sizes. It was this result that convinced even the sceptic Ostwald about the real existence of atoms in 1903. Shortly we shall learn about some of the independent ways that have led into the world of atomic dimensions but, in contrast to our earlier approach, we are not going to follow the methods in the sequence of their historic evolution because it so happens that some of the methods which were applied later are more direct and easier to understand than the older ones.

However, before embarking on any of these studies let us stop for a minute and see what quite simple laboratory observations are able to tell us about atomic dimensions.

It is known that compounds of sodium impart an intensive yellow colour to the Bunsen flame. The phenomenon is extremely sensitive, and using solutions of known concentration it was possible to ascertain that 1/2,000,000 mg of sodium may just be detected in this manner. Therefore the sodium atom must certainly weight less than that and if we calculate the radius of a sodium sphere of this weight we arrive at the value of about 1/100 mm which represents, of course, an upper limit. One of the compounds with the most distasteful smell is ethyl mercaptan, C_2H_5SH. Its smell can be detected even at a dilution where only 1/500,000,000 mg of the material reaches our nose. This amount of substance would be contained in a sphere of about 0·8 micron radius. (A micron is 1/1,000 mm and its symbol is μ.) The mercaptan molecule is composed of nine atoms and if we divide the volume by nine we reach a radius of about 0·4 micron for an average atom, as an upper limit. (The radius of a sphere is proportional to the cube root of its volume; we assume here that atoms and molecules are spherical).

The finest wire drawn from platinum or gold, the so-called "Wollaston wire", has a diameter of about one micron. This must definitely be an upper limit for the diameter of an atom of gold. Gold smoke, that is excessively thin gold foil made by hammering an ordinary foil of gold between pieces of leather is still finer than that, it has a thickness of only 0·1 micron. The foil is transparent to green light at this thinness. Exactly how thin it is can best be calculated from the weight of a known area and the density. We must consider however, that 1/4 of a micron is about the limit to be seen in an ordinary microscope and that many bacteria have dimensions of this order of magnitude. It is clear that we must still be far above the true dimensions of an atom, because after all it seems highly improbable that such a complicated thing as a living bacterium should consist only of a small number of atoms!

We may proceed yet further. An alkaline solution of fluorescein remains visibly fluorescent at a concentration of 1/1,000,000 and the green light of fluorescence is apparently still perfectly homogeneous within a volume of 1 cubic micron, as seen under a microscope. This minute volume presumably contains many molecules. Now as 1 mm³ of water weighs 1 mg, 1 μ^3 (μ = micron) weighs only 10^{-9} mg* and this amount of solution contains only one 10^6th part of fluorescein corresponding to a sphere about 0·006 μ in diameter. We know that the fluorescein molecule contains 37 atoms and so we obtain for the average upper limit of the diameter of an atom 0·002 μ, that is 2 × 10^{-7} cm. We have been reaching lower and lower values for the diameter of an atom and do not know yet where we shall find a lower limit. The only thing of which we are certain is that an atom must be smaller than 2 × 10^{-7} cm in diameter.

A stream of electrons, such as the one that traces the picture upon the fluorescent screen of a television tube, may be made to pass through matter like a ray of light. Instead of the lenses, necessary for light rays, electrons may be deflected by electrostatic or magnetic fields, and by an appropriate combination of such electric or magnetic "lenses" it is possible to build the so-called electron microscopes. Just as the resolution in a light microscope increases with decreasing wavelength of the light (twice as good resolution can be obtained at the violet as at the red end of the spectrum), the resolution of an electron microscope increases with the velocity of the electrons in the beam. This in its turn is of course determined by the voltage used to accelerate the electrons. Now it is possible to work with electrons of so high a velocity that structures as small as 10^{-7} cm can be resolved, and seen or photographed in the electron microscope. Virus particles which pass our finest filters show up as small spheres or rods in these pictures. A virus particle is much more primitive, much less "living", than a bacterium, but still it must surely contain quite a number of atoms. We can now say that the limit for atomic diameters is less than 10^{-7} cm.

The reader will be acquainted with soap bubbles and their beautiful brilliant colours. Probably he will have noticed that before they burst a black spot appears and grows in the midst of these colours. By adding glycerine Dewar was able to lengthen the life-span of such thin bubbles and measure their thickness by elaborate optical techniques. He obtained the result of 6 × 10^{-7} cm for the thinnest of them.

Even thinner layers may be obtained by spreading fatty or oily material on the surface of, say, water. Take a clean surface of water and just touch it

---

* It is time to go over to the usual notation for great and small quantities, the notation based on powers of 10. 10 itself is 10^1; 100, that is 10 × 10 is 10^2; 1,000 is 10^3, etc. Changing to fractions, 1/10 is 10^{-1}, 1/100 is 10^{-2}, 1/1,000 is 10^{-3}, etc. 10^0 is 1. 1 $\mu = 10^{-3}$ mm; 1 μ^3 = 10^{-9} mm³.

with a needle that has been made slightly greasy by touching one's forehead. We can *see* no change whatever in the water, although something very fundamental has happened to the surface. For instance, small pieces of camphor which run hither and thither on a clean surface of water stand

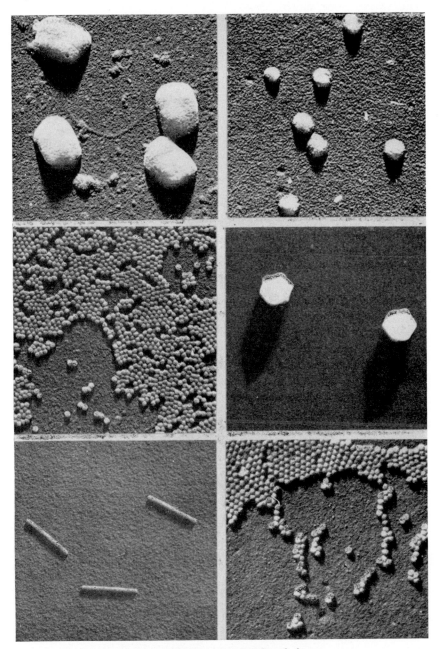

Fig. 33. **Electron micrographs of viruses**

(From Stanier, Doudoroff and Adalberg, *The Microbial World*, 2nd ed., p. 210, Fig. 9-8 (Prentice-Hall, 1963).)

still when it has been contaminated in this manner! How small must this contamination have been, considering that it spread from the tip of an oily needle which may even have been wiped before application!

We are able to demonstrate the effect of the needle in another way too. Let us spread some ignited,* pure talcum powder on a clean water surface before the experiment. Next let us touch the surface with the tip of the needle. The talcum powder moves from the point of contact as if it had been swept away. If we maintain contact for a sufficiently long time, we may sweep all the talcum to the edges of the vessel. But if we remove the tip from the surface, we may halt the process at any moment and then only a round talcum-free area will be seen on the surface of the otherwise talcum-covered water. Something spreads from the needle on to the water surface and its spreading takes time. We may move the vessel or blow on to its surface and thus deform the shape of the clear area, but its magnitude remains unchanged. It can be shown that camphor stands still on the area freed from talcum but moves about on the rest of the surface and it is clear that the fat or oil from the needle has done something to the surface of water, its presence being made apparent by the phenomena it induces.

Oleic acid, long-chain fatty acids, and long-chain amines or alcohols behave in a similar manner on a clean water surface. We find that in general those compounds do so which have a long hydrophobic (water-insoluble) hydrocarbon chain (containing only C and H atoms) in the molecule. But attached to such a chain there must be at least one group which in itself would impart water-solubility to the compound. Thus the short-chain hydrocarbons, methane, ethane, propane, etc., are almost insoluble in water but compounds derived from them and containing an acid group (e.g. acetic acid) or an amine group (ethylamine) or an alcohol group (ethyl alcohol) are readily soluble. The solubility in each of these series decreases with increasing length of the hydrophobic C–H chain. We may safely assume, that the active groups ($-COOH$, $-NH_2$, $-OH$) still try to dissolve in the water and this seems to explain the above experiment. The compounds spread on water because their ends possessing the water-soluble (hydrophilic) group tend to adhere to the water surface whereas the hydrocarbon chains, which resemble the insoluble hydrocarbons themselves, remain aloof from the surface. These considerations were first put forward by *Lord Rayleigh* (1842–1919) and were elaborated in more detail by *P. Devaux, I. Langmuir* (1881–1957) and *W. D. Harkins.*

All this sounds very well, but how can we prove it? Is it possible to deduce some consequences of all we have just said, and compare them with experimental results, if possible in a quantitative manner? It is possible. If our picture was correct, the hydrophilic groups alone are attached to the surface of the water while the hydrocarbon chains stand away from it. Thus all molecules containing the same hydrophilic groups (i.e. all acids, all amines, all alcohols, etc.) should occupy the same area on the water surface irrespective of the length of the hydrocarbon chain to which they are attached,

---

* i.e. previously heated to a high temperature and then allowed to cool.

because these do not contact the water. Therefore we have to determine the area of water occupied by, say, one mole of different compounds belonging to the same series and see whether this area is really the same for each compound within the homologous series. By comparing equal number of moles (a mole is the molecular weight expressed in grammes) we compare equal numbers of molecules, but we just do not know yet how big this number is.

Suppose we choose our example from among the compounds which carry a CO.OH (carboxyl) group on the end of a long hydrocarbon chain. Suitable compounds, in general, are the normal, saturated acids, the so-called fatty acids. (Normal refers to a straight chain, as opposed to a branched one, and saturated means that no double or triple carbon–carbon bonds are present.) One of these occurs in the suet of sheep. It is called lauric acid and has the formula $C_{11}H_{23}.CO.OH$. Two higher members of the series which are to be found in the fat of pigs and cattle are palmitic acid ($C_{15}H_{31}.CO.OH$) and stearic acid ($C_{17}H_{35}.CO.OH$). A much rarer compound is margaric acid ($C_{16}H_{33}.CO.OH$). Let us prepare a solution of each of these acids in "light petroleum ether", a hydrocarbon solvent, of concentration 10 mg (0·01 g) of acid per litre of solution. This is admittedly very dilute, but it can be readily prepared with great accuracy using a chemical balance. Let us convince ourselves that the petroleum ether is pure by confirming that the talcum ring closes up again after a drop of this solvent applied to the surface has evaporated away (Fig. 34a). This should occur in about a minute or so. Assuming that the solvent is found to be pure, we can measure out, with a fine pipette, 0·10 ml of each solution and apply it to a talc-covered water surface. As soon as the petroleum ether has evaporated, the circular, talc-free region no longer changes in area, i.e. no further contraction occurs (Fig. 34b). Now we can measure the area of this circle and for stearic acid we find it to be 5·3 cm². This is the area occupied by the stearic acid that was present in the 0·10 ml of solution, i.e. 10^{-6} (one-millionth) of a gramme, a small quantity often designated by the Greek letter *gamma*, γ. The molecular weight corresponding to the formula of stearic acid is 284, therefore, one mole of stearic acid weighs 284 g. We can easily calculate that the area occupied by one mole of the acid would amount to 150,000 m² of water surface.

Let us now perform the same experiments and calculations with the other three acids. The areas occupied by the amounts of acids contained in 0·10 ml of the solutions are 7·5 cm², 5·9 cm² and 5·6 cm² for lauric, palmitic and margaric acids respectively. These are the areas covered by 1 γ of each of them. But let us consider that their molecular weights are 200, 256 and 270 respectively and then it becomes apparent that one mole of each of them covers the same area, namely 150,000 m², as does the mole of stearic acid. It seems as though our assumptions that the area occupied by a foreign molecule on a water surface would depend only on the end group, is correct.

If this be so, we can calculate the thickness of such a monomolecular layer upon water. The density of stearic acid is 0·94 and hence the volume of 1 γ of stearic acid is $1·06 \times 10^{-6}$ cm³. This amount of stearic acid has spread

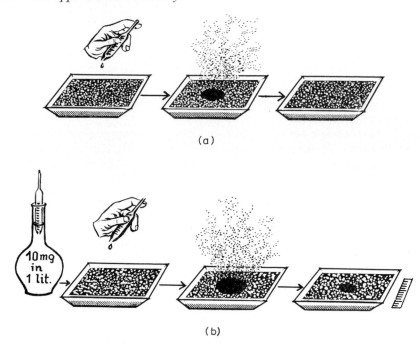

Fig. 34. **Measuring the thickness of a monomolecular surface layer**

(a) A drop of pure petroleum ether pushes aside the talcum powder on a water surface. It soon evaporates completely and the water surface closes in again.

(b) A drop of a 10 mg/litre fatty acid solution in petroleum ether also pushes away the talcum. But after the petroleum ether has evaporated the talcum does not close in completely. The "window" consisting of the fatty acid layer can be measured with a mm ruler.

over an area of 5·3 cm². We obtain the thickness of the layer by simple division: $2·0 \times 10^{-7}$ cm. If we are right, this is the length of the stearic acid molecular chain. (We have, not quite correctly, assumed that the density of the thin film is the same as of the bulk material and that there are no gaps between the molecules).

We may go even further. By calculating the length of a molecule which is just one CH_2 unit shorter than the stearic acid chain, that is the length of a margaric acid molecule, we can find the difference between this and stearic acid and this gives us the length of a CH_2 group along the chain. The length obtained for a margaric acid molecule is $1·91 \times 10^{-7}$ cm and hence the difference attributable to a single CH_2 group is $0·11 \times 10^{-7}$ cm. Knowing from structural chemistry that the two hydrogen atoms are attached across the carbon atom chain, this length should then be due to the carbon atom itself and would therefore be the first reliable measurement of the diameter of the carbon atom we have been able to make! A correction is needed, however, as we have assumed the C—C—C chain to be a straight line. According to Kekulé's theory this is not a legitimate assumption since the valency angle

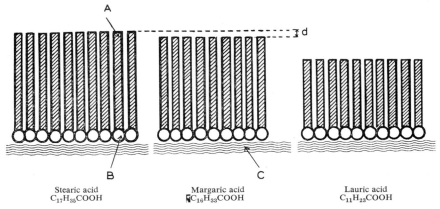

Fig. 35. **Cross-section of monomolecular fatty acid surface
layers on water**

All fatty acids occupy the same surface area on water, irrespective
of their chain length.

A = hydrophobic CH_2—CH_2— chain
B = hydrophilic —CO.OH group
C = water surface
d = length difference due to one C atom

Stearic acid
$C_{17}H_{35}COOH$

Margaric acid
$C_{16}H_{33}COOH$

Lauric acid
$C_{11}H_{23}COOH$

of a regular tetrahedron is about 109·5°. The C—C—C chain should therefore be a zig-zag line with this angle between adjacent bonds (Fig. 36).
Elementary trigonometry allows us to calculate the true C—C distance from
this assumption and yields a value of $0·14 \times 10^{-7}$ cm. Is this, then, more or
less the actual diameter of a carbon atom? Yes! Other more reliable methods

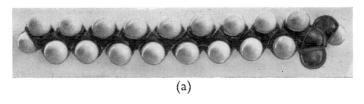

(a)

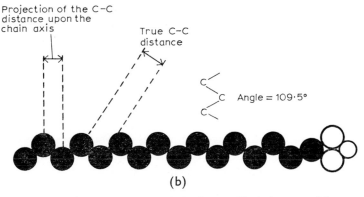

Projection of the C–C
distance upon the
chain axis

True C–C
distance

Angle = 109·5°

(b)

Fig. 36. **Zig-zag structure of the fatty acid chain, caused by
the tetrahedral valency angles**

(a) Realistic model; (b) schematic drawing of the carbon chain.

have given $0{\cdot}154 \times 10^{-7}$ cm for the diameter of carbon atoms in saturated aliphatic chains and we may indeed be very satisfied to have reached so good an approximation with such a primitive method. Remember, we have used nothing except a normal laboratory balance to weigh the acid, a volumetric flask and a pipette to measure the volumes of solution, and a ruler to measure the naked area on the surface. It is really an imposing accomplishment.

Before proceeding, let us define a new unit of length which is very convenient for atomic dimensions. We define 10^{-8} cm as an "Ångström unit", named after a famous Scandinavian spectroscopist, and write for its abbreviation, Å. The result of our measurement thus gives $1{\cdot}4$ Å for the C—C distance instead of the correct value of $1{\cdot}54$ Å. (Our assumption that the density of the acid film is that of the bulk acid was incorrect: the (CO.OH) heads, are broader than the —CH_2—CH_2— tails and prevent their close packing).

We could now try to calculate how many carbon atoms, each of $1{\cdot}54$ Å diameter, are present in one cm^3 of pure carbon, say diamond or graphite. Knowing the weight of one cm^3 of these materials we would arrive at the absolute weight of a carbon atom by simple division. But to do this we should have to know the inner geometry, the building plan of graphite or of diamond, i.e. the way the carbon atoms are arranged around each other. We do not know this at the present stage of our study, and there would be no point in proceeding from arbitrary structural assumptions.

None the less we have broken the magic circle and have at last ascertained the first truly atomic dimensions. The first scientist to accomplish this feat (in 1865) was the Austrian physicist *J. Loschmidt* (1821–1895), who based his calculation on the viscosity of gases. We shall try to give an idea of his reasoning on p. 169. Suffice it to mention here that he was the first to calculate with reasonable precision the absolute weight and diameter of individual atoms (1865). Knowing these absolute weights he had only to divide the conventional atomic weights of the elements by them to find out how many atoms were in each mole (gramme-atom) of any element. We know today that the correct value of this number is $6{\cdot}02 \times 10^{23}$; this is the number of molecules present in $22{\cdot}4$ litres of any gas at N.T.P. (S.T.P.). This factor which connects the world of atoms with the world of our immediate senses is named the "Loschmidt number",* designated by N_L. In Anglo-saxon countries it is named "Avogadro's number", designated by N.

The experiment we have described may be performed in any first-year course of chemistry, the main difficulty being in general the measurement of the area which is seldom strictly circular. In spite of this, it is possible to obtain the length of a fatty acid molecule with an error not exceeding 20 per cent and even from this value one arrives at conclusions similar to those reached above. It is difficult not to be moved by gaining so deep an insight into the structure of matter by such a simple experiment. It should be a part of the curriculum in every laboratory for undergraduate chemists.

---

* It should be noted that the Loschmidt number is sometimes defined as the number of molecules *per cm^3* of a gas at N.T.P.

25. CRYSTALS AND LATTICES

If chemistry had taught us nothing about the atomic structure of matter, the existence of crystals and the fundamental laws which govern their geometry might alone have convinced us that these beautiful gems of nature can only be built from elementary and equivalent building stones arranged in regular patterns. We are now about to become acquainted with the elementary principles of crystallography. Previous chemical studies have given us a qualitative hypothesis of elementary building units within crystals and we may now proceed towards their quantitative measurements. Precisely as a beam of light is diffracted when viewed through the dense fabric of an umbrella cloth, a beam of X-rays is diffracted when passing the dense, regular, three-dimensional structure of crystals and it is this diffraction which allows us to calculate the dimensions and structure of the crystal lattice. At once, the true dimensions of the atoms are obtainable.

The most conspicuous attribute of crystals is the fact that they are bonded by plane surfaces. As soon as man began to measure crystals it became evident that for each substance the angle between particular planes was constant, in spite of differences in magnitude of linear elongation in one direction or another. Fig. 37, for instance, shows us in the bottom row four different shapes which are all combinations of a cube and an octahedron. On the first, the planes belonging to the cube prevail, on the third those belonging to the octahedron. The second shows them just at the stage where they are equally developed, while in the fourth case the whole structure is extended in the vertical direction. But the angles between neighbouring planes remain strictly constant throughout all these variations! This is the sign that we are dealing intrinsically with the same sort of crystal all the time. The crystallization conditions determine which special form will be produced and often several forms occur one beside the other. But we shall never find a crystal of the same material with a different angle, say between the prismatic and pyramidal faces (Fig. 38). This law of constant angles between faces was discovered by *N. Steno* (1638–1686), in 1666.

On crystals of relatively high symmetry we can always find, with relative ease, three directions which may be taken as axes of symmetry. Less symmetrical crystals have axes which are somewhat more difficult to assign but none the less it can always be done (Fig. 39). These three directions are called the axes of the crystal. If we had destroyed the outward appearance of a crystal by, say, carving it with a knife, we should still be able to ascertain the position of the axes by sets of physical measurements or a study of the cleavage planes, because most of the optical, mechanical, etc., properties of a crystal are dependent on the internal direction involved, having in general extreme values along the axes. The ability of a crystal to be cleaved is itself a significant property. Variation of refractive index, light absorption, elasticity constants, etc., can also occur according to the direction of the path involved, through the crystal. Thus in one way or another we are always able to ascertain the axis directions within a crystal.

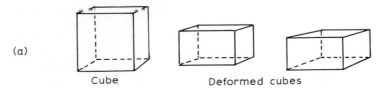

(a)

Cube   Deformed cubes

(*a*) The angles between the faces are the same in all three: right-angles. Right-angles are equally characteristic for the cube and for simple tetragonal and rhombic prisms; all three drawings could represent any of these three crystal forms. Their internal structures would differ.

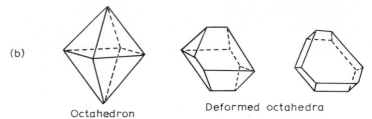

(b)

Octahedron   Deformed octahedra

(*b*) The angles between the faces are the same in all three. Angles between faces of pyramids not belonging to the regular crystal system would differ from the angle of the octahedron, therefore all other crystal systems are excluded. Hence any of these crystals must be classed as an octahedron.

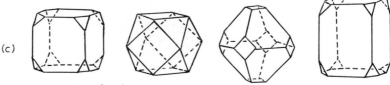

(c)

Combinations of cube and octahedron

(*c*) The angles between octahedral-octahedral and octahedral-cubic faces are characteristic for the regular system, combinations with pyramids of other systems would yield different angles. Therefore all these combinations belong to the regular system and can only be regarded as combinations of the cube and the octahedron. Distortion does not matter as long as the angles remain unchanged.

Fig. 37. **Cube and octahedron: deformed and combined**

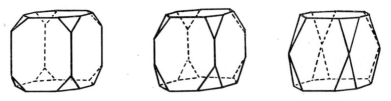

Fig. 38. **Combinations of quadratic prisms with different pyramids**

The crystal on the left is again a combination of a cube and an octahedron. The two others are combinations of tetragonal quadratic prisms with different tetragonal pyramids. The angles between prismatic and pyramidal faces are different in the three figures, therefore they cannot be distorted forms of essentially the same crystal. The angles are determined by the internal structure; different angles mean different atomic architecture within.

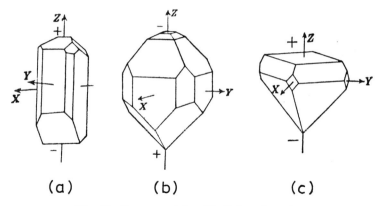

(a)   (b)   (c)

Fig. 39. **Some crystals with their axis systems**

(*a*) Sucrose (cane sugar): monoclinic; (*b*) zinc silicate: rhombic;
(*c*) pentaerythritol: tetragonal.

(From W. Voigt, *Lehrbuch der Krystallphysik* (Teubner, 1928),
Figs. 107, 108, 109. By courtesy of B. G. Teubner Verlagsges.)

The three main axes are the most convenient system of reference coordinates with which to describe the relative position of different crystal faces. In view of the fact that only the orientations of the crystal faces matter whilst their distance apart is of no importance, it is convenient to imagine the faces moved to and fro so as always to be parallel to their original directions until they all coincide at a single point upon one axis at a distance from the zero point of the axis system which we shall take as unity (Fig. 40). It must be emphasized that parallel displacement of faces is not important in crystallography; it occurs whenever the crystal grows more easily in one direction than another.

Now, having imagined all the faces to be brought to coincidence at one point, we may characterize the position of any face by measuring the points of its intersection with the other two axes. This defines its position in space. Parallel planes naturally coincide after having been brought together at one point; they give the same intercepts with the other two axes. Only those faces which are parallel to one or other of the remaining axes fail to intersect them, although mathematically we may say that they intersect at infinity.

R. J. Haüy (1743–1822), a French abbot, was the first to observe that whenever he thus moved together all the faces of a crystal the intercepts on the two free axes obeyed a relatively simple law (1784). He found that these intercepts were small integral multiples (indices) of a length which was characteristic for each of the axes. This is the law of rational intercepts, more often known as the law of rational indices, but what does it mean? It means that a crystal face may not have all of the conceivable orientations in space. From among the infinity of possible directions only those occur in which the two free axes of the crystal are cut at a characteristic distance or at a small integral multiple of it. At first sight one stops aghast at this result. Is there something that prevents a crystal face from having any position whatsoever in space? There *must* be something, evidently. On second thoughts

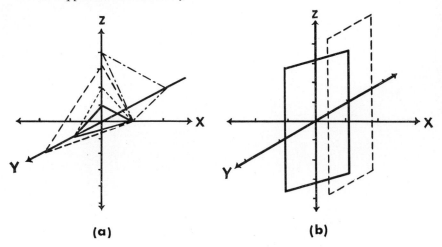

(a)   **(b)**

Fig. 40. **Intercepts of crystal planes with the axis**

All planes are moved parallel to themselves until they intersect
the $+x$ axis in one and the same point. It is then found that their
intercepts on the y and z axes are small multiples or sub-multiples
of two characteristic unit lengths, one on each of these axes (*see*
text on p. 105).

(*a*) Pyramidal faces

———— intersects y at $+1$ and z at $+1$ unit

– – – – intersects y at $+2$ and z at $+3$ units

……… intersects y at $+1$ and z at $+2$ units

—·—·— intersects y at -2 and z at $+4$ units

(*b*) Prismatic faces parallel to the z axis

———— intersects y at $+1$ unit

– – – is also parallel to y

this is not so astonishing at all. The mere fact that matter assumes a regular
outside form with constant angles between plane faces shows us already that
some kind of forces must be operative which work to prevent the crystal
assuming any random shape. This force evidently regulates the permitted
number of orientations too.

Haüy himself had already explained his observation by assuming that
crystals are built from elementary building units according to a given structural scheme (Fig. 41). This idea, however, only received its quantitative
formulation at the hands of *A. Bravais* (1811–1863) who clarified the connexion between the atomic structure of matter and the law of rational
indices (1850). The situation is perfectly analogous to Dalton's having deduced the atomic composition of matter from the laws of the constant and
multiple proportions of combining weights. With crystals as well as molecules we are inescapably encompassed by laws which allow only for ratios
of small integral numbers and in both cases this is explained by the assumptions that "building stones" of definite dimensions combine to form more
elaborate structures, and that only whole building stones exist.

It is not very easy to think in terms of three-dimensional space and to draw in perspective. Therefore it will simplify matters to begin with an analogous set of ideas relating to a single plane of atoms. Naturally, in such a model we are not going to deal with planes as the boundaries of a

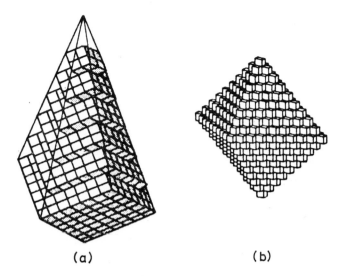

(a)   (b)

Fig. 41. **Haüy's idea of crystals being built from elementary, repeated, units**

(a) Scalenohedron built of rhombohedra; (b) octahedron built of cubes.

(From *Encyclopaedia Britannica*, 14th ed., v. 22, p. 810, Figs. 1, 2).

crystal but with straight lines as the boundaries of the infinitely-thin crystal lamellae which we have substituted instead. Let us assume in accordance with Bravais' hypothesis that the crystals are built from periodically repeated elementary pieces in the three directions of space, that is, in the two directions of the plane of our lamella. Such a generalized configuration is represented by Fig. 42. Suppose we choose the two directions where the density of points is maximal as the two axes, placing one of them horizontally. The

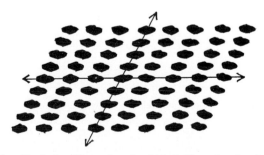

Fig. 42. **A two-dimensional, plane lattice of molecules**

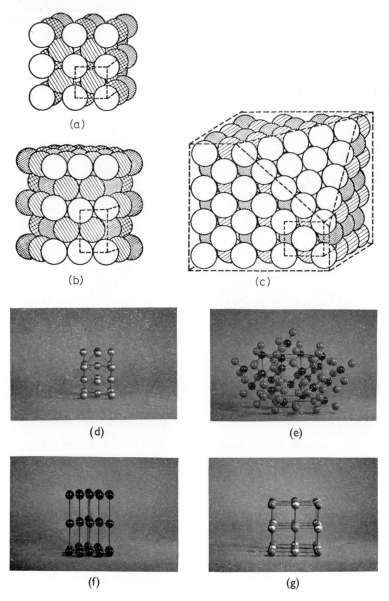

Fig. 43. **Three-dimensional space lattices**

(*upper*) (*a*) Close-packed spheres: cubic; (*b*) close-packed spheres: hexagonal; (*c*) cube broken down along an octahedral face.

The two structures are actually two ways of superimposing identical planes, one above the other. In (*a*) every third plane coincides, in (*b*) every second.

(From *Encyclopedia of Science and Technology*, v. 3, p. 598, Fig. 10 (McGraw-Hill, 1960). By courtesy of the McGraw-Hill Book Co.)

(*lower*) Schematic space lattices with atoms separated.

(*d*) copper; (*e*) quartz; (*f*) graphite; (*g*) common salt.

(From *Encyclopedia of Science and Technology*, v. 3, p. 592, Figs. 2, 3, 4, 10 (McGraw-Hill, 1960). By courtesy of Bell Telephone Laboratories and McGraw-Hill Book Co.)

point density along the two axes will be different and in general the angle between them may have any value. There is no reason why the two axes should not sometimes be perpendicular to each other and the point density along them might even be the same in individual cases. But these special cases correspond to higher symmetry of the two-dimensional crystal. It is evident that this system is fully determined by the angle between the axes and by the two point densities along them; all its other properties can be deduced from this. A three-dimensional space lattice would be fully specified by three angles and three point densities (Fig. 43).

In order that our lamella should have real border lines it is necessary that there should be points of its lattice along these borders. As already chosen the points lie most densely along the lines parallel to the two axes so that any

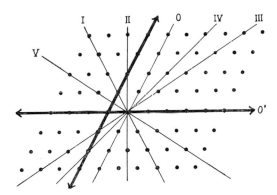

Fig. 44. **Plane lattice with some directions which go across lattice points**

such lines will always be suitable for forming the border. But let us look for other straight lines which contain many lattice points (Fig. 44). Such a line is, for instance, the one which goes through the diagonals of the elementary parallelograms, e.g. the one that goes through the first point on the right-hand side along the horizontal axis and proceeds towards the first upper point of the slanting second axis, and so on (line I). We may draw another straight line (II) across the same point of the horizontal axis but in such a direction that it goes through the second upper point of the slanting one. We can see that the density of points is smaller on line I than on the axes and decreases further for line II. But there are still sufficient points upon I and II to allow them to be regarded as borderlines if they should come to be edges of the lamella. The lines III and IV differ from lines I and II only in so far as they intersect the slanting axis on its lower half and thus form the longer diagonals of the respective parallelograms. The intercepts of these four lines with the slanting axis are evidently in the ratios $1:2:(-1):(-2)$. We are free to draw further lines with larger intercepts, but it is clear that the density of lattice points will go on decreasing along them and they are steadily loosing their significance in the event of being chosen as borderlines of the lamella. There would be hardly any matter along such a border line (Fig. 45), and if we draw circles around the points to represent matter, they either lag behind the line or stand out across it. It could hardly be called a material boundary.

Fig. 44 also contains more complicated directions. One of them (V) again cuts the horizontal axis at the first point to the right, but proceeds between the first and second points of the slanting axis and then passes through the first point to the left of the first upper horizontal row. There are obviously points lying along this line too, though fewer than on the preceding lines. And what about its intercept ratio? It goes through the horizontal axis at unit distance but cuts the other axis at 1·5 units (measured in terms of the

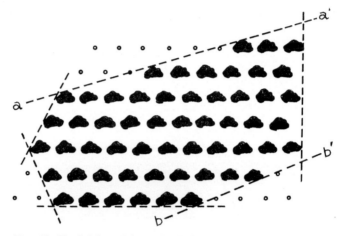

Fig. 45. **Probable and improbable bordering lines of a lamella of a plane lattice**

There would be hardly any material along border-lines like *a–a'* or *b–b'* and the actual border-line would be microscopically rugged.

linear point density of the latter axis). This gives a ratio of 2:3, a ratio of small integral numbers. The periodicity of the lattice provides that whenever we have drawn a line across any two points in it, we shall meet another point if we proceed the same distance further on. But the intercept ratio of such a line will become the ratio of greater and greater integral numbers as the distance of the two neighbouring points increases. Thus the density of matter along a line will decrease as the intercept ratio grows more and more complicated and the line will soon become unsuitable as a boundary of a real material lamella. Very specific conditions must prevail if a more complicated direction such as this really manages to become a boundary line; it could happen, for instance, if the elementary building stones around the lattice points are not circles but have some particular, complicated geometry of their own. This could of course be the case with certain molecules. On the whole, however, it must have become evident by now that the concept of a point lattice in itself practically excludes directions of complicated intercept ratios from becoming boundary lines of a lamella.

Nothing is easier than to expand this set of ideas to the three-dimensional lattice (Figs. 40 and 46). Every crystal face will now be characterized by three intercepts, that is by two intercept ratios and it is clear that the plane of the crystal face will be covered most densely with points, that is with matter, if it is parallel to one of the axes. The density of matter will decrease

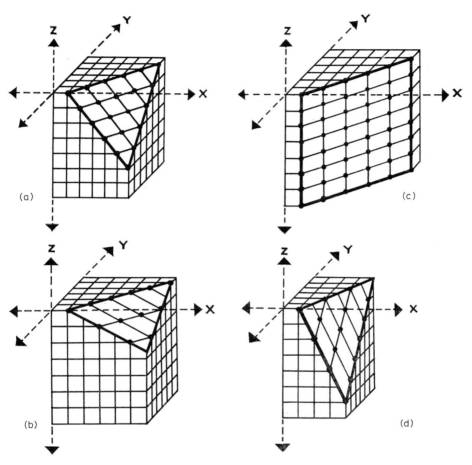

Fig. 46. Space lattice bordered by planes of different directions and material density

(a) Intercepts X:Y:Z = 1:1:1; (b) Intercepts X:Y:Z = 1:1:1/2 = 2:2:1; (c) Intercepts X:Y:Z = 1:1:∞; (d) Intercepts X:Y:Z = 1:2:2.

as we go over to more and more complicated intercept ratios. A crystal face is a plane which must by necessity contain matter, and densely packed matter at that. Hence it is clear that the intercept ratio of real, existing, crystal faces must be ratios of relatively small integral numbers. This is Haüy's law.

We have arrived at a stage of our logical conclusions analogous to that which Dalton arrived at in his atomic hypothesis. It has been made clear that the assumption of discrete material building stones in a periodic space lattice is sufficient to account for the law of rational indices. In the same way the assumption that atoms of given specific mass combine to form compounds was sufficient to account for the laws of stoichiometry. But from Haüy's law we know as little about the actual size of these elementary building stones as we knew about the actual weights of atoms in spite of the enormous number of structural formulae which agreed with the properties of experimentally prepared compounds.

26. INTERFERENCE OF LIGHT

Many of us will have seen the rainbow pattern formed by the cloth of a black umbrella when a street lamp is viewed through it on a rainy night. Physicists say that the light is "diffracted" on the two-dimensional optical

Fig. 47. **Interference of two wave systems generated by throwing two stones into a pond one beside the other**

Observe curves along which oppositely oscillating waves extinguish each other.

(From W. H. Westphal, *Physik*, 2nd ed., p. 137, Fig. 126 (Springer, 1903). By courtesy of Springer-Verlag.) (Photograph: E. Grimsehl.)

lattice of the cloth, the red light being affected more than the blue. In order to understand this phenomenon we must suppose that light is undulatory, that is, that it has the nature of waves. It is not important at this stage what the waves consist of. Interference and diffraction are bound to occur with waves of any type. Think of the surface of a calm pond into which are thrown two stones close to each other (Fig. 47). There will be two circular sets of waves spreading outwards from the places where the stones hit the surface. At a certain distance from the centre these sets of circular waves begin to overlap and we may observe in what manner they interfere with each other. There will be some points in the surface which would be moved upwards by one of the waves, whereas the other wave would move them downwards by the same distance. As a consequence these points will remain at rest. On the other hand there will be other points which are moved upwards or downwards in the same direction by both of the waves and these points will double their elevation or depression. This is the origin of the interference phenomenon of any wave motion.

The distance between two adjacent crests of a wave train is called the wavelength. It is easy to see that the wave-length is going to determine at what distance from each other the neighbouring immobile points of the interference pattern will occur. In the case of light the colour is determined by the wavelength of the radiation and experiments show that the light of any spectrum line is practically monochromatic, that is, it contains waves of a single wavelength, such as the characteristic yellow light from sodium vapour. (As a matter of fact it is composed of two neighbouring yellow wavelengths which are very close together.) Thus, if the lamp in the street referred to at the beginning of this section happens to be a sodium-vapour lamp, the diffraction image across the umbrella will not be a rainbow but a lattice of yellow points on a black background. The really black portion of this pattern comprises the places where light coming through different holes of the cloth interferes and causes extinction, analogous to the immobile points on the pond surface. Its position depends on the wavelength. (The rainbow pattern is simply a superposition of a series of single-coloured patterns with different wavelengths.)

As a matter of fact the distance between points of extinction in an interference pattern is the most basic method of determining the wavelength. Let F in Fig. 48 represent a light source and R an opaque sheet with two fine holes, A and B, in it at a distance d from each other. Light can pass only through these two holes which are a simplified one-dimensional lattice as compared with the two-dimensional lattice of the umbrella-cloth holes. Place a screen S parallel to the sheet R at the distance l from it. For convenience's sake let us use a monochromatic source (e.g. a sodium lamp) or else insert a monochromatic filter in front of the source so that the light with which we work shall virtually contain only one wavelength. In this case we shall be able to observe that there is maximum light intensity on the screen at C, opposite the point in the middle of the distance d which connects the two holes. At C, both trains of waves from the holes have traversed the same distance and are therefore always in the same sense of motion, "in the same phase" as we call it. But on each side of this light-spot the light intensity decreases and at a certain distance a from C there is complete darkness. Proceeding further along this line parallel to the distance d the light intensity increases again on each side to a second maximum, followed by a second total extinction, a third maximum and so on.

What has happened? Every point of our screen obtains light from each hole; why are there points of darkness between the illuminated areas? The answer is because of interference, as in the case of the surface waves on the pond. This experiment proves to us that along the path of the light there is an undulating field. We have darkness at those points where a wavecrest of the light ray coming from one hole coincides with the bottom of a wave from the other hole. This means that in the first dark point from the centre

(*D*) the difference between the two waves must be just one-half of a wavelength. Denoting the wavelength by λ the difference in distance of this first dark spot from the holes *A* and *B* respectively is just $\lambda/2$. Let us denote as *E*, the point on the sheet *R* opposite the point *D*. Then *ED* is *l*, *AB* is

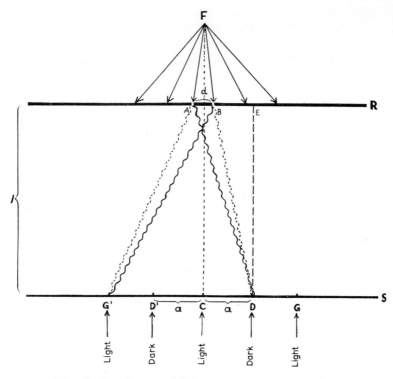

Fig. 48. **Interference of light waves across two openings**

$$AE = (a + d)/2; \quad BE = (a - d)/2$$

Light waves start from *A* and *B* towards the screen *R*. Wherever a crest of the wave from *A* meets a crest from *B* light is maximal. Wherever a crest meets a valley they extinguish each other and a dark region occurs.

d, and *CD* is *a*. The two lengths which interest us, *AD* and *BD* are hypotenuses of the two right-angled triangles *AED* and *BED*. It is clear from the drawing that $AE = CD + \frac{1}{2}AB = a + \frac{1}{2}d$, whereas $BE = CD - \frac{1}{2}AB = a - \frac{1}{2}d$. Applying the theorem of Pythagoras we have

$$AD = \sqrt{\{l^2 + (a + \tfrac{1}{2}d)^2\}} \qquad . \qquad . \qquad . \quad (26.1)$$

$$BD = \sqrt{\{l^2 + (a - \tfrac{1}{2}d)^2\}} \qquad . \qquad . \qquad . \quad (26.2)$$

It is only necessary to insert the measured distances into the formulae, subtract *BD* from *AD* and the difference must be $\lambda/2$. Thus we have determined the wavelength of the light we used. The result can be verified by making similar calculations for the distance between the central and the first lateral maximum, which must yield λ because the two light rays must have

been in phase in order to strengthen each other. We may also use the second
dark point where the crest of one wave must have been counteracted by the
second valley of the other, that is, the difference in path must have been $3\lambda/2$.
All these observations yield the same value for the wavelength, thus confirming our assumption about the nature of the phenomenon and the precision
of our measurements.

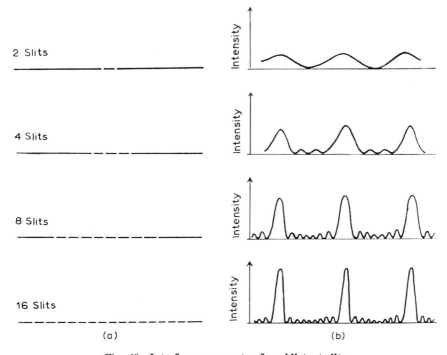

Fig. 49. **Interference on sets of equidistant slits**

(*a*) Arrangement of slits; (*b*) energy distribution in the interference
patterns.

Besides the interference of neighbouring slits, slits further apart
also interfere. Thus slit distances d, $2d$, $3d$, etc., are also active.
Whenever any of these slit distances extinguish the light in a given
direction, the direction remains dark, light is concentrated into
the regions near the original maxima, because this is the direction
in which all possible slit combinations are reinforced. (Slit sizes
are very much exaggerated.)

Evidently the longer the wavelength, the farther away we must go from the
centre of the screen to reach the point where the path difference is half a
wavelength. Experiments show that the spacing of the interference pattern
is broader, about twice as broad, in red light than in blue light. This means
that the wavelength of red light is about twice the wavelength of blue. If
we use red and blue light simultaneously we get first an extinction of blue and
then of red, that is, a two-coloured pattern, and now we understand how a
mixture of different colours of light gives a whole rainbow. The light emitted
from incandescent solids, such as the sun or a tungsten lamp, is, as a matter

of fact, a mixture, a continuous mixture of a great many different wavelengths, as interference experiments have proved.

The effect we are dealing with is by no means so small as to be on the limit of observation. Suppose we measure the wavelength of red light. If the two holes are 0·1 mm apart (this is the "lattice constant" of our system) and the screen is 1 m away from the holes we observe that the first dark point on the axis of our interference image is 3·5 mm from the middle of the image. Using

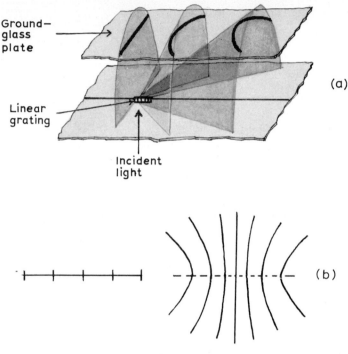

Ground–glass plate

Linear grating

Incident light

(a)

(b)

Fig. 50. **Light emerging, after interference, from a linear grating**

(*a*) Emerging cones trace hyperbolae on the ground-glass plate
(*b*) (*left*) Linear grating; (*right*) interference pattern on the ground-glass plate.

the previous equations we arrive at a wavelength of 7×10^{-5} cm (7,000 Å): we have measured a length of less than a micron with a not very complicated apparatus. Blue light would have given the first dark spot at 2·0 mm corresponding to a wavelength of 4×10^{-5} cm (4,000 Å).

If we had used three holes along a line instead of two, each pair 0·1 mm apart, the interference pattern from the first two points would have been only 0·1 mm displaced from the pattern formed from the second two holes, that is, the two patterns would nearly coincide. The coincidence, i.e. the nearness of overlap of the patterns is greater the finer the lattice, the "grating". But there is additionally a pattern from the first and last holes, with a spacing double that of the pattern from the neighbouring holes, that is, with half the resolution. It can easily be shown, e.g. by graphically adding the amplitudes from all three pairs of holes, that whereas the first two pairs only reinforce each other, the last pair increases the light intensity at its maximum points but decreases the intensity in the region between two maxima, thus enhancing their sharpness. If very many holes or slits are used along a straight line, that is if we really use what physicists call a grating, the whole intensity is concentrated into sharp lines at the maxima (Fig. 49).

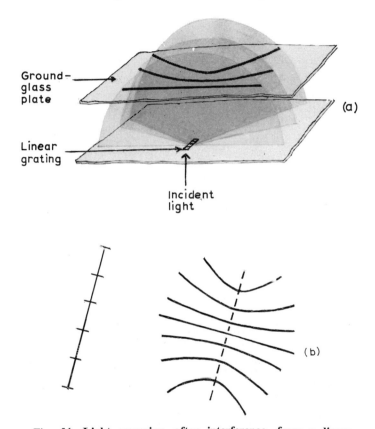

Ground-glass plate

Linear grating

(a)

Incident light

(b)

Fig. 51. **Light emerging, after interference, from a linear grating perpendicular to that of Fig. 50**

(a) Emerging cones trace hyperbolae on the ground-glass plate.
(b) (*left*) Linear grating; (*right*) interference pattern on the ground-glass plate.

If we pay attention to the interference pattern over the whole screen caused by a linear grating we arrive at Fig. 50 where light emerges in the allowed directions all around the grating as an axis. A grating perpendicular to the one in Fig. 50 would give the interference pattern of Fig. 51. It is evident that if the two gratings act at the same time, as in a two-dimensional grating, there can be no light at any point where it has already been extinguished by either of them. There remains only the possibility of having light at the points of intersection of the two primary patterns (Figs. 50 and 51).

Now we are ready to extend our speculations and experiments to two-dimensional lattices, like that of the umbrella cloth with which we began. Regular, two-dimensional arrangements of slits give rise to a pattern as shown

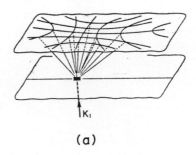

(a)

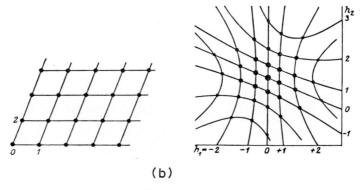

(b)

Fig. 52. **Light emerging, after interference, from a two-dimensional plane grating**

The interference pattern is a combination of those seen on Figs. 50 and 51. Wherever light is extinguished by either combination of the interfering points, as e.g. from the two linear gratings perpendicular to each other, it is permanently extinguished. Light is concentrated into the directions where the two sets of light-cones intersect and produce a two-dimensional interference pattern on the ground glass cutting the cones.

(*b*) (*left*) Plane grating; (*right*) interference pattern on the ground-glass plate.

(*a*) (From P. P. Ewald, *Fifty Years of X-ray Diffraction*, p. 90, Fig. 6-2 (Utrecht, Oosthoek, 1962). By courtesy of A. Oosthoek Publishing Co.)

(*b*) (From P. P. Ewald, *Kristalle und Röntgenstrahlen*, p. 44, Figs. 33, 34 (Berlin, Springer, 1923). By courtesy of Springer-Verlag.)

in Fig. 52. It is built up in the same manner as was the pattern from a one-dimensional grating. One has to sum the intensities from all points of the grating at a given point of the screen, taking account of the relative phases with which the waves arrive. Where the intensities add up there will be light spots, where they annihilate each other there will be darkness.

We have dealt with one-dimensional lattices, and two-dimensional lattices, . . . there is only one step left and that is a three-dimensional space-lattice. It would not be easy to build such a lattice in space from elements separated by less than 0·1 mm or so, but it is not difficult to calculate how such an assembly would behave, using the same principle of summing the

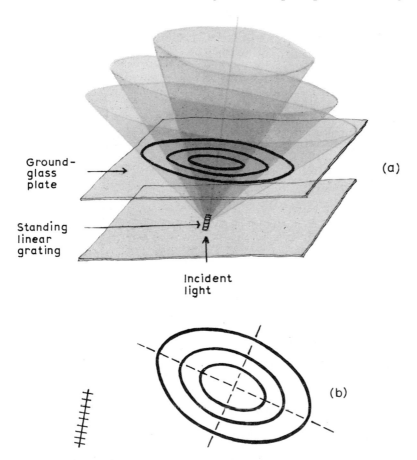

Fig. 53. **Light emerging, after interference, from a linear grating perpendicular to both linear gratings of Figs. 50 and 51**

Emerging cones trace ellipses on the ground-glass plate.
(*b*) (*left*) Obliquely standing linear grating; (*right*) interference pattern on the ground-glass plate from the standing linear lattice.

intensities of the elementary waves according to their phases. We only have to introduce a third linear grating, (Fig. 53) outside the plane of the first two of Figs. 50 and 51. It can be shown that in the general case no light would come through such an assembly at all because the points where light goes through a component of a two-dimensional lattice of the system will not coincide with the points where it can go through the lattice grating of the third direction. Only some fortuitous wavelengths manage to get across such a three-dimensional structure at special angles of incidence and of diffraction, where the diffraction patterns of all three directions happen to coincide. Therefore if we illuminate the space-lattice with white light from a given direction, the lattice will allow the passage of *some* wavelengths of this mixture, that is of light of specific colours, in *some* definite directions. On the other hand, if we illuminate with monochromatic light from all directions, there will be some particular angles in which the diffraction from all three gratings coincides. Thus there will be no transparency only in some specified directions and not in others. These directions will be determined by the spacings of the lattices, by the angles between their three axes and by the wavelength of the light used.

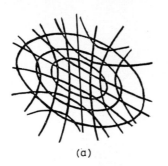
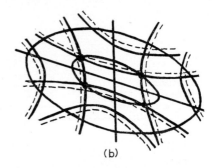

(a)   (b)

Fig. 54. **The three partial interference patterns resulting from three linear gratings which determine a space lattice.**

(Being actually a combination of the patterns in Figs. 50 (*b*), 51 (*b*), 53 (*b*).)

(*a*) In the general case the three sets of interference curves do not meet in such a way that a curve of a third set intersects curves of the other two exactly at the point of their intersection. Wherever light is extinguished from any combination of the gratings it is extinguished for good. Therefore there is no point on the above interference pattern that contains light.

(*b*) In rather special cases, that is for directions at special angles of the incoming beam and special values of wavelength, curves from all three sets intersect at identical points. These, and only these, points of the pattern contain light. Its wavelength is the special wavelength at which this meeting of the curves occurs. The curves marked by broken lines represent slightly different wavelengths: they have no common triple points and are extinguished.

27. CRYSTALS AND X-RAYS

But does not nature present us in its crystals with exactly such space lattices as described in Section 26? In Section 25, we explained the law of rational intercepts by assuming that crystals are regular space-lattices of molecules. If this is true, we should be able to observe diffraction phenomena of the type we have just described, by using crystals as three-dimensional gratings. We may presume from other evidence that atoms and simple molecules will have magnitudes of about 10^{-8} to 10^{-7} cm. What kind of light should we use for such an experiment? Fig. 48 shows us that for perpendicular incidence the wavelength may not be longer than the distance between adjacent lattice points, because the path difference between two rays from two adjacent points can never be greater than this value. The wavelength is equal to it for light diffracted perpendicular to the incident beam, that is along the row of lattice points. Hence ordinary light with its wavelength in the region of 10^{-5} cm is useless. Ultra-violet light has a wavelength not much shorter than that of visible light and is not suitable either. What is needed is light of much shorter wavelength.

M. *von Laue* (1879–1960), the famous German physicist had the idea in 1912, that maybe X-rays were a sort of "light" and that their wavelengths might happen to be of the same order of magnitude as the spacings in a crystal

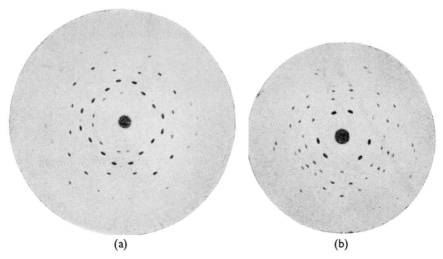

(a)   (b)

Fig. 55. **The first X-ray diffraction pictures of Laue, Friedrich and Knipping, 1912**

Zinc sulphide, (*a*) along the fourfold axis, (*b*) along the threefold axis.

(From P. P. Ewald, *Kristalle und Röntgenstrahlen*, p. 68, Fig. 44, p. 69, Fig. 45 (Berlin, Springer, 1923). By courtesy of Springer-Verlag.)

lattice. At that time it was not clear at all what X-rays really were because no interference phenomena had previously been obtained with them. They did not seem to be waves, although of course, they could have been waves of wavelengths much shorter than would be affected by normal gratings. (If so their interference pattern on normal gratings would lie so as to coincide practically with the central beam going across the grating.) They could have been rays of charged particles like alpha- or beta-rays, but they were not diverted by electric or magnetic fields. Their nature was most uncertain. It is to von Laue's great credit that he combined two different concepts: that of crystals possessing space-lattices and that of the possibility of X-rays being "light" of very short wavelength.

He chose excellent co-workers to try the experiment and in a very short time they had the confirmation in their hands. Fig. 55 shows two of the historically famous first X-ray diffraction photographs. They were obtained by diffraction of X-rays of mixed wavelengths ("white X-rays") by a zinc sulphide crystal, oriented the first time with its trigonal axis and the second time with its tetragonal axis, parallel to the incident beam. The trigonal and tetragonal symmetry, respectively, shows up in the diffraction patterns. The patterns are spaced exactly in the manner calculated from the theory of space lattice diffraction.

Thus two questions had been solved with one experiment. It had been shown that crystals really contained space lattices and that X-rays were waves of light with a wavelength of the same order of magnitude as the elementary spacing in the crystals. We do at least know that spacing and wavelength have similar magnitudes without knowing their absolute values. We have two unknowns and can only calculate one of them if the other is already

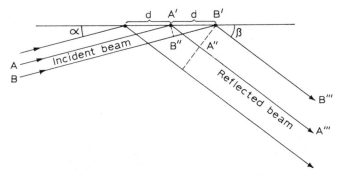

Fig. 56. **Diffraction of slanting X-rays on man-made gratings**
$$n\lambda = d\cos\alpha - d\cos\beta = B'B'' - A'A''$$

(From A. Eucken, *Lehrbuch der chemischen Physik*, p. 961, Fig. 234 (Leipzig, Akademische Verlagsges., 1930). By courtesy of Akademische Verlagsges.)

known to us. In 1912 the real magnitude of atoms was known from other methods and thus Laue's experiments were used to calculate the exact wavelength of X-rays. We should like, however, to be able to arrive at reliable atomic dimensions from X-ray diffraction data and therefore it is very important to know that it is possible to achieve this.

It has become possible by an ingenious extension of the original diffraction method to determine the wavelength of a given radiation using artificial gratings of known spacing. We have mentioned earlier that whenever the wavelength of the incident radiation is very small compared with the lattice spacing the whole interference pattern contracts around the primary beam. This is because a very slight inclination of the primary beam suffices to give differences of the order of a wavelength if the wavelength is short. *A. H. Compton*, the famous American physicist at the University of Chicago proposed in 1928 to circumvent this difficulty by using X-rays of very oblique incidence at the grating (the grazing-incidence technique). *Baecklin* and *Wadlund*, succeeded in performing the relevant experiments. At a very small angle of incidence the X-ray "sees", as it were, a much denser grating than the actual one (Fig. 56) and the emerging rays give an interference pattern which allows the calculation of the X-ray wavelength. Thus we can obtain the wavelength of X-rays from the dimensions of an artificially prepared grating of known spacing between its adjacent lines and from the measurement of the small but definite angles of the incident and emergent X-rays. We have not made any assumption concerning atomic dimensions in the determination of this wavelength. Everything we have used has been of macroscopic dimensions, and so we are now free to use this wavelength to determine the spacings within crystal lattices, once we have measured the diffraction patterns caused by them.

X-rays emerging from an X-ray generating tube are composed of many wavelengths. Now by the appropriate choice of the metal upon which electrons are caused to impinge, producing X-rays, and by controlling the voltage

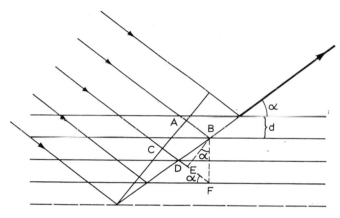

Fig. 57. **Bragg's method for calculating distance of lattice planes from X-ray diffraction angles**

$$n\lambda = 2d \sin \alpha$$

(From A. Eucken, Lehrbuch der chemischen Physik, p. 956, Fig. 229 (Leipzig, Akademische Verlagsges., 1930). By courtesy of Akademische Verlagsges.)

which accelerates the electrons in the X-ray tube, it is possible to influence the wavelength composition of the X-ray beam in any desired manner. If different absorbing metals are employed as "colour filters", it is possible to obtain monochromatic X-rays, that is, all with the same wavelength.

von Laue's method is very useful if one wants to obtain information concerning the symmetry of a crystal lattice in its different directions. But if we are more interested in the spacing between identical lattice planes, parallel to each other within a crystal it is better to use the method of *W. H. Bragg* (1862–1942). Fig. 57 shows a section of a crystal lattice with a set of such parallel planes of identical spaced atoms. The beam of incident X-rays comes from the left and will be reflected from this set of planes if its angle of incidence and its wavelength have suitable values, and "fit into the structure". The figure shows clearly what is meant by fitting into the structure. Waves reflected from planes one below the other must emerge from the lattice in such a way that the emerging wavelets reinforce each other, that is, the difference in their paths must be a whole wavelength or an integral multiple thereof. Thus, for example the path difference between the second and third ray is $CD + DB - AB$ and this must be a multiple of the wavelength, $n\lambda$, where n is an integer. We see that $AB = CE$ and thus $AB - CD = CE - CD = DE$, but DB on the other hand is equal to DF so that

$$n\lambda = DF - DE = EF \qquad . \qquad . \qquad . \quad (27.1)$$

and if α stands for the angle of incidence and d for the distance between adjacent planes we have

$$n\lambda = 2d \sin \alpha \qquad . \qquad . \qquad . \qquad . \quad (27.2)$$

This means that if we go on to vary the angle of incidence we shall in general not satisfy this equation and therefore not obtain reflected X-rays. Only at specific angles given by solving this equation for different integral n values, that is for differences of one, two, etc., wavelengths between successive rays, will reflection occur. Experiments again confirm the theory. If we measure the first, second, etc., angle at which reflection occurs, take the sines of the angles and insert them in the equation we are able to calculate d from the first angle and using this same d value we in fact obtain the integral numbers 2, 3, etc., from the successive measured angles of reflection. The equation is thus verified by experiment and the spacing is determined.

Take rock salt as an example. If we use X-rays of wavelength 1·542 Å and determine the angle between the perpendicular to the cube face and the X-ray beam at which the first reflection occurs, we find 15°53′. Inserting these values into the Bragg equation (27·2) we calculate that the distance between two adjacent crystal planes parallel to the cube face is 2·818 Å. We assume that the external cubic symmetry of the salt crystal is caused by a cubic structure in its interior, as represented in Fig. 43. Then this distance of 2·818 Å between the cube planes is the distance between neighbouring sodium and chlorine atoms (more exactly ions); it is the length of the edge of an elementary cube within the crystal. Thus a neighbouring pair of sodium and chlorine ions occupies two adjacent elementary cubes of 2·818 Å each, that is, a volume of 44·8 Å³ per NaCl unit. 1 cm³ contains this volume 2·23 × 10²² times, therefore 1 cm³ contains this number of NaCl "molecules". From the density of NaCl we know that 1 cm³ weighs 2·164 g; hence one molecule weighs 0·969 × 10⁻²² g. The atomic weight scale tells us that a NaCl molecule weighs 58·0 times as much as a hydrogen atom (Na, 23·0; Cl, 35·5; H, 1·008) and so the absolute mass of the hydrogen atom must be 1·67 × 10⁻²⁴ g. Finally 1·008 g of hydrogen, that is one gramme-atom of hydrogen must contain 6·02 × 10²³ hydrogen atoms. There must be the same number of atoms in a gramme-atom or the same number of molecules in a gramme-molecule, irrespective of the sort of atoms or molecules we are dealing with; the number 6·02 × 10²³ molecules per mole must therefore be a universal constant. It is Loschmidt's number which we encountered during our discussion of the surface-film experiments with fatty acids. Now we have seen one of the methods which gives us the exact value of this fundamental number.

In the preceding paragraphs we have assumed, rather arbitrarily, that the space lattice of NaCl *is* cubic as seen in Fig. 43 and 58. It need not necessarily be so, and the space lattices of Fig. 58 show us other possible structures of high symmetry where two kinds of atoms may be arranged in the ratio 1:1, as in NaCl. The three axes of these structures are also perpendicular to each other and the unit spacing on all three axes is identical, satisfying the definition given by crystallographers to the cubic class of crystals. How do we know that NaCl has the simple cubic structure?

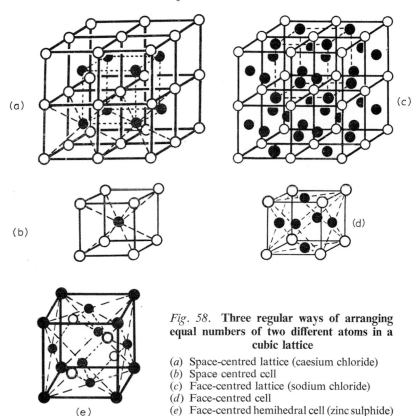

(a)

(b)

(c)

(d)

(e)

Fig. 58. **Three regular ways of arranging equal numbers of two different atoms in a cubic lattice**

(a) Space-centred lattice (caesium chloride)
(b) Space centred cell
(c) Face-centred lattice (sodium chloride)
(d) Face-centred cell
(e) Face-centred hemihedral cell (zinc sulphide)

This is a problem which recurs in the course of every X-ray crystal analysis and nearly always it is much more complicated than in our simple case. It is only possible to sketch the procedure in rough outline. First of all, it is necessary to consider simultaneously all possible sets of parallel planes within a crystal. X-rays permeate the crystal freely and thus all internal sets of planes act simultaneously as possible reflecting planes for the X-rays. On rotating the crystal in the X-ray beam all internal planes which are parallel to a common straight line (all planes of a "zone") will come to reflection in a direction perpendicular to this line ("zone axis") one after the other. Therefore we obtain in one experiment an intermingled set of reflection angles and they must be first untangled by trial and error, sorting out each series by trying to fit the sin α values into the Bragg equations. The angles which can be fitted using the same *d*-value but different integral numbers *n*, belong to the same set of planes and differ only in the number of wavelength differences, in the "order" of the reflection. After some trials, the data are separated and we have obtained a few separate *d*-values which belong to specific sets of internal planes in the lattice. The experiment must be repeated around other zone axes until a complete series of inner plane distances is at our disposal, together with their relative positions within the crystal. Atoms must evidently lie on corners of elementary planes in simple lattices and thus it is possible to reconstruct the position of the atoms as points of intersection along different sets of lattice planes. If we proceed according to this plan for NaCl we confirm our previous *assumption* regarding its internal structure. For another diatomic molecule, zinc sulphide for instance, we should not have confirmed this structure, the site of its atoms would emerge according to Fig. 58 (*e*).

It is by no means necessary that the smallest repeatable unit in a crystal structure, the "unit cell", should be an atom or a molecule. For only slightly more complicated molecules this is no longer the case and the smallest "repeat pattern" of the lattice often contains a number of molecules. Long chain-like molecules such as cellulose on the other hand, may grow through a column of superimposed unit cells (Fig. 59). The determination of structure from X-ray reflection data becomes thus more and more involved. It is not enough to determine the directions of reflection, it is also important to measure their relative intensities because these vary, among other reasons, according to the weight of the diffracting atoms (more correctly according to the number of electrons within them), and according to the interaction of planes parallel to each other but packed to different densities. There is no exact solution of the problem in its most generalized form. The solution must needs make use of all the chemical facts we have about the possible structures of a molecule, and is sometimes nearer to an intuitive art than to science. But if a solution is found it can, and must always be tested. We can always tell what diffraction pattern a particular lattice will give, though we are often not able to tell what lattice has produced a certain diffraction pattern. All directions and intensities of a tentative solution must be compared with those found experimentally and they must all agree in order to prove the structure.

Throughout our search for a lattice we always have to choose between a finite number of types. This is so because it is only possible to arrange points regularly in space according to a finite number of recurrent patterns. As always, it is easier to see this on a plane than in space. We can fill a plane with points according to a pattern of triangles, parallelograms (including, of course, rectangular ones and the square itself) and with regular hexagons. But that is all. You may try forever to fill a plane with regular pentagons or octagons: it is impossible (Fig. 60). The same is true of space. We have crystals which are trigonal, tetragonal or hexagonal but we never see any having the form, say, of a pentagonal prism or a heptagonal bipyramid. We are able to carve these from wood, they do exist in geometry, but they never exist as crystals! The reason is that they cannot be built from an internal space lattice of regularly recurrent pattern, just as we could not fill a

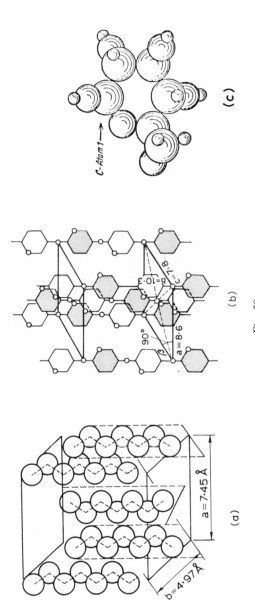

(a)

(b)

(c)

Fig. 59

(a) Schematic view of the crystal structure of the normal paraffin, n-heptane, C_7H_{16}. The unit cell is as high as the length of a molecule and contains two molecules. From the four molecules along the four vertical edges only one-fourth belongs to the central unit cell.

(b) Schematic view of the crystal structure of cellulose.
   Chains of chemically bound glucose molecules grow vertically across any number of identical unit cells.

(c) The atomic structure of a glucose unit. Such units unite linearly in cellulose.

(From K. H. Meyer and H. Mark, *Der Aufbau hochpolymerer Naturstoffe*, p. 43, Fig. 17, p. 60, Fig. 25, p. 109, Fig. 37 (Leipzig, Akademische Verlagsges., 1930). By courtesy of Akademische Verlagsges.)

plane with pentagons. Thus the non-existence of such symmetry elements in crystals is a proof of their regular internal structure, of their being built from atoms and molecules.

Bravais was the first to consider these questions at a time when it seemed utterly beyond hope to gain insight into the real structure of crystals. Following his lead *E. S. Fjedorow* (1853–1919) and *A. M. Schönfliess* (1853–1929) examined all logically possible sets of point assemblies in space and arrived at the conclusion that there are 230, and no more, types of recurring arrangements ("space groups") of elementary units in space. It is necessary to

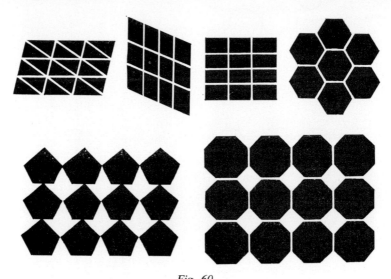

Fig. 60

(*upper*) Polygons that can be fitted together to fill a plane.
(*lower*) Some polygons that cannot be made to fill a plane.

allow that the unit be sometimes asymmetrical because complicated molecules are asymmetrical in general, and they may be units of a crystal. It is an extremely fascinating facet of the history of science that this whole logical theory of possible space groups was completely available at the time when the experiments of Laue opened up the way to use and test it on real properties of matter. The test was successful to the last detail. This is one of the proofs that a seemingly meaningless play with logical concepts may one day become a valuable asset in the practical realm of science or even technology.

Fig. 61 shows us the structure of some other familiar crystals. It is worth mentioning that within the crystal of NaCl, and in some other structures too, we are unable to distinguish individual molecules. Each sodium atom is surrounded by six strictly equivalent chlorine atoms and each chlorine atom by six sodium atoms. Nobody can tell which of them is the "real" molecular partner of the other. There are no molecules as such within these crystals, though they may occur in the gas phase if the solid is vapourized. Other structures clearly exhibit groups containing atoms nearer to each other than to any atom belonging to another similar group, and here the molecular structure prevails. If we so choose, we may call the whole crystal of NaCl a giant molecule because there is no internal boundary in its structure. In between the atomic (or ionic) giant molecular crystal and the crystal built of individual molecules, we have crystals built from layers which are distinct from each other but within which it is again not possible to assign an individual partner to an atom. It is generally easy to separate such solids into thin layers or slices; common examples are graphite, CdS and MoS_2.

28. LET THE NUMBERS SPEAK!

We have now achieved one of the most presumptuous of our aims: we have determined the true masses of the atoms and, on the assumption that they contact each other within the crystals, their diameters also. Our measurements are extremely precise, and we have every reason to be confident in them. The circumstance that the result of X-ray measurements agreed with the order of magnitude that we had already found in the course of our very simple play with oil drops on water, serves to strengthen our confidence. As a matter of fact about ten independent methods have led to exactly the same value of Loschmidt's number N_L and thus we may rest assured that this value is correct. Naturally it amounts to the same thing, whether we determine the number of atoms in a mole or the actual weight of an atom. They are absolutely interconnected because their product must be by definition the atomic weight of the element.

Let us try to visualize some atomic dimensions. We stated earlier that a gold foil may be hammered down to a thickness of about 10^{-5} cm. X-ray analyses of gold crystals have proved that the diameter of a gold atom is $2 \cdot 87$ Å. Hence there are some 350 gold atoms one above the other within the thickness of the gold foil, not a very great number after all!

We know now that the thickness of the soap bubble, just before it bursts, within the "black region" contains just two opposed layers of fatty-acid salt molecules with a total length of some 60 Å. There were 36 carbon atoms across this layer and yet we still managed to handle it without damage! We do not seem to be so far above atomic dimensions after all, surely not so much as we are below the grand dimensions of the universe around us.

The situation changes somewhat if we go over to the weight of an atom, instead of its diameter. This is natural, because weight, like volume is proportional to the cube of the diameter, e.g. the diameter of an atom is about 10^{-8} cm whereas its volume is 10^{-24} cm³. The smallest speck of gold visible in an ordinary microscope contains about a milliard atoms. One cubic millimetre of atmospheric air contains some $3 \cdot 10^{16}$ molecules, in words thirty thousand billions. (For Americans, thirty million billions*.) And in the best vacuum we are able to obtain in our laboratories we still have 300,000 molecules to the mm³.

---

* English billion = 10^{12}; U.S.A. billion = 10^9.

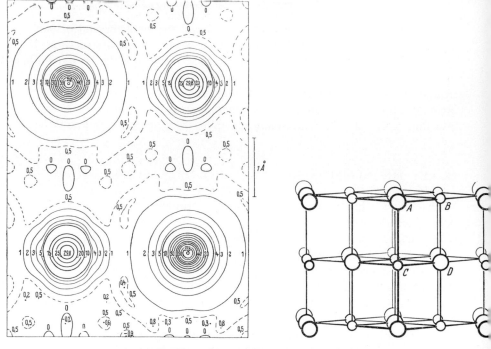

(a) Sodium chloride. The crystal is composed of charged ions side by side, which cannot be arranged into individual pairs.

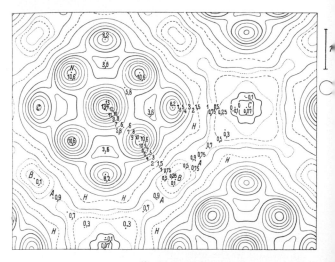

(b) Hexamethylene-tetramine. The cryst

Fig. 61. **Electron density maps and the**

The *left-hand half* of each figure shows an electron density ma constant electron density, as contour lines of a map represer

The *right-hand half* shows a model built from spheres representir clearer picture). The projection brings similar atoms to cove

(From H. G. Grimm *et al.*, *Naturwissenschaften* v. 26, p. 3

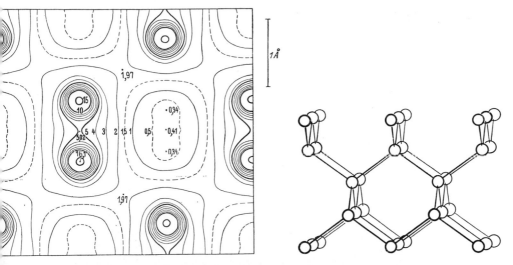

(c) Graphite. The crystal is composed of sheets, one above the other, each in itself a covalently bound molecule.

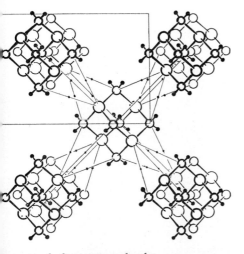

composed of separate molecules.

anslation into a space structure of atoms

nstructed from X-ray diffraction data. Lines represent loci of nstant height.

oms (too small against their distance from each other, to give a ch other.

gs. 1, 2, p. 31 Fig. 3 (1938). By courtesy of Springer-Verlag.)

The average distance between neighbouring gas molecules in normal air is about 30 Å, that is they are about ten times as far apart as in solids and liquids. This is in agreement with the fact that the density of normal gases is about 10^{-3}, a thousand times smaller than that of the liquids into which they condense. In the extreme vacuum mentioned above, this distance

Fig. 62 (a). **Electron micrograph showing the structure of catalase crystals**

The smallest visible structures are the molecules of the enzyme, some 70 Å broad. ($\times$ 500,000.)

(From R. J. Harris, *The Interpretation of Ultrastructure*, p. 271, Fig. 2 (New York, Academic Press, 1962). By courtesy of the Academic Press Inc.)

between molecules would increase enormously, to 15 microns. By comparison, this corresponds to pinheads 50 m apart from each other, which at normal pressure would be separated by only one centimetre.

The resolution of an electron microscope allows us to photograph structures with a breadth of 10–20 atoms! (Fig. 62).

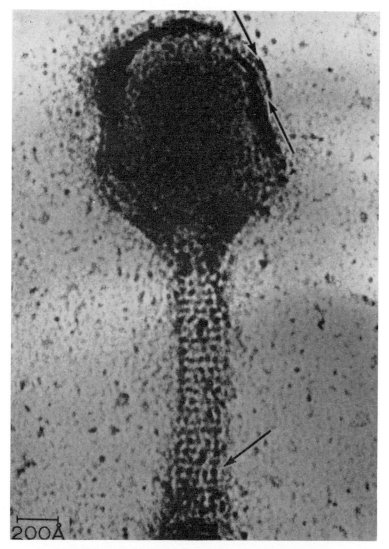

Fig. 62 (b). **Electron micrograph of a bacteriophage**

The small black points are molecules.

(From R. J. Harris, *The Interpretation of Ultrastructure*, p. 419, Fig. 6 (New York, Academic Press, 1962). By courtesy of the Academic Press Inc.)

Part 3

Molecular Movement

29. ARE THE MOLECULES MOVING OR NOT?

The reader must surely be convinced at this point that atoms and molecules really exist and that even their dimensions have been measured. What we do not know as yet is whether these particles are stationery or in motion. Let us discuss some well-known phenomena which resolve this uncertainty.

We know that whenever we connect an evacuated vessel to one containing a gas, the gas flows into the vacuum until the pressure in the two vessels is equalized. The gas has moved, and so by necessity the molecules have also moved. However, we do not know whether they have moved because a

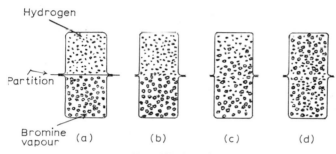

Fig. 63. **Diffusion of gases**

(*a*) Before the experiment; (*b*) at the start of the experiment (partition pulled out); (*c*) after an hour; (*d*) after a day.

repulsive force between them has made them fill as much space as possible or because they were already in spontaneous motion within their original enclosure.

The experiment can be repeated in a slightly different way. Instead of evacuating the second vessel let us fill it with another gas at the same pressure as that in the first vessel. In order to make sure that any motion that occurs is not caused by gravity the vessels should be arranged vertically. Suppose we fill the lower vessel with heavy, brown bromine vapour and the top one with light, colourless hydrogen gas. (Fig. 63). The two containers are initially separated by a sheet of paper which can be carefully withdrawn until the two ground-glass flanges are in contact with each other. If we have proceeded carefully, nothing will appear to have happened and the borderline between the brown vapour and the colourless gas will remain as sharp as before. We must take great care, that the table remains quite undisturbed and that during the experiment no differences of temperature occur in the room causing thermal convection of the gases. But no matter how careful we are the line of demarcation will soon begin to blur. After a few hours it will barely be detectable and in a day or two the bromine and hydrogen will have mixed completely, even against the force of gravity which would have been expected to keep the gases apart. With whatever pair of gases or vapours we repeat this experiment, the result will always be the same, although the velocity of mixing may differ.

An analogous experiment may be performed with two miscible liquids* one of which is coloured. Pour some water into a glass and carefully add a layer of red wine above it. Wine contains alcohol which is less dense than water, and so the red layer will float on the water. No matter how much care is taken to exclude vibrations or temperature fluctuations, the two liquids will have mixed completely in a few days time, red colour, alcohol, water, everything.

If we want to make quite sure that there were no minute vibrations at work assisting the mixing process we can go one step further and work with a jelly. Prepare a suitable jelly by making a 5 per cent solution of gelatine in water, put half of it in a test tube and let the mixture stiffen in the cold. Now dissolve a speck of colouring matter in the remaining gelatine solution and pour this upon the gelatine layer in the test tube. This second, coloured layer is allowed to cool until it has become as stiff as the first. The boundary between the two layers is very sharp, initially, yet becomes unsharp after an hour or so, and in this case also has vanished in a few days.

The same process goes on between, say, gold and silver if pieces of the two metals are held in contact under strong pressure. However, it may take years in this case to get enough silver into the gold, and enough gold into the silver, to detect them by chemical analysis. (Modern radiochemical methods can detect the process after a much shorter time.)

This process, the interpenetration, even against gravity, of two materials which are soluble in each other, without the application of external movement or any other obvious cause is called diffusion. Experience has shown us that the process goes on more quickly at elevated temperature.

In classical chemistry we can only perform this experiment with different compounds, because we cannot follow the process if top and bottom layer are of the same material. But there never has been discovered a pair of compounds, similar though they might be to each other, which do not show the phenomenon of diffusion. Normal propyl alcohol (CH_3—CH_2—CH_2OH) with the OH group on a terminal carbon atom, and isopropyl alcohol (CH_3—$CHOH$—CH_3) with the OH on the central carbon atom diffuse into each other as though they were as different as any of the pairs we have already described. Does it not seem very probable then, that a layer of normal propyl alcohol would also exchange molecules with a superimposed layer of the same substance? Most probably we are only unable to confirm this movement because we are unable to distinguish one molecule of propyl alcohol from another. But remember that in the chapter on inorganic structural formulae we have already mentioned (without having been able to

---

* Although all gases are mutually miscible, the same is not true of liquids, e.g. oil and water will not dissolve in each other to more than a minute extent.

explain it yet) that it is possible today to label certain atoms without changing their chemical properties. The label may be a difference of one or two units in the atomic weight or sometimes the ability to undergo radioactive disintegration. Thus molecules of isopropyl alcohol which contain a heavy hydrogen atom or radioactive carbon may be distinguished from unlabelled molecules of the same compound. By this use of sophisticated techniques it can be confirmed that any chemically homogeneous material does diffuse "into itself". The process is called self-diffusion.

If diffusion is a general property of matter, molecules and atoms *must* be in perpetual movement because they would not be able to change places, to intrude one beside the other without this movement.

30. THE VELOCITY OF MOLECULES

The most beautiful and convincing method which demonstrates molecular or atomic movement and simultaneously permits the measurement of the particle velocities is that devised by *O. Stern* (1888–) in 1920. Let us build a small electric furnace into a chamber which can be evacuated (Fig. 64*a*). There is a slit S_I in the wall of the furnace which contains a metal which evaporates readily in a vacuum, e.g. cadmium. The axis of rotation of the system passes through the middle of the furnace, and along that same radius which connects the axis with the slit S_I we cut a second slit S_{II} in a cylinder C_I, which is halfway between the furnace and an outer cylinder C_{II}. The furnace and the two cylinders are all mounted concentrically and may be rotated together about their common axis. They are contained within the chamber which must be evacuated before the experiment proceeds. When it is heated to 650°C, cadmium, which has an appreciable vapour pressure at this temperature, evaporates and fills the small furnace with its vapour. The slit S_I is closed at this stage but may be opened from outside by using a magnet. The atoms of cadmium (they *are* atoms, because the determination of the vapour density has shown us that cadmium is monatomic in gas phase, i.e. it belongs to the "class containing the smallest number of cadmium atoms from among all cadmium compounds",) are now allowed to escape through the slit of the furnace and will in general reach the first cylinder. Only in one direction, the direction of the second slit, will they be able to go on and reach cylinder C_{II}. We therefore find a Cd deposit on this outer cylinder at the point P which lies on the radius going through both slits. We should not be very much astonished that the atoms have flown right across the apparatus and through the slits in a straight line, like a beam, because there was no reason for them to change their direction, no force was acting upon them. Newton's laws of mechanics demand that all bodies continue in their uniform straight motion if no force deflects them therefrom. (Gravity may be neglected.)

So far the apparatus has been stationary. The result is quite different, however, if we repeat the experiment after the whole system of slits and furnace has been brought into uniform rotation. In such a case the atoms which

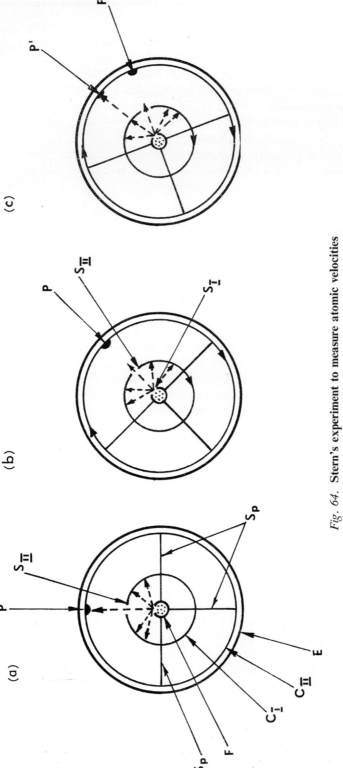

Fig. 64. **Stern's experiment to measure atomic velocities**

(*a*) Rotor stationary; furnace-slit open.
(*b*) Rotor turning; slit opened after full speed achieved.
(*c*) Rotor turning; time so much later than in (*b*) that an atom has flown from S_{II}, to impinge on C_{II}.

E = glass envelope of the vacuum system
F = rotatable furnace
S_I = slit of furnace
C_I = inner cylinder, rotating with F
S_{II} = slit radially aligned with S_I and centre of furnace

C_{II} = outer cylinder rotating with F and C_I
S_p = spokes of the rotor
P = point on C_{II} opposite the two slits
P' = point on C_{II} where the atom arrives in the rotating
  apparatus after passing through S_{II}.

Broken arrows = paths of flying atoms.

would have just reached the second slit on their way from the furnace if the system had been at rest arrive too late: the slit S_{II} has moved round a little during their time of flight. These atoms will therefore reach only the first cylinder and will condense on its wall. Only those atoms are going to pass through the second slit which departed at a small angle ahead of the radius, toward P, just as one has to aim a bit ahead of a flying bird if one wants to shoot it. The atoms must be directed at the point on the first cylinder where the slit S_{II} *will be* at the time of their arrival. Fig. 64*b* shows the same five directions of atomic flight as the preceding figure and we can see that not the third, but the second atom is going to pass the rotating slit. Were the system to come to a standstill at this moment, this atom would reach the outside cylinder at the same point P as before. But the apparatus goes on rotating and the atom will arrive at C_{II} too late to reach the point P, just as its neighbour was too late to reach the slit S_{II} at all. The third figure shows us that by the time this atom reaches the outside cylinder it impinges at a point P' which is somewhat behind P in the sense of the rotation.

The whole phenomenon occurs because the atom needed a finite time to cover the distance between the second slit and the cylinder C_{II}. Had it moved with infinite velocity, it would have reached the point P. But the experiment proves to us, that the atom did not move with infinite velocity since P and P' do not coincide. Knowing the speed of rotation of our system and the distance between the two cylinders it is easy to calculate the absolute velocity of the Cd atoms. Suppose the apparatus rotates at 4,000 revolutions per minute, a speed quite within the range of a centrifuge, and suppose that the radius of the cylinder C_{II} is 10 cm. Then any point on this cylinder moves at the rate of $4,000 \times 2\pi \times 10$ cm/min, that is at 251,000 cm/min or 4,183 cm/sec. After having performed both experiments we can see, using a microscope, that the distance between the centre of P and the centre of P' is 0·574 cm. How long did it take for the cylinder C_{II} to move that distance? Evidently 0·574/4,183 sec, that is $1·37 \times 10^{-4}$ sec. The cadmium atoms flew during this time from the second slit to C_{II} across a distance of 5 cm. Hence they had to fly at a speed of $5/1·37 \times 10^{-4}$, or 36,500 cm/sec. We have therefore measured the speed of flight of cadmium atoms in the vapour under the given circumstances and have found that it amounts to 365 m/sec, about the speed of rifle bullet (0·23 mile/sec).

What do we mean by "under the given circumstances"? The temperature of the furnace. Experiments at different furnace temperatures show that the distance between P and P' decreases with increasing temperature. It is even possible to state, within the limits of experimental accuracy, that this distance decreases with the square root of the absolute temperature of the furnace.

Cadmium may be replaced by another metal in so far as the latter has a practical vapour pressure within the range of experimental temperatures. Potassium could be used; after all it will not oxidize in a vacuum. We find, in this case, that the atoms of potassium move at a higher speed than cadmium atoms at the same temperature. At 650°C, for instance they have a velocity of 622 m/sec. The atoms of mercury on the other hand would lose this race, for they have a velocity of only 273 m/sec at this temperature. In general we find that the higher the atomic weight (K, 39·1; Cd, 112·4; Hg, 200·6), the slower the atom. It is immediately evident that the three velocities (622, 365, 273 m/sec respectively) do not decrease in inverse proportion to the atomic weights. They decrease less than in inverse proportion. However their squares are (approximately): 386,000, 133,000 and 74,000 and we see by simple division that these squares are inversely proportional to the atomic weights. In other words, the product of atomic weights and squares of velocities is a constant at any given temperature.

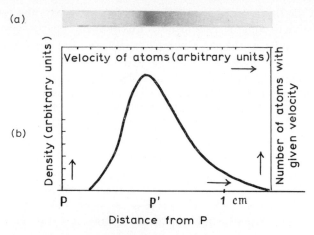

Fig. 65. **Result of the Stern experiment**

(*a*) Density of cadmium deposit.
(*b*) Diagram of the quantitative distribution of cadmium on the second cylinder.

More accurate observation of the spot (or rather line) *P'* under the microscope reveals that it is not so sharp as would be expected from the narrowness of the slits used. We therefore receive the impression that the cadmium atoms leave the furnace with different velocities. Some move at a great speed, others lag behind. The former reach the goal nearest to point *P*, the latter impinge behind it. An accurate determination of the density of the cadmium film around *P'* reveals the distribution of velocities among the cadmium atoms (Fig. 65). We see that there is a maximum on this distribution curve, meaning that there is a velocity which occurs the most often while the number of atoms with greater or smaller velocity decreases.

To summarize: The Stern experiment tells us that atoms (and molecules too, because the same experiment could have been just as easily performed with volatile compounds) move around with velocities of the order of magnitude of that of a rifle-bullet, some hundred of metres in a second. The velocity increases with temperature and decreases with the atomic (or molecular) weight in such a manner that the square of the velocity is proportional to the absolute temperature and inversely proportional to the atomic (or molecular) weight. We have also learned that the particles of a gas possess different velocities distributed around a median value.

31. THE KINETIC THEORY OF GASES

Centuries before the experiments of Stern showed so clearly the movement of molecules, scientists conceived the idea of molecular movement as the result of quite different and more involved reasonings. They advocated the idea that heat in itself was nothing but disordered movement of an aggregate of molecules. Building upon the rudimentary ideas of *D. Bernouilli* (1738), the German physicist *R. J. E. Clausius* (1822–1888) was the first to draw definite conclusions from this assumption (1857). Like Avogadro, he too concentrated on the theory of gases because it was evident that their behaviour was more uniform than that of any other state of matter.

We have seen that the molecules of gases at normal temperature and pressure are so far apart that the forces acting between them can probably be neglected to a good approximation. The following macroscopic experiment serves to prove this point: Whenever a gas of normal pressure expands into an adjoining vacuum, its temperature remains nearly unchanged. If the molecules attracted each other we should have expected the gas to cool, because the energy needed to separate the molecules against this attraction must have been forthcoming from somewhere. On the other hand if there had been repulsion between the gas molecules this would have liberated energy during expansion and the gas should have become warmer. The fact, that the temperature remains constant proves that neither attraction nor repulsion acts between the molecules. That all gases try to expand is thus only due to the fact that their molecules move about the whole time in all possible directions and so can enter any hole they find in an apparatus.*

Let us examine the movement which Clausius himself imagined. We have to suppose that the molecules move about in all directions in complete disorder and only change direction if they hit each other or the wall of their container. When ever they collide they behave as elastic spheres (or more complicated bodies) and according to the relative directions before collision they either increase or decrease each other's velocity. They rebound elastically after a collision with the wall. The molecular velocity depends on the temperature and increases with the latter according to some function to be determined later. According to the above picture the well-known transformation of kinetic energy into heat is microscopically no true transformation at all. If we drop a piece of lead from a tower, the potential energy we imparted to the lead in taking it up will become kinetic energy when it falls. But where is the energy after the lead has struck the earth? Precise measurement shows that it has become heat, the lead and the earth around it being very slightly warmer than before. According to Clausius we ought to say that the kinetic energy remains as such but instead of being one-directional and belonging only to the lead it has lost its order and serves to increase

---

* Actually there *is* a small change in the temperature of gases in course of expansion and this *is* due to forces between their molecules. The Linde process for liquifying air is, for instance, based on this phenomenon. By starting expansion from low pressures, however, even this effect disappears.

the irregular molecular kinetic energy both of the lead and its surroundings. The molecules move more quickly than before: the materials became warmer.

Why do all gases exert pressure upon the walls of their container? According to the kinetic theory, it is because the molecules constantly impinge on these walls and thus press against them, just as the pan of a balance is forced down if we repeatedly bounce small elastic balls off it. Why do gases expand if there is nothing to restrain them? Simply because the molecules go on moving in straight paths until they encounter an obstacle. What is diffusion between gases? This phenomenon may be illustrated by a mechanical analogy. Imagine two sets of balls, white and red, distributed at first in the right and left halves respectively of a box. When the box is rocked the balls move in an irregular manner and the boundary between the white and red group begins to become indistinct. In a short time the boundary has disappeared altogether and the two colours are intimately mixed. They will not separate spontaneously on shaking them further.

What, then, causes the balls, or the molecules of a gas, to fill all open empty space and what makes them mix with each other? Is it a force? No! It is sheer probability. There is no force which would prevent all the gas molecules in a room from clustering together in one small corner. After all they move around in an utterly disorderly manner and by chance they could become assembled anywhere. But we should be very much astonished if this did happen, and the odds are very much against it. Let us try to become acquainted with this "probability" which seems to have such an immense influence in Nature.

Take an open box and divide the bottom into halves with a line through the middle. Put two identical balls into it and shake it, taking care that it always remains horizontal. Stop shaking occasionally and note the positions of the balls. Evidently there are three possible distributions of the balls between the halves of the box. Either both balls are on the right, or both on the left or there is one on each side. This last distribution can occur in two ways: either ball No. 1 is on the right and ball No. 2 on the left or vice versa. (Of course this distinction is only possible as long as there is some observable difference between the balls, a fact which has made it necessary to formulate somewhat different statistics for real elementary particles.) The four different situations are *a priori* equally probable and indeed if we repeat the experiment often enough we shall find that in one-quarter of the observed cases both balls will be on the right or on the left, while in half (= twofourths) of the cases there will be one ball on each side.

Repeat the experiment with four balls (*a, b, c, d*) (Fig. 66). The following distributions are possible: Four balls on the left, none on the right; three on the left, one on the right attainable in four ways (*abc:d, acd:c, acd:b, bcd:a*); two on the right, two on the left, in six possible arrangements (*ab:cd, ac:bd, ad:bc, bc:ad, bd:ac, cd:ab,*). By symmetry there must be four arrangements for the case where one ball remains at the left and four are at the right side, while finally there is again only one way of having all four at the right. This makes altogether sixteen configurations; there are no more. From these

sixteen equally probable configurations there is only one with all four balls
on one specified side of the tray, whilst there are six where they are distributed
equally between the two halves. Thus the probability of finding all the balls
on one given side will be only 1/16 of all cases whereas that of equal distribution will be 6/16. If we wish we may verify this calculation by actual
experiments.

We could go on making calculations and/or experimenting with an ever-
increasing number of balls in the box. The result would be that the probability of finding all the balls on one side would decrease rapidly while the highest
probability would remain that of finding them equally distributed between

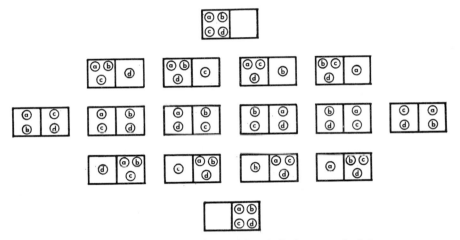

Fig. 66. **Possible distributions of four balls between the halves
of a box**

right and left, i.e. of finding them equally distributed over the entire bottom
of the box. In the same way the molecules of air are distributed with equal
density throughout the whole volume of a room. For any given number of
balls, or molecules, we are always able to calculate a definite number as the
probability of any desired distribution.

In a room we have not four, nor forty, nor even four hundred thousand
molecules but quadrillions of them. From among a quadrillion times a
quadrillion equally probable distributions there would be only a single one
in which all of the molecules meet in one given corner! The winning of the
first prize in a lottery is a trivial accomplishment compared with this event.
This infinitesimally small probability is the *cause* of the fact that we never
observe such a macroscopic bunching of gas molecules. In practice it simply
never occurs.

It would be just as hopeless to wait until the red and white balls have
unmixed again in the course of our shaking the box for any length of time,
however protracted. In the world of molecules this means that the mutual
diffusion of the molecules of two different gases is a process which actually
occurs spontaneously, but its counterpart, the spontaneous demixing of

two gases which were initially completely mixed will never occur. A calculation similar to that above would show that the chance of separating ten red and ten white balls by shaking them in a box is only one in ten thousand. Evidently the chances would be practically zero for a quadrillion of particles.

However, we can imagine other curious situations. It is not necessary for all the molecules in a sample of gas to meet in a given place, it would be quite an achievement if they could manage even to move one beside the other, all in a given direction. What a wind this would be! Well, winds do exist, as we know. But they never arise spontaneously, say within a room of uniform temperature. On the contrary. Whenever we make an artificial wind, e.g. by opening a ventilator, it subsides shortly after the ventilator has been closed. Something has happened which may be likened to children leaving a school at noon. In the neighbourhood of the school there will be more people moving in one direction, away from the school, than towards it because of the outflow of the children. But after a time they will have dispersed hither and thither and the number of people then going in one direction in the street will become equal to the number going in the opposite direction on the average.

It is easy to see that not only the position and the direction of movement of the molecules is going to be equally distributed because of probability, but their velocities are bound to be distributed around some average velocity too, no matter how they were distributed at the beginning. There will be some molecules that move more quickly and some that move more slowly, but the average velocity will, of course, be given by the sum total of the kinetic energy given to them: only this amount of energy can possibly be distributed among them. The cause working for a special distribution of velocities is the great number of recurrent collisions between the molecules. We only need consider some special cases of collision in order to see that colliding particles interchange a part of their velocities.

Consider two identical balls colliding with equal velocities on their head-on run towards each other (Fig. 67*A*). We know that the result of this elastic collision will be that they reverse direction and leave each other with the direction of their velocities interchanged but with their magnitudes unaltered. If a moving ball collides centrally with one that is stationary the first ball comes to a dead stop while the second takes over the magnitude and direction of its movement (Fig. 67*B*). But suppose for argument's sake that two balls collide while moving perpendicularly to one another, one hitting the second exactly against its side (Fig. 67*C*). The resulting motion is obtained from the parallelogram of velocities, and if both balls moved with equal velocity before the collision the one which has hit the other comes to a standstill while the other adds the vertical velocity it has acquired to the horizontal velocity that it possesses already. Pythagoras theorem determines that the enhanced velocity of this ball in the diagonal direction will be $\sqrt{(v^2 + v^2)}$, that is $v\sqrt{2}$. (By the way, energy has been conserved, because we had $\frac{1}{2}mv^2 + \frac{1}{2}mv^2$ in the beginning and $m(v^2\sqrt{2^2})/2 = 2\ mv^2/2$ after the collision.) But all the energy, all the velocity, is now on one of the partners. The other has come to a standstill, just as it would have done if it had hit centrally a standing ball of its own dimensions. Thus we have seen both extremes: one collision whereby the magnitude of individual velocities remained unchanged and another where all velocity, all kinetic energy was transferred from one particle to the other. Of course all intermediate kinds of collision are bound to occur and thus it is clear that even an ordered state of affairs will soon change into a statistically mixed set of molecular velocities, changing from second to second as far as individual molecules are concerned, but

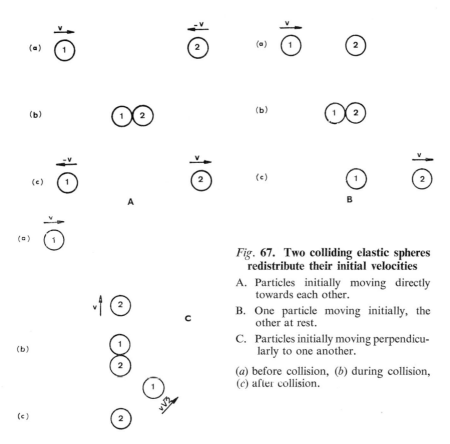

Fig. 67. **Two colliding elastic spheres redistribute their initial velocities**

A. Particles initially moving directly towards each other.

B. One particle moving initially, the other at rest.

C. Particles initially moving perpendicularly to one another.

(*a*) before collision, (*b*) during collision, (*c*) after collision.

retaining the sum total of kinetic energy because energy must be conserved.

J. C. Maxwell (1831–1879), one of the greatest of English theoretical physicists, was able to prove that according to the laws of chance, the distribution of velocities within an assembly of moving balls with a given sum total of kinetic energy remains the same, no matter how individual velocities might change. For each molecule losing a specified velocity there will be on the average another molecule which has just received this velocity in a collision. There is a most probable distribution of velocities and thus of kinetic energy among the molecules and appreciable deviations from this most probable distribution are just as rare as deviations from the homogeneous distribution of gas molecules in space. It was Maxwell who calculated this distribution for each average kinetic energy of the molecules and it is named after him the Maxwellian distribution. The velocity distribution of the cadmium atoms in the Stern experiment (Fig. 65) corresponds exactly with his calculations for the temperature of the furnace. Extremely small and extremely great velocities are very improbable and the maximum of the curve, that is, the most probable velocity, increases with temperature.

Let us remember that the square of the average velocity was proportional to the absolute temperature in the Stern experiment. On the other hand, the square of the velocities of different molecules at the same temperature was inversely proportional to the molecular weight: heavier molecules move more slowly:

$$\frac{m_1}{m_2} = \frac{v_2{}^2}{v_1{}^2} \qquad \qquad (31.1)$$

hence

$$m_1 v_1{}^2 = m_2 v_2{}^2 \qquad \qquad (31.2)$$

or

$$\tfrac{1}{2} m_1 v_1{}^2 = \tfrac{1}{2} m_2 v_2{}^2 = E_{kin_1} = E_{kin_2} \qquad \qquad (31.3)$$

The quantity which is constant for different particles at constant temperature is thus the average kinetic energy.* For different temperatures, on the other hand, it is the square of the velocities, i.e. again the kinetic energy, that is proportional to the absolute temperature. The experiments of Stern thus justify the assumption of Bernouilli and Clausius that the microscopic, kinetic equivalent of the concept "temperature" is nothing else than the average kinetic energy of the molecules. The speed of the molecules is proportional to the square root of the kinetic energy because of the relation $E_{kin} = mv^2/2$ and hence it is proportional to the square root of the temperature also. What does this mean in a practical example? Room temperature is about 300°K; its square root is 17·4. Half of this is 8·7 the square of which is 75. That is, we have to cool a gas from room temperature to 75°K, equivalent to −198°C, in order to make its molecules move half as slowly.

We have reached a conclusion of extreme importance. Temperature has received a mechanical, or rather, a statistical-mechanical meaning. It is disordered movement of particles in contradistinction to ordered movements

---

* The potential energy acquired by a body moving a distance d against a force f is given by the product of these two latter quantities: $E = df$. If the force is removed and the body drops back to its original position then its potential energy is converted into an equivalent amount of kinetic energy possessed by the body solely by virtue of its motion. (Kinetic energy should not be confused with momentum, which is the product of mass and velocity.) The definition of force is the product of the acceleration, a, that it gives to a body of mass m, and the mass: $f = am$. We know from the laws of dynamics that a body moves with uniform acceleration when acted upon by a steady force and so we have $E = d \cdot m \cdot a$ or substituting the increment of velocity by time v/t, which defines uniform acceleration: $E = d \cdot m \cdot v/t$. The body was at rest initially and acquired the velocity v after time t in the course of uniform acceleration, therefore its average velocity must have been $v/2$ during its journey, and the distance d must equal $t \cdot v/2$. Substituting this instead of d into the energy equation we have the kinetic energy expressed as a function of the final velocity

$$E_{kin} = t \cdot v \cdot m \cdot v/(2t) = \tfrac{1}{2}mv^2 \qquad \qquad (31.4)$$

as that of the wind or of the piece of falling lead. From the laws of probability, disorder is always intrinsically more probable than order. Therefore every ordered motion tends to go over into disordered motion, into heat say, by friction.

Disorder, of which temperature is a measure, tends to intersperse matter as much as possible. It causes gases to expand and fill their containers, and counteracts forces which pull or push particles into definite places. At relatively low temperatures, that is, where there is a low degree of disorder, we have the formation of crystals, these monuments of ordered structure which owe their existence to a series of forces which act between particles of matter. We know, however, that even in solids there is diffusion, that is, movement, and the higher the temperature, the quicker diffusion becomes (increasing in general with the velocity of the particles, that is, with the square root of the absolute temperature). Thermal expansion also shows the effect of disordered motion as it loosens the structure of a crystal. At a certain point the fixed, beautiful crystal structure is disrupted and gives place to a disorderly mass of moving molecules, which, however, still remain within each other's field of attraction. Only a few, those which by chance happen to possess high velocities, i.e. those on the right-hand side of the Maxwell velocity-distribution curve, manage to escape from their partners and move away into space. They volatilize. Thereby a liquid is said to have a "vapour pressure", and this pressure goes on increasing with temperature until it becomes equal to the external pressure. At this point the whole liquid boils and forms vapour containing distinct molecules with very small interactions between them, driven about by the disorderly force of their movement, by temperature.

Throughout nature there is a fight between the forces of order and disorder and every thermodynamic equilibrium is a compromise in this struggle. At low temperatures very specific structures become possible which are far away from a random disorder of maximum probability; force prevails over chance. At extreme high temperatures chance prevails and makes all ordered structures practically impossible. In between there is a gradual transition, gradual that is within limits but having points of discontinuous transition between structures of different types as, for example, between a solid and liquid, or between a liquid and its vapour.

The quantitative measure of chance, of the probability of a certain configuration can be given in principle along the lines we discussed when calculating the probabilities of finding balls in one place or another. Statistical mechanics is the branch of physics dealing with such problems. However, the macroscopic results of disorder were evident before this branch of science had evolved. It was evident that certain types of process simply do not occur spontaneously in Nature. As we said, gas molecules do not assemble all together in a corner of a room. A glass of water on the table will not spontaneously separate into a layer of ice and a layer of hot water, in spite of the fact that the energy balance would allow it: the heat from the frozen part could serve to raise the temperature of the remaining liquid. All these facts could be reduced to the empirical statement that heat does not divide spontaneously into heat of lower and higher temperature, because it is extremely improbable that slow and quick particles can segregate. We know this now, but at one time it was simply a law of Nature, the second law of thermodynamics. The first law of thermodynamics is the principle of conservation of energy.

In the hands of the masters these two laws have served as foundation to a vast, logical and beautiful building: to thermodynamics. It so happened that the relevant quantity of all these problems, the probability of a state, occurred in the empirical laws in the form of its logarithm and was named the entropy of a system in the given state. It matters little which we prefer, one increases along with the other: the probability or its logarithm. It was the great achievement of *L. Boltzmann* (1844–1906) to have shown how the two concepts are interrelated, and up to now thermodynamics with its concept of entropy is often simpler to handle than the intricate picture of probabilities (1866–1877). The latter gives on the other hand, much more insight into the structure of Nature.

Boltzmann was able to prove that whenever a particle has the choice of being in one of two states, one where it is attracted by forces and another where it is more free after having paid with the expenditure of the energy E for its freedom, the probability W of finding it in the higher state is

$$W = e^{-E/kT} \qquad . \qquad . \qquad . \qquad . \qquad (31.5)$$

where e is a number, 2·718 . . . (the base of natural logarithms) and k a factor of proportionality (1·380 × 10^{-16} erg/deg per molecule) which has been named the Boltzmann constant. This formula is extremely useful in all problems concerning elementary probabilities in different states.

32. THE GAS LAWS AND KINETIC THEORY

It is worth while to work out a simplified calculation based on the kinetic picture of gas molecules, and which relates the macroscopic concepts of pressure and temperature to the microscopic ones: mass and mean velocity of the molecules.

We have to calculate the pressure as the sum total of impulses of impinging particles on the wall of a container. Let us begin with the force exerted by a single particle rebounding from a wall in a perpendicular direction. Imagine that the wall is held in position from behind by a spring which allows us to measure the force exerted upon the wall during the collision. We shall not be surprised to find that this force is proportional both to the mass and to the velocity of the impinging particle.

Newton's third law of mechanics requires that "action and reaction are equal and opposite." For example, a rifle recoils noticeably as the bullet leaves the barrel. A ball will rebound from a massive wall with its velocity changed only in direction, not in magnitude. The velocity may change say, by $2v$ from $+ v$ to $- v$. Suppose the collision has lasted for time t, the acceleration, i.e. the change of velocity in unit time, is $2v/t$. During this time, t, the ball has acted with a force $m2v/t$ upon the wall. However, we have not to consider a single ball but rather a shower of n balls, striking an area of one square centimetre each second. This means that nt balls have impinged during each time-element t giving a total mean force at any time on a square cm of

$$p = \frac{n.t.m.2v}{t} = 2nmv \qquad . \qquad . \qquad . \quad (32.1)$$

because the force acting on a cm² is the pressure (p) by definition. This is the pressure exerted on a wall by the perpendicular elastic incidence of n balls— or molecules, of course—of mass m, velocity v and density n per second per square cm.

m is a constant of the molecule in question, v is its velocity according to its temperature, but how large is n, and what is the rate of incidence on unit wall area? This quantity is evidently connected with the density N of the molecules in unit volume.* In reality they move about in an irregular manner but in order to facilitate calculation we will arbitrarily assume that they only move in three directions in space, perpendicular to each other, e.g. parallel to the x, y and z axes. There are two senses for each direction of this simplified movement (right left, up–down, forwards–backwards) so that only 1/6 of all molecules are to be regarded as moving towards the wall, the other 5/6 moving either parallel to it or away from it. It can be shown that the result of the calculation is not changed by resolving all possible directions into the above three components. (We have already committed an error of simplification by assuming a mean molecular velocity instead of making a separate calculation for each velocity of the Maxwell distribution.)

Now let us see how the volume density of the molecules, N, determines the density of the impinging shower. We have already agreed that only 1/6 of all molecules is going to reach the wall at all. How many will do so in a second? Evidently as many as are contained in a tetragonal prism which has its base on the one square centimetre of wall under consideration and is v cm long, because the last molecule within the volume of the prism still able to reach the base in a second cannot be farther away from it than v cm, if v cm/sec is its velocity towards the base. Thus we have a tetragonal prism v cm high on 1 cm² area, that is v cm³ in volume, containing vN molecules. But only $vN/6$ are moving in a given direction and this is the number we have been looking for, n. If we insert this expression for n into the equation for the pressure which we derived above, then

$$p = 2mvNv/6 = \tfrac{1}{3}Nmv^2 \qquad . \qquad . \qquad . \quad (32.2)$$

Do not worry about the possibility that the vessel may perhaps be smaller than a prism of length v. In this case molecules flying in the opposite direction will be reflected towards our piece of wall from the other wall; it makes no difference whether two different molecules impinge upon it one after another or the same molecule twice, or more often, in a second.

---

* N normally denotes the Avogadro (Loschmidt) number; here not!

Nor is it necessary to restrict ourselves to a unit volume containing N molecules. We may multiply safely both sides of the equation by any volume V and thus obtain as a general result:

$$p.V = \tfrac{1}{3}V.N.m.v^2 \qquad . \qquad . \qquad . \qquad (32.3)$$

or, if we denote the number of molecules in volume V by N_V,

$$p.V = \tfrac{1}{3}m.v^2\, N_V \qquad . \qquad . \qquad . \qquad (32.4)$$

The product mv^2 reminds us of the mean kinetic energy, $\tfrac{1}{2}mv^2$. We may as well introduce the latter and obtain

$$p.V = \tfrac{1}{3}2E_{\text{kin}} \qquad . \qquad . \qquad . \qquad (32.5)$$

This relationship is the final result of our calculations. It shows that the product of volume times pressure of a gas is in direct proportion to the mean kinetic energy, a quantity we know to be proportional to the absolute temperature, i.e. pV is proportional to the absolute temperature.

Does this fit into what we have learned empirically about the gas laws? Let us see. On page 34 we concluded that the volume, pressure and temperature of a gas are connected by the equation

$$\frac{Vp}{T} = \frac{V_0 p_0}{T_0} \qquad . \qquad . \qquad . \qquad (32.6)$$

p_0 is one atmosphere, T_0 is 273°K and V_0 is the volume of our given gas sample under these conditions. Thus for any given quantity of gas the right-hand side of this equation is a constant. We may rewrite it

$$pV = \frac{V_0 p_0}{T_0}.T \qquad . \qquad . \qquad . \qquad (32.7)$$

pV is proportional to the absolute temperature, as it was proportional to the kinetic energy of the gas in the parallel kinetic argument. Comparing the two equations for pV we see that

$$\frac{V_0 p_0}{T_0}.T = \tfrac{2}{3}E_{\text{kin}} \qquad . \qquad . \qquad . \qquad (32.8)$$

It is customary to write the general equation in terms of one mole of gas, when V_0 becomes 22·4 litres and the constant factor $V_0 p_0/T_0$ is abbreviated as the "gas constant," R. Thus the translational kinetic energy of one mole of gas is

$$E_{\text{kin}}/\text{mole} = \tfrac{3}{2}RT \qquad . \qquad . \qquad . \qquad (32.9)$$

and has the value 33·6 litre-atmospheres at 0°C (i.e. 273°abs). This is the quantity we should have to divide by the Loschmidt (Avogadro) number, N_L, to obtain the mean kinetic energy of the single molecule at this temperature: $5\cdot6 \times 10^{-23}$ litre-atmosphere.*

---

* A litre-atmosphere is the work done by compressing one litre of a gas into a vessel against a pressure of 1 atm. One erg is the work done by moving a body of mass 1 gramme with an acceleration of 1 cm/sec². Taking into account that the unit of one atmosphere is the pressure of a column of mercury 76 cm high, density 13·59, we arrive at the conversion factor: 1 litre-atm equals $1{,}013\cdot3 \times 10^6$ ergs.

Are we going to obtain results consistent with the Stern experiment if we use this energy value to calculate molecular velocities? Of course. One mole of oxygen weighs 32 g and contains N, that is $6 \cdot 02 \times 10^{23}$ molecules. This gives the real weight of an O_2 molecule as $5 \cdot 29 \times 10^{-23}$ g. When we substitute this value for m and the above calculated kinetic energy of $5 \cdot 64 \times 10^{-14}$ erg for E_{kin} into the equation $E_{kin} = mv^2/2$ we obtain 462 m/sec for the mean velocity of an oxygen molecule at 0°C. (Compare Stern's directly measured velocity of the *cadmium* atom, i.e. 365 m/sec.)

Equation 32.4 shows that mean molecular velocities of different gases are inversely proportional to the square root of their molecular weights (temperature being equal). It follows that at 0°C the mean velocity of hydrogen molecules (H_2) is 1,850 m/sec ($1 \cdot 2$ mile/sec), that of helium atoms, 1,310 m/sec while that of the heavier krypton atoms is only 207 m/sec.

33. MICROSCOPIC VERSUS MACROSCOPIC ASPECTS

Compression and expansion of gases. What happens if a volume of gas is enclosed in a thermally insulated cylinder and allowed to change its volume by moving a piston? (Processes occurring within thermally insulated enclosures are called adiabatic; here we deal with adiabatic compression and expansion.)

We can put such weights on the piston that we can attain equilibrium at any given volume and can then cause the volume to change by very slowly taking off or adding extremely small weights (Fig. 68). The molecules are

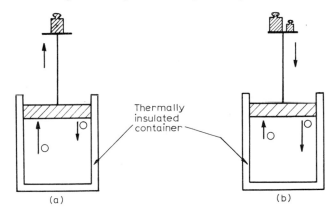

Fig. 68. **Adiabatic expansion and compression of gases**

(a) Pressure of molecules moves weight upwards. The gas performs work and cools down slightly.

(b) The weight does work on the gas. Resultant acceleration of the molecules causes a rise in the temperature.

evidently not going to be reflected in the same way from a receding piston as they would be from a fixed wall. Anybody who plays tennis knows that the ball is slowed down on bouncing off a receding racket. Similarly the molecules lose velocity and thus kinetic energy, too, in the process—the energy necessary to raise the piston with its weights. Thus all molecules recoiling from the piston will be "cooler" than the rest and on mixing into the remaining gas molecules they dissipate this loss of energy among them. The whole gas will have cooled down to the same extent in a very short time.

Exactly the opposite happens if the piston is forced down against the gas pressure: the recoiling molecules will acquire a small amount of additional kinetic energy and the whole amount of gas will become warmer.

An expansion against pressure such as has been described above is utterly different from an expansion into a vacuum where no work is being done. The molecules simply intrude into the new space offered to them and almost immediately strike the wall of the container. But this wall remains stationary and so the molecules will rebound with unchanged energy. The temperature must therefore remain unchanged. We have discussed this case already in Section 32.

Suppose we were to perform a similar expansion or compression using a piston in a cylinder without thermal insulation? In this case heat could flow freely through the walls of the cylinder into or out of the gas. This means that the temperature of the gas will remain constant, because the kinetic energy the molecules give to the piston is refunded to them from the surroundings through the walls. The molecules will not be in Maxwellian equilibrium with the atoms of the walls and are going to receive, on the average, greater impulses from the wall atoms than they give to them during collision. Their kinetic energy is being continuously restored to them and so the temperature of the gas remains constant. Does this mean that it is not the kinetic energy of the gas that has raised the piston? It does, indeed, since the kinetic energy did not change in the process. The energy came from outside the cylinder in the form of heat conducted into the gas and transformed into mechanical work by the ever-reinforced impinging molecules. This is the process called isothermal expansion, in which the temperature remains constant the whole time. Molecules certainly recoil with decreased speed from the piston but others recoil with increased speed from the walls of the cylinder. However, it would have been impossible to transmute this uniform heat energy of the heat reservoir outside the cylinder into mechanical work, had we not increased the entropy (probability) of the state of the gas. We allowed it to expand, to fill a greater volume than it had originally occupied and dispersion throughout a greater volume always means increased disorder, a higher intrinsic probability of a state. Without such a sacrifice the inflowing heat would have never turned into ordered mechanical motion of a piston. The chances would have been as small as those of experiencing a storm in a closed room.

Gases in the field of gravity. Atmospheric pressure concerns the weight of the atmosphere above us. How does it come about that we feel the weight of all air molecules above our heads while we are only in immediate contact with those few molecules that happen to impinge upon us? These are not even in contact with their higher-situated brethren!

Let us begin by considering a single molecule moving about in the field of gravity of the earth. As long as it moves in a horizontal plane its velocity remains constant, but whenever its direction is upwards it looses speed, like a stone thrown up into the air. On a downward course it will gain speed, once more like the stone. Every time the molecule collides with a molecule above it, its speed and thus the magnitude of the impact will be smaller than when collision occurs with a molecule in a lower region. The difference in the kinetic energy of the molecule will be just equal to the difference in height multiplied by the particle mass and the constant of gravity at the given level. As a matter of fact the whole column of gas molecules works in a sort of cascade.

Molecules which possess relatively small upward velocities as a consequence of the Maxwellian distribution will have a chance to return by virtue of gravity, before having met any molecule about them. Thus it looks as if there would be more molecules moving downwards than upwards at any moment. This of course cannot happen in a steady state: if it were to happen at all the molecules would gradually fall to the surface of the earth. Having arrived there, and some of course do arrive there, they would be reflected and proceed upward again. A stationary distribution is only possible if at any moment we have the same number of molecules going both up and down across any horizontal cross-section. The solution is simple. In the state of gravitational equilibrium there must always be more molecules at a lower level than at a higher one. The density of a gas should decrease with the height, as it actually does. The increasing pressure upon us as we descend, say from the top of a mountain, is the result of the greater number of molecules and their increasing total kinetic energy.

The barometric formula. We may put the argument another way round. All molecules within a thin horizontal layer acquire, because of gravity, a component of velocity downwards. This is the origin of the weight of this layer of gas. Each layer thus weighs upon the one beneath it and they successively compress each other until the increased concentration of molecules below causes an increased diffusion upwards into the less densely populated layers. The weight acting upon a unit area is the sum total of the weights of all the molecules above it. It is the pressure which compresses the layer. At a point where we have just half of the atmosphere above us and the other half below us, the pressure is half as great as at the earth's surface. Going upwards halfway between this level and the limit of the atmosphere, the pressure is halved once more, and so on. Halving the pressure means, however, doubling the volume, that is, the third quarter of the atmosphere must be as high as the first half and the next eighth as high again. It takes equal steps to halve again and again the number of molecules above us. The number therefore decreases in a geometric progression for equal difference in height. If h is the height necessary to halve the original density d, then at a height of nh the density will be only $d/2^n$, i.e. it has decreased exponentially (Fig. 69).

What about the height distributions of different gases? They must evidently depend on the molecular weights because according to Avogadro the number of molecules within a given volume is the same for all (ideal) gases at the same pressure and temperature. This means that two different gases with different molecular weights exert the same pressure at equal molecular densities but have different weights per unit volume. Now the whole barometric equilibrium was determined by the fact that the gas pressure within a layer equalled the weight of all the gas above it. If the pressure remains as it was but weight increases in proportion to the molecular weight, it is clear that the heavier the gas, the more quickly will its density decrease with height. It will be necessary to travel up an atmosphere of hydrogen 16 times as far as in oxygen before the density is halved. If gases are mixed, their distributions

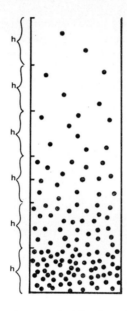

Fig. 69.   **Barometric height distribution**
of gas molecules

The density of molecules is halved after
each *h* increase in height.

are superimposed. The heavier molecules will remain in the neighbourhood of the earth while the lighter ones will reach greater heights. Therefore not only the density but the composition of our atmosphere also changes with the height. At a height of 100 km, where oxygen and nitrogen have sunk already to some 1/100,000 of their original concentration, the concentration of hydrogen is just halved. Although hydrogen represents only 0·01 per cent of the atmosphere at the surface of the earth, it has already become the main component of the atmosphere at the height mentioned. The kinetic reason for this is quite clear: the same kinetic energy, meaning the same temperature, is sufficient to raise a lighter molecule much higher against gravity than a heavier one.

34. THE BROWNIAN MOVEMENT

Let us depart for a while from our gas molecules at different heights in the atmosphere and go over to discuss a seemingly quite different problem. It will become clear that there is a fundamental connexion between the two subjects and we shall gain insight into far-reaching interconnexions which ultimately are going to lead us back to the absolute dimensions of molecules.

The English botanist, *R. Brown* (1773–1858), observed in 1827 that grains of pollen of some flowers never stopped moving in a drop of water when seen under a microscope. The interesting thing was that they simply never became stationary. One might have thought that this was due to local differences in temperature, but the most meticulous care taken to provide uniform temperature did not alter the picture. Neighbouring grains moved in directions absolutely uncorrelated with each other, not as they would, had they all been embedded in the same stream of convected water. Even when every precaution was taken to keep the microscope still and quiet the movement continued unchanged. It was found indeed that if the microscope was deliberately moved about, the drop of water did give rise to currents but that even within these currents neighbouring pollen grains went on moving around in a perfectly unconnected manner. Movement caused by the flow and the strange autonomous movement were easy to distinguish. The strength of the light used to illuminate the microscope stage had no effect either.

This movement was not confined to water. It was observed in all liquids. Its velocity decreased with increasing viscosity of the liquid but that was all; it was violent in ether, and hardly visible but still there, in glycerol. The movement increased with temperature and small grains moved faster than large ones. Only when the grain size was in excess of a critical value was no movement detectable.

The strangest part of this phenomenon was that the movement was incessant. Eventually *Wiener* in 1863 proposed the explanation that the movement was not due to some cause within the particles or outside the drop but was due to movements within the liquid itself, movements characteristic of the liquid state. What could this mean? From all we have already learned it could only mean the thermal movement of the molecules of the liquid—the same movement that was the cause of the diffusion of molecules dissolved within it. Molecules of liquids are in contact and are always acting upon their neighbours and upon molecules or other particles among them, in the course of their disordered thermal motion.

How could the Brownian movement of the pollen grains be envisaged as a consequence of the kinetic movement of the liquid molecules around them? *J. Perrin* (1870–1942), whose beautiful and clearly conceived experiments helped to clear up this whole question (1909–1913) explains the situation by the following comparison. Suppose we are on a cliff at the seashore and gaze out to sea. We do not see the waves at a great distance and it looks as if the sea stands still. But if a small boat is out there, we are able to distinguish the presence of a rough sea from a calm one. We only have to observe whether the boat lies still or whether it is thrown about. It is the same with the molecules. They are too small to enable us to see them directly, even with a microscope. But if a small body passes between the molecules of a liquid they toss it around hither and thither, the more violently the smaller it is. There is a size of these bodies at which they are still small enough to be tossed around by the thermal motion of the liquid molecules but already large enough to be seen under a microscope. A particle jerks to the right or to the left, upwards or downwards according to the direction in which it has just been hit by an energetic liquid molecule. If the particle is too large, the impacts from all directions upon its surface will tend to equalize and the small force which may or may not remain as a resultant will be too weak to give an appreciable impulse to the body. Conversely, the smaller grains may jerk about because at a given moment they are being struck by more molecules in one direction than in another—a small grain is still much larger than a water molecule.

Let us think in more detail about what is going on when the pollen grains dance about in a drop of liquid instead of sinking to the bottom as every well-behaved heavy body is supposed to do. As a matter of fact, pollen is denser than water! An analogous phenomenon occurs when the molecules of a gas fail to surrender to the force of gravity and continue to move around in the atmosphere. Their kinetic energy works to produce disorder in opposition to the external ordering force. According to the kinetic theory there are small molecules, larger molecules and still larger molecules. As a matter of fact, anything that holds together during movement may be regarded as a molecule as far as thermal motion is concerned. It is rather a matter of independently moving particles, than of molecules in the chemical sense. A large body has the same average kinetic energy at a given temperature as a small one and therefore it is going to move about at a smaller speed—kinetic energy being given by $mv^2/2$. A visible speck of dust with mass of say 10^{-9} g will move so slowly that its movement cannot be perceived. We may easily calculate its velocity according to page 151. There we calculated the mean kinetic energy of a molecule at $0°C$ to 5.6×10^{-14} erg. From the above formula for kinetic energy we can calculate that a mass of 10^{-9} g would move at 0.1 mm/sec. In view of the fact that even this slow motion would not be in one direction but would change its direction very frequently, it is clear that we could not observe it. It would be too much to call this speck of dust a gas molecule; it would hardly be able to rise above the ground.

According to what we learned in the last chapter we can estimate the mean height it should be able to attain. The mass of an oxygen molecule is 5.3×10^{-23} g, but our speck is some 2×10^{13} times as heavy as that. The height at which a "gas" consisting of such dust specks would halve its density must therefore be the 2×10^{13}th part of the halving height of the atmosphere, i.e. of 5,000 m. This distance turns out to be as small as 2.5 Å, that is, a negligible fraction of the diameter of the dust particle. In practice, then, the particle will not move about at all, but will remain where it has been placed.

We are able to calculate in the same way that the halving height of particles which are just visible in an ordinary microscope (about 0.3μ diameter) must be some 2×10^{-4} cm, that is, it should be on the limit of observation. Perrin performed his famous experiments on particles of about this critical size but he used a liquid instead of air. Thus, not the whole weight of the particles was tending to draw them downwards, only the difference between their weight and the Archimedean force which tended to make them float. The mastic particles with which Perrin worked had a specific gravity of 1.2 and so, in suspension in water of specific gravity 1.0, the particles behaved as they would have done had they possessed a density of 0.2 in air. Thus in effect he succeeded in making his particles less dense, in the ratio of 0·2 : 1·2, than they really were and gained a factor of 6. We decided above that the halving height would have been 2×10^{-4} cm in air, but now we have extended this to some 10^{-3} cm and are thus well within the range of microscopic observations. Perrin did in fact observe that the population density* of the particles decreased by one-half at this height in an aqueous suspension.

The experiment is so astonishing and so extremely important that it is worth while to stop and look at it in a little more detail. First of all it was necessary for Perrin to obtain small mastic spheres of uniform size. Mastic is a resin soluble in alcohol and can be used to fix bandages on the skin of patients. If one pours this alcoholic solution into excess water one obtains a milky emulsion of mastic in water. The diameter of the particles thrown out of

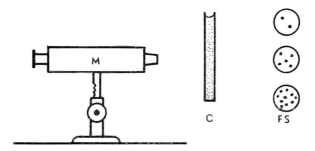

Fig. 70. **Perrin's experiment to determine the vertical distribution of particles in a mastic suspension**

M = microscope
C = cuvette
FS = fields of sight in microscope at different heights.

solution varies of course from below the resolving power of the microscope to visible dimensions. Perrin centrifuged this "milk" at different rotational speeds, that is with different centrifugal forces, and thus separated out first the large particles and then the smaller ones. The fraction he selected he re-suspended in water and centrifuged repeatedly, rejecting smaller and larger "head" and "tail" fractions each time until he saw under the microscope that only spheres of a constant diameter remained. He determined their dimensions carefully by many methods, e.g. by measurement under the microscope; then by counting them as one counts blood cells and evaporating a given volume of the suspension until only the mastic remained. From the weight of this residue, the number of specks and their known density he arrived at exactly the same diameter as by direct observation.

These uniform mastic spheres he now suspended in water again and placed a drop of the emulsion into the vertical chamber of his observing microscope (Fig. 70). On moving the chamber up and down by means of a micrometer screw in front of the objective he observed the density distribution of Fig. 70 among the quickly moving mastic particles. This was exactly what one

---

* i.e. number of particles per unit volume. The density of any single particle remains, of course, constant.

would have expected: the miniature picture of the barometric distribution of gases in the atmosphere. With an ingenious technique of microscopic observation he was able to count the density of his spheres at different heights. For instance, he found at a certain level 100 drops of $0 \cdot 424 \, \mu$ diameter in a cross-section. At a level $30 \, \mu$ above these he counted only 47 drops, going higher by another $30 \, \mu$ their number decreased to 23 and at the next level to 12. Thus a height of $30 \, \mu$ was practically the halving distance. It is in accord with what we said about gases: the density of particles decreases in geometric progression with linearly increasing height. The experimental values $100:47:23:12$ are very near to the halving series $100:50:25:12 \cdot 5$. We should probably have got nearer to it by going somewhat below $30 \, \mu$ per step. But let us say that $30 \, \mu$ is really the "halving height" of these mastic spheres in an emulsion in water. Each sphere has a volume of $40 \times 10^{-15} \, cm^3$ and after subtracting the Archimedean buoyancy force of the water there remains a net force of gravity acting upon a suspended particle of $8 \times 10^{-15} \, g$. The experiment showed therefore, that these particles which were under a force of gravity of $8 \times 10^{-15} \, g$, these "artificial molecules" of this weight, had a halving height of $30 \, \mu$. We know that the halving height of oxygen gas is $5 \, km$; on mountain tops of this height there is only half as much oxygen in the air as at sea-level. Knowing what we do about the halving heights of gases with different molecular weights we are able to calculate the true weight of a molecule of oxygen from these data. The ratio of the halving height of mastic to that of oxygen is $3 \times 10^{-3} : 5 \times 10^5$ that is $1:1 \cdot 67 \times 10^8$. This, then, must be the ratio of the weights of the two and, knowing that a mastic particle weighed $8 \times 10^{-15} \, g$ we find the weight of an oxygen molecule to be $4 \cdot 8 \times 10^{-23} \, g$. The value is not entirely correct; we know, on the basis of X-ray measurements, that the true value should be $5 \cdot 3 \times 10^{-23} \, g$. But the halving height of $30 \, \mu$ was not quite accurate either (*see* above) and the order of magnitude agrees excellently with the expected value. The method was refined later on and yielded even better results.

Perrin repeated his experiments under very varied circumstances. He used other materials instead of mastic and other liquids instead of water. But his results always agreed within the limits of ± 10 per cent. The halving height decreased with heavier particles while it increased with lighter particles, according to the theory. His experiments with glycerol were specially interesting because the particles were less dense than this liquid and thus tended to accumulate at its surface. But just as they did not all sink to the bottom of the cell in the previous experiments with aqueous suspensions, now they did not all rise to the surface. The Brownian movement dispersed them somewhat below it with a distribution which was the inverse of that which we discussed before. In a glycerol medium the deeper we go the fewer particles we find. Effectively, we have a distribution of particles with "negative weight" in the gravitational field. This, of course does not change the calculation since the height and the weight become negative and exactly the same result is obtained as before (Fig. 71). All this surely cannot have been the result of chance. This numerical consistency of the measurements, together with results obtained from other methods based upon the Brownian

movement led W. Ostwald to give up his negative attitude towards the existence of atoms in 1903 and since that time nobody has ever doubted either
their existence or their dimensions.

The agreement between this experiment of Perrin and the kinetic theory
went even further, however. Remember that temperature originates in the
mean kinetic energy of the molecules. This kinetic energy disperses them
against gravity. The greatest height a single molecule of kinetic energy E_{kin}
is able to reach is defined by the maximum potential energy

$$E_{kin} = E_{pot} = mgh_{max} \qquad . \qquad . \qquad . \qquad (34.1)$$

where m is the weight of the particle, g the gravitational acceleration constant and h_{max} the maximum height reached, as in the case of a stone thrown

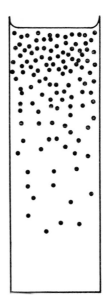

Fig. 71. **Inverse distribution of particles in
glycerine, in which they float, in contrast to
water in which they sink**

"Float" and "sink" refer, of course to the
behaviour of relatively large pieces of material,
unaffected by Brownian motion.

into the air.* Thus the maximum height and with it the whole barometric
distribution is simply extended in proportion to the mean kinetic energy,
that is to the temperature. Perrin repeated his measurements in supercooled
water at −9°C and in water at 60°C. In absolute degrees these temperatures
are 264°K and 333°K, respectively, being in the ratio of 1:1·26. The distribution of heights was in the same ratio to an accuracy of 1 per cent when Perrin
corrected for the density changes due to temperature.

We conclude that the Brownian movement of pollen grains (or other
fine powder particles) fits precisely into the kinetic picture we have

---

* Maybe it is worth while to note that even in a gas containing n particles of a given mean
kinetic energy the maximum height which a particle may reach is limited. It must be so
because evidently no particle can acquire more than the sum of the kinetic energy of all the
other particles. This too can only come about if, by an extremely unlikely distribution of
energies, all other particles remain on the ground with zero kinetic energy, having transmitted all their energy to the one particle. Its kinetic energy will now be nE_{kin} which in its
turn limits the maximum height the particle can attain.

constructed to explain the properties of matter. These small bodies behave exactly as they would if they were molecules some milliard times heavier than the molecules of oxygen. It therefore became possible to calculate the weight of gas molecules as soon as the height distribution of both species had been measured.

Instead of an ordinary microscope the so-called ultra-microscope of Zsigmondy is often used for such measurements. It differs from the ordinary one in that the object is illuminated sideways and no direct light enters the objective lens at all. Only light which has been diffracted upon the particles is able to reach the objective. Spots of light on a dark background are seen through the eyepiece. Particles which are just below the resolving power of the microscope can be made visible in this manner (Fig. 72). A similar phenomenon (Tyndall effect) enables a cloud of cigarette smoke to render a shaft of light visible.

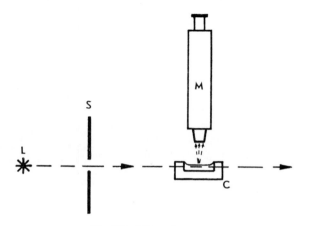

Fig. 72. **Ultramicroscope**

Light from the source L reaches the cuvette C across slit S and would never enter the microscope M if it were not diffracted by small particles suspended in liquid in the hollow cuvette.

35. DIFFUSION AND BROWNIAN MOVEMENT

What was the driving force of the Brownian movement? What made it possible? We say that the particle had to be so small that the impulses which acted on it from different sides did not neutralize each other at a given instant. The movement was determined at each moment by the magnitude and direction of the resultant impulse; *that* was the cause of the remarkable irregularity in the movement of the particle. At the same time the particle had to be large enough to enable us to observe its movements.

But do we really observe the true velocity of the particle? By no means. The path of the particle is completely irregular and hence we do not have the slightest idea where it will be the next moment. We do not know how often and to what extent it has changed its movement and direction in between observations. What we actually observe is only a set of its positions in as short successive time intervals as the human eye is able to register. Whatever happens within these intervals is a secret. The picture we actually see in the ultramicroscope confirms this assertion. A particle seems to be stationary at one moment, to practically disappear in the next, and to reappear again a bit farther on. It is possible to take rapid kinematograms of this motion and to project the picture at reduced speed. We see then, that the shorter the interval of time between two pictures, the more irregularities appear in between two positions which had seemed to be consecutive at longer intervals. Fig. 73 shows the difference between the well-defined continuous movement of,

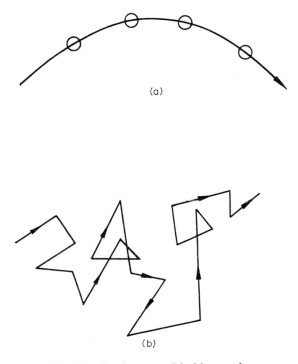

(a)

(b)

Fig. 73. **Continuous and jerking motion**

(*a*) Continuous movement of a flying ball.
(*b*) Jerking movement of a particle in the ultra-microscope (Brownian motion).

say a ball thrown some distance away, and the Brownian movement of a pollen grain. In the former case it is sufficient to observe only some points along the course of the ball. The other points can be fitted in between and are completely predictable by interpolation. The pollen grain, however, moves the more irregularly the shorter the period of observation, the only rule being that on the average it gradually gets farther away from its point of origin. But only on the average. At individual instants it may draw nearer to the origin or even reach it again, improbable as this event may be. We must bear in mind that it is never the true "trajectory" of a grain which we film with our camera, because we are unable to take as many frames per second as would be necessary to do this. What we see is determined by the resolving power in time of our eye, our camera or what ever it may be. If we suppose that the movement of such a particle is absolutely random it is possible to calculate the probable distance travelled from any point of origin in a given time. This calculation is due to *A. Einstein* the famous author of the theory of relativity. His argument runs approximately as follows. Imagine a particle moving around in jerks of some real velocity but which suffers so many collisions that its direction changes from instant to instant. (A refinement allowing for changing velocities gives the same result.) Denote this velocity by v and start from point 0 of Fig. 74 at zero time. After a time interval t_0 the particle suffers its first collision at a distance $\lambda = vt_0$ from the origin. It would be satisfying to know in which direction the particle will proceed from

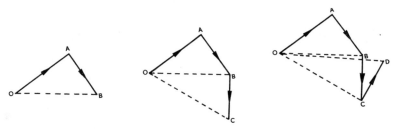

Fig. 74. **Average "random-walk" of a particle that changes the direction of its movement according to pure chance after having traversed a given distance**

$$OB = OA\sqrt{2} = AB\sqrt{2}$$
$$OC = \sqrt{(OB^2 + OA^2)} = \sqrt{(OA^2 + OA^2 + OA^2)} = OA\sqrt{3}$$
$$OD = \sqrt{(OC^2 + OA^2)} = \sqrt{(OA^2 + OA^2 + OA^2 + OA^2)} = OA\sqrt{4}$$

here, but this is a matter of pure chance. Simply because it *is* a matter of chance, we see that on the average it is going to proceed at right angles to its original direction, every direction farther away from this having an equally probable counterpart in the backward direction. This is not to be taken as meaning that any individual particle would be bound to follow this perpendicular direction but that the mean of all occurring directions in a great number of cases will be at a right angle to the initial movement. We repeat the calculation with the mean time interval t_0 in the new direction, obtaining once more a displacement $\lambda = vt_0$. The actual displacement from the origin (d_2) is given by Pythagoras's theorem. It is (Fig. 74):

$$OB = \sqrt{(OA^2 + AB^2)} \qquad . \qquad . \qquad . \quad (35.1)$$

or

$$d_2 = \sqrt{(\lambda^2 + \lambda^2)} = \lambda\sqrt{2} \qquad . \qquad . \qquad . \quad (35.2)$$

After the second collision we again do not know anything about the third direction of the particle except that it will most probably be, on the average, at right-angles (in space, not necessarily in the same plane), to the second displacement. Thus we have to add vectorially BC to OB, and BC on the average will again be λ. This gives us

$$d_3 = \sqrt{(d_2^2 + \lambda^2)} = \sqrt{(2\lambda^2 + \lambda^2)} = \lambda\sqrt{3} \qquad . \qquad . \quad (35.3)$$

The procedure can evidently be continued in this manner giving the most probable displacement after n steps, as

$$d_n = \lambda\sqrt{n} = \lambda\,\frac{\sqrt{t}}{\sqrt{t_0}} \qquad . \qquad . \qquad . \qquad . \qquad (35.4)$$

We must remember that the total time is, of course, $t = nt_0$ and thus $n = t/t_0$. The mean distance between two collisions, λ, and the time between them are governed by the pressure and temperature of the gas, or the density and temperature of the suspension. They are constants under given circumstances, and so the mean displacement from the point of origin becomes proportional to the square root of the time reckoned from the starting time. This is the theory of the "random path."

The result of course only makes sense in a statistical way. Not any one specified particle, but the mean of a great number of particles is going to achieve a mean displacement which will be proportional to the square root of the time. If nothing had stood in their way, the particles would have reached a distance vt equal to $n\lambda$ after this period (t), but owing to their re-peated collisions they only manage to proceed much more slowly and not even for a distance proportional to the time for which they have been moving, but only proportional to its square root (Fig. 74). It would be very improbable but not impossible for some particles to find themselves back for a moment at the point from which they started.

Now the above relationship may be tested by experiments, just as the verti-cal distribution of particles could be tested, using a microscope with a field of view marked by two perpendicular sets of lines or even better by a set of concentric circles. One has to start timing when a particle just traverses the chosen zero point and measure how long it takes to traverse a given distance. The mean result of many readings for each distance confirms the relation that this average distance is proportional to the square root of the time needed to traverse it. Even the scatter of individual points around this mean value fits exactly to a distribution curve of probabilities calculated by Einstein.

36. VISCOSITY OF GASES

Three phenomena are statistically related to the average molecular velocity and the mean free path between collisions of gas molecules. The first is the intrusion of different, or, for all it matters, of similar molecules into each other's domain. This is the phenomenon of diffusion. The kinetic energy determined by the temperature is proportional to v^2. The velocity is thus proportional to the square root of the absolute temperature. Experiments confirm that the velocity of diffusion is in fact governed by this general law. The actual velocity with which one gas spreads out, diffuses, into another (*see* Fig. 63, p. 135) is, however, many orders of magnitude smaller than the molecular velocity of individual molecules. The molecules obviously cannot be moving in unimpeded straight lines: they suffer very frequent collisions with each other.

Instead of considering different kinds of molecules we may take as an example molecules of the same kind but with different amounts of kinetic energy, i.e. at different temperatures. This means that we regard the gas as having at a given time, a sharp difference of temperature along a dividing internal surface. Slow molecules from one side and quick ones from the other will interdiffuse across this boundary in the same way as different kinds of molecules do (*see* Section 29). Kinetic energy is effectively transferred from one half of the gas into the other and we have the phenomenon of heat conduction in a gas. The process tends to establish a Maxwellian distribution of velocities throughout the whole gas. On the other hand if we heat continuously one wall of a container filled with gas while we go on cooling the opposite wall, no homogeneous Maxwellian distribution can be attained and kinetic energy is going to be passed through the gas by virtue of the passage of molecules. This is again the second phenomenon: heat conduction.

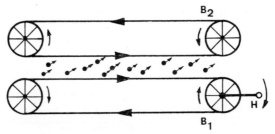

Fig. 75. **Transmission of impulse by a viscous gas**

Belt B_1 is set in motion by turning handle H and the "wind" of the surrounding gas transfers motion to belt B_2.

The third cognate "conductive" phenomenon concerns the conduction of momentum across a gas. Let two conveyor belts be arranged one above the other in an actual experiment with a short distance between them (Fig. 75). The bearings of the conveyor axes must possess extremely low friction to offer small resistance in rotation. If one of the belts is set in motion, the other will be gradually set in motion too by the "wind" generated by the moving belt. Momentum will have been transferred from one belt to the other through the gas. Therefore more energy will have been consumed in moving the first belt than if there had been no gas at all in the system. The internal friction of the gas has acted as a brake on the first moving belt. This internal friction of a gas (or liquid) is called its viscosity. It is defined as the force necessary to move 1 cm² of the belt at a distance of 1 cm from the second one with a relative velocity of 1 cm/sec.

The kinetic cause of this phenomenon is the following. Molecules impinging upon the first belt acquire in addition to their fundamentally irregular velocity a regular component in the direction in which the first belt is moving. (An experiment may be easily performed with a light ball dropped upon a moving conveyor). Molecules from this layer adjacent to the belt diffuse into the layer above them and on the average transmit this directed impulse to the molecules within the second layer. Then molecules from this second layer diffuse into the third, transmitting the impulse in their turn and so on, until finally molecules arrive at the surface of the second belt. The surplus momentum which still persists in the direction of movement of the first belt is going to be transmitted to the second belt and gradually set it moving. We have referred to layers, though no clear-cut layers exist as such in a gas. To make matters easier, we may define as a layer of gas that distance perpendicular to the belts which is equal to the mean free path of a molecule, i.e. to the average distance between collisions, or the distance involved in transference of momentum from one molecule to another.

In principle it would make no difference if we brought the second belt to a standstill with a helical spring and measured the force produced by the moving, viscous gas. The velocity within the gas would in this case decrease linearly from layer to layer, the molecules in contact with the moving belt having attained its average velocity while the molecules at the belt which is stationary move with almost zero velocity, i.e. are also virtually stationary, on the average.

The momentum transmitted by any molecule into the next layer is proportional to the difference of velocity between the two layers. Hence it makes no difference whether the gas between the belt is dense or not. The denser, it is, the more the mean free path will decrease. The layers will become thinner and the difference of velocity between adjacent layers will be smaller. But there will be proportionally more layers with this smaller difference of velocity and containing more molecules each between the belts. The final result will be the same as if the mean free path had extended right through the gas and each molecule had transferred its impulse at one time instead of relaying it via many layers of molecules. The viscosity is therefore seen to be independent of the density (of the pressure) of a gas. It is a paradoxical result but experiments exactly confirm it. J. Maxwell to whom the theory is due, recalculated the result when he first obtained it because it seemed too strange to be true!

To summarize, in a dense gas many molecules are going to impinge upon unit surface area of the stationary belt and they will all transmit some momentum to it. But the momentum transmitted by each molecule will be smaller at a higher density. This is because the molecules will have come through a distance corresponding to a shorter mean free path and have thus come from a layer, the velocity of which differs less from that of the stationary belt than it would have done had the molecule travelled a greater distance in a less dense gas. An exact theory of this process is hardly any more complicated than the kinetic calculation of the gas pressure and it may be simplified as easily as we have simplified the latter.

The result is that the viscosity, which can be regarded here as the momentum transferred, is proportional both to the mean free path and to the density. It is also proportional to the mass of the molecules, because momentum is the product mv. Heavier molecules therefore acquire more momentum than light ones if they acquire the same velocity from the moving belt. This simply implies that not the population density of the molecules but the total mass density of the gas appears in the final formula. Once we have measured the viscosity and the density of the gas, it becomes possible to calculate the mean free path of the gas molecules in gas of given density. We have thus learned something new about gases: besides knowing the weight, diameter, velocity and the average distance between gas molecules, we now know how far they move on the average between successive collisions. This distance is, at atmospheric pressure, about 0·1 μ (1,000 Å), as compared with a typical intermolecular distance of 30 Å and a diameter of, say, 3 Å. Moving as they do with a velocity of about 500 m/sec, they collide on the average some 5,000 million times every second (5×10^9). The same mean free path results if the calculation is based on the corresponding theories of diffusion or heat conduction.

It must be understood, naturally, that this is only a mean value and that much shorter and much longer paths occur within every gas but with decreasing probabilities. It must be appreciated too that the mean free path is not the same for all gases, in contrast to the mean distance between molecules which, according to Avogadro, is the same for different gases at corresponding pressures and temperatures.

37. MEAN FREE PATH AND CROSS-SECTION OF A MOLECULE

What makes the mean free paths of molecules different in different gases? In order to understand the explanation, we must visualize what is really going on within the body of the gas. Molecules must collide with each other during their movement because they are not infinitely thin and so will be in each other's way during their flight. If they were dimensionless mathematical points they would evidently then, and only then, be able to by-pass each other without ever colliding. Thus there ought to be some connexion between the mean free path at a given pressure and the cross-sectional area of individual molecules which are moving around. The finite cross-sections of the molecules cause them to collide, and each kind of molecule will naturally have its own characteristic cross-section.

Two molecules can never approach each other more closely than is required just to produce contact, that is, their centres must be at least as far from each other as the sum of the radii.* If we are dealing with molecules of the same kind this sum is equal to a molecular diameter. Imagine for a moment that all the molecules within a gas remain stationary except for a single one which flies around among them. Let the number of molecules per cm³ be N, the molecular radius r, and the velocity of the moving molecule v. After each collision the latter will change its direction but retain its velocity under these simplified conditions. If it suffers n collisions in a second then its mean free path is $\lambda = v/n$. This is the quantity we dealt with above.

Once we succeed in determining n we have evidently determined λ too, at a known velocity. We again profit by the fact that the situation of the

---

* It is assumed for the time being that atoms and molecules *can* be assigned precise diameters and can be regarded as spherical.

immobile molecules and thus the sites of collision and the total path of the moving molecule are entirely fortuitous. We cannot say anything specific about them and therefore have to reckon with probabilities (Fig. 76).

As often as a stationary molecule comes within the volume of the imaginary cylinder of diameter $2r$ around the path of the moving molecule there will be a collision. Since the stationary molecules are distributed according to chance, i.e. randomly, we may assume that they have a uniform distribution on the average and that any volume V will contain nV of the stationary molecules.

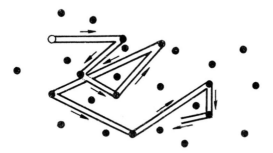

Fig. 76. **The movement of a molecule among stationary molecules**
There is a collision between molecules as often as the cylinder representing the path of the molecule touches one of the stationary molecules. The mean free path is the average distance between collisions.

The mobile molecule moves around with a velocity v and hence the length of the cylinder around its path (though broken at many points) is going to be traversed at the rate of v cm per second. Therefore the volume of this cylinder, swept through per second, is $4r^2\pi v$. The number of stationary molecules in this volume will be $n4r^2\pi v$ and accordingly there will occur this number of collisions per second. Hence the free mean path is simply

$$\lambda = \frac{v}{n4r^2\pi v} = \frac{1}{n4r^2\pi} \qquad . \qquad . \qquad . \quad (37.1)$$

The correction which becomes necessary if we consider that the molecules are not, in fact, all stationary except one is not very important. It is evident that a moving molecule is a greater obstacle in the way of anything than one which stands still because it occupies, as it were, a larger space during the time the colliding partners are near each other. If a molecule does not move it blocks only its own cross-section but if it does move it blocks a cross-section corresponding to the time the two molecules take just to by-pass each other (Fig. 77). It can be shown that on the average the cross-section seems to be increased by this movement by a factor of $\sqrt{2}$ and hence the final formula connecting the cross-section ($4r^2\pi$) of the molecules and their mean free path will become

$$\lambda = \frac{1}{\sqrt{2} \times 4r^2\pi n} \qquad . \qquad . \qquad . \qquad . \quad (37.2)$$

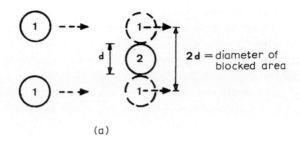

(a)

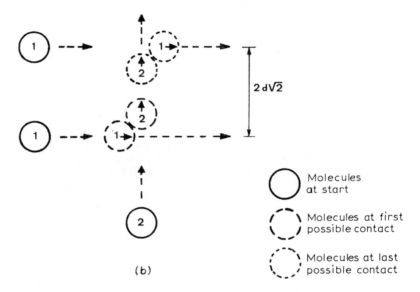

(b)

Fig. 77. **A moving molecule finds more of its path blocked by other moving molecules than by ones which are stationary**

Its mean free path is therefore shorter if the other molecules move than it is if they are stationary.

Suppose all molecules are spheres of diameter d.

(a) Molecule 2 is stationary and blocks an area of diameter $2d$ for all molecules 1 moving towards it in particular parallel directions. A "blocked area" means that the centre of all approaching molecules must be outside the area if they are not going to collide with the stationary molecule.

(b) Molecule 2 moves in a direction perpendicular to that of molecules 1 coming towards it with the same velocity. The first possible contact occurs at the moment when a molecule 1 would just touch 2 at its south-west point and the last possible contact takes place when the touching occurs at the north-east point. The diameter of the "blocked area" for centres of molecules 1 has increased to $2d\sqrt{2}$ (Pythagoras's theorem).

We may also reverse the position. If we have determined the mean free path by measuring, say, the viscosity and density of a gas we are able to calculate the sum total of molecular cross-sections in unit volume of this gas:

$$nr^2\pi = \frac{1}{4\sqrt{2}\lambda} \qquad . \qquad . \qquad . \qquad . \qquad (37.3)$$

Thus we are able to calculate from the approximate mean free path of an oxygen molecule at normal temperature and pressure ($\lambda = 10^{-5}$ cm) that the sum total of the cross-sections of the oxygen molecules in a cm³ is 18,000 cm². Assuming for simplicity that the molecular cross-section is circular (which is surely not true for a molecule O_2!) and that the molecules are arranged in a close hexagonal pattern in a plane, we can calculate that they occupy an area of some 20,000 cm² because of the empty interstices between them. One *mole* of the gas (22·4 litres would occupy in this way 22·4 × 10³ as much area, that is 45,000 m²!

Let us compare this crude result with what we measured as the molar area of fatty acids on the surface of water (p. 99). This amounted to 150,000 m², we have therefore arrived at the same order of magnitude for molar cross-sections of different molecules, in spite of the fact that we have only performed an approximate calculation. Historically, Loschmidt followed this complicated reasoning using mean free paths derived from viscosity measurements. We chose the simpler and more direct experiment with the drop of oil. Nevertheless the result of the historical approach is satisfactory.

38. LOSCHMIDT'S DERIVATION OF N_L

We have derived the sum total of the cross-sections of the molecules in a given volume of gas. If by some means we were able to measure the sum total of their volumes we should be able to calculate the radius of an individual molecule. The calculation is straightforward, as we shall see.

Now how can we obtain the sum total of molecular volumes? There is more than one way of doing this but we shall proceed using the simplest assumptions. Let us suppose that condensed gases at very low temperatures are practically close-packed. For oxygen the experimental density at its boiling-point, $-182°C$, is 1·12 g/cm³ but oxygen contracts still further with falling temperature. After freezing at $-227°C$ its density has already reached 1·27 g/cm³. Below this temperature, at $-252·5°C$ (20° Abs.), the density has increased yet again to 1·43 g/cm³. Let us base our calculation on this last density, hoping that it will not increase much more down to the absolute zero of temperature. The total volume of 32 g, that is of one mole of oxygen of this density is evidently 32/1·43 cm³ that is 22·3 cm³. This same amount of oxygen had a sum total of cross-sectional area amounting to 45,000 m² according to the last section. The volume divided by the base area will give the thickness of this monomolecular oxygen layer: 22·3/(4·5 × 10⁸) cm that is 5·2 × 10⁻⁸ cm or 5·2 Å. Once more we have reached the molecular dimensions with which we are acquainted. Now how many molecules make a mole? One molecule of diameter 5·2 Å has the cross-sectional area πr^2 = 21 Å² and we have seen that the sum total of molecular cross-sections in 1 cm³ of oxygen amounts to 18,000 cm². The corresponding area for one mole of gas, is therefore $18,000 \times 22,400 = 4 \times 10^8$ cm². Dividing this by the cross-sectional area of a single molecule which we have just calculated we obtain $1·9 \times 10^{23}$ molecules per mole. Admittedly, this is only one-third of the true value as we know it today, but its order of magnitude is correct. As oxygen is a diatomic molecule our assumption as to its sperical shape is certainly unfounded and we really do not know to what extent the molecules at 20° Abs. actually touch each other, to say nothing about the question of whether a true touching of molecules can be defined at all. On calculating the area from the mean free path we operated with the distance of interaction between fast-moving molecules, whereas we calculated the total volume from the stable arrangement of molecules moving (swinging around equilibrium positions) with much smaller kinetic energy. Furthermore we do not know yet whether they are "rigid" or not.

Be it as it may, this was the first method by which man ever arrived at true molecular dimensions; it is the method of Loschmidt. Later on he used more refined corrections throughout his calculation and arrived at values which approximate the modern value more closely. This victory over the unknown certainly deserves that we should have glanced at his line of argument even if we have only done so in a rather sweeping manner.

39. MOLECULAR MOVEMENT A *PERPETUUM MOBILE*?

Once again we are at the end of a series of important questions. It is proper that we should review our results. In the first part of our treatise we asked whether matter can be subdivided infinitely or not. We found weighty arguments for assuming that it cannot, and a further series of investigations even revealed to us the true dimensions of these smallest particles of the elements, the atoms.

We then proceeded to ask whether these atoms and molecules are supposed to stand still in so far as we cannot observe their macroscopic motion or whether they have an invisible motion of their own. We first made it seem only plausible that the latter case was true but presently proceeded to give exact proofs of this assumption. We have even determined directly the molecular velocities, have shown how the pressure of gases is caused by this motion, and have found that the mean kinetic energy is a measure of what we know as the absolute temperature. The Brownian movement has shown us the transition between the movement of invisible atoms and that of minute particles just visible in the ultramicroscope. The similarity between the height distribution of such minute particles in the field of gravity and the same distribution of gas molecules in the atmosphere allowed us once more to calculate the weight of individual atoms. Finally, we discussed phenomena which could only be understood by assuming that the molecules of gases collide very often, this assumption in its turn implying that the molecules must have a finite cross-section. The concept of molecular collisions explains why diffusion, conduction of heat and of momentum proceed relatively slowly. Finally, the relation between the sum total of cross-sections and volumes in the same amount of gas gave still another method, historically the first, of calculating the dimensions of the molecules.

Before embarking upon the relations of electricity, light and magnetism to matter, let us glance once more at the dimensions of the world of atoms. The molecules of simple gases have diameters of a few Ångström units, more complicated ones are correspondingly larger. At room temperature and atmospheric pressure these molecules are at a mean distance of a few times 10 Å from each other and collide on the average after having flown a mean free path some 100 times their diameter. They move around at a speed of a bullet, the heavier ones more slowly than the light ones.

Remember our enlarged picture on page 133. The molecules there were represented by dots the size of pinheads, on the average 1 cm apart and moving on the average 30 cm between collisions. If we retain this number of collisions per second we should ascribe to the molecules an increased velocity amounting to some 4 million km/sec.

But it would be highly undesirable to leave these problems without inquiring into the relation between molecular motion and a *perpetuum mobile*. After all, we have supposed that molecules and atoms at finite temperature keep moving all the time and what is this movement if not perpetual motion in itself? And was it not just the existence of a *perpetuum mobile* that was forbidden by the first law of thermodynamics?

We must be quite clear about this point. The first law of thermodynamics, the law of conservation of energy, sums up a long series of negative experiments and proclaims that it is impossible to create movement, electrical energy, heat, light or any other kind of energy without simultaneously sacrificing an equivalent amount of energy of a different kind. Nothing prevents an array of molecules from moving about with undiminished kinetic energy so long as they have not performed mechanical work by so doing or have not given rise to any other type of energy by their movement. As we saw earlier, as soon as a gas moves a piston in a cylinder it has to cool down, it has to lose some of its kinetic energy if no heat—that is no new kinetic energy—is allowed to feed into it through the walls of its container. The first law demands that energy should neither be created nor destroyed. Heat is conserved as kinetic energy within matter at constant temperature so long as it is not used to create another kind of energy. Everything is all right so far.

Another long series of negative experiments has shown that it is impossible to construct a device which enables us to gain work by simply extracting energy from a body of constant temperature and cooling it down, without changing anything else. In the example above we changed the volume of the gas! Such a device would have respected the first law, and work would be derived from heat.

But it is impossible to derive work from heat alone without changing another parameter if there is no difference in temperature at hand. This is one way of stating the second law of thermodynamics. The device which is hereby proclaimed non-existent is a *perpetuum mobile* of the second type, one which would yield work by changing nothing in the world except the lowering of the temperature of some body. Just imagine how convenient it would be to extract work from the sea merely by cooling it a little! We have seen that the intrinsic reason why this will not occur is based on the enormously small probability of molecules of different velocities separating themselves spontaneously.

The Brownian movement makes it easy to understand how this comes about. Imagine a microscopic living being who has just decided to build himself a house and to make use of the Brownian movement to this end. He is sitting on a ridge within the fluid and just waits for the kinetic movement to bring up from below those small particles, say pollen grains, that he has envisaged as bricks for his building. Eventually he just manages to catch one and is glad to have it at his feet. But by the time a second brick appears, this first one is far, far away: it simply had to continue its irregular movement.

A still smaller being would experience a similar fate if he wished to entrust his locomotion to the kinetic movement. Certainly he would be tossed around and indeed might reach great distances. But this would not be a journey towards a preset goal since it is transportation by wild chance. The probability that he would arrive at any given point during finite time is so small that it is practically zero and even if he did by some chance reach his goal he would immediately be tossed on and on, Heaven knows where. If he really wanted to reach a given place, he would have to expend energy, to "swim" by himself. Similarly his colleague the bricklayer would have had to expend energy if he really wanted to accumulate and fix his particles into a structure.

The molecular movement is truly perpetual motion in the sense that it goes on for ever without friction, because the latter is the process which converts macroscopic movement into the irregular movement of molecules, into heat. There is no friction, no second-order heat, into which molecular motion itself can be transformed! But this perpetual motion cannot be used to perform any useful work and will not help to achieve any premeditated aim because its very essence is irregularity, the totally irregular assembly of molecular positions and velocities.

If we say that the water in the upper part of a glass of water at 50°C is not going to boil at the expense of another part freezing, this is the same as saying that the faster molecules from among so-and-so many billions are not going to separate spontaneously from their slower neighbours. Not that this is impossible, but it is so extremely improbable that it is practically certain never to occur. Nor will all the molecules in a room convene a meeting in one of the corners at a given date. Extreme improbability and the impossible are practically the same for macroscopic dimensions.

Part 4

The Atoms of Electricity, Light and Magnetism

40. IONS

In Section 1, we have already used the word electrolysis in connexion with the electrolytic decomposition of water. The fact that such a decomposition takes place is commonly known and so we did not go into details at that time. Then again, in the chapter dealing with acids, bases and salts we went into some detail as far as the separate existence and movement of ions in solution was concerned. Remember for example that the blue copper ion and the yellow chromate ion migrated in opposite directions in an electric field (p. 64). Now, however, we shall discuss the relationship of matter and electricity in more detail and shall be interested to see how far electricity exists separately from matter.

The fact that a great number of compounds are decomposed by an electric current was discovered by the great Englishman, *M. Faraday* (1791–1867), and his then employer, *H. Davy* (1778–1829), whom he served as a laboratory assistant and servant in his younger years. While Davy prepared, for example, the free alkali metals (1807) and alkaline-earth metals from their molten compounds, Faraday made quantitative and systematic examinations of the ratio of electricity consumed to matter transformed.

Let us recapitulate somewhat. We know already that certain melts and and solutions conduct electricity (salts, acids, bases) whereas others do not (sulphur, iodine, sugar, oil). Metals of course conduct electricity, and so do some compounds and elements called semi-conductors (graphite, silicon, cuprous oxide, germanium, grey selenium), though to a smaller extent. There is, however, an important difference between the salt-acid-base and the metal-semiconductor groups. Whereas electricity passes through the latter without displacing matter and without any particular change occurring at the places where the current enters or leaves (electrodes), the salt-acid-base group always shows movement of matter on passage of current (e.g. the copper-chromate system, page 64) and chemical reaction, mostly decomposition, at the electrodes (e.g. with acidified water). As far as heat evolution is concerned, both groups evolve heat on passage of current.

We have explained the process of material movement by the assumption of Arrhenius that some atoms or radicals carry electric charges with them and therefore move in an electric field. We have called them ions (Greek ἴων = migrant). Positive ions moving with the positive current are named cations, negative ones anions. Insulators do not conduct electricity and do not seem to have any mobile charged particles within them, whereas metals and

173

semi-conductors must contain something which has electric charge but no mass and no chemical matter attached to it.

We have seen that those solutions which conduct a current deviate from Raoult's law concerning the change in vapour pressure and in melting- and boiling-point. They deviate in the sense that they would do if they contained a greater number of molecules per unit volume than we suppose. The deviations occur because these phenomena are not connected with the number of *molecules*, as originally supposed, but, as W. Ostwald has pointed out, with the number of independently moving *particles* within the solution. If a molecule is split into ions, it is the number of these that counts. We can hazard, on this basis, a rough kinetic picture of vapour pressure lowering. As long as the solvent is pure its surface contains only its own molecules and by the laws of chance it will happen that one or another of these will accumulate enough kinetic energy to break away from the attractive force of its partners and escape into the gas phase. On the other hand such vaporized molecules have a chance, proportional to their concentration in the gas phase, of striking the liquid surface again and being recaptured by it. Dynamic equilibrium will be established when the number of molecules breaking away and the number recaptured from the same surface are equal. It is evident that the higher the temperature, the more molecules will acquire the energy necessary for evaporation and therefore vapour pressure should increase with the temperature. This effect is observed experimentally.

If the liquid contains dissolved particles, these will occupy a part of its surface, to a first approximation in proportion to their number and fewer solvent molecules will be able to pass into the gas phase in the same time as before. Hence the vapour pressure of the solvent should decrease, as it is actually found to do. Whether the solute molecules are able to leave the solution (e.g. alcohol from water) or not (sugar) depends upon the forces of cohesion between them and the solvent molecules. According to this the solute is or is not going to build up a vapour pressure of its own above the solution. But this has nothing to do with the vapour pressure of the solvent, which has decreased in any case. Thus we begin to understand why an ionized molecule has a greater effect upon the vapour pressure of a solution than has an undissociated molecule.

Melting points, etc., are in causal relationship to vapour pressure, correlated through the laws of thermodynamics, but as always, it is not difficult to visualize the kinetic aspect of such a process. A melting crystal must be in kinetic equilibrium with its parent liquid, just as the vapour above a solution. The number of molecules leaving a crystal surface and passing into the melt and the number being built back into the crystal must be the same at the melting point, at the equilibrium temperature between crystal and melt. If the melt, however, contains other species of molecules (solutes), the crystal will lose more of its own molecules than it receives back in exchange because some of the molecules impinging on its surface are "strangers", solute molecules, which cannot be built into the crystal structure. Therefore the crystal melts—at a temperature above that at which it would have melted in the absence of solutes. Accordingly the system must cool down until the number of molecules leaving the crystal has decreased and become equal to the number of molecules *of its own species* impinging upon, and being built into, its surface in the presence of impeding solute molecules.

Let us return to our ions. It was very difficult in the time of Arrhenius to understand that individual particles of *sodium* should be moving about in a solution of common salt. Sodium reacts with water! It had to be made clear and understood that a sodium atom and a sodium ion are by no means the same thing. The presence of an electric charge upon a group or an atom

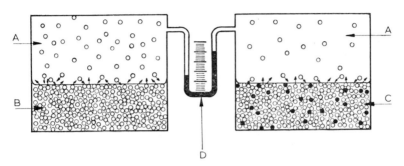

Fig. 78. **The vapour pressure of a solution is smaller than that of the solvent**

Equilibrium vapour pressure means that the same number of molecules leave the liquid surface in a given time as recondense on it. If part of the liquid surface is occupied by non-volatile solute molecules, the number of evaporating solute molecules decreases and the gas-phase concentration, that is the pressure, must decrease until the number of molecules recondensing has also decreased to the same extent. In the figure the decrease amounts to 25 per cent.

A = solvent vapour
B = pure liquid solvent
C = liquid solution of non-volatile solute
D = differential manometer
○ = solvent molecule
● = solute molecule

matters a great deal. Indeed, the energetic reaction of sodium with water is in part the transformation of a sodium atom into the charged sodium ion Na^+. Today we have grown accustomed to this difference between neutral particles and charged ones but at that time the idea was very unpalatable indeed. A change in charge involves a change in most of the properties of a particle. Copper is a red-brown metal; with one positive charge the copper atom becomes the colourless cuprous ion whereas with two positive charges it changes into the well-known blue cupric ion.

The Coulombic force between ions in solution decreases slowly, only inversely with the square of the distance between the ions, similarly to gravitation. Hence it is not adequate to consider ions in solution as being kinetically independent of their neighbours, except at extreme dilutions when the interionic distances are relatively very great. Neighbouring ions of opposite sign attract, those of the same sign repel each other and so it is certain that on the average each negative ion will have more positive ions in its neighbourhood than negative ones, and vice versa. This leads into the rather complicated modern theory of ionic solutions which is certainly beyond the boundaries of our present discussion.

41. THE DISSOCIATION EQUILIBRIUM. THE LAW OF MASS ACTION

In those cases when the deviation from Raoult's law for an electrolyte solution was less than would have been expected according to Arrhenius's hypothesis of complete dissociation into ions, the classical assumption has been that only a fraction of the molecules is dissociated in solution. This now seems to be more or less true for weak acids and bases, e.g. acetic acid or ammonia in water. The vapour pressure depression of their concentrated solutions is not much more than would be expected from their molecular weights, i.e. there is but little dissociation. On diluting the solutions, the depression of course also decreases, but more slowly than does the number of molecules per unit volume. At extreme dilutions the depression is very small of course, but it has decreased so slowly that it now amounts to double the depression calculated for the given number of undissociated molecules per volume. It seems that the molecules have by now completely dissociated into ions.

This was the original assumption, and for weak acids and bases it is still held to be true. It is sometimes possible to follow the change of visible colour or of the equally real, but invisible ultra-violet or infra-red "colour" of undissociated molecules and of their ions throughout this process of dilution. The "colour" of the molecules fades away as that of the ions intensifies.

The electrical conductivity of electrolyte solutions also undergoes analogous changes, but in order to interpret them systematically we have to compare the conductivity of the same amount of electrolyte at different dilutions. In Fig. 79 are illustrated two metal plates as electrodes in a trough. Let us pour a concentrated solution between them, so that it covers only their lowest part, apply a voltage and observe the current. For practical reasons concerned with decomposition at the electrodes it is necessary to work with alternating current, so that anodic and cathodic processes quickly cancel. Now pour some water into the trough and mix the solution: the current will increase. It goes on increasing with further dilution until gradually a limiting conductivity is reached. It seems reasonable to assume that this is the dilution at which practically all the molecules have undergone dissociation: there are as many ions in between the electrodes as can possibly be formed by the given quantity of acid (Ostwald, 1889).

The process of dissociation is itself, of course, a kinetic phenomenon. It does not happen that a molecule once dissociated remains so for ever while those not dissociated at a given moment, have to wait for their turn on further dilution. In the case of acetic acid solutions a hydrogen ion may encounter an acetate ion irrespective of whether it was its original partner or not and may then reunite to form a molecule while at the same time other molecules undergo dissociation. Equilibrium will be attained when the same number of molecules are dissociated as are formed in unit time from ions.

Now the probability of a given number of molecules dissociating per unit volume and time is evidently proportional to their concentration. If there are twice as many, twice as many will dissociate in the course of their intra-molecular vibrations. But the probability of a hydrogen ion meeting an acetate ion is proportional to the number of hydrogen ions there are in the volume and to the number of acetate ions there are to meet them! The association process is thus proportional to the concentrations of both types of

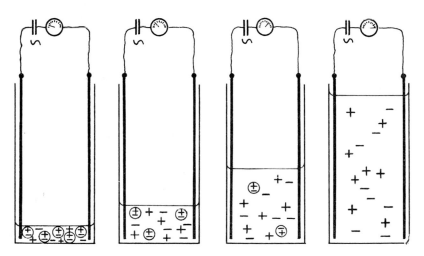

Fig. 79. **The molar conductivity of weakly dissociating electrolytes increases with dilution**

A tall, parallel-walled, vessel contains two electrodes connected to an a.c. source and ammeter. It contains a rather concentrated solution of, say, a weak acid. Most of the molecules are associated: ⊕. On gradual dilution more and more molecules dissociate into ions, the solution gradually fills the space between the electrodes. The current accordingly increases; its limiting value is attained when *no molecules are left*, only ions.

ion and hence to the product of their concentrations. Let us denote the probability of dissociation by P_d and that of association by P_a and the constants of proportionality, the "velocity constants" of the events, by k_d and k_a respectively, so that we have (concn. $\equiv$ concentration)

$$P_d = k_d \times \text{concn.}_{\text{acid}} \qquad . \qquad . \qquad . \quad (41.1)$$

$$P_a = k_a \times \text{concn.}_{\text{H}^+} \times \text{concn.}_{\text{acetate}^-} \qquad . \qquad . \quad (41.2)$$

At equilibrium the two probabilities must be equal, the same number of molecules decomposing as are formed:

$$k_d \times \text{concn.}_{\text{acid}} = k_a \times \text{concn.}_{\text{H}^+} \times \text{concn.}_{\text{acetate}^-} \quad . \quad (41.3)$$

Since Ostwald's time it has become customary to denote concentration of a species, usually in g-ions or g-moles per litre, simply by writing its chemical symbol in square brackets: concentration of hydrogen ions becomes $[\text{H}^+]$ and concentration of acetic acid $[\text{CH}_3\text{COOH}]$, etc. This makes the writing of such equations much easier. Using these symbols and after a trivial transformation we arrive at

$$\frac{k_d}{k_a} = \frac{[H^+].[CH_3COO^-]}{[CH_3COOH]} \qquad . \qquad . \qquad . \quad (41.4)$$

The ratio of the two constants on the left is naturally a constant as well. It is the constant characterizing the dissociation equilibrium, the equilibrium represented by the chemical equation:

$$CH_3COOH = H^+ + CH_3COO^- \quad . \qquad . \qquad . \quad (41.5)$$

All this is true, not only for electrolytic dissociation, but for all chemical reactions as well. They may always be imagined as being reversible and the position of the equilibrium will be always at the point where the reaction and its reverse proceed with equal speed. The velocity of either process will be proportional to the probability of the partners involved meeting each other, that is, to their respective concentrations. The velocity of one process is thus proportional to the product of concentrations of the disappearing species and that of the reverse reaction to the product of the concentrations of the species formed. Whenever more than one type of molecule or ion is involved in either reaction, it is assumed for simplicity that the appropriate number of species must actually meet in one place. For instance let us consider the complete dissociation of dibasic sulphuric acid:

$$H_2SO_4 = 2H^+ + SO_4^{--} \quad \text{or} \quad = \quad H^+ + H^+ + SO_4^{--} \quad . \quad (41.6)$$

Hence the reverse reaction, insofar as it takes place in a single step, is only possible if we have a triple collision* of two H^+ ions with an SO_4^{--} ion, the relevant probability becoming

$$P_a = k_a\,[H^+][H^+][SO_4^{--}] \quad \text{or} \quad k_a[H^+]^2[SO_4^{--}] \qquad . \quad (41.9)$$

At equilibrium we must have

$$\frac{[H^+]^2[SO_4^{--}]}{[H_2SO_4]} = \frac{k_d}{k_a} = K \qquad . \qquad . \qquad . \quad (41.10)$$

The species which occur *n*-fold in the equation have their concentrations raised to the *n*-th power in the equilibrium equation because they occur *n* times as a product (or reactant).

This kinetic derivation of chemical equilibria is due to two Swedish chemists, *C. M. Guldberg* and *L. P. Waage* (1867) and is valid throughout chemistry, with the restriction that it is not exact for strong solutions of electrostatically interacting electrolytes. It is called the law of mass action.

---

* The need for multiple collisions is obviated if we assume that dissociation occurs in more than one step:

$$H_2SO_4 = H^+ + HSO_4^- \qquad\qquad (41.7)$$

$$HSO_4^- = H^+ + SO_4^{--} \qquad\qquad (41.8)$$

Similarly for the association process. Writing the mass-action equation for both processes and eliminating (HSO_4^-) yields eqn. 41.10 again.

It is so important that we must look somewhat deeper into its implications. It may be restated by saying that the product of the concentrations involved in the forward reaction is proportional to the product of the concentrations involved in the back reaction. This means for example, for the dissociation of acetic acid, that the concentration of the undissociated molecules must rise if either the concentration of H^+ or of CH_3COO^- is by some means or other increased in a given solution. Evidently this is so: the chance of ions meeting each other has increased. But how can this be realized in practice? Very easily indeed. We may increase the concentration of hydrogen ions by adding a strong acid, or that of acetate ions by adding sodium acetate. In this special case all the components are without visible colour, but they may differ in other properties, e.g. their spectra in the ultraviolet are different. and the resulting increase of undissociated acetic acid molecules can thus be ascertained.

Concomitantly the number of acetate ions will decrease if acid is added, while if we add acetate, the concentration of free hydrogen ions, which give rise to the acidity of our solution, must necessarily decrease because they combine with acetate ions to form undissociated acid molecules. This can be shown by using an appropriate colour indicator for acidity. For instance, the colour of methyl orange changes from orange towards yellow if sodium acetate is added to an acetic acid solution.

If the salts are coloured and the colour of the undissociated molecule differs from that of the ion as in the case of $CuBr_2$, the process can be observed directly. $CuBr_2$ is brown, Cu^{++}, as we know, is blue. A concentrated solution of this salt is brown at first and changes gradually to blue on dilution. But addition of potassium bromide to this dilute solution increases the concentration of Br^- and thus reverses the dissociation: the colour begins to turn brown again.

Even so simple a process as the solubility of compounds which undergo electrolytic dissociation is governed by the law of mass action. The easiest way to see how this comes about is to assume that an ionic crystal is in dynamic equilibrium with its undissociated molecules within a solution, e.g. a crystal of sodium chloride is in equilibrium with the (extremely few) undissociated NaCl units.* The number of these is governed by the equilibrium:

$$NaCl = Na^+ + Cl^- \qquad . \qquad . \qquad . \ (41.11)$$

that is by

$$K_{diss} = \frac{[Na^+][Cl^-]}{[NaCl]} \qquad . \qquad . \qquad . \ (41.12)$$

The concentration of undissociated molecules must therefore increase whenever the concentration of Na^+ or Cl^- is increased in the solution. But if the original solution was saturated with respect to NaCl this means that any increase in the concentration of NaCl molecules is bound to oppose

---

* These are to be regarded as dipolar ion-pairs Na^+Cl^-, rather than neutral molecules, NaCl.

the "dissolving tendency" of the crystals and the excess salt in solution is going to crystallize on to the existing crystals until the original concentration of undissociated molecules in the solution is restored. For any given temperature and solvent this equilibrium concentration of NaCl units is constant. Hence we may introduce a new constant, S, instead of the product K_{diss} [NaCl] and so we have

$$S = [Na^+][Cl^-] \qquad . \qquad . \qquad . \qquad . \qquad (41.13)$$

This new, combined constant is called the solubility product of the ions in question and in reality it is this product that is important in cases where the same type of ion may come from different sources into a solution.

For example, we may precipitate common salt from its concentrated solution either by adding concentrated hydrochloric acid or concentrated sodium hydroxide. In one case we have raised $[Cl^-]$, in the other $[Na^+]$ and therefore as much of the opposite ion is bound to disappear as will maintain the resultant ionic product $[Na^+][Cl^-]$ at its original value. If the original solubility, that is the solubility product, was low such a procedure often serves to precipitate a salt quantitatively from a solution. As an example we may note that lead sulphate is a very slightly soluble salt, it dissolves at room temperature to the extent of 0·04 g/l. that is to a $1·3 \times 10^{-4}$ molar solution. (Its molecular weight is 304.) The product of ionic concentrations in this concentrated solution, the solubility product, is thus $1·3 \times 10^{-4} \times 1·3 \times 10^{-4} = 1·7 \times 10^{-8}$. If we add sufficient free sulphuric acid to the solution, so that the latter becomes N/1 with respect to sulphate ion, the constancy of S demands that the Pb^{2+} concentration fall to $1·7 \times 10^{-8}$, that is $5·2 \times 10^{-6}$ g/l. which is practically zero for most analytical purposes. We have quantitatively precipitated lead as its sulphate.

In reality the great majority of salt molecules are mostly dissociated into ions and therefore the process of crystallization does not really proceed through the molecules as we assumed above. It rather proceeds by sodium ions impinging upon free chloride ion sites on the crystal surface, and by chloride ions on free sodium ion sites. Thus, not molecule by molecule but ion by ion, is the crystal built up or dissolved. But the result is the same. If we increase the concentration of one ion we flood the surface of the crystal with it and the oppositely charged sites are going to be covered by them very quickly indeed. Thus the crystal face acquires a net $+$ or $-$ charge as the case may be and attracts a greater number of ions of the opposite charge so that the solution loses a great many of these opposite ions too. Thus first one, then the other kind of ion has become attached to the crystal: the salt precipitates out just as though molecules had become attached to the crystals. The process goes on until equilibrium is restored. What kind of equilibrium? Suppose we have added chloride ions to the saturated sodium chloride solution, using hydrochloric acid (HCl) for instance. The crystal surfaces have acquired a negative charge and sodium ions have thereby been attracted. Hence there are now in proportion even more chloride ions in solution, compared with sodium ions, than there were originally. The surfaces receive proportionally more impacts from Cl^- and retain even in equilibrium a negative charge. By virtue of this charge the Cl^- ions from the solution will be repelled and fewer of them will be able to reach the crystal and locate themselves at sites next to Na^+ ions on the surface. The Na^+ ions of the solution, though fewer, are attracted to the surface and those within the lattice will find it more difficult to leave it. Equilibrium is the situation when just as many Na^+ and Cl^- ions are able to leave the crystal as become attached to it in the same period of time. The negative electrical charge of the surface makes it possible that the equilibrium should be main-tained in spite of the much higher concentration of chloride ions in the solution. It can be shown, that the above process leads quantitatively to the same relation, as would be obtained if the surface were in dynamic equilibrium with undissociated sodium chloride molecules (similarly to eqns. 41.7, 41.8, 41.10).

This, by the way, is a very general principle of equilibria. Whenever we find a way by which an equilibrium can be established, necessarily all other possible ways must lead to the same equilibrium. Otherwise there would be no equilibrium at all! The equilibrium position of a balance is, for example, independent of the side from which movement to attain it begins. Therefore one always looks around for the most easily understandable way in which a given equilibrium can be attained, uses this as a basis for calculation and there-by achieves simply a result obtainable by other, more complicated routes which finally lead to the same equilibrium.

That the surface charge of crystals varies with the ionic composition of the adjoining solution is no hypothesis, it can be easily demonstrated. Small crystals can be caused to move in a solution if an electric field is applied and this movement, electrophoresis, can be observed in a microscope or some-times even without it. By changing the ionic nature of the solution the velocity of this movement alters, showing that the charge has changed. At a given value of ionic composition the migration ceases: there is now no net charge upon the particle and by changing the ionic composition further the direction of motion is reversed since the particles have acquired an opposite charge. The point of zero charge is called the isoelectric point of a suspension or emulsion. The effect is not peculiar to crystals; any particle, even a solute molecule which has some sort of affinity towards ions of different sign may absorb them to different extents and thus become charged. A change in the ionic concentrations then brings about the same changes in movement as referred to above. Proteins, macromolecular organic com-pounds which have acidic and basic groups in their molecules and their molecular building stones (the small amino acids, e.g. $NH_2.CH_2.COOH$), show this reversed "electric" migration to a great extent and their isoelectric points at a given acidity are fundamental for the stability of the biochemical structure of living matter.

42. IONIC ACTIVITIES

We ended Section 40 by saying that there are two different approaches to the behaviour of electrolyte solutions. The first was the dissociation equilibrium

with which we have just dealt. It assumes that a molecule is either undissociated or has dissociated into its constituent ions, the equilibrium being
describable by the law of mass action. If a molecule is able to fall apart
into three or more ions, these steps may, and in general do, take place one
after the other as in eqns. (41.7) and (41.8):

$$H_2SO_4 = HSO_4^- + H^+ \qquad . \qquad . \qquad . \quad (42.1)$$

and

$$HSO_4^- = SO_4^{--} + H^+ \qquad . \qquad . \qquad . \quad (42.2)$$

We have, however, already mentioned the fact that the electrostatic forces
between ions in solution decrease slowly with their distance and therefore
even separated ions should not be looked upon as entirely independent units
within a solution. They are bound not to one specific partner, but to a whole
assembly of counter-charged ions to an extent depending upon their mean
distance apart, that is upon the concentration of the solution. This is the
second aspect of ionic dissociation, due to *P. Debye* (1884–) who gave
a quantitative first approximation of the situation together with *E. Hückel*
(1923). It has become increasingly clear that salts are nearly always quite near
to complete dissociation in a solution. We know of weak acids and weak
bases whose solutions at normal concentrations are rather poor conductors of
electricity. But we do not know of weak salts; their solutions are always
good conductors.

From what we have said about acids and bases in Section 17 it is clear
that their dissociation equilibria are dependent on the loss of or addition of a
hydrogen ion:

$$HCl = H^+ + Cl^- \qquad . \qquad . \qquad . \quad (42.3)$$

$$NH_3 + H^+ = NH_4^+ \qquad . \qquad . \qquad . \quad (42.4)$$

$$H_2O = H^+ + OH^- \qquad . \qquad . \qquad . \quad (42.5)$$

The hydrogen ion is a part of the smallest atom, hydrogen; thus it is even
smaller than a hydrogen atom. It is the smallest "heavy" particle of chemistry. No wonder that it is able to get nearer to a given partner than any other
ion and thus is attached more strongly than other ions are.* This is the cause
of its being held so firmly that sometimes it is not easily removed from the
parent molecule, in which case we are dealing with a "weak" acid, dissociated
to a small extent. A weak base on the other hand is a molecule, for instance

---

* The uniqueness of H^+ lies not only in the small size (that of a proton) which means there
is an intense change of field in its vicinity, but also in the fact that it has no electrons
around it (p. 270) and so no repulsion against other atoms or ions which are all surrounded
by orbital electrons. This is why H^+ cannot exist as such, but is always solvated.

ammonia, which does not attract a hydrogen ion with sufficient force to unite easily with it and to form a positive ion.†

Coming back to salts and strong acids or bases which for practical purposes are all entirely dissociated, the activity theory of Debye pictures their ions as going farther and farther from each other with increasing dilution. As long as they are close, ions are attracted to their nearest partners of opposite charge and are therefore not so free to move around as they would be at infinite dilution, where they would have practically no attracting neighbours. They are not as free as might be expected so far as their mobility in an electric field is concerned because their partners tend to drag them back by electrostatic attraction. They are likewise not so free, i.e. available, for any equilibrium in which they become involved. They are, in a word, less "active" than they would be, without neighbours, at infinite dilution. It is not their concentration, not the moles per litre, that alone decides their chemical or electrochemical activity (their mass action) but a corrected concentration which takes into account this electrostatic interaction of ions. Formally this is arranged by multiplying the concentration by a so-called "activity coefficient" which, however, is a complicated and not always easily known function of all the ions within a solution, called the "ionic strength" of the solution. It is evident that this electrostatic interaction is much greater if ions of higher valency, that is higher electrostatic charge are involved.

From what we have considered up to now one should think that the activity of an ion can only be smaller but never greater than its concentration. Alas, even this is not true. The reason is simple. As soon as the solution becomes so concentrated that the amount of water cannot any longer be regarded as infinite, the forces between ions and water molecules decrease the activity

---

† As long as we are dealing with solutions in water, which will suffice at present, we should always be aware of the role that the solvent water molecules play in these processes. A hydrogen ion is so small that it will always combine with something near to it. Thus the dissociation of an acid is in reality not a simple splitting-off of a hydrogen ion from a molecule, but its transmission to a molecule of water

Not $\qquad$ $HA \rightleftharpoons H^+ + A^-$ ("A" is any anion) . . (42.6)

but $\qquad$ $HA + H_2O \rightleftharpoons H_3O^+ + A^-$ . . . . (42.7)

where a water molecule has added a hydrogen ion, behaving as a base and giving rise to the so called oxonium ion, H_3O^+. In the same way a base, e.g. ammonia, reacts with water and extracts therefrom a hydrogen ion, because it is more "basic" than water:

$$NH_3 + H_2O \rightleftharpoons NH_4^+ + OH^- \qquad . \qquad . \qquad . \qquad . \qquad (42.8)$$

Whether a species is an acid or a base in a solvent depends upon its being able to transmit or to accept a proton from this solvent. NaOH is a base, because its OH^- ions are going to bind some of the free H^+ ions always present even in the purest water as represented by

$$H_2O \rightleftharpoons H^+ + OH^- \qquad . \qquad . \qquad . \qquad . \qquad (42.9)$$

an equilibrium which has a constant ionic product, at room temperature, of 10^{-14}. Hence in pure water we have 10^{-7} H^+ ions and the same number of OH^- ions per litre. Addition of one mole of NaOH per litre makes $[OH^-] = 1$ and to keep the ionic product constant $[H^+]$ has to decrease to 10^{-14}. The OH^- ions have bound H^+ ions from water because they are more basic than water. A molecule of HCl on the other hand transmits a H^+ ion to water because it is more acidic ($HCl + H_2O = H_3O^+ + Cl^-$).

of the latter. Water molecules are often bound to ions which are small and/or strongly charged and so are not as free to move about as they might otherwise have been. Thus the activities of both solute and solvent have decreased simultaneously and the solute is dissolved in a less active solvent, in a smaller amount of solvent as it were. This might go so far as to *increase* the relative activity of the solute.

Qualitatively the dissociation and activity theories are similar. A solution does not conduct electricity as well as it could because only a fraction of its molecules are dissociated, according to the dissociation theory. The activity theory claims that the molecules are all dissociated but that their ions are held within bounds through electrostatic forces acting between them. The quantitative formulations differ. As far as weak acids or bases are concerned, the dissociation theory yields good agreement with experiment. However, when we deal with "strong electrolytes" the activity theory takes over, complicated though it may already be for solutions of even medium concentration.

43. ELECTROLYSIS

We have come to know something about the movement of ions in solution during electrolysis and we know that ions react at the electrodes to give rise to new chemical species. They simply unite with electricity of the opposite sign and "discharge", become neutral species, and either remain as such or react further with the solvent, the electrode or each other as the case may be.

Faraday performed quantitative experiments on the relation between the amount of electricity that passes through a solution and the amount of chemical reaction occurring at the electrodes. He found that the extent of the reaction was always proportional to the amount of electricity used, amount of electricity being the product of current and time, because current is defined as amount of electricity going through a circuit in unit time. (In so-called practical units electricity is measured in coulombs, current in amperes, one coulomb being one ampere-second). But something much more than this simple proportionality emerged. Faraday also discovered that the amount of matter evolved at the electrodes by a given quantity of electricity was always proportional to the equivalent weight of the discharged ion. This is Faraday's law of electrolysis (1833).

Let us consider this result which is illustrated in Fig. 80. Suppose we fill a number of beakers with solutions of the following substances: sulphuric acid, sodium hydroxide, common salt, cupric sulphate, ferrous chloride and ferric chloride. Immerse electrodes in the beakers and connect them all in series in one circuit and let current from a battery pass through all of them. Evidently the same current, and thus in any time the same quantity of electricity, has passed through all the beakers. An ammeter in the circuit shows us how many amperes have passed and the time for which the current flows can be read on a watch. We can carefully and quantitatively collect any product

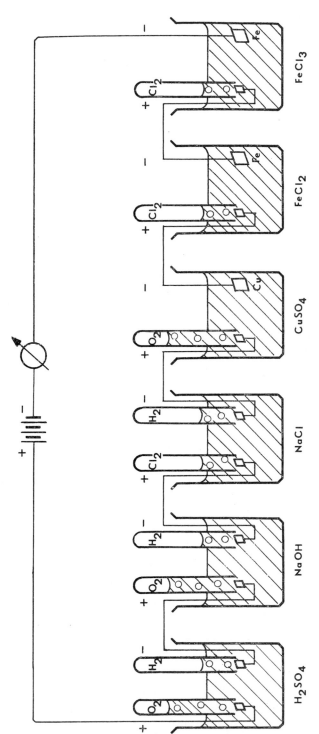

Fig. 80. **Faraday's law of electrolysis**

The same quantity of electricity passing through different solutions liberates at the appropriate electrodes the same number of equivalents of the positive or negative ions. In the case of gases this can be found simply from the volumes: the volume of divalent oxygen liberated is half that of monovalent hydrogen or chlorine (all three are diatomic). But all the *weights* of the gases, liquids or solids of the liberated material are in proportion to the equivalent weights of the respective ions. The weights of iron deposited from ferrous and ferric solutions are therefore in the proportion of 3:2.

which is formed at electrodes. If it be an insoluble gas we collect it in an inverted tube over water; if it be a solid deposited on the electrode, we weigh it. Suppose we have passed a current of one ampere for ten minutes that is, for 600 seconds. This means that 600 coulombs of electricity have passed through the circuit. We find the following results: 70 ml of hydrogen and 35 ml of oxygen are evolved from the sulphuric acid at cathode and anode, respectively. The same result is found with the sodium hydroxide solution. At the cathode of the common salt solution, we obtain again the same amount of hydrogen and at its anode 70 ml of chlorine gas. The copper sulphate gives 0·198 g of copper and once more 35 ml of oxygen. The ferrous chloride solution yields 0·174 g of iron and 70 ml of chlorine, whereas the ferric chloride solution yields only 0·116 g of iron but 70 ml of chlorine. (The volumes are reduced to S.T.P.)

We see that whenever H_2, O_2 or Cl_2 were evolved their volume was always 70, 35, and 70 ml respectively, or in grammes: 0·0062, 0·050 and 0·221. If we now write in a row the equivalent weights of hydrogen, oxygen, chlorine, cupric copper, ferrous and ferric iron we have the ratios 1·0 : 8·0 : 35·5 : 31·4 : 27·2 : 18·6. This is exactly the series of ratios we have found between the decomposition products at the electrodes. This is what Faraday's law implies.

The same fact may be expressed in other terms. In order to decompose a ferrous iron solution to yield the same amount of iron as a ferric iron solution we need only two-thirds as much electricity. This is just the ratio of the valencies: ferrous is divalent, ferric trivalent. The same amount of electricity is necessary for each *unit* of valence! Naturally so, because equivalent weight is simply atomic weight divided by valence. In order to decompose a divalent ion we need twice as much electricity as for a monovalent one, and three times as much is necessary for a trivalent ion.

We have until now been so engrossed with Faraday's law that we have paid no attention to the chemical nature of the electrode products themselves. Some are evidently the expected product, as for instance hydrogen, when it is evolved from sulphuric acid, and copper, iron and chlorine when they are derived from their respective ions. Here the ions have merely discharged, become neutral and emerged as the free elements.

But what about hydrogen evolved from sodium hydroxide or salt solutions and oxygen evolved from sulphuric acid? Are these natural end products also? Well, we know that metallic sodium decomposes water giving hydrogen gas and sodium hydroxide:

$$2Na + 2H_2O = H_2 + 2Na^+ + 2OH^- \qquad . \qquad . \quad (44.1)$$

Thus if sodium metal is the primary product at the cathode, it is no wonder that it instantaneously yields hydrogen. If it happens to be a solution of sodium hydroxide which is undergoing electrolysis it remains in effect unchanged, and only water is decomposed in the end.

The case of the sulphate ion discharge is analogous. We have no evidence for the free radical SO_4 without negative charge, which can be regarded as the primary product at the anode in the electrolysis of sulphuric acid. Thus we are not able to say, as in the case of metallic sodium, that we know that SO_4

can decompose water. But evidently it does so: octavalent sulphur is never encountered. Therefore it is quite logical to assume that SO_4 as a neutral compound is at least as labile as sodium in contact with water and we may write

$$2SO_4 + 2H_2O = 2H_2SO_4 + O_2 \qquad . \qquad . \qquad . \quad (44.2)$$

Again, only water is decomposed, and the sulphuric acid remains unchanged.

All this is true as long as the electrodes do not react with the products, if for example they are made of platinum. If the anode were of a less noble material, it would react. Copper would dissolve as copper sulphate at the anode if a sulphate ion were discharged on it. As a matter of fact electrolytic refining of copper proceeds in just this way: The copper in an impure sample is dissolved at the anode and deposited in pure form at the cathode. An iron anode would be dissolved as ferric chloride ($FeCl_3$ which ionizes) if chloride ions were discharged on its surface. Lead oxide as a cathode would be reduced by discharging hydrogen to metallic lead: This is what goes on at the cathode when an accummulator is charged. In general: the precise nature of an electrode process is determined by the ions *and* by the composition of the electrodes.

Let us see finally, how great is the amount of electricity necessary to decompose one gramme-equivalent of any material. The experiment described above provides the answer. 600 coulombs of electricity liberated 0·00622 g of hydrogen, and so 96,490 coulombs will liberate 1·0 g. This 96,490 coulombs is the fundamental quantity in electrolysis. It is called the faraday and is denoted by F. It is the ratio between electricity and matter which governs all electrolytic decompositions, for instance all electro-deposition processes.

44. CHEMICAL FORCE—ELECTRIC FORCE?

Ever since the structure of modern chemistry emerged from the mists of alchemy in the 18th century, the riddle of the bonding forces which combine atoms into molecules has puzzled the greatest chemists. It seems natural that chemists should have tried to find a connexion between the electrical forces with which they became fairly well acquainted and these mysterious forces constituting the chemical bond.

It was the great Swedish analytical chemist *J. J. Berzelius*, (1779–1848) who first took sides in this controversy (1814), soon after Dalton's fundamental publication appeared. He tried to enforce the hypothesis that all chemical forces originated in electricity. Electrolysis was already known and this gave support to his ideas. He distinguished between positive and negative atoms and tried to explain all chemical facts by this hypothesis. In the opinion of Berzelius positive elements always combined with negative ones and the primary compounds which arose this way combined further with each other to yield secondary and tertiary super-structures. Of course he was unable to divine that atoms in themselves were neutral and that only their ions carried charges.

It was easy to fit many simple inorganic compounds, especially the electrolytes, within the framework. Salt, hydrochloric acid and water, are examples of Berzelius's primary polar compounds. Sulphuric acid, built from water and sulphur trioxide, is an example of the binary super-structure. But there are difficulties. Is sulphur a negative or a positive element? It is negative with respect to hydrogen in H_2S but it also combines with oxygen in SO_2 and SO_3, and oxygen itself was negative with respect to hydrogen! There seems to be no way out of the dilemma without ascribing sometimes negative, sometimes positive character to elements according to the compounds of which they are a part. We have seen that ferric iron "combines" with 3/2 times as much electricity as does ferrous iron. Must we change even the *sign* of the charge in the case of sulphur? It very much seems so. We are unable to confirm the existence of the positive charge upon the sulphur atom in SO_3 by electrolysis but we are apt to suppose it is there and that it holds together the oxygen atoms. We shall see much later on (Section 66) how the concept of the homopolar or covalent bond helps to elucidate such questions.

Always bearing in mind the fact that we are working with a more or less crude approximation, it is quite interesting to see how far one reaches among simple inorganic compounds by assuming that they are built from electrically charged atoms, ions as we now call them. We are going to follow the arguments put forth in beautiful simplicity by W. Kossel in 1916. He and later van Arkel and de Boer went as far on the road of Berzelius as it is likely that one can go at all.

Let us concentrate on compounds which contain oxygen and hydrogen apart from one central atom and see whether we are able to explain why some of them are acids and other are bases. We will assume that oxygen is present as a divalent negative ion, hydrogen as a monovalent positive ion and that the central atom is a positive ion of some particular valency. We are going to characterize them by their charge and by their radii, determined say, in the course of X-ray crystal structure determinations.*

The first group contains alkali metals, for example sodium and potassium. The radii of alkali-metal ions are relatively large in comparison with some others (Fig. 81). Therefore the centre of the negative oxygen ion will be relatively much nearer to the hydrogen ion than to the centre of these alkali-metal ions and, owing to the fact that electrostatic attraction decreases with the square of the distance between charges, it is evident that oxygen and hydrogen will be held together much more strongly than oxygen and the alkali-metal ion. The latter will break away from the molecule with relative ease, leaving a hydroxyl ion. We are therefore dealing with a base. Experiments always confirm that solutions of alkali hydroxides contain cationic alkali-metal ions and negative hydroxyl ions. This is a characteristic property of caustic alkalis.

Within the series of the alkali hydroxides the smallest (lithium) ion is

---

* H atoms are not directly detected by X-rays.

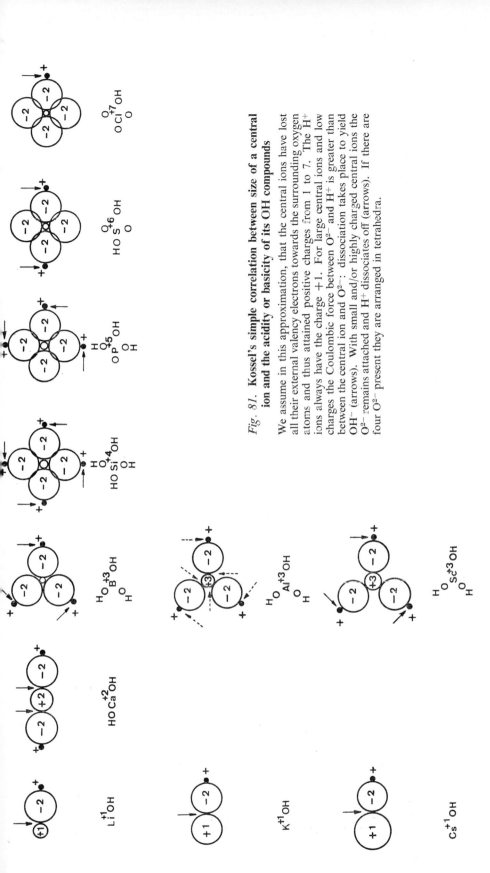

Fig. 81. **Kossel's simple correlation between size of a central ion and the acidity or basicity of its OH compounds**

We assume in this approximation, that the central ions have lost all their external valency electrons towards the surrounding oxygen atoms and thus attained positive charges from 1 to 7. The H^+ ions always have the charge $+1$. For large central ions and low charges the Coulombic force between O^{2-} and H^+ is greater than between the central ion and O^{2-}: dissociation takes place to yield OH^- (arrows). With small and/or highly charged central ions the O^{2-} remains attached and H^+ dissociates off (arrows). If there are four O^{2-} present they are arranged in tetrahedra.

bound most strongly to oxygen, the largest (caesium) ion least strongly of all. Lithium hydroxide is a weaker alkali than caesium hydroxide, in accord with the experimental facts.

The next example is the hydroxide of an alkaline-earth metal, say calcium. The figure shows us that the radius of the calcium ion is smaller than that of potassium. Besides this, it has a two-fold positive charge. Hence it is bound to the oxygen ion more strongly both by reason of its charge and by reason of its smaller radius. The bond between metal ion and hydroxyl will be stronger than it was with the alkalis, the dissociation involving this bond will be weaker and we must be dealing with a weaker base, as indeed we are. The hydrogen ion is now much nearer to the central atom and the Coulombic repulsion of the latter must be more effective at the hydrogen site, although it is not sufficient to cause the hydrogen ion to dissociate from the oxygen at this stage.

Our next example concerns trivalent aluminium, with a radius even smaller than that of calcium. Its three positive charges and the much reduced distance between it and the oxygen ion work together to strengthen the bond between these two. Furthermore the repulsive force of the central ion acting across the oxygen against the hydrogen ions is again increased. Thus we should expect that the probability of a rupture between aluminium and oxygen will have decreased and between oxygen and hydrogen will have increased. Aluminium hydroxide is really a so-called amphoteric compound, able to react either as a base with free positive aluminium ions or effectively as an acid with free negative $Al(O)_3^{3-}$ ions according to whether it finds a more acidic partner, for example hydrochloric acid, or a more basic one such as sodium hydroxide. Both compounds $Al^{3+}Cl_3^-$ and $AlO_3^{3-}Na_3^+$ are known, soluble electrolytes (though $AlCl_3$ is rapidly hydrolysed), whereas the hydroxide itself, $Al(OH)_3$, before "having made up its mind" as to whether it will act as an acid or a base, is insoluble in water. Electrolysis reveals that aluminium migrates to the cathode in acidic solution and to the anode if the solution is alkaline.

It has to be kept in mind, that all these equilibria are dynamic. The molecules of the parent compound, $Al(OH)_3$, undergo both kinds of dissociation with the utmost rapidity, some one way, some the other. The relative proportion of molecules undergoing scission between aluminium and oxygen to that undergoing scission between oxygen and hydrogen is determined by the probability of these events, that is, according to the Maxwell–Boltzmann relation, by the energy necessary in each case. The reverse process, the attachment of free H^+ or OH^- ions to the ionic "residue" is proportional to the concentration of H^+ and OH^-, respectively, in the solution. Hence if there are many hydrogen ions present, the dissociation into H^+ and $(AlO_3)^{3-}$ will be suppressed to a great extent, while hydroxyl ions being necessarily very scarce in this acidic solution, the dissociation into Al^{3+} and $3OH^-$ will be able to go ahead. This is the kinetic picture of $Al(OH)_3$ dissolving to yield Al^{3+} ions in acid solution. The reverse is true if the solution is alkaline. That both kinds of dissociation may occur is due to the fact that the magnitude of the Coulombic forces is nearly equal in the Al—O and O—H bonds. The further complication that free H^+ immediately adds to H_2O to give the oxonium ion only adds a new term to the energy of the whole process.

This equality in the two different bonds does not hold true for the lighter and smaller sister element of aluminium, boron. The B^{3+} ion is much smaller than the Al^{+++} ion and is attached much more firmly to its surrounding oxygen atoms while repelling the three H^+ ions more strongly at this reduced distance. It thus comes about that $B(OH)_3$ always behaves as an acid. It splits only at an O—H bond and never at a B—O bond. $B(OH)_3$, although it is an acid shows only feeble acidic properties.

A heavier sister element of aluminium, the relatively rare scandium has a larger atom than does aluminium and therefore it behaves in the opposite manner: the Sc—O bond is weaker than the O—H bond. $Sc(OH)_3$ splits off OH^- accordingly and is a base, though a weak one.

Proceeding in the atomic weight range of aluminium along the row of elements of increasing maximum valency (*see* Section 55), we have next to discuss the oxy-acids of silicon, phosphorus, sulphur, and chlorine. Their maximum valency in their oxides is 4, 5, 6 and 7 respectively and the radii of these ions decrease from silicon to chlorine. Therefore it is clear that the oxygen ion is bound with increasing strength to these central ions while their repulsion across to the hydrogen ion increases at the same time. Accordingly the hydrogen ions dissociate with increasing ease and we are dealing with a series of acids whose strength increases as the valency of the central ion increases. In $HClO_4$ perchloric acid, we reach the maximum of acid strength attainable in practice. A more detailed discussion should, of course, include the repulsion between the oxygen ions as well, and should include another effect with which we shall become acquainted when we arrive at the electronic representation of the chemical bonds.

The formulae of the acids referred to above are: $Si(OH)_4$, $PO(OH)_3$, $SO_2(OH)_2$ and $ClO_3(OH)$. We write them thus to emphasize that, at least in the pure state, they are hydroxy compounds. One normally writes them H_4SiO_4, H_3PO_4, H_2SO_4 and $HClO_4$ which gives a better division between cations and anion. It is remarkable that throughout this series the number of oxygen atoms around the central atom remains constant at four whereas it has increased steadily from one to four between the alkali metals and silicon. The sister acid of silicon, carbonic acid has the formula H_2CO_3, with one molecule of water less than H_4SiO_4. Carbon is smaller than the silicon atom. There do exist organic carbonates which are derived from the hypothetical acid H_4CO_4 (orthocarbonic acid) but no such inorganic salt is known. On the other hand the much larger sister of chlorine, iodine has an acid of the formula H_5IO_6 with a central heptavalent iodine atom, that is, only one molecule of water is missing from the maximum $I(OH)_7$.

We cannot help believing that the number of oxygen atoms around a central atom is mainly determined by the space available. Very small central ions are only able to accommodate two or sometimes three, while larger central ions can accommodate four which can be tetrahedrally arranged with high symmetry; still larger ions are surrounded by six and not five oxygen atoms because six atoms can form a highly symmetrical body, the regular octahedron whereas there is no regular solid body with five corners. Thus spatial and symmetry relations determine the degree of hydroxylation of these hydroxy-oxy compounds. Factors based on electron distributions are at work, too, but in the ultimate analysis these are also concerned with symmetry. It is the harmony of the spheres within the atom, a harmony which might easily be one of the most important factors determining the structure of matter.

It might be worth while to glance again at the oxygenated acids of chlorine in this connexion. We have discussed them to some extent in the section on inorganic structural formulae (p. 54). They are: hypochlorous acid, $H^+(Cl^+O^{2-})$; chlorous acid $H^+(Cl^{3+}O_2^{4-})$; chloric acid $H^+(Cl^{5+}O_3^{6-})$ and

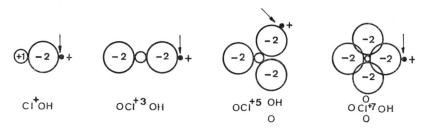

Fig. 82. **Kossel's simple correlation applied to the four oxy-acids of chlorine**

We assume in this approximation that the central Cl atom loses 1, 3, 5 and 7 electrons respectively, to the oxygen ions beside it. Hence the central Cl^- ion contracts because it gradually loses all its valency electrons and its formal charge increases so that the electrons that remain are attracted even nearer to the nucleus. Decreasing size and increasing positive charge of the central atom decreases the resultant Coulombic attraction between O^{2-} and H^+ so that the acids become stronger along the row from left to right. (The four O^{2-} in $HClO_4$ are actually in the corners of a tetrahedron and do not overlap.)

perchloric acid $H^+(Cl^{7+}O_4^{8-})$. The series is interesting for two reasons. First, the acid strength increases with the positive valency of the central chlorine ion from the very weak hypochlorous acid to the extremely strong perchloric acid. This agrees with Kossel's picture of the Coulombic attraction between the chlorine and oxygen ions and the repulsion of H^+ by the increasing charge of the central chlorine ion. Secondly the number of oxygen atoms attached to chlorine lags behind that maximum which is theoretically possible as $Cl^{n+}(OH)_n$ except in HClO and never exceeds the tetrahedric configuration (ClO_4), though we could well imagine at least $H_5(ClO_6)$ as in periodic acid. There is not enough space (Fig. 82).

After all that has been said above we can come nearer to understanding the true structure of, for example, a sodium chloride crystal. It is built from ions, not from atoms, as we know already. X-ray analysis has revealed that each sodium ion is surrounded by six chlorine ions as nearest neighbours and vice versa (Fig. 43, p. 108). The force holding this lattice together is electrostatic in nature. We have to introduce a new, repulsive force which prevents the attracted ions of opposite sign from penetrating each other.

This so called "Born" repulsion (named after the famous German physicist *M. Born*) is responsible for the fact that ions behave to a great extent as if they were rigid spheres, which cannot come nearer to each other than a minimum distance. From these forces it is even possible to calculate the mechanical strength of such a crystal, and under very special experimental conditions, if surface faults are carefully excluded, this strength may indeed be realized.

Thus we could gain the impression that the idea of Berzelius has conquered the field. But it is not so. By means of it we should be able to explain the bonding force in all kinds of molecules: in the diatomic elementary gases H_2, O_2, N_2, Cl_2, etc., and the forces which act in the chains and rings of organic and some inorganic compounds, not to mention some finer details within the realm of ionic chemistry which we have just reviewed. The partisans of the electrostatic bond theory have done their best to extend the theory in these directions. They have made advances, but in the extreme cases they have failed. The reader must rest assured that we have not yet told the whole story of the chemical bond. What we have discussed is called the electrostatic or heteropolar bond between atoms. In due course we shall have to deal with the so called homopolar or covalent bond (Sections 68–74) which, for instance, is present in the diatomic molecules of the elements; the very special bond which is characteristic of metals (Sections 79–83); finally those weaker forces which are responsible for the cohesion of liquids, even for the liquid states of the very inactive rare gases (Section 67).

45. THE CHARGE ON THE ELECTRON

It is a consequence of Faraday's law that an ion may only be associated with a specific quantity of electric charge. Knowing the Loschmidt (Avogadro) number we have only to divide by it our 96,400 coulombs (*see* p. 187) to know how large this charge on a univalent ion really is. It works out at $1 \cdot 6 \times 10^{-19}$ coulomb.

However it still remains an open question whether this amount of electric charge corresponds to some "atom" of electricity or whether it is only the amount of electricity that "finds a place" in some manner or other within an ion. It could be imagined that electricity in itself is divisible without any limit and each ion provides itself with this "fluid" by an amount given by its valency. On the other hand, if we conclude that atoms of electricity really exist, only because the total amount of charge carried during electrolysis across a solution can be divided by the number of ions transported, we argue as someone who, seeing that a belt of buckets in a well always conveys the same amount of water per bucket, would say that water in the well is made up of indivisible water-units of the size of the buckets. However, the well could be full of tennis balls instead of water and in this case the appearance of one ball per bucket would reflect at the same time the discontinuity of the balls *and* the capacity of the buckets. Thus we have to decide between these alternatives as far as electricity is concerned. Should our investigations prove that electricity in itself is composed of electric atoms and should we be able to measure their charges, we should then be in a position to reverse the above calculation and determine the Loschmidt number from the faraday (96,400 coulombs) and from the magnitude of the elementary charge. As a matter of fact, this has turned out to be the case and the value obtained for the Loschmidt number is the most accurate one of all.

There is more than one method of determining this elementary charge. In this section we are going to follow the experiments and the reasoning of the famous American experimental physicist, *R. A. Millikan* (1910). The method is easy to visualize and in principle it is based on simple experiments. However, a great deal of experimental skill was necessary to overcome seemingly minor difficulties and to attain the ultimate precision of the results. We are all familiar with atomizers and sprays which disintegrate fluids into extremely tiny droplets. These drops are so small that they remain dispersed in the air for a considerable time. If the dispersed liquid is not volatile, the drops will not lose weight by evaporation and will sediment very slowly. In a vacuum, of course, they would be falling with the velocity of freely-falling bodies because of gravitation, but in air they are slowed down by friction, lose their acceleration very quickly and thereafter fall with a small, constant velocity which is determined by their weight, their radius and the viscosity of air. Calculations as well as experiments proved that this terminal velocity is proportional to the radius of the small droplets and inversely proportional to the viscosity of the medium through which they are falling.

Millikan dispersed heavy oil very finely as droplets in air and used a microscope to determine the velocity of their downward movement. The vessel in which the particles were falling was of course closed against draughts and held at a uniform temperature. At the top and bottom of this glass vessel was a metal plate; any desired electrical potential could be made to exist between these plates (Fig. 83). The remarkable result of Millikan's measurements was that the drops of oil settled with constant velocities as long as he did not apply the electrical field, but they immediately changed their behaviour if the field was switched on. Then some of the drops continued to settle with unchanged velocity whereas others began to move with different velocities. If the size of the droplets was the same these new velocities fell into discrete, different groups with no intermediate velocities.

The diameter of the droplets was of the order of a few microns; the distance between the top and bottom plates was 1 cm and the applied voltage was about 10,000 volts. Owing to friction between the droplets and the air the uniform velocity of sedimentation of these droplets was between 0·1 and 0·5 mm per second according to their diameter. It seemed an obvious conclusion that those particles which changed their velocity in the electrical field were electrically charged. This would by no means be abnormal: friction between a dispersing fluid and air nearly always gives rise to a charge of frictional electricity on the resulting droplets. This phenomenon, for example, can be observed near waterfalls. Furthermore the action of strong light, and also especially of X-rays, on matter often liberates electrical charges as we shall learn in Section 49. Thus it seemed quite in order that a series of droplets should have been charged. It was of interest that the magnitude and in some cases even the sign of these charges seemed to change after a time, the evidence being that some of the drops would abruptly change their velocities and sometimes even proceed in the reverse direction. The important point was that this change occurred *abruptly* from one moment to the next, and that it always occurred in steps. Only special, discrete values of velocities occurred for each droplet. This, then, was the indication that electrical charges must be composed of discrete units and that a droplet could only gain or lose a whole unit (or several units), of this elementary charge.

The course of the experiment is as follows. The observer chooses a particular droplet on which to make measurements and determines the time

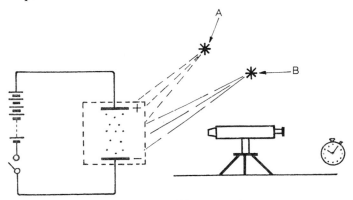

Fig. 83. **The Millikan experiment**
Oil droplets irradiated with X-rays in air, suddenly change their velocities from time to time in a vertical electric field. The change is discontinuous and is caused by their electric charge changing by one or more multiples of a unit, elementary, charge. The velocity is measured with a stop-watch and a microscope.

A = X-ray source   B = light source

which elapses between the droplet's passing an upper and a lower mark in the field of the microscope. Then he switches on the electric field, choosing its direction so as to make the droplet ascend. Its velocity of ascension is measured in the same way as its previous descent, provided it does not suffer any of those abrupt changes in velocity in the meantime. As often as the droplet reaches the upper mark, or whenever its velocity, dependent upon its charge, changes abruptly, the field is switched off again and the time necessary to reach the lower mark is determined once more. In this manner the experiment may be repeated on the same drop any number of times, thus increasing the accuracy of the determination of its velocity. A droplet will always descend without the electric field with the same velocity, but the velocity of its ascent will differ according to the abrupt changes of charge which it has suffered between different determinations.

What we are interested in is the electrical charge on a droplet in the course of its different ascents. We know that the electrical force acting upon a charged body in a given field is proportional both to the charge and to the strength of the field. The field gradient of course is the greater the higher the voltage between the plates and the less the distance between them. In our example the field was 10,000 volts per cm. The electric force between two parallel plates is independent of the distance from either one of them because the nearer one goes to one plate the further one is from the other. The force acting upon a charge e in a field of strength E is hence proportional to eE and by appropriate choice of units can be made equal to this product. Acting against this force we have the weight of the droplet. This can be determined either from its radius, measured with the aid of a microscope and its density, or more accurately from the uniform downwards velocity with which it sediments in the absence of an electric field. The smaller the droplet, the smaller its velocity.

We have mentioned elsewhere that a force acting upon such small particles in a viscous medium is very soon equalized by the force of friction so that the terminal velocity becomes uniform and it is this velocity, not the acceleration, which becomes proportional to the driving force. Therefore, if we manage to determine the terminal velocities of a droplet when it carries different charges, we have only to subtract these velocities from each other to obtain the difference in the driving force which will be proportional to this difference in velocity in each case. Now, the results of such measurements have shown that *all upwards velocities differ from the primary downward velocity by an integral multiple of a fixed unit of velocity*. For instance, suppose that a droplet was moving downwards without the field with a velocity of 0·01 cm/sec. When the field was switched on we should observe one of the following effects: either the original downward movement would remain unchanged or the velocity would become directed upwards amounting to either 0·02, 0·05, 0·08 or 0·11 cm/sec, etc. The upward velocity of 0·02 cm/sec exceeds the original downward velocity by 0·03 cm/sec because they differ in sign. It will be noted that all the upward velocities differ from each other by this same amount of 0·03 cm/sec. No velocities in between these values are ever found, and so there must be no charges whose magnitude lies in between those successive charges on a droplet. It seems evident that these series of upwards velocities correspond to one, two, three, etc., elementary charges on a droplet.

We may now proceed to calculate the numerical value of the elementary charge e. Terminal velocities are, as we know, proportional to the acting resultant forces. Hence the force of gravity which pulls a droplet downwards is in the same proportion to the force which pulls it upwards when it has the smallest possible charge, as the downward velocity is to the smallest upward velocity. Let us denote the original downward velocity by v_0, the weight of the droplet by p, the smallest upward velocity by v_1, and the unit elementary charge by e. Then the upward force will be the difference of the electrical force eE and the weight which counteracts it, p. This gives the relation

$$v_0 : p = v_1 : (eE - p) \qquad . \qquad . \qquad . \qquad . \quad (45.1)$$

solving for the elementary charge, we have

$$e = \frac{1}{E}\left(\frac{pv_1}{v_0} + p\right) \qquad . \qquad . \qquad . \qquad . \quad (45.2)$$

The mean result from many thousand of Millikan's extremely careful measurements yielded the value of $1 \cdot 59 \times 10^{-19}$ coulomb for this elementary charge, in agreement with our calculation from the faraday and Loschmidt's number at the beginning of this chapter. Conversely, we may calculate the Loschmidt number by division of 96,400 coulombs by $1 \cdot 59 \times 10^{-19}$ and obtain $6 \cdot 05 \times 10^{23}$ once more, independently of previous determinations. Perrin's experiments on mastic particles and the diffraction of X-rays in crystals were really fundamentally different both from each other and from Millikan's experiment. Yet all of them have yielded almost exactly the same number for the number of molecules in a mole! There must surely be the truth behind these very different hypotheses.

Of course, any determination of this fundamental value permits the calculation of connected fundamental values. From the Loschmidt number which we have just determined from Faraday's and Millikan's experiments we can immediately deduce the absolute mass of an atom: we have only to divide the appropriate gramme-atomic weight by the Loschmidt number N_L. If we desire to know how large atoms are, the density of crystals of known composition, and a knowledge of N_L will enable us to make the calculation. We can work out the number of atoms in a given volume of the crystal and thus the volume of a single atom within it. The mass, the volume and the number of atoms per mole are all interconnected.

Once more, it was the occurrence of small integral multiples of a given value in Nature which revealed the existence of atomic structure. For example, the law of multiple stoichiometric proportions led to Dalton's atomic hypothesis. Then the law of rational crystal indices led to the assumption of atomic structure within crystals. Finally, the integral multiples of a velocity in Millikan's experiment have led to the conclusion that electricity itself has a discrete structure.

46. THE FREE ELECTRON

Ever since men of science have been concerned with electricity it has been a recurrent question to ask "about its true nature". Is electricity matter, is it a state, is it one "*fluidum*" or two? One has to be very careful with such questions because they try to verbalize too fundamental concepts. Do we really know "the true nature" of anything at all? Hardly. Many things may be broken up into components, and we are able to classify these and tell how they were put together. In the end we arrive at seemingly primary indivisible entities and here we can do no more than describe all their properties, their whole behaviour. Maybe we succeed in breaking up a component further, but this only shifts the difficulty back another step. "The true nature" of something is not physics anymore, it is beyond it—metaphysics.

As physicists we may ask, however, whether it is possible to encounter this elementary charge of electricity all by itself, not bound to ions or droplets of oil? Has it a mass of its own? Anticipating the answer, we may affirm that such an elementary electric charge does exist, that it is only stable if it is negative and that it has a mass of its own which is however much less than that of the lightest atom, hydrogen. For a long time it was supposed that positive electricity only existed in the positive ions of the elements, bound to chemical matter. *C. D. Andersen* (1905–) succeeded in detecting the positive counterpart of this elementary charge (1936), which, however, combines almost immediately with one of the negative charges and so disappears, (p. 434). The free negative electric charges are now called electrons. Owing to the above-mentioned instability of the positive electron the name electron is generally used to designate the negative elementary

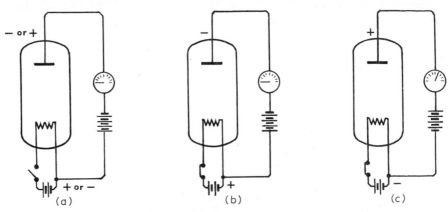

Fig. 84. **Two electrodes in a vacuum tube (diode valve)**

(a) If the lower electrode is not heated there is no current between the electrodes, irrespective of their sign of charge.

(b) If the lower electrode is heated but is positive with respect to the upper electrode there is still no current between them.

(c) If the lower electrode is heated and is negative with respect to the upper electrode, a current of negative electricity flows between them.

charge while the positive one is often termed a positron for distinction. Its existence was theoretically predicted by Dirac.*

The easiest way to observe free electrons is to utilize evacuated tubes similar to radio valves or television tubes. Let us evacuate a container to such an extent that the mean free path of the relatively few atoms or molecules which remain exceeds the dimenisons of this container. This means that the molecules left do not very often collide with each other but in general bounce from wall to wall. A metal can be brought to incandescence in such an evacuated container by electrical heating, for example the filament of a lamp. (It is best to employ a tungsten filament containing a small amount of thorium, or nickel with a very small amount of barium on the surface, although other metals can also be used.) This incandescent metal acts as one electrode while another piece of metal with leads through the insulating wall of the container (say glass or silica) forms the other electrode (Fig. 84). On connecting a battery and an ammeter between the electrodes

---

* It has been called a "donkey" electron because in an electric field it would be expected to move in the *opposite* direction to that taken by a normal (negative) electron.

something very strange happens. If the incandescent electrode is made negative a current flows in the circuit, but if the voltage is reversed there is no current! It seems that negative electricity is emitted from the glowing surface and is attracted by the positive electrode. If both electrodes are cold there is no current in either direction. This phenomenon lends itself immediately to practical application: it is possible to build current rectifiers along these lines. Only those peaks of an alternating current which make the incandescent electrode negative are able to traverse such a device. This is one of the basic processes of rectification which converts alternating to direct current.

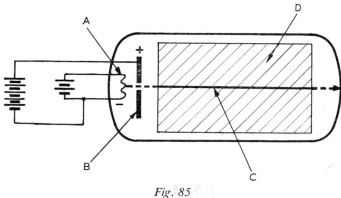

Fig. 85

The electron beam continues in a straight line from the cathode through the slit in the anode and along the length of the tube.

    A = incandescent cathode
    B = anode
    C = electron beam
    D = fluorescent screen, nearly parallel to the electron beam.

If the voltage between the electrodes is sufficient it is even possible to see where this negative electricity impinges on, say, the glass wall. There are many compounds which, like zinc sulphide, emit light if such a stream of electricity hits their surface, invisible though the stream itself may be. This happens as a matter of fact at the front face of a television tube. An apparatus like that in Fig. 85 allows us to perform some interesting experiments with this streaming electricity. The anode is placed near to the incandescent cathode and it has a small opening in the middle. Behind this anode there is a slanting plane surface covered with some fluorescent material, nearly but not quite in line with the connecting line of the cathode and the slit. A part of the streaming negative electricity goes through the slit and touches the fluorescent surface. We observe a line of light on this luminescent screen, a straight line which is the continuation of the line drawn from the cathode through the slit to the screen. The electricity carries on in a straight line from its source, just as a ray of light, or a ray of atoms in the Stern apparatus (p. 138). Such an electrical beam is often called a cathode ray. It does not change its position, no matter how we turn the tube around.

A ray of light is not deflected from its path by an electric or a magnetic field. (Only a very strong gravitational field is able to bend its path, in accord with Einstein's theory of general relativity.) But our cathode ray can be deflected both by electric and by magnetic fields. Fig. 86 shows practically the same apparatus as Fig. 85 but it also contains a pair of metal plates parallel to, and on both sides of the original cathode ray. If we connect the positive pole of a battery to the lower plate and the negative pole to the upper one the ray bends downwards. It has to do so as it is negative electricity and we always find that unlike charges attract each other. The electrons—and we seem to be dealing with these—behave in this electric field like a heavy body in a gravitational field. They move downwards along a parabola, because the longer they are acted upon by the field the greater their velocity

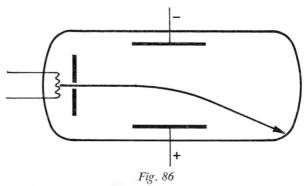

Fig. 86

The cathode ray describes a parabola between the two charged, deflecting, electrode plates, like a jet of water in the gravitational field of the earth.

in the downwards direction. They either cannot be very heavy, or else must possess immense velocity because the force of gravity alone is unable to deflect them.

Now what about a magnetic field? Suppose we bring the north pole of a magnet from the front, from the side of the reader, towards the cathode ray. The ray bends downwards in a circle, the more sharply the nearer or the stronger the magnet. A south pole causes bending in the opposite direction. This is, of course, the same process which we should observe if a flexible wire carrying a current had been subjected to a magnetic field; it is simple electromagnetism as is utilized in an electric motor. Magnetic fields do not act upon stationary electrical charges but they do act upon moving charges, that is when there is a current. Thus its behaviour towards magnetic fields shows once more that a cathode ray is a current of negative electricity moving from the cathode towards the anode and even through it via the slit.

It is possible to use these two deviations—in an electric and in a magnetic field—to determine the ratio of charge to mass of the electrons, and also their velocity in the cathode ray. We will follow the argument through in a qualitative manner. As we have seen, the action of the electric field upon the ray was like that of gravitation on matter. The electrons within the ray are projected through the slit and proceed in a straight line with uniform velocity as long as no force acts upon them. But between the two horizontal, charged plates they are acted upon by the electric field between the plates. Remember the

Millikan experiment. The field between parallel plates is homogeneous, the field strength being the same everywhere between them. But whereas the droplets studied by Millikan fell in air, and were damped by viscous forces into uniform motion, here we are working in a vacuum and so a constant force does give rise to a constant acceleration, as when a body falls to earth. The vectorial addition of the horizontal, inertial motion and the vertical accelerated fall gives a parabola, just as in case of a stone which has been thrown horizontally. The direction of acceleration of the electrons is towards the positive plate. The more quickly the electrons move forward from the slit, the less time they will spend between the plates and they will accordingly have less time for vertical acceleration. For a given field, the parabola will be flatter, the greater the original horizontal velocity, or, more exactly, the horizontal momentum (velocity times mass). On making the calculation we thus obtain an equation which connects the known strength of the deviating field, the measured length of deviation from the horizontal across a measured horizontal length, with the unknown mass, charge and velocity of the electrons.

The laws of electromagnetism teach us that the force which a magnetic field exerts upon a current is perpendicular to both and is constant for a given field and current. Now it is clear from mechanics that whenever a constant acceleration is enforced upon a body, in uniform movement, perpendicular to its path, the body must acquire motion along the circumference of a circle. Compare a stone swung around on a rope, when the centripetal force through the rope is constant. The circle will be smaller, the greater the force and we know from electromagnetism that it is proportional to the current as well as to the magnetic field. But the circle will be the larger, the greater the momentum of the moving body, because of the inertia which has to be overcome when bending the motion out of a straight line. We again obtain an equation between the known strength of the magnetic field, the measured radius of the circle on the fluorescent screen and the unknown mass, charge and velocity of the electron.

From these two equations we may calculate two unknowns, though alas we have three with which to deal. But it happens that the charge and mass of the electron always occur as a ratio in the equations and so we are at least able to determine this ratio as one unknown and the velocity as the other, by the following procedure.

Let the electrons first pass the electrically charged plates of Fig. 87. It can be found by measurement that they deviate distance b from the horizontal while passing through a horizontal distance a. If their velocity is v, they will fly through distance a in time $t = a/v$. The field strength E acts upon their charge e with force eE, and owing to the fact that they possess mass m, they will acquire a vertical acceleration eE/m. A freely falling body under gravitational acceleration g moves through a vertical distance $\frac{1}{2}gt^2$ in t seconds. Now the distance the electron has "fallen" in time $t = a/v$ is, in the same way

$$b = \frac{Ee}{m} \cdot \frac{t^2}{2} = \frac{Eea^2}{2m \cdot v^2} \tag{46.1}$$

The electron next arrives at the magnetic field which must be obtained from large pole pieces (in order to have it homogeneous), giving the magnetic field-strength H,

Here it is forced to move along a circle of radius r. The force acting upon a current is proportional to the current i and the magnetic field H and is, as we have said, perpendicular to both. Thus it is, using appropriate units, equal to iH. But how can we express i for moving electrons? Simply. A charge e moving with velocity v is nothing but a current $i = ev$. Therefore the magnetic force which acts upon the electron is Hev. This pulls it towards the centre of the circle while the centrifugal force mv^2/r works against it, trying to throw it out radially. In the equilibrium circle

$$Hev = mv^2/r \tag{46.2}$$

Eliminating from these two equations first e/m and then v, we have

$$v = \frac{Ea^2}{2Hrb} \tag{46.3}$$

and

$$\frac{e}{m} = \frac{Ea^2}{2H^2r^2b} \tag{46.4}$$

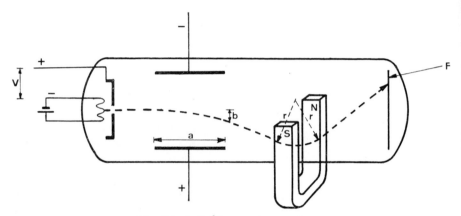

Fig. 87. **J. J. Thomson's experiment**

Cathode rays are deflected in a parabola by an electrostatic field and then in a circle by a magnetic field. Their total deviation is measured on a fluorescent screen F. The two fields are applied one after the other and the deflexions measured. From the field strengths and the deflexions the velocity of the electrons and their specific charge, e/m, can be determined.

The results are as follows. It is apparent that the velocity of the electrons, v, depends only on and may be determined by, the accelerating potential difference between cathode and anode. The kinetic energy of electrons which have fallen through a potential difference V is $eV = mv^2/2$ and so their terminal velocity is proportional to the square root of the accelerating potential difference. In a similar way the terminal velocity of a freely falling body is proportional to the difference in height it has dropped through. These two phenomena are analogous and this fact may help the reader to understand the situation. An electron moves in a constant electric field as does a body in a constant gravitational field. The terminal velocity of the electron in the cathode-ray tube is very great indeed. After passing, or "falling", through 100 V it is $5·9 \times 10^8$ cm/sec, somewhat less than one-fiftieth of the velocity of light! As we have stated this velocity varies with the square root of the potential difference, that is it increases tenfold at 10,000 V and decreases to one tenth at 1 V.

However as far as the ratio e/m, the ratio of charge to mass of the electron, is concerned, this remains absolutely constant up to very great velocities (J. J. Thomson, (1856–1940) in 1897). It is 1.77×10^8 coulomb-grammes. In view of this constancy we may conclude that the cathode rays are really composed of identical units. The experiment of Millikan showed that the charge of an electron is constant. Now we have found that the ratio of this charge to the electronic mass is a constant too, and so the mass must necessarily be constant. The charge was 1.59×10^{-19} coulomb, e/m was 1.77×10^8 coulombs and therefore the mass of the electron must be 9×10^{-28} gramme; very small indeed, only 1/1837th part of the mass of the hydrogen atom! It is the smallest mass of any particle known to the present time.

We have qualified our statement as to the constancy of e/m by the words "up to very great velocities". This means that up to an accelerating potential of say 100,000 V the ratio is practically constant but it begins to decrease at higher velocities. At the same time the velocities cease to increase proportionally to the square root of the applied accelerating potential difference. They increase more slowly than that and tend asymptotically towards the velocity of light ($c = 3 \times 10^{10}$ cm/sec) as an ultimate limit. Taking both facts together, we gain the impression that the mass of the electron increases as its velocity approaches that of light and it is just the inertia of this increased mass that finally prevents the particle from reaching the velocity of light. As it comes nearer and nearer to the velocity of light its mass increases without limit and it cannot be accelerated any further.

This phenomenon is a consequence of a deep-rooted fundamental principle of physics. After the greatest experimental physicists, first among them *A. Michelson* (1856–1931), had failed to find any difference in the velocity of light according to whether the observer moved towards the source, away from it, or parallel to it (1887) *A. Einstein* (1905) announced it as "a principle of relativity" that the velocity of light is the greatest of all existing velocities and that no other velocity can be added to or subtracted from it. This removed the last hope of ascertaining whether one or another of many systems which moved relatively to each other with uniform velocity "was really at rest". There is no mechanical experiment which allows us to distinguish between a "stationary" seashore and a "moving" ship. You may play tennis or perform any other movement of balls, objects or yourself upon either: there will be no difference. The man on the ship may claim as well as the man on shore that the other is moving. The only certain statement is that they do move relatively to each other. Now if there had been an ocean of something, called at one time "ether"* in which light is propagated independently of material bodies it would have been possible to determine absolute motion relative to this ether by determining the different velocities of light in different moving systems. But no differences were found experimentally and thus, between uniformly moving systems, only relative motion remained.

---

* Not to be confused with the class of compounds called "ether" in organic chemistry.

Our whole mechanical outlook is modified to its foundations by the principle of relativity—but the latter is experimentally true. We have been used to considering that velocities may be simply added to and subtracted from each other if they are in the same direction. But this is only true so long as they do not approach the velocity of light. It is quite impossible here to go into further details of relativity, but the experiments on cathode rays seem to offer at least another approach to the same paradoxical end. It seems that whenever the velocity of something as seen from a given system (here the laboratory with its cathode-ray tube) approaches the velocity of light, its mass begins to increase infinitely as far as this system of observation is concerned and this makes it impossible for the moving object to exceed, nay even to reach, the velocity of light. No finite force is able to accelerate an infinite mass. The cathode ray tube was a very convenient instrument with which to observe this phenomenon because experimentally available accelerating potential differences are able to accelerate the tiny electron into the neighbourhood of this limiting velocity. The immense particle-accelerating devices of today, cyclotrons, betatrons, cosmotrons and others, all work in a range of velocities where further force serves to increase only the mass of electrons but is practically unable to increase their velocity. e/m decreases and v is very near to its limit, c.

47. ELECTRON WAVES: THE PRINCIPLE OF UNCERTAINTY

We have concluded from experimental facts that electrons have a constant charge, a constant mass at non-relativistic velocities, suffer deviation in an electric or magnetic field, and so we assume that they are surely corpuscles. Something like ions, perhaps, but two thousand times lighter than the lightest of ions (H^+). Great was the astonishment and consternation of physicists when *C. J. Davisson* and *L. H. Germer* in 1928 and *G. P. Thomson* a year later, clearly established the fact that electrons sometimes behave exactly as if they were waves: they show the phenomenon of interference that we have learned about in connexion with light (p. 114).

Davisson and Germer, in the laboratories of the Bell Telephone Company, directed a beam of uniform-speed electrons on to crystals and obtained the same sort of interference figures which we learned to know with X-rays (Fig. 88). A fine pencil of parallel electrons is only reflected in special directions from the internal planes of crystals and the quantitative relation between reflections is the same Bragg relation which we described on page 123. Using this equation and crystals of known structure, it became possible to determine the wavelength of the electrons. Interestingly enough, this wavelength was not a characteristic constant of the electrons but was uniquely related to their momentum. Not taking into account a small correction due to the potential difference at the surface of the crystals, this relation was very simple indeed. It was found that the wavelength was inversely proportional to the momentum, to the product of mass and velocity of the electrons. The faster the electron as a corpuscle, and the larger its relativistic mass, the shorter its wavelength in diffraction experiments! We have previously mentioned this property on page 96 when we described the electron microscope. The very intense magnification of this instrument is due to the extremely short wavelength which electrons can attain by sufficient acceleration.

Recapitulating: series of reflections of an electron beam from a crystal face are represented by the Bragg equation

$$\lambda = n2d \sin \alpha \qquad . \qquad . \qquad . \qquad . \quad (47.1)$$

where n is an integer, d the distance between internal planes and α the angle of deflection.

Fig. 88. **Electron diffraction across a mica lamina**

A magnetic field deflects the whole picture, proving that it arises from charged particles.

(From W. H. Westphal, *Physik*, 2nd ed., p. 536, Fig. 487 (Berlin, Springer-Verlag, 1930). By courtesy of Springer-Verlag.)

The values of λ calculated from any such series depend only on the momentum, determined by the accelerating potential of the electrons in the beam. If we denote an important constant of Nature by h (*see* p. 208), it is

$$\lambda = \frac{h}{mv} \quad . \qquad . \qquad . \qquad . \qquad . \quad (47.2)$$

Working with electrons of mixed velocity one obtains a diffraction image like the Laue photographs with mixed, that is with non-monochromatic, "white" X-rays (Fig. 55). The same pattern, the same calculation, is encountered in both cases.

We have seen that a series of fundamental properties of the electron speak for its being a corpuscle. It has a fixed charge and moves according to the laws of mechanics if acted upon by an external force, according to the laws of relativistic mechanics, it is true, but this applies to any other particle also, as far as its greater mass can be accelerated near c, the velocity of light. This can easily be done today in particle accelerators. Another set of precise quantitative experimental results suggests on the other hand, that electrons are waves. A similar dualism is going to come our way as soon as we begin to discuss the processes in which the energy of light waves is converted into any other sort of energy: we shall soon see that these waves behave as if they consisted of lumps of energy with a given impulse. Energy and impulse are connected with the wavelength again by simple relations containing the fundamental constant h (eqns. 47.3, p. 209 and 47.4, p. 210).

At first sight the situation seems remarkable, but not as tragic as it really happens to be. Actually it involves a contradiction which cannot be bridged within the boundaries of classical physics and even the fundamental concept of causality hardly survives it. We shall see shortly how this comes about. In classical physics all bodies which move without a force acting upon them, continue moving in a straight line with uniform velocity. Knowing this velocity we are able to predict the position of the body at any time we please, if we know its position at one time. On the contrary, any point on a wave is

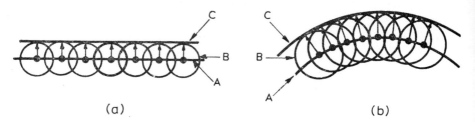

(a)   (b)

Fig. 89. **Huygens's principle of elementary waves**

(a) Progressing plane wave; (b) wave front progressing from a centre.

A = original wave front
B = elementary wavelets
C = new, external wave front.

itself the origin of a new wave system. The interaction of all these micro-wave systems determines the actual wave-front according to a principle discovered by *Huygens* (1690), the great Dutch physicist. A plane wave is the external surface of parallel elementary wavelets, a spherical wave the covering surface of concentric wavelets (Fig. 89). By confining a part of these wavelets as may be seen for example on the photograph of water waves passing a slit on Fig. 47 the slit becomes the visible centre of the new progressing wave system. To put the difference in plain language: one may hide behind a wall from a bullet because it flies in a straight line, but one cannot evade its sound because this propagates by a wave system and finds its way around the corner. What, *now*, is the nature of an electron? Can one hide from it behind a corner? The old experiments with slits in the cathode-ray tube seemed to answer "Yes": the shadows cast by a slit were sharp ones. But now we know that this was due only to the high velocity of the electrons and their concomitant short wavelength. As with X-rays, only slits of atomic dimensions yielded diffraction images of appreciable magnitude. Thus an electron is able to go round a corner without the help of any external force. But *where* will it be in fact? The interference pattern around a hole acting the part of a slit has, of course, circular symmetry and it exhibits a set of concentric circles on a screen. Circles of maximum "electron intensity" alternate with circles where no electrons are arriving. Where, on any one of these circles, will a single specific electron arrive in fact? It cannot spread out, according to our macroscopic notions, into the whole circumference of the circle and still conserve its unity as a particle with a fixed charge. Whenever it gives up its energy in an observable process, as a source of light or of electrons (*see* later) it is observed at *one* specific point on these circles. From this point of view it is behaving as if it were a particle. But a particle should not have gone round the corner in the first instance! It seems at present impossible to predict just at which specific point on an interference pattern an electron is going to be caught after it has traversed such a seemingly simple system as a small hole in the path of the beam! Inability to predict where the particle will be, after all its describable parameters such as direction and velocity have been given, implies that we are unable to predict effects from causes.

Now, in most macroscopic experiments we are dealing with a great number of electrons at any one time. The sum of the individual behaviours determines the interference pattern, and what we observe in these cases may be only the distribution of a great number of electrons, according to some law of probability, over a predetermined pattern. We may position a photographic plate instead of the screen and so photograph the diffraction rings, obtaining the whole distribution with intensities determined by the probability of arrival from each direction. But the intensity of the electron beam may be reduced further and further. Assuming that the sensitivity of scientific photographic plates is already such that they can respond to the impact of single electrons, the pattern of concentric circles is then going to disintegrate into points scattered here and there along the circles. Just where, then? We do not know and we cannot tell.

The new, basic theory of physics, the theory of quantum mechanics declares that it is fundamentally impossible to answer such questions. It allows us only to make statements about the probability of finding a small particle somewhere at a given moment. As long as we are making observations on a great number of electrons, or other small particles—their actual distribution is going to follow some pattern of probability—a hundred out of a million are going to be at a place where their probability amounts to 0·0001, and so on. This probability can be calculated according to the fundamental, axiomatic, wave equation of quantum mechanics.

E. Schrödinger (1887–1961) followed up an extremely interesting idea of *L. V. de Broglie* (1892–) in 1926 concerning a possible analogy between mechanics and optics. Both branches of physics state an important law of Nature in terms of a "minimum postulate". The path of a light-ray through any series of media with any refractive indices is defined by the postulate that the time necessary for the light-ray to reach a given point from any of its previous points should be a minimum among all paths which can be imagined. On the other hand we learn in mechanics that the real path of a material body in any field of force is defined by the postulate that the integral sum of the impulse at all parts of its path multiplied by the length of the elementary paths shall be a minimum between any two points of the path.

There is a far-reaching formal resemblance between these two minimum postulates. de Broglie's and Schrödinger's idea was to extend this analogy. In optics it was already known that there are really no "rays of light" because light consists of waves, "it will go round a corner." It will only look like a "ray" as long as its wavelength is short in comparison to the dimensions we are looking into. Maybe the same is true for mechanical motion? de Broglie found an expression for the wavelength of such "material waves" by an ingenious combination of relativity theory and a new postulate of his own. The result was equation 47.2 which was soon verified by Davisson and Germer. Schrödinger went a step further than the wavelength: he found a differential equation describing mechanical motion as waves, exactly as Maxwell's equations describe the wave of electro-magnetism. The equation connects spatial and temporal derivatives of a function with the potential energy and kinetic energy of a particle at any point in space, just as Maxwell's equations connect spatial and temporal derivatives of electric and magnetic field strength with dielectric constant and magnetic susceptibility at any point of space.

But what is the "function," the derivatives of which he has combined in an equation? It is a very strange entity; it is related to the *probability* of finding the body at any given point of space at a given moment. And here we are back to probability again. The Schrödinger equation cannot be deduced from classical mechanics. It is an independent, new axiom, even contradicting causality as we shall see. But it works. Throughout the huge domain of the physics of elementary particles, solutions of this intuitively postulated wave equation yield the experimentally determined probability of finding a particle here or there at a given time.

The equation is a wave equation and waves have amplitudes. Amplitudes can be positive or negative. But probabilities are essentially positive quantities, so it is clear that the amplitude of the wave function cannot represent a probability. The situation is similar to that of the electromagnetic equations: there the amplitudes cannot represent energies, because energy is essentially a positive quantity. In Maxwell's equations the amplitudes are the field strengths, and the energy at any given point and time is proportional to the square of the field strength. In Schrödinger's equation too, the probability is proportional to the square of the amplitude of the function. (When complex numbers are being used in the calculations, as is the case here, it is not exactly the square of the amplitude but the amplitude multiplied by its conjugate complex number. Essentially we may think of it as being the square.) The function itself is the square root of the probability of something that has a meaning in physics. It has no other meaning in itself. Therefore it is only designated by a Greek letter: ψ (psi) and is known as this everywhere in quantum mechanics.

As far as constants are concerned the equation contains the mass of the body we are dealing with and the universal constant h, which we have already encountered. This has a very small value in C.G.S. units: $6{\cdot}55 \times 10^{-27}$ and its dimensions are rather strange, ergs times seconds (i.e. erg-seconds). This is energy times time or what amounts to the same thing, impulse times length: g cm² sec⁻¹. We have encountered it in the minimum postulate of mechanics when a sum of products of momentum times length has to be a minimum. It is called "action" and it is very difficult to visualize what it amounts to. But it *is* exactly defined and it seems to be fundamentally important in Nature.

Historically, *h* first entered theoretical physics at the hands of the great German physicist *M. Planck* (1858–1947) when he found an equation describing the wavelength distribution of the light energy that is emitted by a black body (say a small hole in a box) at given temperatures (1900). Since light consists of waves of coupled electrical and magnetic fields, physicists naturally hoped that this experimentally well-measured distribution function could be derived from a combination of electromagnetic theory and thermodynamics (the theory of heat and work, based on the laws of the conservation of energy and the fact that no "second kind of *perpetuum mobile*" is possible (Section 39). However, the combined theories yielded results in flat contradiction to experiments. They demanded that the maximum of energy emitted should be at an end of the total wavelength spectrum, whereas experiments showed and we all know it, that it is somewhere in the middle, tending towards shorter wavelengths with increasing temperature. A body at room temperature emits infra-red (heat) waves. As its temperature is increased it begins to glow, first dark red, then red, yellow and even blue in the case of the hottest stars. An "end" of a wavelength spectrum is an absurdity in itself: it is non-existent and a formula locating maximum energy in a non-existent region is absurd too. Planck found a solution by making an assumption that had no basis whatsoever in the theoretical physics of his time. He postulated that the amount of electromagnetic energy radiated by a rotating or oscillating charge cannot vary continuously but only in jumps, in "quanta" which are proportional to the frequency of the movement.

$$E = h\nu \qquad (h \text{ is the constant of proportionality}) \qquad . \quad (47.3)$$

By introducing this strange assumption into the theory, the resulting equation fitted the experimental results excellently. With this formula the notion of discontinuous quanta of energy and the constant *h* entered the domain of physics to become more and more important ever since, and to overthrow even the idea so natural to us that all happenings are causally determined.

The wave equation of Schrödinger only gives us probabilities of where to find a body, but on the other hand it explains automatically the existence of energy quanta for certain states of motion. It so happens, that it often has no solution at all, save for cases where the constant representing the total energy of the system has a series of discrete, well-defined values. For all energy values in between there is no meaningful solution; this is a mathematical property of this equation. There was no need to introduce the discontinuity as a postulate of physics. The permitted values of energy are, so to speak, the "*own* energy values" of the equation for given boundary conditions. It has become customary to use the German word for "own" in most languages, among them English. The "own value" is called the "eigenvalue" and the corresponding, permitted psi (ψ) function is an "eigenfunction."

Differential equations which can only be solved when one or more of their constants have definite values are not new in physics. An analogy from normal macroscopic physics may help to clarify the situation a little. A string of a violin cannot vibrate in all modes which may at first seem possible. First of all it can of course vibrate with its maximum amplitude in its middle. But every violinist knows that it can also be made to vibrate in "flageolets" by gently touching it with a finger at 1/2, 1/3, 1/4, etc., of its total length. In these cases it is going to vibrate with one, two, three, etc., nodal points equidistant along its length and it is going to emit sound of twice, three times, etc., its original frequency (octave, second quint, quart, etc.). No other modes of vibration are possible for the whole string. In two dimensions an elastic plate can be made to vibrate by rubbing it with a bow at given points while holding down other points upon it. Different notes are emitted and the mode of vibration can be demonstrated by putting some powder on the plate: it will accumulate in the "nodal lines" where the plate surface remains at rest, just as did the violin string at its nodal points (Chladni figures). The possible vibrating states of this plate are now characterized by two sets of numbers, a set for each main axis of the plate. The string had only one set of characteristic numbers, cutting it into a certain number of equal sub-lengths. Something capable of vibration in three dimensions should, by analogy, be characterized by three sets of numbers (p. 261).

We possess in the Schrödinger equation an invaluable tool for predicting the probability of finding particles at specific points at given times, and we can also obtain given, discrete values for their energy in a series of their discrete stationary states. Only a few months before Schrödinger, a young German physicist, *W. Heisenberg* (1901–   ) attacked the same problem from an entirely different starting point. In his fundamental principle, the axiom called the principle of uncertainty, he demands that no two physical quantities which yield a quantity of the dimension of "action" (p. 209; $g\,cm^2\,sec^{-1}$) when multiplied by each other, should be simultaneously accessible to measurement beyond a certain limit of accuracy. Thus the position and the impulse of a body (which when multiplied yield action) or the energy and the duration of a process (which also yield action when multiplied) can never *both* be measured, simultaneously, with infinite accuracy. Heisenberg postulated that the uncertainties of these quantities (denoted by the symbol Δ), multiplied by each other, should always yield Planck's constant, h. (According to how we define the uncertainty of a measurement it is h or $h/2\pi$). Thus for the position coordinate x and the impulse coordinate mv we obtain the fundamental relation of Heisenberg:

$$\Delta x\,\Delta(mv) = h \qquad . \qquad . \qquad . \qquad . \quad (47.4)$$

(For velocities which are not near to the velocity of light we need not take account of relativistic effects and therefore we may regard m as being constant, so that the uncertainty of measuring impulse actually means the uncertainty of measuring velocity.)

It was a glorious feat of Heisenberg to have shown that his theory and the wave equation of Schrödinger are equivalent: one can be deduced mathematically from the other. Both are axioms and both introduce the concept of microscopical uncertainty, of probability, instead of an exact and binding relation between cause and effect. The approach of Heisenberg is probably more beautiful as a logical structure, but the wave equation of Schrödinger lends itself better to the usual methods of differential calculus. Actual problems are therefore generally attacked from the side of the Schrödinger equation.

Let us see now the quantitative consequences of the principle of uncertainty. These are, of course, determined by the magnitude of the constant h in equation (47·4) and naturally in the wave equation of Schrödinger too, which is not reproduced in this book because it is devoid of meaning for those not acquainted with partial differential equations. As stated already h has the value $6·55 \times 10^{-27}$ g cm^2 sec^{-1}. This was the value that helped Planck to fit his formula to known experimental values of the energy emission curves of black bodies. It is also the value that we shall encounter when dealing with the photoelectric effect in Section 49.

How does the magnitude of h affect two measurements of different orders of magnitude? First take a body of mass 1 mg, that is about the magnitude of a pinhead, and let it move extremely slowly, with a velocity of say 1 mm in 100 sec. This amounts to a velocity of 0·001 cm/sec (10^{-3} cm/sec). Since its mass is 10^{-3} g, its momentum (mv) is 10^{-6} g cm/sec. If we divide h in equation (47·2) by this value we arrive at the "wavelength" as a "material wave". We obtain $6·52 \times 10^{-21}$ cm, a billionth of an Ångström unit. This means: we are certain about its location. There is no crystal lattice so fine as to come near to diffracting such a minute wavelength. As a second example take a molecule of hydrogen. We know its mass to be $3·3 \times 10^{-24}$ g and its mean velocity at normal temperature $1·7 \times 10^5$ cm/sec. Thus its momentum is $5·6 \times 10^{-19}$ g cm/sec. Divide h again by this value: $1·2 \times 10^{-8}$ cm is the result. Thus we have had to descend to the lightest molecule in order to come into the Ångström unit range for its wavelength. As a matter of fact it has really become possible to obtain diffraction phenomena by using molecular beams of the very lightest molecules. The momentum of an electron is even some two thousand times smaller than that of a hydrogen atom and therefore its wavelength is going to be longer in this proportion at the same velocity.

This is the reason why the wave nature of matter did not appear in macroscopic physics. The objects were too large as far as their mass times velocity (their momentum) was concerned.

Fifty years ago it seemed trivial that it is enough to know the position and velocity of a body together with all the forces acting upon it in order to predict its path exactly in the future. The future was thought to be determined by the present. Now we come to learn that the path of the same small particle under the same external circumstances, say during diffraction on the same crystal, will lead into an endless series of different places in accordance with a mysterious rule of probabilities. The present does not determine the future any more. It is only possible to say that a given future situation follows with a certain given probability if the present situation is given.

But is a situation ever "given" at all? There are at least two serious objections to this being the case. First of all every body interacts with all other bodies in the universe and the whole situation is only described completely if this virtually infinite number of interactions is taken care of, which is evidently impossible. Thus one could always say that the different result of "the same" repeated experiment is due to the simple fact that the experiment has not been the same, we have not taken care of all this enormous series of influences.*

But we do not need to go far out into the universe. As we shall see presently, atoms themselves are composed of many electrons and other particles and hence the situation within a given crystal may change enormously between two processes of diffraction if these electrons move around in some way or other, as they were postulated to do in the Bohr theory of the atom (p. 215). Thus it might be that two seemingly identical situations of the lattice are not identical at all, and that we have neglected important parameters which we were unable to observe. It would be a situation similar to one encountered in the kinetic theory of gas molecules: we do not predict the path of a molecule because it is going to interact with so many others in such a complicated manner. We only draw some statistical conclusions of probability as to the pressure exerted by a great number of molecules on a wall, their mean free path, rate of diffusion, etc. Attractive as this solution to our dilemma seems, it has been refuted by the great mathematician *J. v. Neumann* (1903–1961). However, we have arrived at present-day science with living controversies. J. v. Neumann's theorem is accepted by the majority of living scientists specializing in this field. But one of the fathers of the quantum wave theory, *L. de Broglie*, together with *D. Bohm*, is not convinced by his arguments and still hopes for a solution of the basic problem by invoking neglected parameters for the fundamental uncertainty.† The struggle for more understanding is going on before our eyes; though one party has increased its influence, the other is not vanquished as yet. It is very interesting that other great physicists of our times, *A. Einstein* and another father of the quantum wave theory, *E. Schrödinger*, were also among those who were unhappy about the present situation and hoped that out of this transient confusion a clearer understanding of fundamentals would arise.

The majority of physicists seem to be quite happy with the present state of affairs. The author has met great scientists who have tried to convince him

---

* For forces like gravity, which only decrease with the square of the distance from the particle in question, it would not even help us to argue that distant bodies can be disregarded as sources of influence. Supposing that the density of matter is the same everywhere in the universe, the amount of matter within a spherical hull surrounding our particle will *increase* as the volume of this hull, that is *with the square of its distance* from the particle. The intensity of gravitational attractive force, for example, *decreases with the square of this distance*, so that the effect of a spherical hull, no matter how far away, will always remain the same and add up to infinity—or to the limit of the universe.

† The author regrets that he is unable even to sketch the outlines of v. Neumann's argument and the counter-arguments because they far exceed his understanding. The reader would be well advised to learn the elements of quantum mechanics, also, from a more authoritative source.

that there is no dilemma at all, and that a statistical statement is just as good as a non-statistical one. They have said that determinism is conserved. Maybe the difference lies in what we call determinism. For the author, personally, it is certainly not the same whether he knows that 50 per cent of slowly incoming neutrons would blow up a given atomic bomb or that a single neutron which is going to reach it presently will have the same effect or not. But in fact it may be that common sense is not sufficient to cope with the situation.

There is a second, non-controversial cause besides neglected parameters which may be behind this fundamental uncertainty. It is not necessary to look for a multitide of unknown parameters in order to explain an inability to predict the path of a particle. We have said that with a knowledge of the position, the velocity and forces acting upon a particle, classical mechanics should enable us to predict its path in the future. In the preceding paragraphs we have discussed the question whether "all forces" acting upon a particle can possibly be known. But how about its position and its velocity? In order to measure anything we must have instruments. To arrive at a reasonable accuracy of measurement it is necessary that the measure should be "finer" than the thing to be measured. A table can be measured satisfactorily in centimetres and not in metres. However, what are we going to use to measure electrons? They are the lightest stable particles known to us and for all we know they may really be the lightest in existence—and the "smallest", too, so far as a meaning can be attached to the concept of the dimensions of an electron.

If this is true, we have no smaller or lighter entity at our disposal to measure any property of an electron than an electron itself. There remains the possibility of making some measurements with "immaterial" electromagnetic waves, very short-wavelength waves, beyond the X-ray region even. Such so-called gamma rays do exist as we shall see in a subsequent section (p. 404). But we have already promised to show that the energy of light is concentrated for all purposes of interaction with anything else, into small bundles, "quanta" (= photons) as they are called. We shall learn more about them in the next chapters. These quanta are localized, at the moments of interaction with anything, and carry momentum. Experiments made in 1933 by *A. H. Compton* (1892–1957) on the interaction of electrons and X-rays showed definitely that an electron moved away after such an encounter as it would have done had it collided with an elastic ball of the energy and momentum of the light quantum.

The energy, as well as the momentum, of a light quantum, have been found to be inversely proportional to its wavelength (p. 223). Thus the shorter the wavelength we choose for performing measurements upon an electron, the greater the momentum of this light quantum is going to be. Now let us suppose that we are going to locate the position of an electron by means of a very short-wavelength (light) quantum, in some hypothetical instrument we could well call a γ-ray microscope. The theory of the optical microscope shows that structures smaller than one-fourth of the wavelength of the light used for viewing cannot be sharply seen and hence they cannot be accurately located. Therefore we have to "look" at an electron with light of the shortest wavelength at our disposal and this is the reason why we are trying to use gamma (γ) rays. But short wavelength means high frequency and high frequency means high energy of the photons which will interact with the electron like one swiftly moving ball with another. We may have determined the electron's position at the moment of collision with the gamma-photon but because of the Compton interaction we have changed its velocity. Therefore we cannot measure its original, undisturbed, velocity by, say, determining its position again after a short time interval and dividing the length of its displacement by the time that has elapsed.

Worse than this, not even the direction of the velocity of an electron can be determined, because this would mean that we should have to observe which way it is going to pass through two consecutive, narrow, slits. But we have seen that an electron can go round a corner! Its path will not necessarily be defined by the straight line joining the slits since it can be bent away from it as in the experiments of Davisson and Germer. Its being found here or there around the corner is governed by probability, not by causal law.

Many clever experiments can be devised for measuring the position and momentum of an electron (or of any particle with small momentum) but all will fail because of the Compton interaction or because of the experimental fact that electrons are diffracted like waves.

These experiments cannot *prove* by classical methods that the principle of uncertainty must prevail in physics, because they use experimental facts such as the Compton effect or the wave aspect of particles which in themselves are against the principles of classical mechanics or electromagnetism. The electro-magnetic theory of light knows nothing of photons with impulse in a given direction and carrying a defined quantum of energy. Classical mechanics knows of particles which proceed along straight lines as long as they are not acted upon, and waves which go round corners, but of no hybrids between the two. What these imaginary experiments *can* prove is only that *new facts* of experimental physics exclude simultaneous exact measurements of position and momentum and thus, having excluded the possibility of knowing exactly and simultaneously where a body is at a given moment, and with what velocity it is just moving in which direction, we have lost the hope of predicting its future from its present. This is because we can never really know its present. And thus, as physics stands today, the principle of causality has fallen a victim to experimental facts.

The constant h entered physics at the hands of Planck exactly at the beginning of our century. Discontinuous "quanta" of energy, together with this "elementary quantum of action" were needed to explain the continuous curves which have been found to describe the emitted energy from a black body as a function of wavelength, at different temperatures. However, light emission of *gases* is already in itself a discontinuous phenomenon: a few wavelengths of light are emitted by glowing gases and that is all, the rest of the spectrum being black, empty. Their light absorption is similar. As we shall see in Section 64, it took thirteen years (1913) after Planck's breakthrough before another giant in physics, *N. Bohr* (1885–1962) explained the structure of the simplest spectrum known, that belonging to the hydrogen atom, by combining the constant h with new and *ad hoc* assumptions concerning stable states of electrons revolving in an atom, and emission of energy in discrete light quanta. Since then the whole science of spectroscopy has become unified by quantum mechanics and Bohr's *ad hoc* assumptions appear as logical consequences of the principle of uncertainty.

Not only electrons but all matter is subject to this extension of mechanics. Whenever momentum (impulse) is small, and whenever the space to which a particle is confined is small, so that the product of the space coordinate with its connected ("conjugate") coordinate of impulse becomes commensurate with h, we have to resort to quantum mechanical calculations. Not only electrons have been diffracted by crystals. Bundles consisting of neutrons (p. 427), and positive ions of hydrogen and helium (*see* Section 97) also suffer diffraction, and their effective wavelength is determined by the same equation, 47.2.

The science of quantum mechanics has had to be built up in such a way that, on one hand, its results fit the experimental results of the submicroscopic world of the smallest particles, while going over, on the other hand, into the well-established laws of classical macroscopic physics, as the momenta of the systems in view increase. It is one of the many immense merits of the great Danish physicist, N. Bohr, that he proclaimed this "principle of correspondence" in 1920, in the first years of the then new quantum theory. He demanded that whatever the new laws of quantum mechanics were eventually going to be, they must be such as to go over continuously into the laws of macroscopic classical mechanics with the increase of momenta. This demand proved to be of great benefit to those who searched for exact formulation of the new equations.

There is no doubt that quantum mechanics has disturbed "normal" physical thinking even more fundamentally than the theory of relativity. It demanded newer and much more abstract concepts and arguments. But nothing would be more false than to imply that physics has collapsed. Physics, like all natural science, is based on observations, and arranges these into a system. From time to time we learn that descriptions of these systems are only valid within certain limits: on the side of great velocities relativity, on the side of small momenta quantum mechanics take over from normal classical mechanics. But there are no boundaries in between them: the new equations are such that they include the old ones within themselves for the normal range of momenta and velocities. The unity of the system has been conserved, only unwarranted extrapolations have been replaced by others which are born out by experiments. Causality became one of the casualties in this evolution.

Just to show how far-reaching the consequences are even outside the fields of science and technology, the author would like to take the liberty of digressing for a short moment into ethics. Determinism or indeterminism is an old problem not only of science but also of ethics. If everything is predetermined, there is no personal moral responsibility, and what remains is rigid predestination. Good or evil may remain according to an established or proclaimed or rational code of morals, but a bad man is no more to blame for his action than is the blond man for the colour of his hair. And there seems to be very little point in doing anything one does not wish to do, because maybe it is not determined that it should be done anyway. So why do it? The arguing around determinism has gone on for thousands of years. The net result is nil. Science seemed to teach us in the last centuries that everything in the material world is governed by strict causal relations, that is, it is predetermined. So far as I am able to judge the behaviour of my fellow beings, nobody has ever acted upon this principle in making decisions of any kind and very few have rejected moral standards because of predestination, divine or physical as the case may be. Perhaps the Calvinists were at least consequent determinists in assuming that God created bad men for a special reason of His own which exceeds our intelligence and that He created their predetermined damnation as well. Communists become furious if one tries to defy their *moral* arguments by accepting their materialistic *determinism*. They are not to blame for not having solved a problem which has been there since the beginning of history. But everybody should be honestly aware that he lives as if he possessed a free will while often believing on scientific grounds that no such free will can exist.

Has this situation now changed? Is one hundred per cent determinism really dead and are we free to hope that free will, moral responsibility and a consciously built future is going to emerge from this revolution in physics? I surely do not know and if anybody does profess to know I shall instinctively disbelieve him. But it *may* well prove to be the case after all.

We still live in the turmoil of these physical revolutions. The author is alas not in the position to explain them more comprehensively and probably stands as baffled before them as most of the readers. It would evidently be better for the readers to consult somebody who is really acquainted with these new domains in order to gain more insight into them. In the meantime the advance continues and we have every reason to look with anxious interest into the future.

48. X-RAYS: LIGHT FROM ELECTRONS

When the electrons of a cathode-ray beam impinge upon matter many things happen. The body upon which they impinge becomes hot: the kinetic energy of the electrons is transformed into heat, in the same manner as the kinetic energy of a falling stone when it strikes the earth. This heat may be sufficient to melt even highly refractory metals and it is often used for this purpose. These vacuum-melted metals may attain a very high degree of purity. The cathode-ray electrons are also able to transmit momentum. If a very light wheel with blades protruding beyond the rim is mounted in the way of the beam so that only the upper or the lower blades intercept the electrons, the wheel begins to turn as soon as the beam is switched on. The momentum of the electrons has been transferred to macroscopic matter.

But a quite different phenomenon may occur, apart from these thermal and mechanical effects. *W. K. Röntgen* (1845–1923) and *P. Lenard* (1862–1947) observed in 1895 that a then new kind of radiation emerged from that part of the tube upon which the electrons impinged. It was invisible to the naked eye but blackened a photographic emulsion and caused gases to ionize and conduct electricity when it traversed them. This radiation passed through matter, the more easily the lighter the atoms of which the matter was composed. Thus it could permeate the human body. The fleshy parts being composed mainly of the atoms H, O, C, N were more translucent to it than the bones with their heavier Ca and P atoms. The best way to screen off the radiation was to introduce the heavy lead atoms into its path. The penetrating power of the rays increased with the kinetic energy of the original cathode-ray electrons which produced them. They are now called Röntgen rays in continental Europe and X-rays in Anglo-saxon countries. We have already referred to them in Section 27, where we learned how v. Laue managed to prove that they had the nature of waves and how it became possible to determine their wavelength by diffracting them through crystal lattices. X-rays never suffer deviation in electric or magnetic fields, however strong. Thus it was concluded that they are electromagnetic waves like normal and ultraviolet light but that their wavelength must be some three orders of magnitude shorter.

The interference, or diffraction, experiments revealed that the wavelength of X-rays becomes shorter the greater the velocity, and thus the kinetic energy, of the generating cathode-ray electrons. The kinetic energy in its turn is of course determined by the voltage drop across the X-ray tube, because it is this potential difference which accelerates the electrons (p. 202). Now the X-rays emerging from an X-ray tube are not monochromatic, that is consisting of a single wavelength, any more than is the light emitted from an incandescent lamp. They consist of X-rays having a range of wavelengths. The long wavelength limit to X-rays begins with those rays which are just not absorbed by the glass or other window of the tube, after which the radiation has a decreasing wavelength until a well-defined experimental limit is reached. No rays whose wavelength would be shorter than this limit are emitted. The limit depends *only* on the kinetic energy of the generating electrons; as a matter of experimental fact it is inversely proportional to this energy. The numerical factor of proportionality is $1 \cdot 234 \times 10^{-4}$ if we measure the wavelength in centimetres and the energy of the electrons by the voltage that was necessary to accelerate them. Thus we have:

$$\lambda = \frac{1 \cdot 234 \times 10^{-4}}{\text{voltage}} \qquad . \qquad . \qquad . \qquad . \quad (48.1)$$

This experimental equation may be transformed a little and in doing this we shall encounter a constant with which we are already acquainted. First of all we will use the frequency of light instead of its wavelength. In the case of a wave of wavelength λ, travelling with a given velocity, c, it will have travelled c centimetres in a second and in this distance there will be c/λ waves, of course. This number of waves must pass through any point in the direction of travel in each second. There will therefore be c/λ maxima or minima of the waves passing a given point in a second and this number is the frequency v of the waves:

$$v = \frac{c}{\lambda} \qquad \cdot \qquad \cdot \qquad \cdot \qquad \cdot \qquad \cdot \qquad (48.2)$$

Now let us change over from the units of accelerating volts to real kinetic energy of the accelerated particles. We have seen on p. 201 that a charge e acquires a kinetic energy Ee after having "fallen" through a potential difference E, just as a falling body near the surface of the earth acquires the energy mgh after having fallen through a difference of levels h (g is the constant of gravitational acceleration at the earth's surface.) Inserting the appropriate constants for the charge of the electron and for the potential in volts we obtain $1\cdot59 \times 10^{-12}$ erg of energy for an electron that has "fallen" through 1 volt and hence $v \times 1\cdot59 \times 10^{-12}$ erg for one that has fallen through v volts. Therefore the voltage that yields the kinetic energy E_k is

$$V = \frac{E_k}{1\cdot59 \times 10^{-12}} \qquad \cdot \qquad \cdot \qquad \cdot \qquad (48.3)$$

Inserting this value for the volts, and c/v for λ in equation (48.1) we arrive at

$$\frac{c}{v} = \frac{1\cdot234 \times 10^{-4} \times 1\cdot59 \times 10^{-12}}{E_k} \qquad \cdot \qquad \cdot \qquad (48.4)$$

and remembering that c, the velocity of light, is 3×10^{10} cm/sec we finally have

$$6\cdot55 \times 10^{-27}v = E_k = hv \qquad \cdot \qquad \cdot \qquad \cdot \qquad (48.5)$$

The constant, $6\cdot55 \times 10^{-27}$ is an old acquaintance of ours; it is Planck's constant which we denoted by h in the preceding section. It is the same constant that connected the momentum of an electron with its wavelength. Now it connects the energy of the electrons with the shortest wavelength (or highest frequency) of X-radiation which they are able to generate.

We shall see that this relation holds whenever energy is converted into light or vice versa. It is a very remarkable relation, not at all in harmony with classical concepts of an electromagnetic radiation process. According to classical theory, the frequency of an electromagnetic wave should be determined by the frequency of the movement of a charge along an axis, as in the case of the movement of an alternating current in an antenna. Now we begin to learn that when we reach the dimensions of elementary particles the frequency is determined by the energy of the particle whose energy is transformed into light by some process!

In an entirely different experimental set-up due to *J. Franck* (1882–1964) and *G. Hertz* (1887–) the same relation emerges for light near the visible range of wavelengths (1914). A beam of electrons is accelerated through a potential difference of a few volts in a gas at low pressure, say mercury vapour. At a certain voltage the mercury vapour begins to emit ultraviolet light and it can be ascertained that at this moment the electrons lose all their kinetic energy to the vapour (*see* detailed discussion on page 296). The voltage was 4·9 V and the wavelength of the u.v. light which was emitted was 2,537 Å, that is 2.537 × 10⁻⁸ cm. Again converting voltage to electron energy by multiplication of v by 1·59 × 10⁻¹² we may determine the ratio of the energy to the frequency ($v = c/\lambda$). Once more we find that the ratio is exactly 6·55 × 10⁻²⁷ erg, that is, is h. This ratio is always found to be h whenever energy is converted into light.

49. PHOTOELECTRICITY: ELECTRONS SET FREE BY LIGHT

X-rays are produced when fast electrons collide with matter. In Franck and Hertz's experiment slow electrons collided with matter and produced u.v. light. Now let us consider the reverse of these phenomena. Can electrons be

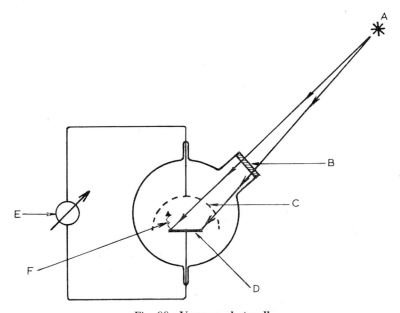

Fig. 90. **Vacuum photocell**

When the metal cathode is illuminated, negative current flows from it towards the wire-mesh anode.

A = light source   D = photo-cathode, e.g. alkali metal
B = quartz window   E = micro-ammeter
C = wire-mesh hemisphere anode   F = path of ejected electron.

produced from matter by the action of light? They can. (*W. Hallwachs,* 1888).

Figure 90 shows us an apparatus called a vacuum photoelectric cell. It is a bulb of glass or silica containing a metal plate coated with a layer of say, sodium metal. A metallic lead through the bulb provides an electric connexion to this sodium film while another lead connects to a wire-mesh electrode which is positioned opposite to the sodium layer. The two electrodes are connected through a microammeter. As long as the cell is in the dark nothing happens, but the moment we allow light to fall upon the sodium we observe a current. Its direction reveals that electrons must have left the sodium surface and have reached the other electrode. Application of an external potential so that the sodium becomes negative and the grid positive (photocathode and anode respectively) increases the current to a limiting value for any level of illumination. No matter how much we increase the potential, the current will not exceed this value, but it does increase, and in direct proportion, with the *intensity* of the light. This is the reason why photocells are often used to measure the intensity of light with great precision.

However, if we vary the wavelength of the incident light towards longer and longer wavelengths we observe that at a well-defined limit the whole photo-electric phenomenon disappears. This limiting wavelength is characteristic of the metal used within the photocell; in our case, sodium. The limiting wavelength is very sensitive to small contaminations of this metal surface.

It was not imperative to apply a voltage across the photocell. During illumination electrons left the sodium spontaneously and arrived at the anode quite by themselves, but some, of course, went astray when there was no positive voltage on the anode to attract them. They must have traversed the space between cathode and anode with some particular velocity. This velocity can be determined by applying a reversed electric field to the photo-cell, making the sodium positive and the anode negative in order to repel the electrons. By slowly increasing this reversed voltage from zero the current of electrons gradually decreases and vanishes entirely at some characteristic voltage. The value of this voltage depends both on the metal *and* on the wavelength of the light used in the experiment. The work of *Lenard* (1899) revealed that the maximum kinetic energy, eP, of the electrons (i.e. the last electrons in the experiment, those which were forced back into the cathode at the limiting potential P) was linearly related to the frequency of the light:

$$eP = h\nu - \text{const.} \qquad . \qquad . \qquad . \qquad (49.1)$$

The coefficient of the frequency is once more h, 6.55×10^{-27} erg. sec, Planck's constant! This time it governs the transformation of light of fre-quency ν into kinetic energy of electrons! This is exactly the reverse of the phenomenon that we discussed in the preceding section.

What about the subtractive constant in the above equation? Remember that there was a limiting wavelength above which it was impossible to detach any electrons at all from the given metal surface. This wavelength corre-sponds of course to a limiting frequency through the relation $\nu = c/\lambda$, below which there is no effect. Denoting this limiting frequency by ν_0 we arrive at the subtractive constant once more by multiplying by h; it is $h\nu_0$.

The equation of the maximum kinetic energy of the emerging electrons thus becomes

$$E_k = eP = h\nu - h\nu_0 = h(\nu - \nu_0) \qquad . \qquad . \qquad (49.2)$$

The graphs in Fig. 91 show this experimental relation.

The meaning of the limiting frequency ν_0 is fairly self-evident. It represents, on multiplication by h the energy deficit of the emerging electrons compared

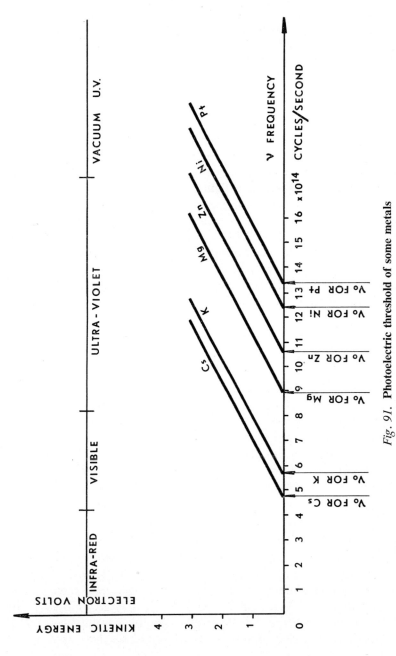

Fig. 91. **Photoelectric threshold of some metals**

For each metal there is a characteristic frequency of light below which no photo-electrons emerge. Above this threshold the kinetic energy of the photo-electrons increases linearly with the frequency of incident light.

with such electrons as might be imagined to emerge with the total energy of the light quantum $h\nu$. This difference is the energy the electron has to use up in leaving the lattice of the metal, in analogy to the heat of vaporization which a molecule has to pay when it leaves the surface of a liquid. It is called the work function of the metal.

Surveying the alkali metals and after them all other metals which are photo-electric (that is all metals), we conclude that photoelectrons leave metals the more easily the less noble they are. Caesium loses electrons even when irradiated with infra-red light. Rubidium follows next and then all the other metals in roughly the same order as the potentials necessary to decompose their compounds in electrolysis. This is to be expected. Reactive, non-noble metals part easily with their electrons and yield ions, be it a matter of chemical reaction, e.g. transferring the electrons to oxygen (*see* Section 60), or dissolution at an electrolytic anode when the electron goes into the external current, or by the action of light.

Fig. 91 shows us the photoelectric energy–frequency plots of some metals: the lines are of course all parallel to each other, the slopes being equal to h because of the relationship $E_k = h(\nu - \nu_0)$. They only differ in their intercepts $h\nu_0$ which are a measure of their "nobility". Electrons may be ejected from non-metals and even from gas molecules also, provided the energy of the light quantum is greater than their energy of ionization; in general for non-metals (and many metals) this occurs in the ultraviolet region of the spectrum.

50. PHOTOCHEMISTRY

The photoelectric effect is not the only effect of light on matter. There are some chemical reactions which proceed only in light or which proceed in light at appreciably lower temperatures than they would without it.

On mixing hydrogen with chlorine gas nothing will happen as long as the mixture is in the dark at not too high a temperature. Red light will not affect the mixture; the latter will, however, explode violently on irradiation with blue light. Hydrogen chloride is the product.

$$H_2 + Cl_2 = 2HCl \quad . \qquad . \qquad . \qquad . \quad (50.1)$$

The dependence of photoresponse on wavelength is also found with the silver halides used in photography. If they are not specially sensitized by dyes they are stable in red light and only begin to decompose if light of shorter wavelength than corresponds to the green falls upon them. Silver chloride turns a violet or black colour, the bromide only grey but they both acquire the property of being capable of photographic development, that is reduction to silver by certain solutions which would not have reduced the halide without illumination. Wavelengths longer than blue-green are not absorbed by these silver halides, this is why they do not act upon them. This is the first universal law of photochemistry: only light which is absorbed by a body can act upon it (*Grotthus*, 1820). By adsorbing special dyes on to the surface of the silver bromide crystallites, they are made to absorb, for example, red light also, and transmit its energy to the halide, thereby bringing about the primary reaction which makes the halide capable of development. These are the "panchromatic" emulsions.

There is, however, a wavelength limit in all photochemical reactions above which no wavelength causes reaction, just as there was a limit in the photoelectric effect. The mixture of hydrogen and chlorine reacts only when irradiated with light quanta of sufficient energy to dissociate the chlorine molecule (blue-green). This energy is known from thermochemical measurements also, which may be used to recalculate by means of the $E = h\nu$ relation, the frequency and thus the wavelength minimally necessary to initiate the reaction. This calculated wavelength agrees with that determined photochemically, except for a small difference which can be accounted for.

What happens, however, when a photochemical process does begin to take place? There are two main possibilities. Either the reaction proceeds simultaneously with the light absorption and in proportion to the product of the light intensity and time of irradiation. Or a minimum amount of light suffices to *initiate* a reaction which then proceeds by itself. Such is the reaction of hydrogen with chlorine. It is not even necessary in this case that the whole mixture should have been exposed to light; exposure of a small region is sufficient to initiate the reaction. In reactions of this type it is evident that light only serves to overcome an initial inertia of a reaction which is able to proceed by itself, as a fuse detonates an explosive.

Here we are more interested in the former case. From careful measurement of the amount of light absorbed, of its wavelength (or frequency) and of the amount of matter that has reacted photochemically, a very interesting result emerges which was first announced by the founder of relativity, A. Einstein (1905). Let us compare the amounts of substance reacted for equal total energy absorbed, at different frequencies. The amount decreases with increasing frequency, but in the direction of lower frequencies there is a definite limit. At lower frequencies, that is at longer wavelengths than this limit there is no reaction at all. We have already mentioned this fact in connexion with the chlorine-hydrogen reaction, but that belonged to the explosive type. Why does the amount reacted decrease in the non-explosive reactions with increasing frequency? Einstein said: because the energy of light is not available for chemical reactions in a continuous manner but can only be made to act in bundles, in "light-quanta" or "photons" of given magnitude. The magnitude of a quantum of light is given, according to Einstein by $E = h\nu$, and is therefore proportional to the frequency; h is of course again Planck's quantum constant.

We are dealing with the same equation that determined the wavelength of X-rays as a function of the generating electrons and which inversely determined the velocity of a photoelectron ejected by X-rays. (In this case the energy of the work function (p. 220), $h\nu_0$, is negligible beside the enormous frequency, that is energy, of the X-ray quanta. Therefore the equation $E_k = h(\nu - \nu_0)$ simplifies for X-rays to $E_k = h\nu$). We shall soon conclude that in all three cases we have come to deal with the "atomic structure of light."

But let us return first to the photochemical consequences of the Einstein relation.

Einstein supposed that one light quantum is absorbed by one molecule and in so far as the energy thus absorbed was sufficient for the given chemical reaction, this reaction would occur with that molecule. Thus for every light quantum, $h\nu$, absorbed, there is one molecule that reacts. In view of the fact that the energy of a quantum is given by $h\nu$ it is clear that a given total energy corresponds to a smaller number of light quanta the higher the frequency of the light. Thus, necessarily, the number of reacted molecules is also smaller, as has actually been observed. The optimum case evidently occurs when the energy, the frequency, of the light corresponds exactly to the minimum energy necessary to enable a molecule to react: a given sum total of energy will then activate the maximum number of molecules. If the energy of a quantum is greater than necessary the surplus will only serve to impart kinetic energy to the product of reaction: it will appear as heat. But the moment the frequency falls below the optimum value, the magnitude of a quantum is not sufficient to bring about a reaction of the molecule and there is no photochemical reaction at all.

This law of Einstein is generally called the law of photochemical equivalence. As often as the difficulties of measuring all the necessary data have been overcome, this law has always been verified for primary reactions, in the case of the direct photochemical decomposition of silver halides, for instance.

But the law also explains more complicated relations. For example, experiments showed that for every quantum of ultraviolet light absorbed, two molecules of hydrogen iodide are decomposed according to the equation

$$2HI = H_2 + I_2 \quad . \quad \quad . \quad \quad . \quad \quad . \quad (50.2)$$

Thus it seems to be sufficient if one of the two reacting hydrogen iodide molecules has absorbed light energy. It dissociates into a hydrogen and an iodine atom and because the bond energy between the two hydrogen atoms in the hydrogen molecule is much greater than between the hydrogen and iodine atoms in hydrogen iodide a free hydrogen atom reacts with a non-irradiated hydrogen iodide molecule to yield molecular hydrogen and a free iodine atom. This eventually combines with the iodine atom liberated by the original act of light absorption.

$$HI + h\nu = H + I \quad . \quad \quad . \quad \quad . \quad \quad . \quad (50.3)$$

$$H + HI = H_2 + I \quad . \quad \quad . \quad \quad . \quad \quad . \quad (50.4)$$

$$I + I = I_2 \quad . \quad \quad . \quad \quad . \quad \quad . \quad (50.5)$$

Thus one quantum of light causes the decomposition of two hydrogen iodide molecules. The assumption as to the reaction of the free hydrogen atom with a hydrogen iodide molecule can be proved directly by allowing atomized hydrogen to react with the latter—it is no *ad hoc* hypothesis.

In some cases there are deviations from Einstein's law. Fewer molecules react than have absorbed light quanta. This comes about when the activated or dissociated molecule has only a small chance of finding the necessary partner for the second step of the reaction series and has sufficient time to lose its energy of activation or to recombine before it reacts. Thus there are deviations from Einstein's law in both directions: one in the direction of induced explosions or successive polymolecular reactions, the other in the direction of deactivation. But both types of deviation generally lend themselves to well-founded explanations and many cases have been analysed completely, including much more complicated ones than the photodissociation of hydrogen iodide. The law of Einstein holds good for the *primary* action of light on matter; the successive steps of the normal chemical reactions which ensue must be elucidated in each case.

In general, photochemical reactions are not influenced by the temperature of the reactants, whereas the velocity of normal chemical reactions increases considerably with temperature. Why? To answer this question let us calculate the energy of a single light quantum of, say, the wavelength of the yellow sodium light. This wavelength is $4 \cdot 8 \times 10^{-5}$ cm. Dividing the velocity of light by this number according to eqn. (48.2) we obtain the frequency $5 \cdot 2 \times 10^{-14}$ waves per second. Multiplying with h we obtain the energy of a single such quantum: 34×10^{-13} erg. Let us compare this energy with the average kinetic energy of a molecule at room temperature (p. 150) which is about $0 \cdot 56 \times 10^{-13}$ erg. The light energy is about 60 times the greater! Therefore it would be necessary to raise the temperature to 60 times the room temperature (on the absolute scale) in order to have energies commensurate with that of a yellow light quantum. This would mean a temperature of $60 \times 300 = 18,000°$ absolute. It is clear that under these circumstances temperature cannot compete with light energy. This is the enormous advantage in using photochemical reactions: in a single, well-aimed stroke one can introduce an enormous amount of energy into a single molecule, often into a single bond, which is then able to undergo a specific transformation, without exposing the whole mass of matter to temperatures which would surely destroy all but the most stable of molecules. The environment of the photoreacting molecule remains relatively cold.

51. PHOTON AND LIGHT WAVE

No matter whether light is transformed into chemical or kinetic energy or these energies are transformed into light, we always find that the Einstein relation between the energy involved and the frequency of the light in question is valid.* Perhaps the reader has noticed that a fundamental difficulty is implicit in this simple relation. It flatly contradicts the assumption that light has the nature of waves.

According to the wave theory, light should be continuous in space and time, whereas according to the experiments which have led to the Einstein

---

* Chemical energy is converted into light in chemiluminescence. We have not dealt with this phenomenon, but it is also governed by the Einstein equation.

equation it is essentially discontinuous as often as its energy changes into any other sort of energy. It can only do this in discontinuous units of energy, in individual quanta of light which are proportional to its frequency.

Let us consider this problem in conjunction with the photoelectric effect or any photochemical reactions. Assume we are using a light source the colour (wavelength, frequency) of which can be chosen freely and the intensity of which can be adjusted as necessary. Place a photographic plate and a photoelectric cell alongside each other opposite to the light source and at a given distance from it. As long as the frequency of the light does not reach the limiting frequency of the plate or of the photocell, that is as long as it is on the "red" side of this limit nothing will happen, no matter how intensely and for how long we illuminate our instruments. However, the moment that the frequency has passed the respective limits the reaction starts. We may now weaken the intensity of the light to any degree. It will be only a matter of time until a silver atom is set free here or there on the plate or an electron emerges from the photocathode.

What would have seemed natural according to the wave theory of light? Something entirely different. Imagine the surface of a calm lake which is set into wave motion by a piston moving up and down somewhere below its surface. The waves encircle this point of origin and progress with constant velocity but decreasing amplitude as they spread out on the surface. Now take pairs of balls and bind each pair together with a very fine, weak thread. Let them float on the surface of the lake, around the origin of the waves. They are to be regarded as the models of diatomic molecules, held together by binding forces. When the waves start, all "molecules" within a well-defined circle are going to break apart because of the forces exerted upon them by the waves of appreciable amplitude. Outside this limiting circle, however, the amplitude of the waves becomes insufficient to break the threads: the ball-molecules will be safe here. The frequency of the waves makes no difference. The pairs of balls are unable to absorb energy from the wave system and accumulate it until enough is present to break their threads. Only in the very rare case when the frequency of vibration of a pair of balls against each other (e.g. as might occur if they were connected by elastic threads) is the same as the frequency of the approaching wave crests, could it happen that they might accumulate energy from a series of wave crests. The reason for this is that successive waves would always arrive in the same phase with the relative swinging motion of the balls. This is resonance, the same phenomenon that makes the string of a musical instrument begin to vibrate if it is subject to sound waves which are of its own frequency. But in the general case our ball-molecules will not accumulate energy in this way and their threads will *not* be broken below a certain level of wave amplitude. Beyond that level *all* threads will be broken.

For real molecules the intensity of the light waves is irrelevant, it is only their frequency that counts. It is as though it is only a question of resonance between the wave and the molecule. This picture of ours with water waves and ball-molecules does not seem to give a good analogy with photochemical disintegrations or the liberation of photoelectrons. But let us go straight forward to the next difficulty. *Which* silver chloride molecule from among all those within the emulsion is going to be decomposed if the total energy of light which has fallen on the plate is only sufficient to break up a single one? Or *which* sodium atom from among all the atoms in the surface of the photocathode is going to emit an electron if the total energy hitting the cathode is only enough to liberate a single one? According to the theory of waves the energy is spreading out equally on growing spherical surfaces across the whole space! The same amount of electromagnetic energy penetrates through each surface unit of a given wave sphere around the origin.

We are unable to tell where, but only *one* silver chloride molecule will be split and only *one* electron will be emitted from an unpredictable point of the cathode surface. The classical concepts are only statistically true, but in reality the energy of a light wave cannot be spread out in a continuum, but has to remain concentrated in compact bundles of energy, of light quanta, each containing the amount of energy hv. They are to be regarded as almost like particles in their compactness, and usually they are called photons instead of quanta to emphasize this particle analogy. When light and a molecule meet it is as if two molecules have met. Molecule and photon have to arrive simultaneously at the same place in order to react, if not they have just missed each other. Matter is not swimming around embedded in an ocean of light waves, it is moving about amidst a shower of photons. Whenever molecule and photon meet, the energy of the photon must be above a minimum value to be sufficient to bring about a reaction—no matter whether we consider a chemical reaction or the emission of an electron. If the photon's energy is smaller than this minimum, nothing will happen, no matter how many photons may be at our disposal. The presence of many photons corresponds to intense illumination, but the intensity of illumination in itself will never make a reaction go if the individual energy of single photons is insufficient. Many small photons are not as good as a few larger ones. Furthermore the energy of the individual photons is linked in some mysterious manner with the frequency of a wave motion which statistically governs their distribution, for instance in a diffraction experiment.

We have arrived at exactly the same difficulty that confronted us in Section 47 when we learned about the wave nature of the electrons. There, an entity which we thought was a well-defined particle suddenly revealed a partial wave nature. Here, a phenomenon we attributed to waves suddenly manifests itself as being due to bundles, particles, of energy! Two vast domains of Nature approach each other here in a strange and incomprehensible dualism: matter, particles on one side and continuous waves on the other seem to be interconnected more intimately than we have hitherto supposed.

We have to remember, however, that the great *Newton* (1642–1727) thought of light as composed of particles moving very rapidly through space and that he was able to deduce all the laws of geometrical optics from this assumption. But physical optics, the phenomena of diffraction, interference and polarization were explained better by the wave theory of Huygens and because the results of geometrical optics followed also from the wave theory, it was generally accepted instead of Newton's approach.

As far as we can see today, both theories have to be applied simultaneously. The propagation of light, interference and diffraction are described by the theory of waves. The same is true of particulate matter if the corresponding wavelength is taken as being inversely proportional to the momentum in accordance with de Broglie's relation (eqn. 47.2 on p. 205). But whenever we arrive at saying: this is going to be the intensity of light or the number of impinging particles at a given position, this assertion will only be valid in the statistical sense, as a probability valid for great numbers. We are never able to locate less light than a single photon and the true path of this photon will remain as uncertain as that of any particle in quantum mechanics. The principle of uncertainty emerges again and we could repeat every word of what we have said about causality in the chapter concerning the wave nature of electrons (p. 212).

52. MAGNETISM AND THE ELECTRON

According to the macroscopic laws of electrical and magnetic fields which have been codified by *J. C. Maxwell* (1831–1879) in 1873 as a result of the experiments of a number of great physicists including Faraday, a magnetic field is generated whenever electrical charges are in motion or the intensity of an electrical field changes. On the other hand these magnetic fields act upon moving charges and generate electrical fields in their turn whenever their intensity changes.

So far we have only asserted that electrons can behave as charged particles and that they are subject to these laws of electromagnetism during their movements, while creating magnetic fields of their own. As we have not yet discussed the structure of atoms it would be premature at this stage to discuss the magnetic implications of electrons moving within an atom or a molecules; we will engage in this discussion in Section 61.

However, we cannot very well leave the subject of the individual electron without at least mentioning the fact that it seems to behave as an elementary magnet in itself.

Let a beam of electrons pass through a very thin film of iron magnetized in one direction within its plane. Absorption of the electrons in the iron will of course reduce the intensity of the beam. But something else will happen. If the same beam is made to traverse a second parallel film of magnetized iron it will be important how the direction of the magnetic field in this second film is oriented with respect to the first, The electrons will pass most easily if the field directions coincide, whereas they pass in smaller numbers the more the two directions differ from each other. There must have been some magnetic interaction between the fields and the electrons which allows for some kind of orientation of the electron relative to the field. Thus the electron cannot be represented merely by a simple sphere of undefinable radius charged with negative electricity. It must have some vectorial, directed property as well which causes it to be oriented in a magnetic field. In the simplest picture we can imagine, the electron corresponds to a rotating sphere of negative electricity, which accordingly would comprise a circularly moving charge and thus a magnetic moment according to the laws of electro-magnetism. It would have an axis and two poles of rotation and the axis could be oriented in specified directions.

The idea of the spinning electron was proposed by *S. A. Goudsmit* and *G. F. Uhlenbeck* (1925) to explain some complicated features of atomic spectra before these direct experiments of electrons moving through a magnet were carried out. The spectral measurements are much more refined and the whole theory of "spin" has been developed from spectroscopic observations. It involved a mechanical spinning angular momentum, which obeyed the conservation laws of mechanics but was also subject to quantiza-tion as far as allowed orientations in an external magnetic field were concerned.

The experiment with the two consecutive parallel magnetic layers set at different angles of magnetization to each other corresponds to some extent to the experiments on light polarization (p. 75). It was a great success of theoretical physics when *Dirac* (1902–   ) was able to deduce the spin of the electron from a relativistic theory of electron waves (1928). This question unfortunately lies above the level of this book and of its author.

Part 5

Parts of the Atom

53. PROUT'S HYPOTHESIS

Atoms were named atoms ($\overset{\text{'}}{\alpha}\tau o\mu o\varsigma$ = indivisible) because they were considered to be the smallest particles into which matter could be subdivided. At the beginning of modern chemistry the number of known elements, that is of new sorts of atoms, increased rapidly. Later on it became more and more an event if a new element was discovered. At last, with the best modern methods some 90 natural elements have become known. One might perhaps have expected that these would have been taken as the ultimate building stones of Nature.

However, it somehow does not suit the unifying tendency of human nature to acquiesce in such a large number of ultimate particles. Hardly had the heroic period of chemistry begun when *Prout* (1786–1850), an English physician, ventured to propose the hypothesis that all the atoms of the heavier elements are composed of atoms of hydrogen (1816), in some indeterminate manner. He based his argument on the empirical fact that, especially among the lightest elements, atomic weights are very often nearly integral multiples of the atomic weight of hydrogen, (Table III). If every single sort of atom were an ultimate building stone of Nature there should be no reason for their weights to have any simple relation to each other. If any such relation is found it must be taken as a sign that they are composed of some common component, as a multiple thereof. True, there were many exceptions to this empirical rule. As a matter of fact there were more exceptions than examples in its favour, especially among the heavier elements. But the number of fits was so high that it could not be ignored. It had to be hoped that some explanation would eventually be found for the heavier elements not fitting into the picture. It was thought, that perhaps redeterminations of their atomic weights—crude as they were at that time—would show a way out of the dilemma.

But for a long time the trend was in the opposite direction. The better the determinations of atomic weights, the more certain have the discrepancies become. Only a few decades ago did we learn the true reason for non-integral atomic weights (*see* Section 91 on "isotopes"). Until then the increasing accuracy of atomic weight determinations discredited the hypothesis of Prout, which, however, managed to survive in textbooks of chemistry as a curiosity without explanation.

Table IV of atomic weights (p. 235) shows how far the average atomic weights were from integral numbers, that is from being integral multiples of the atomic weight of hydrogen. However, it is worth while to extract from that table the 16 lightest elements and to mark with an asterisk those whose atomic weights lie within $\pm 0{\cdot}10$ of an integral number, i.e. within the range of $0{\cdot}2$ unit about an integral number (Table III).

Disregarding carbon which is taken as a standard for the definition of relative atomic weights we have compared 15 elements and of these the atomic weights of 12 of them lie within $\pm 0{\cdot}10$ unit of integral numbers. On the basis of pure chance only one-fifth of the quoted elements should have had atomic weights falling within a range of $0{\cdot}2$ (that is of $1/5$) of a whole number; this would have amounted to only three elements. But we have found 12! It does not look like mere chance. Among the elements following those in Table III in order of increasing atomic weight the number of those

TABLE III

H	1·008	*
He	4·00	*
Li	6·96	*
Be	9·02	*
B	10·82	
C	12·00	standard for definition
N	14·01	*
O	16·00	*
F	19·00	*
Ne	20·18	
Na	23·00	*
Mg	24·32	
Al	26·97	*
Si	28·06	*
P	31·02	*
S	32·06	*

with atomic weights near to integral numbers decreases, but for a while it remains appreciably above the chance value of $1/5$ of the elements. Only among the heaviest elements is this frequency of $0{\cdot}2$ attained.

The situation is very curious indeed. We are confronted by a higher frequency of an event that can be due to chance alone but we are not confronted with a law. No wonder that chemists at his time finally rejected Prout's hypothesis. After all, either all elements are made up from the common ancestor hydrogen in which case their atomic weights must be *exactly* a multiple of that of hydrogen, or else they cannot be made up from hydrogen. Experiments at that time showed that they are not.

On the other hand there can be no doubt that the data in Table 3 do need some interpretation. It is hard to conceive of any other explanation than that there is some sort of dependence between the atomic weights of different atoms; surely nothing short of this could be the reason for the observed regularity among the light elements. It was about a century before the necessary explanation was found, but this shall be discussed later.

54. PERIODIC REGULARITIES

If it were true that the atoms of all the elements are independent ultimate units of Nature, their properties should be entirely uncorrelated. The most we could expect in such cases would be that they all agreed in some properties and then we could say that these properties are characteristic for matter as such. But experience has taught us otherwise. It has been known ever since the identities of the true elements emerged in chemistry that there are groups of similar elements. In the chapters dealing with inorganic chemistry we have already mentioned these groups and have used their collective names. Thus we have the alkali metals which are all monovalent and highly electropositive; they are soft like wax, of low density and are immediately oxidized in air. A similar coherent group of elements are the alkaline-earth metals. They are all divalent, slightly less positive than the alkalis, of low density although not to the same extent as the former group, and they too are prone to oxidation by the atmosphere. We also know of six monatomic gases which hardly form any compounds at all: the rare or noble gases. Evidently these also form a group of their own.

Let us proceed further. We are acquainted with the halogens: fluorine, chlorine, bromine and iodine. They form diatomic molecules, are either gases or are easily volatilized at room temperature and all of them attack the lungs and the mucous membranes. They all combine with hydrogen to form the corresponding hydracids HX and the latter are strong acids in aqueous solution (X denotes a halogen atom). Their salts and their compounds with oxygen are very similar also.

We have only mentioned the most striking examples of such groups. A great number of the elements known today were unknown when *J. W. Döbereiner* (1780–1849), the inventor of platinum sponge, remarked that the difference of atomic weights between three triads of such related elements is nearly constant:

Li 6·94; Na 23·00; K 39·10; (difference ca. 16)

Ca 40·07; Sr 87·63; Ba 137·37; (difference ca. 48·5)

Cl 35·46; Br 79·92; I 126·92; (difference ca. 46)

He found other similar regularities in the table of atomic weights but his observations remained isolated.

Only much later, in 1864, were these regularities taken up again, this time by the German chemist-physician *J. L. Meyer* (1830–1895) and in 1869, by the great Russian chemist *D. I. Mendeleev* (1834–1907) of the Academy of St. Petersburg. They took great trouble to go through the whole table of elements and found that it was full of interesting regularities (Fig. 92). To a first approximation we might formulate all these regularities into a rule by stating that if we write the elements in the order of their atomic weights, the elements which belong to characteristic groups always recur in the same order, one element of one group appearing after another element of another group. It was especially Mendeleev who went into details of several physical properties and even managed to predict the discovery of then unknown elements from missing ones in this system.

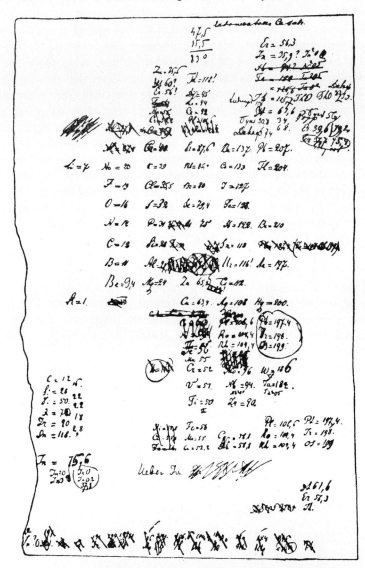

Fig. 92. **Mendeleev's original drafts of the Periodic Table**
(From O. N. Pissershevsky, *Dmitri Ivanovich Mendeleev* (Moscow, Verlag für Fremdsprachige Litteratur, 1954).

Let us look at Table IV (p. 235). Helium is the first rare gas. It is followed by lithium, the first alkali metal, then by beryllium which resembles very much the alkaline-earth metals. The next rare gas is neon, eight places after helium in order of atomic weights. One place before neon we find fluorine, the first halogen and one place after it, that is once more succeeding a rare gas, we have the next alkali metal, sodium. This is followed again by an alkaline-earth metal, magnesium. We must then proceed again by eight elements after fluorine to arrive at the next halogen, chlorine. This should be again succeeded by a rare gas, argon. In fact, it is succeeded by the alkali metal potassium of atomic weight 39·10 rather than by argon (39·88), but

TABLE IV. Chemical Atomic Weights

Based on $^{12}C = 12\cdot00000$ as unit (*see* p. 39)

Element	Symbol	Atomic Weight	Element	Symbol	Atomic Weight
Hydrogen	H	1·00797	Tellurium	Te	127·60
Helium	He	4·0026	Iodine (Jodum)	I (J)	126·9044
Lithium	Li	6·939	Xenon	Xe	131·30
Beryllium	Be	9·0122	Caesium	Cs	132·905
Boron	B	10·811	Barium	Ba	137·34
Carbon	C	12·01115	Lanthanum	La	138·91
Nitrogen (Azote)	N	14·0067	Cerium	Ce	140·12
Oxygen	O	15·9994	Praseodymium	Pr	140·907
Fluorine	F	18·9984	Neodymium	Nd	144·24
Neon	Ne	20·183	Promethium	Pm	145
Sodium (Natrium)	Na	22·9898	Samarium	Sm	150·35
Magnesium	Mg	24·312	Europium	Eu	151·96
Aluminium	Al	26·9815	Gadolinium	Gd	157·25
Silicon	Si	28·086	Terbium	Tb	158·924
Phosphorus	P	30·9738	Holmium	Ho	164·930
Sulphur	S	32·064	Erbium	Er	167·26
Chlorine	Cl	35·453	Thulium	Tm	168·934
Argon	Ar	39·948	Ytterbium	Yb	173·04
Potassium (Kalium)	K	39·102	Lutetium	Lu	174·97
Calcium	Ca	40·08	Hafnium	Hf	178·49
Scandium	Sc	44·956	Tantalum	Ta	180·948
Titanium	Ti	47·90	Tungsten (Wolfram)	W	183·85
Vanadium	V	50·942	Rhenium	Re	186·2
Chromium	Cr	51·996	Osmium	Os	190·2
Manganese	Mn	54·9381	Iridium	Ir	192·2
Iron (Ferrum)	Fe	55·847	Platinum	Pt	195·09
Cobalt	Co	58·9332	Gold (Aurum)	Au	196·697
Nickel	Ni	58·71	Mercury		
Copper (Cuprum)	Cu	63·54	(Hydrargyrum)	Hg	200·59
Zinc	Zn	65·37	Thallium	Tl	204·37
Gallium	Ga	79·72	Lead (Plumbum)	Pb	207·19
Germanium	Ge	72·59	Bismuth (Wismut)	Bi	208·980
Arsenic	As	74·9216	Polonium	Po	210
Selenium	Se	78·96	Astatine	At	210
Bromine	Br	79·909	Radon (Emanation)	Rn (Em)	222
Krypton	Kr	83·80	Francium	Fr	223
Rubidium	Rb	85·47	Radium	Ra	226
Strontium	Sr	87·62	Actinium	Ac	227
Yttrium	Y	88·905	Thorium	Th	232·038
Zirconium	Zr	91·22	Protactinium	Pa	231
Niobium			Uranium	U	238·03
(Columbium)	Nb	92·906	Neptunium	Np	237
Molybdenum	Mo	95·94	Plutonium	Pu	242
Technetium	Tc	97	Americium	Am	243
Ruthenium	Ru	101·07	Curium	Cm	247
Rhodium	Rh	102·905	Berkelium	Bk	247
Palladium	Pd	106·4	Californium	Cf	249
Silver (Argentum)	Ag	107·870	Einsteinium	Es	254
Cadmium	Cd	112·40	Fermium	Fm	253
Indium	In	114·82	Mendelevium	Md	256
Tin (Stannum)	Sn	118·69	Nobelium	No	254
Antimony (Stibium)	Sb	121·75	Lawrencium	Lw	257

the difference is slight and we feel inclined to correct the observed order in atomic weights to conserve the chemical sequence. Later on (p. 302) we shall see why we are justified in doing this three times in the system of the elements. After argon we have, as expected, an alkaline-earth metal, calcium, eight places after magnesium. Let us now find the next halogen element. It is bromine and this time we find it 18 elements after chlorine. It is followed in the "normal" sequence by a rare gas, an alkali and an alkaline-earth metal: krypton, rubidium and strontium.

It would be possible to go further in this manner and it is astonishing how many interesting sequences can be found. It was found to be more or less possible to arrange the whole series of elements in a system according to increasing atomic weights such that some sort of regularity prevailed throughout. As we have already mentioned, it becomes only necessary to change the sequence of three pairs. The first is the K-Ar pair we have already encountered. The other two pairs are I-Te and Ni-Co. The arrangement of the elements with their atomic weights increasing throughout the system, but in a table where groups of similar elements are arranged to lie near each other, is the so-called periodic system of Mendeleev and L. Meyer.

Before embarking upon the details of this system we must emphasize that the sheer existence of it speaks in favour of the assumption we mentioned when discussing Prout's hypothesis: there must be some connexion, some fundamental connecting principle, between the atomic weights of different elements. It seems very improbable that such regularities would exist between unconnected ultimate building stones of Nature. And it is hard to conceive any connexion between different atoms without crediting them with some sort of a structure which is built in a regular sequence. "Ultimate particles" should have no "structure," a structure involves more elementary parts within an atom. We become rightly suspicious that the periodic recurrence of characteristic qualities in the row of atoms is due to some fundamental principle according to which they are built up. Hence they probably cannot be ultimate particles at all.

55. THE PERIODIC SYSTEM

We will next construct the periodic system of the elements according to our knowledge of today. In the times of Meyer and Mendeleev the number of known elements was of course much smaller than it is now; for instance the whole group of rare gases had not then been discovered. It was one of the most striking features of the system that it was possible to foretell many properties of elements which still had to be discovered and which would not have been felt to be missing in the absence of such a system. Gallium and germanium (the now famous element of transistors) were among the most conspicuous examples of this kind. Mendeleev proclaimed in 1869 that in his opinion elements were missing below aluminium and silicon which he provisionally called Eka-aluminium and Eka-silicon. He was able to predict the range of atomic weights he thought they would possess and described some of their most probable properties. *L. de Boisbaudran*, some six years later actually discovered gallium while *Cl. Winkler* isolated germanium in 1886. Atomic weights, chemical properties and mineralogical relationships fitted well with Mendeleev's prediction.

A similar but somewhat more intricate argument led *G. v. Hevesy* to predict in 1923 that it was not cerium, with the long series of adjoining rare earth metals, that should be placed below zirconium in the fourth column of the periodic system, but a new, unknown tetravalent metal with an atomic weight between 175 and 180. Using the then new method of determining the wavelength emitted in the X-ray region by elements (*see* p. 302) he and his collaborators soon succeeded in finding a new characteristic X-ray wavelength in the minerals they examined. This was followed by isolation of the new element hafnium (named after Kjobenhaven where he worked) which was tetravalent and had the atomic weight 178·6. It was so similar to zirconium, that up to then it had never been separated from it.

Now let us examine the system itself (Table V). The lightest element, the one that Prout thought was the building stone of all the others, is hydrogen, and is located at the beginning of the system. It is monovalent and in most of its compounds it is electrochemically positive. Therefore it has some electrochemical resemblance to the alkali metals. But in the last forty years during which we have been finding out about hydrides it has been known that the hydrogen in them is anionic, migrates towards the anode and is evolved there during electrolysis of the melts. Since in these compounds the hydrogen is monovalent and negative like the halogens, should we place it above lithium or fluorine? It is uncertain. Better let it stand as it is, a singular, fundamental element at the beginning of the periodic system.

The next element in the order of increasing atomic weights is helium. It has no stable compounds, it is a monoatomic gas and the lightest member of the very characteristic family of rare gases. The other members are, in order of increasing atomic weights neon, argon, krypton, xenon and radon (radium emanation).* If we build our system by writing similar elements below each other there can be no doubt about placing all these in the same column. Owing to the fact that there is an interesting relationship between the vertical columns and the maximum valency of elements belonging to them, this column of elements, practically without chemical valency, is sometimes denoted as the zeroth column.

The third heaviest element in the system is lithium. It is of low density, soft, easy to oxidize, positive and monovalent. We already know that it has many brethren among the elements and that this most positive group of metals is called the alkali metals. They are so similar to each other that in this case, too, there can be no doubt but that they belong to one column; sodium, potassium, rubidium, caesium and the very rare (radioactive) francium

---

* Until 1962 no compounds of any of these gases were known, apart from the He_2 molecule which was identified by its spectrum in gas discharges as an extremely short-lived species. Now some stable fluorides of xenon and krypton have been prepared which yield hydroxy-compounds on hydrolysis (p. 278).

TABLE V. The Long (Werner) Form of the Periodic System

Unit: $^{12}_{6}C = 12 \cdot 00000$

Period	Ia	IIa	IIIa	IVa	Va	VIa	VIIa	VIII	VIII	VIII	Ib	IIb	IIIb	IVb	Vb	VIb	VIIb	0
1	1 H 1·00797																	2 He 4·0026
2	3 Li 6·939	4 Be 9·0122											5 B 10·811	6 C 12·011	7 N 14·007	8 O 16·000	9 F 18·998	10 Ne 20·183
3	11 Na 22·990	12 Mg 24·312											13 Al 26·982	14 Si 28·086	15 P 30·974	16 S 32·064	17 Cl 35·453	18 Ar 39·948
4	19 K 39·102	20 Ca 40·08	21 Sc 44·96	22 Ti 47·90	23 V 50·94	24 Cr 52·00	25 Mn 54·94	26 Fe 55·85	27 Co 58·93	28 Ni 58·71	29 Cu 63·54	30 Zn 65·37	31 Ga 69·72	32 Ge 72·59	33 As 74·92	34 Se 78·96	35 Br 79·91	36 Kr 83·80
5	37 Rb 85·47	38 Sr 87·62	39 Y 88·91	40 Zr 91·22	41 Nb 92·91	42 Mo 95·94	43 Tc 97	44 Ru 101·07	45 Rh 102·91	46 Pd 106·4	47 Ag 107·87	48 Cd 112·40	49 In 114·82	50 Sn 118·69	51 Sb 121·75	52 Te 127·60	53 I 126·90	54 Xe 131·30
6	55 Cs 132·91	56 Ba 138·34	57* La 138·91	72 Hf 178·49	73 Ta 180·95	74 W 183·85	75 Re 186·2	76 Os 190·2	77 Ir 192·2	78 Pt 195·09	79 Au 196·97	80 Hg 200·59	81 Tl 204·37	82 Pb 207·19	83 Bi 208·98	84 Po 210	85 At 215	86 Rn(Em) 222
7	87 Fr 223	88 Ra 226	89† Ac 227															

	IVa	Va	VIa	VIIa	VIII			Ib	IIb	IIIb	IVb	Vb	VIb	VIIb
* 58–71 Rare earth metals	58 Ce 140·12	59 Pr 140·92	60 Nd 144·29	61 Pm 145	62 Sm 150·35	63 Eu 151·96	64 Gd 157·25	65 Tb 158·92	66 Dy 162·50	67 Ho 164·93	68 Er 167·26	69 Tm 168·93	70 Yb 173·04	71 Lu 174·97
† 90–103 Actinide metals	90 Th 232·04	91 Pa 231	92 U 238·03	93 Np 237	94 Pu 242	95 Am 243	96 Cm 247	97 Bk 242	98 Cf 249	99 Es 254	100 Fm 253	101 Md 256	102 No 254	103 Lw 257

follow one below the other. All are monovalent, their column is termed the first column of the system.

There are eight elements from helium to neon and again from neon to argon. There are eight from lithium to sodium and from sodium to potassium. But there are eighteen elements from argon to krypton and from potassium to rubidium and eighteen from krypton to xenon and from rubidium to caesium. Furthermore there are thirty-two elements from xenon to radon and from caesium to francium. Thus the length of periods between similar elements is not constant and increases with increasing atomic weight. The sequences between like elements are called periods or rows of the periodic system. Remembering that hydrogen and helium form a very short row at the upper end of the table, this represents the first row and the number of elements in the succeeding periods increases thus: 2, 8, 18, 32. It was soon found that these numbers conform to a rather simple formula: $2n^2$ if n is given the value of the integers 1, 2, 3, and 4. This rather mystical, cabbalistic relationship was only explained much later when quantum rules of atomic structure became known (p. 60).

Since there are more and more elements in the rows of heavier elements it is clear that one cannot find a place for all of them below elements in the shorter rows above. Many an effort was made to solve this problem of equating 8 to 18 and even to 32, but no satisfactory scheme was devised. The consequent procedure is to go on building the system from columns which are as clearly coherent as those of the rare gases and the alkali metals. After reaching a point in the longer periods where no sister element is to be found above them, one simply discontinues the short periods for a while. But after the alkalis this is not yet the case.

Following lithium, the third heaviest element, we find a divalent metal, beryllium. Looking for similar divalent, electropositive metals in the systems we soon find the whole group: magnesium, calcium, strontium, barium and radium. They are more or less easily oxidized and their hydroxides are bases, though not so strong as those of the alkalis. When their compounds are volatilized in a flame, they impart specific colours to it, as do the alkali metals (magnesium needs a spark instead of a flame). But looked at in a spectroscope these colours are found to be made up of a more complicated mixture of wavelengths than are those of the alkalis. From calcium onwards the sulphates of this group are sparingly soluble in water. However, we are not studying detailed inorganic chemistry at the moment and it would lead us too far to enter upon all the chemical similarities between the elements of a column. Suffice it to say that one nearly always manages to interpolate the chemical characteristics of an element correctly if those of the element above it and below it are known. The periodic system is truly the genealogical tree of the chemical elements.

The fifth element in the sequence of increasing atomic weights is boron, a trivalent metal, or rather a semiconductor. Eight elements further on we have another trivalent metal, aluminium, and after another eight, eighteen and thirty-two places, respectively, there follow other trivalent metals. The oxide of aluminium being one of the most common components of soil-forming minerals, and the higher elements of this group being often admixed with it, the group was named the earth metals. They are, after aluminium: scandium, yttrium, lanthanum and the radioactive actinium. Scandium is one more example from among the elements predicted by Mendeleev and later found to exist accordingly. In the period of lanthanum we find on the side of increasing atomic weights from the latter a sequence of fourteen rare metals, very similar to each other and all with a principal valency of three. Some of them, for example cerium and neodymium, have acquired considerable industrial importance. They are so similar to each other and to lanthanum that they are called the group of rare earths. In the period above them we do not encounter their like: scandium stands above lanthanum, and zirconium, as we have seen, above hafnium. So logically we cannot help but break the row of elements after scandium to continue again above hafnium with zirconium. In practice this would mean a lengthening of the table by fourteen places and would make it somewhat unwieldy. Therefore for convenience one generally puts an asterisk after lanthanum and lists the series of rare earth metals separately, beneath the general table. However, we should know that they do belong to this place and have not been left aside for any logical reason. This is one of the cases in the system where not only series of elements in a column below each other, but series of neighbours, with increasing atomic weights show very pronounced similarity. In every case when such a similar group of neighbours appears, there is an extension of the length of the horizontal period.

The sixth element of the system, the right-hand neighbour of boron is carbon. We already know it as the mother of organic compounds. It is nearly always tetravalent, as we have seen, and it is equally capable of accommodating four positive or four negative elements around itself, having a "coordination number" of four. Silicon, the element eight places beyond carbon and appearing below it in this fourth column of the system is also capable to some extent of forming —Si—Si— chains analogous to the carbon-carbon chains, —C—C—. There exists a limited silicon chemistry which resembles to some extent the chemistry of carbon chains—with the very important difference, however, that these compounds are much more liable to oxidation than the majority of those encountered in organic chemistry. Silicon is more electropositive than carbon and this implies that it combines with oxygen more readily than does carbon. We shall return later to the phenomenon of increasing electropositive character as one descends any of the vertical columns. Germanium, the element whose existence was predicted by Mendeleev, still has some chain-structured compounds with hydrogen, but they are even more labile than the "silanes" as the silicon hydrides are called. But on jumping from silicon by chemical analogy to germanium we have trespassed over a border line: we did not by-pass eight elements to find an analogue as we did from aluminium to scandium, but eighteen instead! The eighth element after silicon is titanium. It is a metal somewhat similar to aluminium and scandium and with ever-increasing technical importance as a strong and light metal. But it is not similar to silicon, except in so far as its maximum valency is four. It also often exists in the trivalent state.

As in the row of the rare earths, we encounter the phenomenon of neighbours becoming more similar rather than elements which are eight or eighteen places from each other. This, then, must be the place where we have to cut through the short, eight-element periods in order to accommodate the other ten elements which are included in the new eighteen-element row. We write scandium below aluminium and interrupt the aluminium period while titanium, vanadium, chromium, manganese, iron, cobalt, nickel, copper, zinc and gallium appear in the lower row after scandium. These, the so-called first "transition" metals are all decidedly metallic in character, and neighbours in the series are in general quite similar to each other. Only after gallium do we encounter germanium, which, as we mentioned before is similar to silicon. Thus, above germanium, silicon and carbon occur and there are no elements above the transition metal series.

Below germanium, at a distance of eighteen and thirty-two elements respectively, we place the two metals tin and lead, both with the maximum valency of four. They are more electro-positive than their sister elements above them and join germanium in that they also often occur with the lower valency of two.

Looking at the period of rubidium in the neighbourhood of the third and fourth column we observe that a series of ten transition metals occurs here also. This must be so because this row contains eighteen elements as does the preceding one. The series extends from zirconium to indium, containing only metals, and neighbours resemble each other again. The next, the longest, period which begins with caesium contains the fourteen lanthanide rare earths, but after their completion there begins once more the "transition" series of metals from hafnium to thallium. We are going to deal with these transition groups separately a little later on. The last period is incomplete. It contains only radioactive, that is, as we shall see in Part 7, unstable, elements, many of them having been made only in the laboratory. The fact that the period is incomplete, with fewer elements than the preceding one is evidently caused by the instability of its elements. It took some time before scientists agreed where to place the elements from actinium onwards, but now it seems pretty certain that we are dealing with a group similar to that of the lanthanide rare earths and it seems most satisfactory to write them below the latter.

We will not consider at the moment elements of the transition or actinide groups but proceed to the seventh element in the system, to nitrogen, and try to find its sister elements. At a distance of eight elements from it we find phosphorus, after eighteen more arsenic, after a further eighteen places, antimony and then, after another thirty-two places, bismuth. The maximum valency in combination with oxygen or halogens is five for elements of this column and the hydrates of the oxides act as acids, the more so the higher they lie in the column. Remembering the Kossel theory of hydroxy-compounds (p. 189) we understand why this must be so; high valency of the central atom attracts the oxygen and repels the hydrogen atoms. Increasing atomic diameter on the other hand weakens these coulombic forces as we descend within the column and thus also weakens the acid character of the hydroxides. Nitrogen and phosphorus are non-metals, though one of the phosphorus modifications resembles the metallic state. Arsenic has both a non-metallic and a metal-like modification; from antimony onwards all are metals. This, by the way, is a general rule in the system. The metallic character of the elements increases hand-in-hand with increasing electropositive behaviour as we descend the columns (p. 240). In this, the fifth, column we encounter a new characteristic phenomenon as far as valency is concerned. The hydrogen compounds of these elements only contain three hydrogen atoms as in our old acquaintance, ammonia (NH_3), in PH_3 and in the increasingly labile hydrides of the heavier elements. Maximum valency in combination with (positive) hydrogen differs from that with negative elements as for instance with oxygen or halogens. In the fourth column this was not so and the valency was four in both directions. Some short-chain hydrogen compounds exist also in this group but they are rather labile and cannot compare with anything like those of organic chemistry. The hydrides are gases or volatile liquids.

The eighth element of the periodic system is oxygen and there can be no doubt about sulphur, selenium, tellurium and the radioactive polonium standing below it in the sixth column of elements. This time it is easier to decide the matter on inspection of the hydrides which all conform to the general formula H_2X and are either gases or volatile liquids. There seems to be no point in speaking about oxides of oxygen, but as far as the other elements are concerned they go up as far as the trioxides (valency six) which form acids with water, and their maximum valency when combined with fluorine is six also. Their lower oxidation state is four with negative elements. Conceding the normal increase in electropositive and metallic character on descending the column these are indeed fairly homologous elements. Tellurium is nearly a metal and polonium is metallic. The strength of their acids decreases down the column because of the increasing atomic diameter, as explained by Kossel's theory.

The last element at the beginning of the system before the second rare gas neon, is fluorine. It thus brings the first period of eight elements to an end. Its homologues are of course chlorine, bromine, iodine and the unstable artificial element astatine. The natural elements of this column are well known, either gases or else very volatile liquids, with agressive smell and physiological behaviour. Their hydrogen compounds have the formulae HX and are gases, very soluble in water, in which they act as strong acids. Their salts are water-soluble with very few exceptions (Ag, Pb, Hg).

It is now high time to consider the elements we rejected previously as belonging to the so-called transition series. First of all we simply write them one beside the other as they occur, in order of increasing atomic weights (remembering the three exceptions discussed on p. 236). There are ten of them in each of the periods where they occur, so there is no difficulty in forming them into ten columns. Now let us see whether any regularities emerge. The fact that horizontal neighbours in these series are rather similar to each other has already been referred to.

Titanium, zirconium and hafnium constitute the first of these columns.

They are tetravalent. For a long time thorium was added at the bottom, because its maximum valency was four like that of its predecessors, but many fine details of chemical behaviour and even more the novel understanding of what lies behind the regularities of the system have convinced us that it is better to place it among the heavier rare earths, the actinides. The nitrides and carbides of titanium, zirconium and hafnium belong to the most refractory solids we know: they melt at temperatures around or above 4,000°C. Zirconium and hafnium are so alike in their chemistry that the latter was not found until Hevesy made a special search for it: then it proved to be not so rare at all. It was merely an undetected companion of zirconium.

The next column contains vanadium, niobium (called columbium in the U.S.A.) and tantalum. As with many of the transition metals in the first columns they may be alloyed with steel, improving its character. Their maximum valency is five but they can be reduced to lower valency states, vanadium, as a mater of fact, showing all possible valencies up to five. Salts of their lower valency states are deeply coloured, and paramagnetic (*see* p. 284). Their highest oxides are anhydrides of not very strong acids, their lower oxides are bases, conforming to the Kossel theory.

Chromium, molybdenum, tungsten (= wolfram) form the next column with uranium now transferred to the actinides, irrespective of its maximum valency of six which would make it fit into this column as well. They too are important steel-alloying components. Their maximum valency is six in the acid-forming trioxides. Compounds in the lower valency states, and for chromium even in the highest, are deeply coloured, and the lower oxides of the elements form bases. It is hard to find anything to distinguish them from the preceding group and from the next group containing manganese and rhenium (as well as the artificial element technetium) except their maximum valency. In the next group, that of manganese, this rises to seven but in every other respect this group is quite similar to that of chromium. The dark purple potassium salt of permanganic acid, which contains heptavalent manganese is very well known as a disinfectant (potassium permanganate) and is used in industry also. The analogous perrhenates are less well known because rhenium is a rather rare element; it was discovered only thirty years ago.

On trying to continue the building of vertical columns we fail with the next three sets of three elements in each period. We have iron, cobalt and nickel in the upper row, ruthenium, rhodium and palladium, three noble metals below it, and finally osmium, iridium and platinum at the bottom. These nine elements are generally gathered within a single group of the system, those in a row being closer to each other than those in a column. Perhaps the ability to absorb great quantities of hydrogen gas is the only property which holds together the column of nickel, palladium, and platinum. Even the maximum valencies fail to do so. Osmium and ruthenium have tetroxides which are volatile and which do not form acids; in these compounds we have arrived at the maximum valency we know: eight. Following the column of manganese with a valency maximum of seven, this is what we should expect. But iron cannot go beyond a maximum of six, cobalt beyond three and nickel scarcely beyond two. The upper three elements are steel-forming metals *par excellence*, while the two lower triplets belong to the series of the most noble metals. They have a highly developed complex chemistry, sharing this property with their predecessors in the chromium and manganese family.

Passing on to copper after nickel the vertical analogy improves to some extent: copper, silver and gold do have some similarity to each other. But their relation to their left-hand neighbours remains very intimate indeed: nickel and copper, palladium and silver, platinum and gold have many properties in common. This new column contains the elements of great electrical conductivity. The metals are coloured, even the very pronounced white colour of pure silver being impressive. In thin layers they are transparent: blue-green, blue and green respectively. All three can be monovalent, but only silver is regularly so. Copper is often divalent, gold trivalent and even di- and trivalent silver compounds are known. They belong to the least electropositive, that is, most noble, metals of the system.

To the right of copper we find divalent zinc, below this divalent cadmium and further down mono- or divalent mercury. They melt at relatively low temperatures, mercury melting even below $0°C$; their sulphides are insoluble in water and even in acetic acid, while their sulphates are easily soluble in water, in contradistinction to those of the other divalent metal quartet of the periodic system: Ca–Sr–Ba–Ra.

With the column containing gallium, indium and thallium we close the series of transition metals. Gallium melts not much above room temperature and herein resembles mercury, but the melting points of the whole group are relatively low, similar to those of the preceding one. Their volatile salts impart strong colours to flames. They have a maximum valency of three and this is their most frequent valency.

It is best to disregard here the last natural elements of the system up to uranium and the following artificial elements because these are best dealt with in the chapters concerning radioactivity. As we have noted already, they form a chemical group of their own, the actinide group corresponding to a certain extent to the group of the lanthanide rare-earth metals.

We have now made a cursory review of all the elements and mentioned some of the most characteristic relationships which can be seen in the periodic system. But we have written the system in its modern form which dates back to A. Werner, the father of modern chemistry. The original Mendeleev form of the table is given in Table VI. The transition elements are not clearly separated from the others in this arrangement. We can make use of the fact that the maximum valency—in combination with oxygen—increases among the transition metals in the same manner as among the non-transition elements, for instance in the series Ti–V–Cr–Mn from 4 to 7, and thus serves as an extension of the increasing-valency series A–K–Ca–Sc which precedes it. After manganese (Mn) we arrive at the inseparable triplets Fe–Co–Ni which are lumped together in an "eighth" group. As we have seen, only ruthenium and osmium attain the valency of eight, but these at

Unit: $^{12}_{6}C = 12 \cdot 00000$

Period	I a	I b	II a	II b	III a	III b	IV a	IV b	V a	V b	VI a	VI b	VII a	VII b	VIII	0
1	1 H 1·00797															2 He 4·0026
2	3 Li 6·939		4 Be 9·0122			5 B 10·811		6 C 12·01115		7 N 14·0067		8 O 15·9994		9 F 18·9984		10 Ne 20·183
3	11 Na 22·9898		12 Mg 24·312			13 Al 26·9815		14 Si 28·086		15 P 30·9738		16 S 32·064		17 Cl 35·453		18 Ar 39·948
4	19 K 39·102	29 Cu 63·54	20 Ca 40·08	30 Zn 65·37	21 Sc 44·956	31 Ga 69·72	22 Ti 47·90	32 Ge 72·59	23 V 50·942	33 As 74·9216	24 Cr 51·996	34 Se 78·96	25 Mn 54·9381	35 Br 79·909	26 Fe 55·847 27 Co 58·933 28 Ni 58·71	36 Kr 83·80
5	37 Rb 85·47	47 Ag 107·870	38 Sr 87·62	48 Cd 112·40	39 Y 88·905	49 In 114·82	40 Zr 91·22	50 Sn 118·69	41 Nb 92·906	51 Sb 121·75	42 Mo 95·94	52 Te 127·60	43 Tc 97	53 I 126·9044	44 Ru 101·07 45 Rh 102·905 46 Pd 106·4	54 Xe 131·30
6	55 Cs 132·905	79 Au 196·967	56 Ba 137·34	80 Hg 200·59	57* La 138·91	81 Tl 204·37	72 Hf 178·49	82 Pb 207·19	73 Ta 180·948	83 Bi 208·980	74 W 183·85	84 Po 210	75 Re 186·2	85 At 215	76 Os 190·2 77 Ir 192·2 78 Pt 195·09	86 Rn(Em) 222
7	87 Fr 223		88 Ra 226		89† Ac 227		(104)		(105)		. . . .					

* 58–71 Rare earth metals

58 Ce 140·12	59 Pr 140·907	60 Nd 144·24	61 Pm 145	62 Sm 150·35	63 Eu 151·96	64 Gd 157·25	65 Tb 158·924	66 Dy 162·50	67 Ho 164·930	68 Er 167·26	69 Tm 168·934	70 Yb 173·04	71 Lu 174·97

† 90–103 Actinide metals

90 Th 232·038	91 Pa 231	92 U 238·03	93 Np 237	94 Pu 242	95 Am 243	96 Cm 247	97 Bk 247	98 Cf 249	99 Es 254	100 Fm 253	101 Md 256	102 No 254	103 Lw 257

least do belong to the group. The next triplet of transition elements in this eighth column, Ru–Rh–Pd contains one of these two while the triplet one period below them, Os–Ir–Pt, contains the other.

With the exception of this eighth column, the Mendeleev system places the transition elements into subgroups, sub-columns of the periodic system. Copper, silver and gold form the subgroup of the monovalent alkali group, sharing with the alkalis the not-quite obligatory valency of one. We have remarked on the high electric conductivity of this group of noble metals and now find that in this respect they are comparable with the alkalis which have the highest conductivities within the periodic system.

The divalent transition metals zinc, cadmium and mercury, appear beside the divalent alkaline-earth metals, though they have not much in common with their chemistry. However, the phosphates of magnesium and zinc do show some similarity, and an extensive comparison of the chemistry of the main and sub-groups would reveal some other similar properties. On the whole, however, the similarity between neighbours outweighs the similarity of vertical columns among these metals. Trivalent Ga–In–Tl are alongside the trivalent earth metals Sc–Y–La; quadrivalent Ti–Zr–Hf alongside quadrivalent Ge–Sn–Pb; pentavalent metals V–Ni–Ta alongside pentavalent metalloids and metals As–Sb–Bi, which often occur in the trivalent state also. But note that the vanadium group of metals also have their own lower states of valency.

The difference between metals and non-metals increases in the sixth and seventh groups between members of the sub-group and the main group: Cr–Mo–W occur next to Se–Te–Po, and Mn–Tc–Rh alongside the halogens. Only the maximum valency and the formal composition of the respective compounds remains similar between group and sub-group.

On the whole, the Werner structure of the periodic system is more realistic than the original Mendeleev structure, but the latter has become firmly established and does serve to show a few interrelations between groups and subgroups, so that it cannot possibly be ignored completely.

Before taking leave of the chemical interrelations within the periodic—or natural—system of the elements we have to comment once more on a very important rule concerning valencies. The maximum valency of elements combined with oxygen or with the even more electronegative element fluorine, is nearly always equal to the number of the column or sub-column to which the element belongs. It begins with the mostly zerovalent rare gases and ends with the eighth group where ruthenium and osmium win the prize with the maximum valency of eight. But in the right-hand upper corner of the system we find elements forming gaseous compounds with hydrogen. The limit of this region is marked with a heavy line in the Werner system. The maximum valency of these elements in their hydrides is different from that in their oxides or fluorides. We have the halogen hydracids HF, HCl, HBr, HI in the seventh column, and the water group, H_2O, H_2S, H_2Se, H_2Te and H_2Po in the sixth column. The fifth column contains ammonia and its homologues: H_3N, H_3P, H_3As, H_3Sb and H_3Bi. All are in order of decreasing stability throughout the columns if we proceed downwards.

Only in the fourth column with CH_4, SiH_4, GeH_4 and SnH_4 is the valency in the hydrides the same as that in compounds with the electronegative elements (e.g. CO_2, CF_4).

On closer inspection we conclude that the sum of the maximum valencies in compounds with the negative elements and with hydrogen is eight in each of these columns: e.g. $7 + 1$ in the case of Cl, $6 + 2$ in the case of S, $5 + 3$ in the case of N and $4 + 4$ in the case of C, and so on. It is once again apparently a piece of cabbala which we shall understand better as we proceed to investigate the structure of atoms. Every regularity of this sort is an extra argument for the existence of such a structure. Real ultimate particles, without some common structure could not be expected to yield such remarkable relations. Nor would they yield the whole periodic system, of course.

56. PERIODICITY OF PHYSICAL CONSTANTS

We have often referred to the size of atoms. Some methods by which their magnitudes may be ascertained are already known to us: X-ray diffraction measurements, the mean free path of gaseous molecules and the volume occupied by a mole of the condensed element. For many purposes this latter method is quite adequate; we ascertain the smallest volume to which an element can be compressed at low temperature, and dividing by the Loschmidt (Avogadro) number we obtain the atomic volume.

Now let us see how these atomic volumes change in the course of the periods of the periodic system. In Fig. 93 we have the atomic volume as ordinate and the place within the periodic system as abscissa, for each element. It is immediately evident that this curve is periodic and has as many maxima as we have periods in the periodic table. On the top of each wave we find an alkali metal and in the valley between two peaks are the transition metals. The lower they lie on the curve, the smaller their volume, and in view of the fact that their atomic weight of course increases from left to right it is obvious that their density must increase in the same direction. We find the elements of highest density in the valley at the right of the curve. The regularity is so impressive that once more we can hardly doubt that a structural cause within the atoms underlies it.

Where are the metals on this curve? They begin at the peaks with the alkalis and continue downwards into the valleys until the curve begins to ascend again. Here we find the metals which are not so very metallic after all, at least as far as their electrical properties are concerned, e.g. B, Si, As, Se, Te. Next to them we gradually enter the realm of true metalloids, especially at the beginning of the system. To the left below the alkali metals we find the rare gases and below these, again to the left, the halogen elements.

A second characteristic physical property of the elements is their electrical conductivity. This, together with the related properties of high thermal conductivity and metallic light reflection is the most characteristic property of metals and we have already seen that metals occur on the left side and in the lower part of the Werner system. In order to compare the electrical

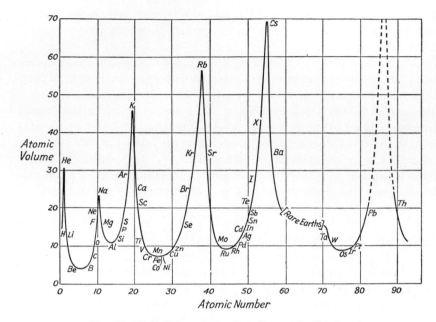

Fig. 93. **Periodicity of atomic volumes of solid elements**

(From A. Eucken, *Lehrbuch der chemischen Physik*, p. 646, Fig. 135
(Leipzig, Akademische Verlagsges., 1930). By courtesy of Akademische Verlagsges.)

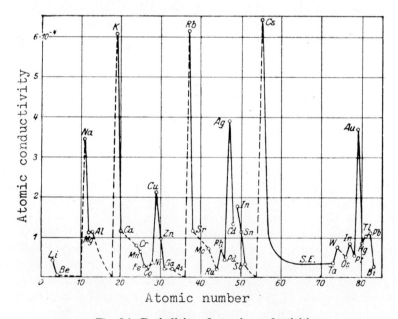

Fig. 94. **Periodicity of atomic conductivities**

(From A. Eucken, *Lehrbuch der chemischen Physik*, p. 648, Fig. 136
(Leipzig, Akademische Verlagsges., 1930). By courtesy of Akademische Verlagsges.)

conductivity of different elements on an atomic basis we must somehow compare the conductance of an equal number of atoms. The most natural way to do this is to think of one mole of every element arranged in a cube (as large as its atomic volume) and to compare the conductivity of these gramme-atomic cubes with each other. This is what we call the atomic conductivity, in contradistinction to ordinary conductivity in which we compare the conductivity of equal volumes of substances and thus not of equal numbers of atoms.

Fig. 94 shows these atomic conductivities as a function of the sequence of elements in the periodic system. Once more there is a periodic curve, but this time it is more complicated than before. We see now twice as many maxima: not only the alkali metals, the metals of the first column, are on peaks but also the copper–silver–gold family, the first sub-group within the Mendeleev representation. True, these peaks are somewhat below the peaks of the alkalis, but there they are. By the way, it is generally not known that the alkali metals have less electric resistance, i.e. greater conductance, per atom than these noble metals. Owing to the fact that the alkali metals are very easily oxidized and very soft, their high conductivity is of no technical use for making wires, but it does exist.

This curve is very instructive in so far as it reveals a similarity between columns and sub-columns of the periodic system, at least in the neighbourhood of the peaks. Thus not only chemical valency but some physical properties also, connect column with sub-column and thus give some foundation to the Mendeleev version of the system of elements.

It is known that the volatile salts of many elements give colour to flames: alkali and alkaline-earth metals are well-known examples. In cases where the temperature of ordinary flames is not sufficient to evoke this characteristic light emission, a hotter flame or an electric arc will need to be used. This coloured light can be analysed with a spectrograph into its component wavelengths or frequencies. Now, emission spectra can be measured with the utmost precision. They contain a great number of spectral lines even when they are relatively simple, and thus very detailed and accurate quantitative knowledge has been gathered from them. We shall discuss this topic very shortly in Section 64. Here we only wish to emphasize that the build-up of atomic spectra reveals far-reaching periodic regularities also. There is, for example, much similarity between the spectrum of the monovalent hydrogen atom and the spectra of the monovalent alkali metals. There is great similarity between the spectra of the alkaline-earth metals, and so forth. The higher the number of the column, the more complicated the spectra of its elements. The elements in the first three sub-groups also impart colour to flames.

57. WHAT IS WITHIN THE ATOM?

We must first sum up a few experimental facts which we have already encountered in order to draw some conclusions as to what there might be within atoms.

We have seen that electrons emerge from metals, in fact from matter in general, if light of sufficiently high frequency (quanta of sufficient energy) impinges upon it. This is the phenomenon of photoelectricity. Then we have also observed that electrons emerge from very hot metals, the phenomenon of thermoelectric emission. Flames have been found to conduct electricity and so thermal energy must have generated electrically charged particles in the hot flame-gases. Cold gases are insulators. It has also been observed that once a gas has become conducting its conductivity can be increased by the application of a high voltage across electrodes in the conducting gas: we obtain an arc or a spark according to circumstances. In these cases the primary charged particles are accelerated by the voltage and they proceed to break up other atoms or molecules into charged fragments.

It seems to be of no importance how we introduce energy into matter: with light, heat or the kinetic energy of accelerated gas-ions. Electrically charged particles and sometimes free electrons are the result. Free electrons can always be characterized by determining the ratio of charge to mass as described in Section 46. Negative electricity has a charge:mass ratio which either corresponds to that of a free electron or is some thousand or so times smaller and thus belongs to an electrically charged ion. Positive charge was never observed in these experiments, other than with a charge-to-mass (e/m) ratio corresponding to that of an ion.

Summing up we may conclude that electricity seems to be contained within originally neutral matter because it can be made to separate from it by the action of different kinds of energy. This electricity within matter—that is within atoms, because matter consists of atoms—must be in part composed of negative electrons since these are sometimes observed to emerge from matter. There must be some sort of positive electricity within atoms, too, to neutralize the electrons because originally atoms are all neutral and also because positive ions have actually been observed.

Another long series of experiments has given information about what is definitely *not* within atoms. There is no continuous matter within them. This we know from the fact that high-speed elementary particles, e.g. electrons and the alpha-particles with which we shall soon become acquainted in Section 47, are able to go through thin layers of matter without suffering very much deviation in most cases. This could not be so if atoms were spheres, of the diameter given by X-ray or other experiments, all filled with solid material. Surely particles could not pass through a thickness of, say, a thousand layers of such solid atoms.

Molecules of solids and liquids are practically in contact with each other. Even if some channels are present between them owing to special forms of their arrangement in space, these would only exist in quite special directions and could not be the general rule. Usually it is quite safe to say that even if a particle managed to slip across holes in a few molecular layers, rather sooner than later it would be bound to collide with a molecule in a subsequent layer. It might of course be imagined to force its way through them. But this should involve a permanent deformation of the material after a certain number of particles have gone across it, and in general such deformations are not observable. Thus we must begin to suppose that atoms are not those rigid spheres of solid matter we thought them to be according to their behaviour within crystals and in the kinetics of gas molecules. If they are shot at by appropriately small-sized particles, they reveal their loosely-built structure. A fence of trellis-work is effectively a solid wall so long as one kicks at a football at it. But on trying one's luck with table tennis balls it may soon become apparent that the fence contains more holes than wood. Molecules are like the football and fence with respect to each other and thus they do not interpenetrate in crystal lattices or on collision in gases or liquids. It seems though, that electrons or the alpha particles referred to are much smaller than molecules are and thus manage to pass through the open spaces within the molecule's structure.

P. Lenard (1962–1947), the famous physicist who fought all his life for the credit of having discovered X-rays when an assistant to *W. C. Röntgen* to whom this credit was given, made a very meticulous investigation concerning the penetration of electrons through matter. He determined the changes in velocity and direction suffered by them on passing through layers of different thickness of all kinds of material. He found that a beam of electrons broadens after a while, showing that changes in direction have occurred and that the electron velocities which were homogeneous at the beginning are in many cases slowed down to cover a wide range of values after having traversed a number of layers. The effect was the more pronounced the slower the electrons were at the beginning, and the more matter they had passed through. However, the most interesting point in his results was that a beam of electrons managed to emerge after passing through a macroscopic film, for instance a foil of aluminium, and still remained a more or less defined beam. Consider, that the electrons must have gone through thousands of layers of irregularly placed aluminium crystallites!

It did make a difference from which element the layer was made. Elements with low atomic weight disturbed the beam of electrons much less than did heavier atoms. Lenard found that to a first approximation the disturbing effect of elements upon electron beams was proportional to the atomic weight of the element. He explained this result by assuming that there are many points of concentrated matter within the atoms, the rest of the atomic volume being empty space. The number of these concentration regions—dynamides, as he called them—would have had to be proportional to the atomic weight of each element. In spite of the fact that it was known that electrons must somehow be accommodated within atoms, Lenard did not identify his dynamides with electrons because he was aware of the fact that positive electricity and of course ponderable matter itself must also find their place within the atoms. Only later did experiments reveal the fact that Lenard's dynamides were for the most part the electrons in the external "shells" of the atom.

The mass of an electron is very small in relation to its electric charge. Hence an electron easily suffers deviation if acted upon by an electric field. It is easier to obtain further information on the internal structure of atoms by using heavier particles as probes to traverse them because these particles will not be deflected as easily by electric fields. Such a heavy particle might find its way across a loosely built atomic structure as easily as does an electron, but would only suffer appreciable deflexion if it encountered electric fields of considerable intensity. The ratio of charge to mass of alpha particles, which are emitted by some radioactive elements, have been determined in the same manner as that of electrons, using electric and magnetic fields for their deviation (Sections 46, 87). Their individual charges were determined by collecting a known number of them on a capacitor. Thus it was shown that they carried twice the charge of an electron, but that the charge was positive instead of negative. Their mass was roughly eight thousand times that of an electron. It is therefore to be expected that the same electric field which deflects the path of an electron completely, will not considerably

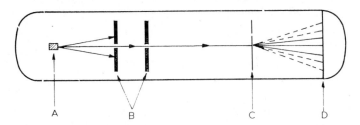

Fig. 95. **Deflexion of alpha rays in a vacuum after passing through a thin sheet of an element**

The pencil of alpha rays is defined by the slits and diverges somewhat on passing across a thin sheet of the element to be examined. The number of scintillations on the screen decreases rapidly with the angle of deflexion.

A = radium preparation   C = thin sheet of element
B = lead slits   D = fluorescent screen.

influence an alpha particle. If, for instance, an alpha particle comes near to an electron (which has only 1/7,548 of its mass) the electric force between the two will only influence the alpha particle in this proportion, as compared with the effect on the electron. This is the principle of Newton's action and reaction in ordinary mechanics. Very many collisions of this sort with electrons would eventually combine to deviate to some extent even a particle as heavy as an alpha particle. But as a matter of fact it was observed that a small number of alpha particles suffered considerably greater deviation than could be accounted for even by this cumulative process.

What does one see in such an experiment? First of all we must have a source of alpha rays. Any compound of radium, or a compound of many of the radioactive elements may serve as a source. Alpha particles which are emitted from the source in all directions must be sorted out so that only a well-defined, narrow beam of them remains. This may be accomplished by using holes in two sheets of a heavy metal, say lead, which is opaque to alpha rays (Fig. 95). Finally, it is necessary to have some means of observing them. One of the simplest devices to this end is a layer of fluorescent material, e.g. zinc sulphide or the luminescent powder of a television tube. Every time such a luminescent layer is hit by an energetic alpha particle it begins to emit light. It emits a continuous glow of light at the spot where the main beam of particles hits it. But around this point one observes an irregular flicker of very short, very intensive light pulses (scintillations), diminishing quickly in number as one proceeds away from the ray's main axis. These miniature lightning flashes are nothing less than the impinging of individual alpha particles upon the luminescent layer! (p. 395).

Now we happen to be interested in just these stray alpha particles which have lost their way on traversing the foil of the element with which we happen

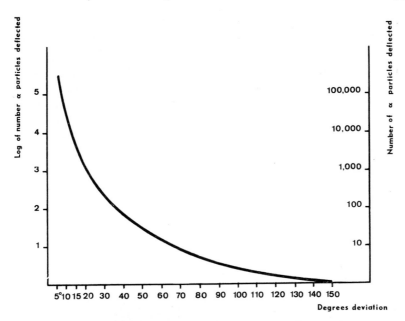

Fig. 96. **Number of alpha particles deflected after passing a layer of gold, as a function of the angle of deviation**

to be experimenting. How can it be that while the overwhelming majority of particles simply cross this foil without any visible deviation, a few unfortunate ones are thrown out of their path to a great extent? It is not an interference effect since the intensity of the out-of-bounds particles shows no periodic changes on going farther and farther away from the original direction. The number of such particles decreases regularly and very quickly at that. Fig. 96 shows their distribution which is quite different from any known diffraction pattern.

58. ATOMIC NUCLEI

We owe these fundamental experiments to *E. Rutherford* (1871–1937) who performed them with his brilliant team of scientists, *H. Geiger, E. Marsden* and *J. Chadwick*. The University of Cambridge, where he ultimately settled down has remained ever since an international centre of this fascinating branch of science. In 1911 Rutherford astonished the world by his fundamentally novel explanation of these experimental results.

According to Rutherford, the whole positive charge *and* the whole ponderable mass of an atom is concentrated in a single, very small particle within it, in the atomic nucleus. Thus it becomes possible that an alpha particle which happens to traverse an atom in the vicinity of this nucleus suffers an appreciable deviation, while those alpha particles which do not come into this vicinity go on practically undisturbed. The *a priori* probability of an alpha particle traversing an atom at any given point on its cross-section is of course the same. Let us draw circles of increasing radii around a spot we consider to be the nucleus. The probability of a perpendicularly moving alpha particle passing through any of these circles is akin to that of a bullet fired with closed eyes striking into a ring around a target. The probability of hitting any ring is simply proportional to the area of the ring. The repulsive force of the nucleus upon the alpha particle is Coulombic, that is it increases inversely as the square of the ring's radius. Thus the deviation increases with decreasing ring radius. But the ring area, that is the probability of coming near to the nucleus decreases in proportion to the radius. Great deviations must therefore be very improbable indeed.

Fig. 97 (*a*) serves to illustrate the situation. The rings must be imagined to lie in a plane perpendicular to the plane of the paper, what we see of them are only the points of intersection with the drawing. Across them we see the lines of flight of some alpha particles; the nearer these pass to the nucleus, the greater their deviation will be.

The paths in this figure have been drawn on the assumption that the repulsion between the two particles obeys the law of Coulomb, that is the force is inversely proportional to their distance apart, and that the laws of classical dynamics may be applied. The momentum of the alpha particle is large enough to make the introduction of wave mechanics unnecessary, while if its velocity should approach that of light to a great extent this would lead us into the domain of relativity. On these assumptions, the paths of the alpha particles are hyperbolae, like the paths of some comets. We assume that the charge and mass of the alpha particle are already known, as indicated above and as discussed in a later section (p. 396). According to Rutherford's assumptions the mass of the nucleus should be virtually equal to the total mass of the atom and therefore should be known too. There remains a single unknown quantity in the whole experiment: the charge on the nucleus. This determines the angle of deflexion of any alpha particle which passes a nucleus at a given distance.

It was necessary to count, in the scintillation apparatus of Fig. 95, the number of alpha particles as a function of the angle of deflexion. The experimental curve (Fig. 96) had to be compared with the curve calculated on the basis of the aforementioned assumptions, and the only free variable, the charge of the hypothetical nucleus, had to be adjusted until the best fit was obtained. The great proof of Rutherford's assumption lay in the fact that no matter which element was used as foil, there was always some particular charge which made the whole experimental curve fit the one calculated on theory, using that charge value. In view of the fact that the whole curve between say 5° and 150° deviation could be made to fit by choosing a single parameter, there was no doubt about the validity of

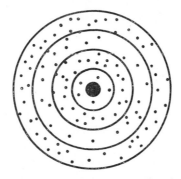

Fig. 97 (a). **Alpha rays pass around a nucleus in the same way as bullets aimed at a target by a blindfolded gunner**

Only the nearest particles suffer appreciable deflexion by the nucleus.

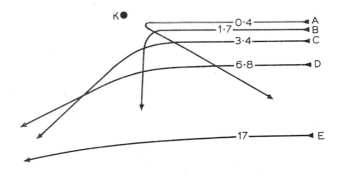

Fig. 97 (b). **Deviation of alpha rays due to Coulomb repulsion of the nucleus, when nearing the nucleus at different distances from it**

(From A. Eucken, *Lehrbuch der chemischen Physik*, p. 695, Fig. 147 (Leipzig, Akademische Verlagsges., 1930). By courtesy of Akademische Verlagsges.)

Rutherford's assumption. The number of deflected particles varied in this range in the ratio of $1:200,000$!

Rutherford was thus able to calculate the positive charges of the nuclei. Experiments on a series of elements showed at first that this charge was proportional to the atomic weight, but later on an even simpler relation emerged from more precise measurements. Taking the charge of the electron as a unit it turned out that the positive charge upon a nucleus was so many times greater than this unit, depending upon the position of the element in question within the periodic system. Platinum for instance occupies the 78th place in the system and Rutherford obtained the best fit of his curve by assuming a nuclear charge of 77·4 electron units. Silver is the 47th element and its charge turned out to be 46·3 units, while that of zinc with atomic number 29 was 29·3. From preliminary, not very accurate results this relation had already been predicted by *van der Broek* in 1913, and more exact measurements amply confirmed this view a few years later.

A glance at Fig. 97 (*b*) reveals that each angle of deflexion corresponds to a definite distance of the alpha particle from the nucleus. Therefore after having found the electrical charge which gave the best fit of the curve one could proceed to calculate the smallest distance between the two particles, the distance corresponding to the greatest angles of deflexion. Of course these extreme deflexions are rare events, but still they provide an upper limit for the radius of an atomic nucleus. The result was of the order of magnitude 10^{-12} cm, varying of course with the atomic—that is nuclear— weights. We may therefore be sure that atomic nuclei are not greater than this size: we have now learned about the dimensions of a new set of building stones of matter. Nuclei are some ten thousand times smaller than their atoms!

Alpha particles are able to approach closer to the nuclei of light atoms than to those of heavy ones because the charge upon light nuclei is smaller and thus the Coulombic repulsion is less. But the experimental curve for the number of deflected particles versus angle of deflexion ceases to agree with the theoretical curve for very great deflexions. That is, for cases when the particle was very near to the nucleus. One or more of our assumptions concerning the process of deflexion must have lost its validity. The most plausible explanation seems to be that it is no longer permissible to represent the nucleus by a point-charge. Its extension in space and possibly asymmetric electrical structure must be allowed for and hence the simple Coulomb relation between point-charges becomes invalid. Calculations showed that $5 \cdot 10^{-13}$ cm is about the distance between particle and nucleus when the above factors must be taken into account and this represents a lower limit for the radius of the nucleus. More recent measurements of the utmost importance (p. 451) revealed that an attractive force begins to act between the particles as the distance further decreases and this attraction then increases very quickly as they continue to approach each other. This force belongs to the kind which holds the nucleus together.

Let us recapitulate. We concluded that electrons must exist within atoms because we often succeed in breaking them loose from the atoms. We also concluded, that atoms are by no means solid, compact spheres as we had previously thought, but on the contrary, they are very loose structures indeed and only a very small part of their overall volume really contains anything. This "anything" must be concentrated partly in the electrons, partly in the positively charged nucleus which neutralizes the charge of the electrons and concentrates nearly all of the mass in itself. The number of positive charges in the nucleus (in electron units) is the same as the number of the element in the periodic system. The radius of the nucleus is some 10,000 times smaller than that of the atom and hence occupies only about one millionth of a millionth of its total volume. The atomic sphere of page 133 which we represented by a pinhead is not so solid a sphere after all; it has proved to be easily permeable to our subminiature projectiles, the alpha particles. All the mass of the pinhead would be concentrated in a sphere of about one micron diameter, if we wish to retain our picture.

59. ELECTRONS AROUND THE NUCLEUS

As soon as Rutherford's hypothesis concerning atomic nuclei was thus verified, one had to answer the question as to how the negative electrons were arranged around the nuclei. Owing to the fact that atoms as a whole are neutral, there must be as many negative electrons within each atom as there are units of positive charge upon the nucleus, that is, as many electrons as the ordinal number of the element in the periodic system. We understand only now what arranges the different atoms into a system at all: it is the steadily increasing number of electrons within them.

Hydrogen, the lightest of elements, has a unit positive charge on its nucleus and one negative electron outside it. The nucleus of helium has a double charge and two outer electrons, accordingly. And so on. Moving from one element to the next in the system, the positive charge on the nucleus always increases by one electron unit and the complement of outside electrons by one electron. At last we gather hope that we may understand the properties of elements within the system. But we have not come nearer to understanding that remarkable regularity concerning the atomic weight of the light elements which made Prout announce his hypothesis that all elements are built from hydrogen. We have only learned something about the charge upon the nuclei and about the fact that all the atomic mass is practically concentrated within them. We do not know why the mass of an atom is as great as it is. Only much later on, when discussing the structure of the nuclei themselves are we going to shed some light on this important question (*see* p. 459).

Let us remain content now with the problem of arranging the outer electrons around the nucleus—we shall see that this is a very intricate problem in itself. The main difficulty is to imagine negative electrons surrounding a positive nucleus without falling into this nucleus by pure Coulomb attraction. Lenard's experiments showed that there is only free space between the "dynamides," the free electrons, of the atom and that these occupy a very small fraction of its total volume. The experimental justification of this was that the majority of the small, light electrons that he shot through the layers of matter hardly suffered any change in velocity or direction.

Soon after the decisive experiments of the Rutherford laboratory, in 1913, *N. Bohr* (1885–1962) in Copenhagen laid the foundations of the theory of atomic structure. And though today we are much less inclined to represent atoms by such rigid models as did Bohr in his first venture, Copenhagen has ever since remained one of the spiritual centres of atomic theory. Bohr tried to circumvent the difficulty that would arise from Coulomb attraction between nucleus and electrons, by assuming that the electrons moved around the nucleus and that their centrifugal force counteracted the electric attraction. This was the picture of a miniature planetary system. It was only necessary to assume appropriate radii for these electron orbits and to assign appropriate velocities to the electrons in order to balance the forces.

But a new difficulty emerged. According to the laws of electromagnetism an electric charge which moves must needs become the source of electromagnetic waves. If we choose the radii of electron orbits to be of the order of magnitude of those of the atoms themselves, the rotational frequency of the electronic motion turns out to be of the same order of magnitude as the frequency of visible light waves. This corresponds in short, to the number of rotations per second which makes electrostatic attraction and centrifugal force equal. Hence electrons rotating around nuclei should emit electromagnetic radiation of this frequency *all the time*, And because they emit this energy, the electron's own energy should decrease, their velocity should accordingly decrease too, and the centrifugal force would become insufficient to balance the attraction. In almost no time at all, electrons would spiral down towards the nucleus and fall into it.

The atomic model of Bohr had to fight right from its inception against this difficulty and it never really won the battle. On looking back to those days one does not understand how it was possible that anyone could explain the very large amount of experimental knowledge on the basis of this model by deliberately ignoring this difficulty. Bohr made the sweeping assumption that certain orbits exist which—in flat contradiction to electromagnetic theory—do not radiate. He choose these orbits by combining "planetary" atomic theory with Planck's quantum constant h. Those orbits were assumed not to radiate in which the momentum of the electron was in a specific relation to h or one of its integral multiples.

Bohr himself and somewhat later *Sommerfeld* (1868–1951) in Munich, performed wonders with this electron-astronomy and, developing it in incredible detail, managed to account for the experimentally observed spectral lines emitted by the lighter elements of the periodic system. The fundamental assumption was always that radiant energy could only be emitted when an electron "fell" from an outer, non-radiating orbit into one closer to the nucleus and with less total energy. The difference in energy would then be emitted as radiation and the frequency of this radiation was determined by the Einstein relation $E = h\nu$. However, this procedure did not always work. Not only because calculation became insurmountably difficult if a system of as few as four or five electrons had to be considered. This would have simply meant the difficulties of the many-body problem of ordinary mechanics and could probably be solved today, with good approximation, by electronic calculating machines. But the method did not work in other aspects either, in cases where the mathematical difficulties had been overcome. Slight but significant differences between theory and experiment remained.

It is not at all easy to explain these difficulties on an elementary level. We shall describe only one of them and even this in a very crude manner. If we believe that a hydrogen atom is built from a nucleus and one electron rotating around it, its shape would be roughly disc-like. Its cross-section in the plane of rotation should be much greater than perpendicular to it. Now an electron revolving around a nucleus constitutes an electric current; it should have a magnetic moment and should be capable of orientation by an external magnetic field. Thus oriented, it should diffuse more readily in the plane of its orbital than perpendicularly thereto because of the smaller cross-section. Experiments have been devised and performed which are in principle equivalent to the idea set forth above and in all cases the result was entirely negative: no experiment whatever showed any deviation of the hydrogen atom from spherical symmetry. It is not a disc.

The deep inherent inconsistency of the planetary model was only over-come in the twenties, when Schrödinger and Heisenberg first took into consideration the wave nature of electrons. Or to be more precise: when they first postulated this wave nature which was confirmed only a few years later by experiments of Davisson, Germer and Thomson. The experiments were expressly devised to prove this point (*see* p. 204).

The fact that it is impossible to discuss position and momentum of an electron simultaneously beyond a given limit of "quantal" precision led to the consequence that the all too concrete picture of planetary orbits had to be abandoned. Either the position or the velocity of a planet in its orbit can always be determined, but we are unable to say *exactly* where an electron is going to be found if we want to say anything about its velocity (exactly: its momentum) *at the same moment*, and vice versa. The wave nature of the electron becomes all important and we are only able to locate it and to speak about its velocity in terms of statistical probabilities (*see* p. 207).

Looking at the problem from this aspect obviates the necessity of *ad hoc* assumptions about special, non-radiating orbits. An electron is unable to fall into the nucleus, because if it did so, we would have defined its momentum within the limit of $h/10^{-13}$ that is, say, 6×10^{-14} g. cm sec^{-1} (h being $6\cdot55 \times 10^{-27}$ g cm^2 sec^{-1}). Taking into account that the mass of an electron is 9×10^{-28} g this means that its energy would have to be 2 ergs*, an incredibly enormous amount of energy for a single electron, corresponding to a velocity of 7×10^{13} cm sec^{-1}. The potential energy of the electron at this distance from the nucleus (e^2/r) would be $(4\cdot79 \times 10^{-10})^2/10^{-13} = 2\cdot29 \times 10^{-6}$ erg, quite insufficient to keep it there. It could not remain in the nucleus with this kinetic energy. Therefore it does not fall into it at all.

In a somewhat crude manner we may imagine this enormous kinetic energy as throwing the electron out, away from the nucleus into a more expanded region within which it is localized "somewhere." The greater this region of uncertainty, the less the uncertainty of momentum and thus of

---

* Momentum is mass times velocity, $I = mv$. Kinetic energy is mass times the square of velocity over 2, $E_k = mv^2/2$. Hence kinetic energy is the square of momentum divided by twice the mass $E_k = I^2/2m$.

kinetic energy and of thrust. The most probable location of the electron with respect to the nucleus according to this picture is on the surface of a sphere around the nucleus. The radius of this sphere is determined by the fact that the total energy of the electron is, on the average a minimum at this distance. Farther away, energy increases because work has to be performed against the electrostatic attraction. At closer distances the energy increases again because of the increase of mean kinetic energy with decreasing volume of localization explained above.

This strange sort of kinetic energy is quite new in physics and is characteristic of quantum mechanics. It is called zero-point energy, because it is independent of thermal motion: cooling electrons, or for all it matters, hydrogen, to any temperature in the neighbourhood of absolute zero cannot decrease this zero-point energy. It is inherent in matter and owes its existence, as we have stated, to the fact that all matter and energy has the nature of waves, and cannot be localized precisely without introducing an uncertainty into its momentum. The smaller the volume in which a particle is enclosed, the less we know about its momentum, its velocity. Thus its velocity cannot be zero, because zero motion is just one fixed velocity, and fixed it cannot be. It will have a statistical distribution and the mean value of this zero point velocity combined with the mass determines the zero-point kinetic energy of the particle.

This has all been a very qualitative treatment of a difficult situation. The exact treatment is based on Schrödinger's wave equation (p. 208) which is the relatively simple expression of the fact that, though matter and light are diffracted like waves, the mass and the energy respectively of these waves is concentrated within wave packets. The simplest two-body problem for this equation is that of the hydrogen atom with one central nucleus and one indeterminate electron surrounding it. This equation has an immense advantage over the orbit equations of Bohr: there is no need to introduce *ad hoc* quantum conditions for non-radiating orbits. The equation itself only makes sense, it only possesses sensible solutions, for discrete, well-defined amounts of total energy (p. 209). For each of these discrete energy values, a probable spatial and momentum distribution can be constructed with places of maximum probability for the electron. If we so wish, we may designate these positions of maximum probability as orbits of the electron, and for the hydrogen atom they really coincide with Bohr's orbits. But it must be clear that any other position is possible also, with some smaller probability—a probability which can be exactly represented by the solution of the Schrödinger equation for the given total energy. Now Schrödinger's equation operates with waves in three-dimensional space for a single electron and it happens to be characterized by three sets of numbers which between them determine the permitted states, the permitted distributions of the electron around the nucleus. Each solution has a specified content of total energy. The sets of numbers are called the quantum numbers and they completely characterize each state of the electron-wave system (*see* "Eigenvalues" on p. 209).

That is, nearly completely. We should now remember the fact (Section 52) that electrons are elementary magnets and have thus an axis of their own. This has the important consequence that it is not sufficient to determine their probable distribution in space, it is also necessary to know how the magnetic axis is oriented, against either a given magnetic field or against the axis of a second electron. It can be shown from the study of mathematical physics (though it cannot be shown here) that the orientation of this magnetic axis cannot be chosen to lie in any given directions, but is "quantized" also, and only discrete orientations are possible solutions of a more complicated Schrödinger equation which has been set up by *Dirac*. The so called "spins" of the electrons may be either parallel or antiparallel to each other, i.e. can point in the same or the opposite direction. Hence for every solution of the original Schrödinger equation, which was characterized by three quantum numbers, we have two real solutions for the spinning electron (or for any other particle which happens to possess the same spin).

The calculation involves a great deal of mathematics. But its results can be plotted on diagrams. Curve (i) of Fig. 98 (*a*) shows the radial probability distribution of a single electron around a nucleus for different quantum numbers. It is a mathematical property of the wave equation that its "principal" quantum numbers, the quantum numbers belonging to the first set, determine the total of quantum numbers possible in the second set, and these two between them determine the possible number in the third set. The spin quantum number of the electron can always have only one of two possible values. Now, if the principal quantum number has its lowest value, the next two have only a single value; they only can gradually attain more and more separate values as the principal quantum number rises. Thus there is only one single state of two possible spins related to the lowest principal quantum number. This "density" of probable electron positions is represented in curve (ii). We see that this curve has its maximum *in* the nucleus (!) and decreases spherically around it. Thus, from among all possible *points* in space, the nucleus is the most probable site of an electron in this lowest, ground state.

But it is not going to be *found most often* at this zero distance, because *all* points lying on a sphere with radius *r* around the nucleus have the same chance to accommodate an electron. *All* points of the sphere are, of course, equidistant from the nucleus. Therefore the total probability of finding an electron at a distance *r* from the nucleus will be proportional to the product of finding it at any of the points of the sphere (given by the ordinate of curve (ii) at the abscissa *r*) and the surface of the sphere $4\pi r^2$. The result of these multiplications is plotted in curve *c* which represents the probability of finding the electron in any direction at the distance *r*. In this plot the probability of the electron being in the nucleus is zero because, so to say, the nucleus represents only "one point" from among the infinite points of space, whereas all spherical shells represent a large number of "points," proportional to the surface of each shell. Curve (iii) has a definite maximum at the distance where the influence of intrinsic probability (curve (ii)), which decreases with a function of *r*, is just compensated by the fact that the surface of the shells increases with r^2. This is the distance from the nucleus where the electron will be found more often than at any other particular distance. It happens to be identical with the first Bohr orbit, but it was found without recourse to any magic quantum relations as the simplest possible solution of the wave equation.

Consider now solutions of the equation which belong to principal quantum numbers larger than one. As stated above, the lower quantum numbers

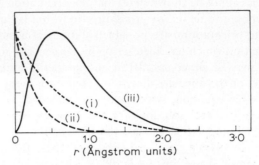

Fig. 98 (a). **Distribution of an electron around the proton in the lowest state of the hydrogen atom according to the Schrödinger equation**

Curve (i)   Radial distribution function of ψ, the variable of the equation.

Curve (ii)   Radial distribution of ψ^2 which represents the relative probability of finding the electron at a given point of space.

Curve (iii) ψ^2 multiplied by $4\pi r^2$, the surface area of the sphere of radius r. The product represents the probability of finding the electron in any place on the surface of a sphere with radius r. The maximum of this probability is at the radius calculated by Bohr in his first, rather *ad hoc* theory.

Fig. 98 (b). **Picture visualizing the density distribution around the nucleus**

(From G. M. Barrow, *Physical Chemistry*, p. 224, Fig. 9.2 (New York, McGraw-Hill, 1961). By courtesy of the McGraw-Hill Book Co.)

may then increase, according to a set of rules which are purely mathematical and not physical. They follow logically from the mathematical form of the wave equation. However, they need not necessarily increase, they may equally well retain their lowest value, the one they had when the principal quantum number was 1. Calculation shows that the solution of the wave equation remains spherical in these cases, in other words, the function ψ depends only on the distance from the nucleus and is independent of the direction. However, its dependence on the radius is more complicated than it was for the principal quantum number 1. It has as many maxima as the numerical value (n) of the quantum number itself. The curves $n = 2$ and $n = 3$ are shown in Fig. 99. The highest maximum is always the one farthest away from the nucleus. We are very far from knowing where our electron "really" is, even the probability of finding it at a given distance from the

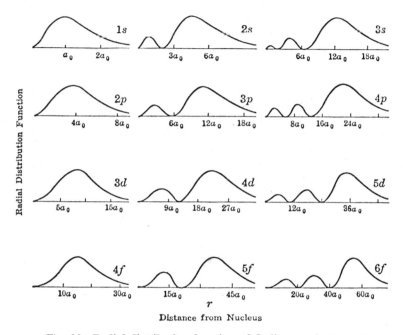

Fig. 99. **Radial distribution function of finding an electron at a given distance from the nucleus for higher main and secondary quantum numbers**

Units in a_0 = the lowest Bohr radius of the H atom. The representation is that of Fig. 98a (iii). The numbers on the curves are the main quantum numbers, the letters s, p, d, f characterize the secondary, azimuthal, quantum states. The quantum number 1 must always be an s state, 2 may be an s or p and so on. Only the s states are spherically symmetrical, *see* Fig. 100.

(From S. Glasstone, *Theoretical Chemistry*, p. 59, Fig. 5 (Princeton, N.J., van Nostrand, 1944). By courtesy of the van Nostrand Co.)

nucleus has maxima and minima! But each solution represents a quite definite content of total energy. And the difference of two such possible energy values is emitted in electromagnetic waves—as we shall see—according to the Einstein relation $\Delta E = h\nu$. And this relation holds good.

There is a fundamental change in the situation as soon as we allow for an increase of the second quantum number in cases where the main quantum number n is greater than 1. The mathematical solution of the distribution function loses its spherical symmetry. This means that places of equal probability do not lie any more on spheres but are arranged in different directions in space. It can be shown that for the next stage of the second quantum number three solutions along the three space axes, respectively, are obtained—more are impossible. Fig. 100 gives some idea of this distribution. For $n = 3$, the second quantum number may have three values: the lowest with spherical symmetry as shown in Fig. 100(*a*), the next with the propeller shapes of Fig. 100(*b*) and a last series of five different distributions as represented to some extent by Fig. 100(*c*). For purely historical reasons concerned with experimental spectroscopy it has become conventional to designate these distribution types of the second quantum number by the small italic letters: *s*, *p*, *d*, *f*. The spherical distribution corresponds to the letter *s*, the double cigars to *p*, the more complicated ones of Fig. 100 to *d*, and so on.

60. ELECTRON SHELLS

Possessing the Schrödinger equation, it should be possible to calculate the electron distribution for all elements in the periodic system. Proceeding from one element to the next, the charge on the nucleus will increase by one positive unit and the number of electrons around the nucleus will increase in the same way. Accordingly, the number of coordinates to be determined will increase by three each time, because each new electron involves three new cordinates.* It is easily seen that a differential equation with such a quickly increasing number of variables does not lend itself to exact solution—purely because of technical difficulties, as are encountered, for example, in the many-body problem of classical mechanics. But many an astronomical problem has been solved by appropriate approximations, even before computer calculation was invented. Very much has been, and is going to be done by approximations in atomic and molecular structure calculations also. Mathematical skill has to be combined with the intuition of the good physicist to find the shortest and most efficient approximation, to feel which part of the electronic system may be simplified into a cloud of negative electricity, or to decide what kind of statistical assumptions are most helpful when a great number of electrons are concerned. All this is going on in the present day.

As far as we are concerned, it is best to go once more through the periodic system and see how experimental facts arrange themselves if we look at them from the point of view of quantum mechanics. However, before embarking on this project it is necessary to insert one further principle into the theory. This very fundamental principle is not inherent in any assumptions we have made up to now as far as quantum and wave-mechanics respectively are concerned. It is an independent postulate of physics, announced by the lately deceased theoretical physicist *W. Pauli* (1900–1958) and is known as the "Pauli principle." It states that it is not possible to

---

* Because of the uncertainty principle we can only determine exactly the three space coordinates of an electron if we have given up saying anything at all about its impulse coordinates or vice versa. This is the fundamental difference between quantum and classical mechanics: in classical mechanics three space and three momentum coordinates should determine the situation. Now we have only half as many coordinates as on classical grounds.

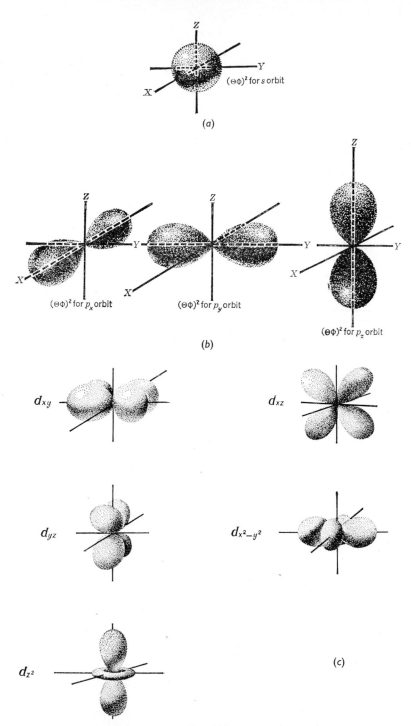

(a) for *s* orbit

(a)

$(\Theta\Phi)^2$ for p_x orbit $(\Theta\Phi)^2$ for p_y orbit

$(\Theta\Phi)^2$ for p_z orbit

(b)

d_{xy} d_{xz}

d_{yz} $d_{x^2-y^2}$

(c)

d_{z^2}

Fig. 100

(a) Probability sketch of the *s* state of an electron around a nucleus.
(b) Probability sketch of the three *p* states of an electron around a nucleus.
(c) Probability sketch of the five *d* distributions of an electron around a nucleus.

(From G. M. Barrow, *Physical Chemistry*, p. 192, Fig. 8.3, p. 242, Fig. 9.12 (New York, McGraw-Hill, 1961). By courtesy of the McGraw-Hill Book Co.)

TABLE VII. The Grouping of

The data at the beginning of this table are based on exact quantum
never occurs as a free, single atom and thus the theoretical grouping
spectroscopic and chemical evidence and on analogy

Element	Number of electrons	K	L		M			N				O				P			Q
		s	s	p	s	p	d	s	p	d	f	s	p	d	f...	s	p	d...	s...
H	1	1																	
He	2	2																	
Li	3	2	1																
Be	4	2	2																
B	5	2	2	1															
C	6	2	2	2															
N	7	2	2	3															
O	8	2	2	4															
F	9	2	2	5															
Ne	10	2	2	6															
Na	11	2	2	6	1														
Mg	12	2	2	6	2														
Al	13	2	2	6	2	1													
Si	14	2	2	6	2	2													
P	15	2	2	6	2	3													
S	16	2	2	6	2	4													
Cl	17	2	2	6	2	5													
Ar	18	2	2	6	2	6													
K	19	2	2	6	2	6		1											
Ca	20	2	2	6	2	6		2											
Sc	21	2	2	6	2	6	1	2											
Ti	22	2	2	6	2	6	2	2											
V	23	2	2	6	2	6	3	2											
Cr	24	2	2	6	2	6	4	2											
Mn	25	2	2	6	2	6	5	2											
Fe	26	2	2	6	2	6	6	2											
Co	27	2	2	6	2	6	7	2											
Ni	28	2	2	6	2	6	8	2											
Cu	29	2	2	6	2	6	9	2											
Zn	30	2	2	6	2	6	10	2											
Ga	31	2	2	6	2	6	10	2	1										
Ge	32	2	2	6	2	6	10	2	2										
As	33	2	2	6	2	6	10	2	3										
Se	34	2	2	6	2	6	10	2	4										
Br	35	2	2	6	2	6	10	2	5										
Kr	36	2	2	6	2	6	10	2	6										
Rb	37	2	2	6	2	6	10	2	6			1							
Sr	38	2	2	6	2	6	10	2	6			2							
Y	39	2	2	6	2	6	10	2	6	1		2							
Zr	40	2	2	6	2	6	10	2	6	2		2							
Nb	41	2	2	6	2	6	10	2	6	3		2							
Mo	42	2	2	6	2	6	10	2	6	4		2							
Tc	43	2	2	6	2	6	10	2	6	5		2							
Ru	44	2	2	6	2	6	10	2	6	6		2							
Rh	45	2	2	6	2	6	10	2	6	7		2							
Pd	46	2	2	6	2	6	10	2	6	8		2							
Ag	47	2	2	6	2	6	10	2	6	9		2							
Au	48	2	2	6	2	6	10	2	6	10		2							
In	49	2	2	6	2	6	10	2	6	10		2	1						
Sn	50	2	2	6	2	6	10	2	6	10		2	2						
Sb	51	2	2	6	2	6	10	2	6	10		2	3						
Te	52	2	2	6	2	6	10	2	6	10		2	4						
I	53	2	2	6	2	6	10	2	6	10		2	5						
Xe	54	2	2	6	2	6	10	2	6	10		2	6						

Orbital Electrons of the Elements

mechanical calculation for free atoms—though carbon, for instance,
of its electrons cannot be checked. The rest of the table is based on
to the results of exact calculation at the beginning.

		K	L		M			N				O				P			Q
Element	Number of electrons	s	s	p	s	p	d	s	p	d	f	s	p	d	f...	s	p	d...	s...
Cs	55	2	2	6	2	6	10	2	6	10		2	6			1			
Ba	56	2	2	6	2	6	10	2	6	10		2	6			2			
La	57	2	2	6	2	6	10	2	6	10		2	6	1		2			
Ce	58	2	2	6	2	6	10	2	6	10	1	2	6	1		2			
Pr	59	2	2	6	2	6	10	2	6	10	2	2	6	1		2			
Nd	60	2	2	6	2	6	10	2	6	10	3	2	6	1		2			
Pm	61	2	2	6	2	6	10	2	6	10	4	2	6	1		2			
Sm	62	2	2	6	2	6	10	2	6	10	5	2	6	1		2			
Eu	63	2	2	6	2	6	10	2	6	10	6	2	6	1		2			
Gd	64	2	2	6	2	6	10	2	6	10	7	2	6	1		2			
Tb	65	2	2	6	2	6	10	2	6	10	8	2	6	1		2			
Dy	66	2	2	6	2	6	10	2	6	10	9	2	6	1		2			
Ho	67	2	2	6	2	6	10	2	6	10	10	2	6	1		2			
Er	68	2	2	6	2	6	10	2	6	10	11	2	6	1		2			
Tm	69	2	2	6	2	6	10	2	6	10	12	2	6	1		2			
Yt	70	2	2	6	2	6	10	2	6	10	13	2	6	1		2			
Lu	71	2	2	6	2	6	10	2	6	10	14	2	6	1		2			
Hf	72	2	2	6	2	6	10	2	6	10	14	2	6	2		2			
Ta	73	2	2	6	2	6	10	2	6	10	14	2	6	3		2			
W	74	2	2	6	2	6	10	2	6	10	14	2	6	4		2			
Re	75	2	2	6	2	6	10	2	6	10	14	2	6	5		2			
Os	76	2	2	6	2	6	10	2	6	10	14	2	6	6		2			
Ir	77	2	2	6	2	6	10	2	6	10	14	2	6	7		2			
Pt	78	2	2	6	2	6	10	2	6	10	14	2	6	8		2			
Au	79	2	2	6	2	6	10	2	6	10	14	2	6	9		2			
Hg	80	2	2	6	2	6	10	2	6	10	14	2	6	10		2			
Tl	81	2	2	6	2	6	10	2	6	10	14	2	6	10		2	1		
Pb	82	2	2	6	2	6	10	2	6	10	14	2	6	10		2	2		
Bi	83	2	2	6	2	6	10	2	6	10	14	2	6	10		2	3		
Po	84	2	2	6	2	6	10	3	6	10	14	2	6	10		2	4		
At	85	2	2	6	2	6	10	2	6	10	14	2	6	10		2	5		
Rn	86	2	2	6	2	6	10	2	6	10	14	2	6	10		2	6		
Fr	87	2	2	6	2	6	10	2	6	10	14	2	6	10		2	6		1
Ra	88	2	2	6	2	6	10	2	6	10	14	2	6	10		2	6		2
Ac	89	2	2	6	2	6	10	2	6	10	14	2	6	10		2	6	1	2
Th	90	2	2	6	2	6	10	2	6	10	14	2	6	10	1	2	6	1	2
Pa	91	2	2	6	2	6	10	2	6	10	14	2	6	10	2	2	6	1	2
U	92	2	2	6	2	6	10	2	6	10	14	2	6	10	3	2	6	1	2
Np	93	2	2	6	2	6	10	2	6	10	14	2	6	10	4	2	6	1	2
Pu	94	2	2	6	2	6	10	2	6	10	14	2	6	10	5	2	6	1	2
Am	95	2	2	6	2	6	10	2	6	10	14	2	6	10	6	2	6	1	2
Cm	96	2	2	6	2	6	10	2	6	10	14	2	6	10	7	2	6	1	2
Bk	97	2	2	6	2	6	10	2	6	10	14	2	6	10	8	2	6	1	2
Cf	98	2	2	6	2	6	10	2	6	10	14	2	6	10	9	2	6	1	2
Es	99	2	2	6	2	6	10	2	6	10	14	2	6	10	10	2	6	1	2
Fm	100	2	2	6	2	6	10	2	5	10	14	2	6	10	11	2	6	1	2
Md	101	2	2	6	2	6	10	2	6	10	14	3	6	10	12	2	6	1	2
No	102	2	2	6	2	6	10	2	6	10	14	2	6	10	13	2	6	1	2
Lw	103	2	2	6	2	6	10	2	6	10	14	2	6	10	14	2	6	1	2

arrange two particles in such a manner that *all* their quantum numbers (including spin) are identical (1925). In classical physics it seemed quite natural that two objects could not occupy the same place at the same time. We have to be much more sophisticated now. We are not so certain about the "place" of a particle as we used to be and even the concept of two events being "simultaneous" has suffered some restrictive clarification at the hands of Einstein. So our trivial-looking statement about two objects not being simultaneously at the same place has lost quite a bit of its meaning. Perhaps it is a consolation to look at the Pauli principle as the quantum mechanical equivalent of that primitive statement.

We have already spoken about the hydrogen atom because it is the simplest system built from electrons and nuclei. It has the lightest known nucleus with a single electron "around" it. The nucleus itself carries practically the whole mass of the atom, $1·66 \times 10^{-24}$ g and its charge is equal to that of the electron, but positive. We have discussed the possible probability distributions of the electron in the field of this nucleus, have found a spherically symmetrical state of lowest energy for this atom and several states of higher quantum numbers—some relating to spherical states, others not—which possess more energy than the ground state but are capable of falling back into the latter by emitting their surplus energy, e.g. as light waves. The frequency of these radiations fits exactly into the scheme calculated from the energy differences by the Einstein relation $\Delta E = h\nu$.

The second element in the system is helium. It belongs to the family of the rare, noble gases which we thought for a long time to have no chemistry at all, i.e. they remain monatomic and have no compounds. According to its place in the system the helium nucleus must carry twice the charge of the hydrogen nucleus, or proton as it is generally called, because it is the first, the most fundamental of all the nuclei. And of course there must be two electrons around the helium nucleus. We are now dealing with the quantum mechanical three-body problem, on the verge of exact mathematical solutions. Calculation shows that the state of lowest energy in this system corresponds to a field of probability for the electrons around the nucleus which is spherical and very similar to the field of probability for finding the electron of the hydrogen atom somewhere around the proton. But what about the Pauli principle? The solution we spoke about assigns to both electrons the lowest quantum numbers in all three sets of these numbers. The only possible way of doing this without violating the Pauli principle is to assign opposite spins to the two electrons. The magnetic interaction of these two oppositely oriented elementary magnets even liberates a certain amount of energy in this combination, adding to the main energies of electrostatic attraction between electrons and nuclei and reducing repulsion between the electrons.

All in all something has come to an end with this structure of electrons. It will evidently be impossible to add a third electron to the system in the same lowest set of quantum numbers, because both possible spin directions are already filled. A third electron—say the electron which we have to dispose of for the third element, lithium—will have to be accommodated in a state, with a probability distribution around the nucleus which is characterized by at least a principal quantum number of $n = 2$. This amounts to a greater average distance from the nucleus, as shown in the curve for $n = 2$ in Fig. 99. There will not even be a great difference between the electrostatic force acting upon this third electron and that acting upon the electron of the hydrogen atom in its $n = 2$ state, because the positive charge of 3 on the lithium nucleus is very nearly shielded by the two negative electrons occupying the inner $n = 1$ distribution, and so yields the effective field of charge $3 - 2 = 1$. This seems to be the simplest way a lithium atom can be built if the Pauli principle is to remain valid. The third electron has been expelled with great probability into a sphere farther away from the nucleus than that of the electron of the hydrogen atom. The positive charge acting upon these electrons is nearly the same in the two cases. Hence the more distant third electron of lithium must be bound with much less energy to the remaining core of the atom which consists of the nucleus and the twin electrons in the lowest energy "orbital".

How does all this fit into what we know from chemistry? Lithium is a monovalent metal. This means that it is prone to lose one of its electrons and yield a monovalent positive ion. It is a metal, that is it readily conducts electricity in its condensed state, probably because it is prone to lose this same electron in this case too, and to allow it to move hither and thither between the resulting positive ions within the crystal lattice. Chemical facts and the electronic structure of the lithium atom therefore seem to agree excellently.

As a matter of fact they did agree well in the case of helium. We shall see later on that in one way or another the electrons of an atom, at least the electrons at the surface of an atom, are responsible for the different kinds of chemical bond. In order to go into chemical combination with another atom these surface electrons have to be loosened to some extent, if not given up entirely, as by lithium when giving rise to the Li^+ ion. Or an electron of another atom has to be attracted into the sphere of interest of the first atom. In the case of helium it is pretty clear that neither of these events is very likely. Both of its electrons are bound with high, spherical symmetry into the lowest possible sphere around the nucleus. To remove one electron by ionization or even to excite one into a higher quantum level would mean expenditure of more energy than could conceivably be gained by entering into chemical combination with another atom. The energy *can* be calculated of course, and it turns out to be excessive. Only in electrical discharges, in fields of high energy density may it happen now and then that a helium atom is excited to such a level that a transient combination with an atom of another element or of helium, might come about. Spectra obtained from such discharges actually do reveal the existence of these short-lived molecular species and they agree with the spectra calculated for such species on the basis of their possible states of energy. But these are no stable, ordinary chemical compounds, they disintegrate immediately. The other possible way of forming compounds, the combination with an electron from another atom is again impossible for helium because it could only accept this electron into its second quantum sphere, far from the nucleus, where the electrostatic attraction available to bind it is nil: the two electrons around the nucleus shield its field completely. This is the reason why helium is a noble gas.

But what about the chemistry of hydrogen? Its atom has a single electron in the lowest quantum state and there is the possibility of accommodating therein a second one without violation of the Pauli principle. The nuclear charge is only half that of helium. This has the effect that in the hydrogen atom the electron will be farther away from the nucleus, as mentioned, and hence its binding energy will be smaller for two reasons: by virtue of the smaller charge and the greater distance. Thus a hydrogen atom will be able to lose its electron much more easily than does a helium atom. On dealing with acids we have become familiar with the fact that there are hydrogen compounds which easily lose a positive hydrogen ion. Now we know already that this ion is the proton itself, a bare nucleus of unit charge, the only *bare* nucleus in chemistry. (It therefore always combines with the solvent as in H_3O^+ or with other species as e.g. in NH_4^+.) All other ions, radicals, atoms and molecules possess electrons around their nuclei. These species which are prone to lose a proton we have termed acids while that other class of chemical species which readily add a proton we have termed bases (Section 17). Thus it seems that a bound hydrogen atom is capable of dissociating itself from its original electron-partner and of associating itself with others, in accordance with what we have said about its electronic structure. Compounds which contain positively ionizing hydrogen are acids. Hydrogen is evolved at the cathode on their electrolytic decomposition. But we know of another class of compound where hydrogen is evolved at the anode. It must have been present as a negative ion in these cases and so must have added at least one electron to achieve this. The best known of these compounds is LiH, a salt-like solid which can be electrolysed after melting. Of what is it composed? Lithium migrates to the cathode, and so must have been present as a positive ion as always, after simply letting lose its third electron which, as we have seen, is not bound very strongly. This solitary electron must have added itself to the incomplete lowest quantum state of the hydrogen atom, compensating the spin of its single electron. Thus a very special electronic structure results. Li^+, as well as H^-, has two electrons around its nuclei. The structures of the two ions must be extremely similar, but with one important difference however: the nuclear charge acting upon these electron pairs is $+3$ for Li^+ and only $+1$ for H^-. The former must be, and is, substantially smaller than the latter.

Besides these extreme cases of parting entirely with one electron or permanently acquiring one, there are all kinds of intermediate situations for the various elements and for hydrogen also. The great majority of organic compounds containing carbon and hydrogen for instance are bound, as we shall see later on, by bonds of this type. It should, however, always be borne in mind that hydrogen is unique throughout the whole periodic system, because when it has given up its electron, a bare nucleus remains. This bare nucleus may be drawn into the electronic system of its chemical partner. It is the only nucleus which is capable of doing so, because it has absolutely no electrons left around it to establish a sphere of repulsion against foreign electron systems. This is the reason why the concepts of acids and bases had to be evolved.

The fourth element, beryllium, has four positive nuclear charges and four electrons—three which we had to dispose of for lithium and the fourth which is allowed to occupy the same state as the third electron but with opposite spin. The two interior electrons are exposed to the full attraction of the $+4$ charges at a small distance and are thus strongly bound. They do not leave the region of the nucleus to enter chemical bonds. The two others are farther away because their principal quantum number, n, is 2. Practically, two charge units of the nucleus are shielded from them by their interior partner and they are farther away because $n = 2$; as a result they are bound to a lesser extent and do enter into chemical combinations. However, the resultant charge acting on them is $+2$ as against $+1$ in lithium and hence they are bound more firmly than the valency electron of lithium. Beryllium is divalent because of these two electrons but less electropositive than lithium because of this higher charge. We now understand; these experimental facts are now explained on a logical basis.

With boron a new situation arises. The next, the fifth, electron is prevented by the Pauli principle from entering the same quantum state as the two outer beryllium electrons. But the solutions of the Schrödinger equation allow different secondary quantum numbers for the main quantum number $n = 2$. They permit three such solutions, that is, they allow for six electrons if we double each solution of the equation to allow for the two possible electron spins. These are the first non-spherical "p orbitals" we have already mentioned. The last electron of boron can only enter one of these latter states.

It has become customary to write electronic formulae for atoms. The principal quantum number is written first, then a letter s, p, d, etc., stands for the next quantum numbers while a number as a right-hand superscript to this letter shows how many electrons are in the state thus described. For example we write the symbols for the first five elements, i.e. for those we have discussed up to now as: $H: 1s^1$; $He: 1s^2$; $Li: 1s^2, 2s^1$; $Be: 1s^2, 2s^2$; $B: 1s^2, 2s^2, 2p^1$. (The superscript 1 is often omitted.)

Boron has two strongly bound inner electrons ($1s^2$) and three less strongly bound ones in the $n = 2$ region. It is trivalent with respect to these three disposable electrons but less prone to lose them, less positive than is beryllium because the effective pull upon these three electrons is about equivalent to $+3$ charges (two of five are practically shielded). Owing to this increasing charge these three electrons are pulled closer towards the nucleus and the atom becomes smaller in accordance with the curve of atomic volumes in Fig. 93. Its smaller radius accounts for the fact that it is able to bind, e.g. oxygen, more strongly than the preceding elements can: its hydroxy compound is an acid as we have seen in discussing the ideas of Kossel on acids and bases (p. 189).

We now reach carbon, the essential element of organic chemistry. Evidently its electronic formula must be $1s^2$, $2s^2$, $2p^2$. However, there are three p

states with two spins each. Where has the new electron gone: into the same p state as the first, compensating its spin, or into one of the other two of the three, cigar-shaped, mutually perpendicular "p orbitals"? Such questions cannot well be answered from the chemical facts. Spectroscopic and magnetic evidence has shown, however, that in such cases the different "p orbitals" are occupied one after the other and spin compensation within occupied orbitals only takes place when there is no free orbital of the same low energy. This rule was announced by *Hund* as a result of spectral observations which we are unable to follow at this level. But the meaning of Hund's rule ("maximum multiplicity") can be made quite clear. Electrons naturally occupy the states of lowest energy accessible to them. The question we have posed above amounts to asking: which state has the lower energy, the one where two electrons are in the same spatial distribution of probability with opposite spin or where they evade each other, as far as possible, in different "orbitals"? The fact that Hund's rule prevails shows that the latter is the case. It means that the electrostatic repulsion of two negative electrons makes them go out of each other's way as long as they are able to do so, as long as orbitals of the same kind are available, that is unoccupied. The magnetic energy which can be and is won by spin compensation is much smaller than this electrostatic repulsion, and therefore only comes into play when the former repulsion cannot be evaded any more.

The above electronic structure for the carbon atom shows us that it really has four electrons in a "second sphere" which could account for its tetravalency. However, they are not quite equal, two of them being of the spherical symmetry "s" type while two occupy cigar-shaped, perpendicular "p" orbitals. True, we see a slight chemical evidence for this difference in the chemical fact that carbon forms two oxides, CO and CO_2, the two pairs of electrons are apparently utilized for bonding, one after the other. But in general we have seen that the facts of organic chemistry are best explained by an absolute tetrahedral symmetry around the carbon atom. It has been shown by elaborate quantum mechanical calculation that such a tetrahedral symmetry is possible for the four $n = 2$ electrons of carbon. The Schrödinger equation has certain permitted solutions for such a tetrahedral distribution as idealized in Fig. 101. In such a case the orbitals are equally shaped and point towards the four corners of a regular tetrahedron. Since they consist of a mixture of originally s and p state electrons they are termed "sp hybrid" states. It is a question of energy content whether the $2s^2$, $2p^2$ or the $2sp^3$ state will be the more stable, and it has been shown that the latter can be the stable arrangement for compounds of carbon. Similar but sometimes more complicated "hybrid" solutions occur in many higher elements too, and play an important role in inorganic complex chemistry. It is always important to remember that we are searching for possible stable solutions of a wave equation, in analogy to the Chladni figures of a vibrating plate. Given the equation, the form and number of permitted solutions, together with their characteristic energy contents, is a question of mathematical technique.

Nitrogen, oxygen, fluorine and neon go on to complete the set of six p electrons belonging to the main quantum number $n = 2$. With neon each of the three cigar-shaped distributions contains its full complement of two spin-paired electrons, and as the mathematical theory of the Schrödinger equation shows, no more electrons can be accommodated without increasing the principal quantum number. Surely it is the completion of this $n = 2$

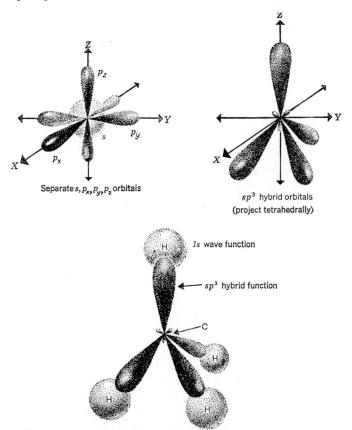

Separate s, p_x, p_y, p_z orbitals

sp^3 hybrid orbitals
(project tetrahedrally)

$1s$ wave function

sp^3 hybrid function

Fig. 101. **The sp^3 orbitals that can be formed from linear combination of the s and the three p orbitals and the overlapping orbitals that lead to bonding in methane**

The angular part of the probability function is shown and has been distorted for clarity.

(From G. M. Barrow, *Physical Chemistry*, p. 232, Fig. 9–5 (New York, McGraw-Hill, 1961). By courtesy of the McGraw-Hill Book Co.)

series* which is responsible for the chemical stability, lack of reactivity, of this second rare gas. Helium, the first noble gas has a complete K shell, neon, the second one, a complete L shell. It can be shown by addition of the three p probability functions that the three double cigars exactly add up to a

---

* Instead of saying "the shell of electrons belonging to the main quantum number 1, 2, 3, etc.," one often uses the capital letters K, L, M . . ., etc and speaks of helium, for instance, as the element with a complete K-shell of electrons.

sphere once more: a very compact, highly symmetrical arrangement has been reached once again. The sphere of electrons shields the nucleus completely: there remains no force to attract an electron from another atom (compare helium). As with helium the removal of an electron from this highly symmetrical structure into a higher quantum level would mean that a great amount of electrostatic energy must be expended. Owing to the fact that we are now dealing with L electrons ($n = 2$), farther away from the nucleus than those of helium, and to the fact that we are now dealing with eight electrons instead of two so that they repel each other to a greater extent, it is easier to remove an electron from neon than from helium. The minimum velocity of an electron beam necessary to ionize neon is less than for helium; the ionization potential of neon is lower (21·5 V against 24·5 V). Still, it is high enough to prevent neon from forming chemical compounds.

The repulsive force between the electrons of the L shell makes the radius of the atoms increase in the second half of the period. In the first half, from lithium to carbon it decreases because of increasing effective nuclear charge; after carbon the repulsion between the increasing number of electrons outweighs this effect, and the radius of the L shell and thus of the atom itself increases (*see* Fig. 93, atomic volumes). This phenomenon is repeated throughout each successive period and explains the shape of the curve in Fig. 93.

Increase of volume means more space to accommodate electrons and within this increased space their position is less exactly determined. According to what we have said about zero-point energy (p. 260), this means that the uncertainty of velocity and thus of energy are less as the volume increases and therefore these configurations are favoured from the point of view of energy content. Is this really so? It is. Beginning with carbon every element in this period forms hydrides (or alkali compounds) in a way that the number of its own L electrons together with the number of electrons of the added hydrogen atoms is exactly eight: CH_4, NH_3, OH_2, FH . . . and Ne with no hydrogen at all. The configuration of the rare gas, with its complete set of p electrons seems to be favoured (Kossel). The space available for an electron has increased in going over from H to the other atom and its zero-point energy has decreased accordingly. Hence the process does take place and the compounds are formed. The remaining protons accommodate themselves within this larger system of electrons of high symmetry. We shall see on discussion of the chemical bond that each hydrogen atom is bonded by two electrons to its partner within this structure (p. 324). This means that all the electrons of carbon are used for these bonds in CH_4 while one, two and three pairs respectively of electrons remain unbonded in NH_3; OH_2, and FH_1 giving rise to electric asymmetry. They are prone to be attacked by a free, positive, hydrogen.

The case of boron shows clearly, that two electrons are desirable to bind a hydrogen atom. Otherwise there would be no reason to discontinue the series and not write BH_5 with eight L + H electrons. No such compound exists: eight electrons are insufficient to bind five H atoms! The chemistry of the B–H compounds is extremely complicated for this very reason, even those boron–hydrogen compounds which do exist have less than two electrons per bond (*see* Section 69).

On the other hand the elements at the beginning of the period are prone to lose their electrons to elements at the end of it. Lithium and beryllium give positive ions Li^+ and Be^{++} with a complement of electrons which once more belongs to a rare gas, but to the next lower one, helium in this case. Boron forms halides of the general formula BX_3 and their structure is certainly near to one where the electrons of boron have entered the outer sphere of the halogen atom. The oxygen compounds are similar and increasing continuously in valency (p. 247): Li_2O, BeO, B_2O_3, CO_2, N_2O_5. The electron of lithium is probably completely lost to the oxygen atom while the higher we go in this series, the more the electrons are *shared* between the two atoms of the bond. This will be discussed at length when we come to deal with the chemical bond. But as a very crude approximation we may look at the matter as if the more positive partner has released its L electrons to fill the L shell of the more negative partner, until the complement of 8 L electrons is reached. This is the electronic explanation of Kossel's rule (p. 247) that the sum of the maximum valency of elements in combination with hydrogen (where they accept electrons until they have 8) or in combination with oxygen or fluorine (where they more or less lose all their electrons from the L shell which can hold 8), is eight throughout the periods. The fine structure of the L shell appears, for example, in the fact that nitrogen need not necessarily lose all five of its L electrons. If it retains its two 2s electrons and loses three 2p electrons we have N_2O_3, nitrous acid HNO_2 and the nitrites. Nitrogen is able to form oxides with less simple electronic structures too (N_2O, NO, NO_2) but what matters now is its maximum valency in combination with oxygen.

After the rare gas neon the whole story repeats itself once more. The Schrödinger equation has no further solution as long as its constant representing the principal quantum number remains 2. When $n = 3$ we encounter the M shell of electrons in the same manner as we encounter the L shell when $n = 2$: the next electron takes up the 3s state. The real charge on the nucleus is $+11$ but the 10 electrons in the K and L shells screen approximately ten of these units. Only $+1$ remains, effectively. Hence the attraction at the increased distance of the M shell is relatively small. We have once more an alkali metal atom; very electro-positive because of its loosely-bound electron, and monovalent because only one electron *is* so loosely bound. We are dealing in fact with sodium, Na. (Its name is natrium in many languages.) It is more positive than is lithium because the effective charge which attracts its outer electron is the same, $+1$, but the distance from the nucleus is increased because $n = 3$ (Fig. 98). This resemblance goes on throughout the period. In each case the element in this, the third, period is similar to the others but slightly more positive or less negative in character than the corresponding element in the period above. The reason for this is that the outer electrons, the electrons of chemical valency, are farther away from the nucleus but are attracted to it by practically the same effective charge. We accordingly have the pairs: Li–Na, Be–Mg, B–Al, C–Si, N–P, O–S, F–Cl.

After chlorine we arrive at the next rare gas: argon. Its complement of $3p$ electrons is again filled, and it has altogether 18 electrons arranged according to the formula $1s^2$, $2s^2$, $2p^6$, $3s^2$, $3p^6$. The argon atom is somewhat larger than the neon atom, just as neon was of course larger than helium because the maximum of the most probable electron densities in the M shell ($n = 3$) are farther away. The same reason accounts for argon being more easily ionized than neon but still difficultly enough to bar it from any chemical combination.

The next electron, that of potassium, K (its name is kalium in many languages), has a choice. The solution of the Schrödinger equation for $n = 3$ allows for the existence of five so-called d states apart from the s and p states which are already filled. Doubling these five states for the two possible permitted directions of spin, we have room for 10 further electrons in the M shell. But these d orbitals do not *yet* begin to be occupied. What we actually observe is the occurrence of a typical alkali metal, potassium. Its chemical, physical, and spectral properties all agree so perfectly with those of sodium that no doubt can arise concerning its electronic structure. The last electron of potassium must be in the N shell ($n = 4$), but evidently only because it so happens that the energy of the whole system of electrons is less when this particular electron goes into the s state of a higher shell than when it goes into the d state of the preceding one. Potassium is again somewhat larger and hence in all respects more positive than was sodium. The next element is a typical divalent metal, calcium, analogous to magnesium in having two electrons in the N shell. Following is a trivalent element similar to aluminium, namely scandium. It is somewhat larger and more electropositive than the former. Spectroscopic evidence suggests that it also must possess only two N electrons, while the newly added electron joins the $3d$ subshell: $3s^2$, p^6, d; $4s^2$. The $3d$ electron is lost easily, together with the two $4s$ electrons so that there is no chemical evidence of the new situation.

But titanium (Ti), though usually tetravalent, is not analogous to silicon. And the next element, vanadium (V), is most certainly not a sister element of phosphorus! Both belong to the steel-alloying series of the metals called transition metals (Section 55). Now it becomes evident to the chemist that something novel has happened. In view of the fact that a series of 10 elements begins which have no sister elements in the period above, it is fair to conclude that we have arrived at the point where it has become more favourable, as far as energy is concerned, to build the next 9 electrons into the d quantum states of the M shell which were by-passed starting with potassium. The Schrödinger equation of a nucleus and 21 electrons (scandium) does not of course, lend itself to exact solution. We could not have calculated *a priori* that at this stage and nowhere else are the $3d$ electrons going to appear. But we *were* able to predict that they will appear somewhere in this region, and all the properties of these transition metals point to the fact that these electrons *do* appear at this point.

The five d states are occupied one after the other according to Hund's rule, and a given state only houses more than one electron through spin coupling if there is no other empty state available for occupation. Thus the number of electrons with uncompensated spin increases from scandium to the fifth element in this transition series, manganese (Mn). Five electrons with uncompensated spin are quite a lot. Taking into account that only *s* states are so symmetrical that they do not possess a magnetic moment in themselves (orbital momentum of the planetary electron in the Bohr model!) we have to add up the magnetic spin and orbital vectors of five *d* electrons for this element (*see* following section). The result is an atomic magnetic moment of quite appreciable magnitude. Do not forget, we are speaking about manganese, the neighbour of iron! The atomic magnetic moment decreases in elements on each side of manganese, to the left because there are fewer *d* electrons in all and to the right because the sixth *d* electron compensates the spin of one of its neighbours. At the end of the transition series the atomic magnetic moment becomes zero. Around iron we have manganese, cobalt and nickel: known magnetic metals.

As far as valency is concerned the transition metals show less regularity than those of the main series. Titanium generally loses its two *d* electrons together with the two electrons in its N shell, but it can retain one more electron because we know it can also be trivalent. Vanadium at the most loses all three of its *d* electrons together with both those in the N shell, and attains the positive valency of five. But it may retain any number of these five electrons and so can show all the possible valencies 1, 2, 3, 4, and 5. Chromium has a maximum valency of six, and manganese seven. So far, all M*d* and N*s* electrons can be abandoned. But from iron onwards this is no longer the case. The maximum valency of iron is six in the rare ferrates, of cobalt and nickel only three, of copper and of zinc two.

The valency problem of the transition elements is complicated by the fact that they yield the typical, complex-forming metal ions. We shall have to discuss complex bonding later on in connexion with the chemical bond. It seems that these ions are able to acquire from other molecules electrons which are used for a special bond formation and can thus complete their incomplete electronic systems entirely (or, in some cases, only partially). It is interesting to note how the magnetic moment of the ions changes according to the number of foreign electrons they have inserted into their electron systems (p. 286).

With zinc (Zn), the $3d$ subshell is completed and the two N electrons, $4s^2$, act as regular valency electrons above a filled M shell. Gallium introduces a third N electron, germanium a fourth, and from now on we are among sister elements of the former main period: arsenic, selenium and bromine are real homologues of phosphorus, sulphur and chlorine. They form characteristic simple hydrogen compounds from germanium to bromine (*see* Section 55) with decreasing numbers of hydrogen atoms in their molecules, so as to increase their complement of $4p$ electrons to 6, to the number of electrons in the next stable structure of the system, the rare gas krypton (Kr).

Here we have to pause for a moment to show that even the simple inorganic chemistry of elements is not as finished and petrified as we are sometimes inclined to suppose. For nearly sixty years we learned and taught that the atoms of the rare gases, the "noble" gases, do not combine with other atoms, that they have no compounds. Then, in 1962 *N. Bartlett* and *D. H. Lohmann* prepared the first fluoride of xenon and a research team of the Argonne Laboratory soon showed that it is relatively easy to prepare three different xenon fluorides and also fluorides of krypton. It was easily shown that the rare gases are positive in these compounds whereas the most electro-negative of all elements, fluorine remains negative as always. Most of these compounds are quite stable. They can be hydrolysed to hydroxy-compounds which are much less stable, but can be well characterized also.

The theory of these compounds is simple. We noted that the energy neces-sary to extract an electron from neon is less than the energy to extract one from helium; the ionization potentials were 24·5 eV respectively. Because of the increasing distance of the outer electrons from the nucleus this energy of ionization decreases continuously through 15·7 eV for argon and 14·0 eV for xenon. This is still quite high, but not higher than the ionization potential of the oxygen molecule, for instance. Therefore it is no wonder that the atom of fluorine, the atom which attracts electrons into its outer shell more strongly than any other atom in the periodic system, is able to react with these two noble gases and change them into positive ions.

Coming back to the N period (N represents here the fourth period of the system of elements and not the element nitrogen!) which ended with krypton we can sum up as follows. Everything said for the previous period (M) remains valid, the respective sister elements resemble each other, but the elements of the N period are larger and more electropositive (less electro-negative) than the elements one period above them.

The element following the rare gas is once more an alkali metal, rubidium (Rb). Hence it must have a single electron around the symmetrical electronic structure of krypton in the O shell ($n = 5$). Its electron is further away from the nucleus and therefore rubidium is more positive than potassium. Then follows strontium (Sr) as a typical alkaline-earth metal with two O electrons, after which a new set of ten transition metals begins. The electrons have the chance to go into the d sub-shell of the N shell, which has been by-passed by rubidium as the d subshell of the M shell was by-passed by potassium one period before. Y–Zr–Nb–Mo–Tc (artificially produced technetium)–Ru–Rh–Pd–Ag–Cd come one after the other, and only with divalent cadmium is the d subshell filled to completion. Indium is a regular trivalent metal, succeeded by the main series Sn–Sb–Te–I until the next rare gas with its complete set of $5p^6$ electrons, namely xenon (Xe). The electropositive charac-ter again increases in comparison with the sister elements of the preceding period: tin (Sn) is no doubt a genuine metal and antimony (formerly called stibium, Sb) is practically a metal, too. Tellurium (Te) is rather more of a metal than not and even iodine crystals have metallic reflection, though of course iodine is a typical halogen. Hydrogen compounds exist as before, but are more labile because the elements are more positive. The energy of ionization of xenon is again lower than that of krypton: its newly discovered compounds form more easily.

After xenon we find an alkali metal once again: caesium (Cs). Thus the next electron must have entered the $n = 6$ shell, the so called P shell. The

alkaline-earth metal barium (Ba) follows all right as the homologue of strontium but with lanthanum (La) we embark upon a new interim series of
elements (*see* Section 55), that of the rare earth metals. They number 14
and are very similar to each other. As in the two preceding transition metal
series, we are dealing with a series where the two P electrons remain in their
place but the next ones find it more practical (that is of lower energy!)
to enter another subshell which has been by-passed in the mean-time. This
is the $5d$ subshell which opens with lanthanum. But the solution of the
Schrödinger equation provides for the possibility of accommodating seven
so-called f electrons beside the s, p and d electrons in the complement of the N
shell. This opportunity has been by-passed by the whole series of the fifth
period and the first three elements of the sixth. But it seems that, at this stage,
a lower state of energy can be attained by entering this by-passed f subshell
than by going on in the regular manner. Each of the seven f states can accommodate two electrons of opposite spin and hence we have 14 elements in
this rare earth group. Once again, beginning with cerium, the f states are
filled according to Hund's rule, that is without spin compensation as far as
possible, and therefore we find the seventh element, gadolinium, again at a
maximum of atomic magnetic moment with seven uncompensated electrons.
The valency of these metals is mostly three, owing to the loss of the $5d$ and
$6s^2$ electrons, the deep-sited f electrons mostly taking no part in ionization,
but small deviations do occur around this valency value.

Only after this f subshell has been completed does the transition metal
series of this sixth period continue. The d subshell of the fifth period (O
shell) fills its complement of $10d$ electrons in the series, La . . .–Hf–Ta–W–
Re–Os–Ir–Pt–Au–Hg and reaches completion with the last element. Osmium
and ruthenium of the former period, are the only elements which reach the
octavalent positive state in combination with oxygen and fluorine. The
resemblance between the transition elements which occur below each other
in the three periods, is great enough and has been discussed in Section 55.
It results from their analogous electronic structures. The main series with
completion of the $6p$ electrons goes on from thallium (Tl) through
Pb–Bi–Po–At (synthetic radioactive halogen, astatine) to end once more with
a rare gas of $6p^6$ configuration: emanation (Em), also called radon (Rn).
The fact that all elements above lead (Pb) are, or may be, radioactive is very
important of course, but does not represent a property of the electronic
structure. It is an instability of the nucleus itself as we shall see in Part VII
of this book. Thus, as far as chemistry is concerned, we need not concern
ourselves whether the elements are radioactive or not. If they live long
enough for their chemistry to be established, established it will be.

This brings us to the end of the periodic system, to the last, radioactive
period after emanation. Once more it begins with an alkali metal but this
is already radioactive and was only discovered during the second world-war:
francium (Fr). Next comes the famous sister element of barium: the alkaline-
earth metal radium (Ra). All this, so far as the chemistry is concerned, is in
perfect analogy to previous periods. The first two s electrons of principal
quantum number $n = 7$ (Q shell) have appeared. From here on, after actinium

(Ac), a new more deeply seated *f* subshell is beginning to be filled and the last elements of the natural system as well as the artificially produced elements which follow them are probably within a group analogous to the rare earths, the group of the "actinide" metals.

61. MAGNETIC MOMENTS

We discussed the spin of the electron and the magnetic moment associated with this spin in Section 50, but postponed the discussion of magnetic properties of atoms and molecules until we knew more about their internal structure. Now we are able to do so.

In macroscopic physics we know of three kinds of magnetic behaviour. A material may be ferromagnetic, paramagnetic or diamagnetic. Ferromagnetic materials can be permanently magnetized, for instance if they are placed along the axis of a solenoid fed by direct current. After removal from the solenoid, the rod made of ferromagnetic material is found to have become a magnet: it attracts or repels the magnetic poles of solenoids or of other permanent magnets, and yields two shorter permanent magnets when broken in two. It increases the magnetic field of the solenoid by several orders of magnitude.

Paramagnetic materials also increase the pole strength of a solenoid when they are placed within it. A rod of unmagnetized paramagnetic material is attracted by both poles of a permanent magnet and therefore turns towards the poles, into the line connecting them, when placed between the poles.

Diamagnetic materials weaken the field strength when placed within a solenoid, and are repelled by the poles of a permanent magnet, so that they turn away perpendicularly to the line connecting its poles. As far as actual strength is concerned, ferromagnetism is much stronger than paramagnetism and diamagnetism is by far the weakest of all.

Let us begin with the electronic explanation of *diamagnetism*. According to the laws of electromagnetism, all moving electric charges are set in circular motion when they are in a magnetic field that changes its intensity with time. The direction of this induced circular motion is such that the magnetic field generated by it (the circular motion of a charge is a ring current) tends to counteract the increase or decrease of the inducing field.

Atoms and molecules are composed of light electrons and relatively heavy nuclei, and the circular motion of the light electrons will become stronger or weaker if they are brought into a magnetic field. This change in their rotation will be such as to counteract the field of the solenoid that is influencing them or it will be such as to counteract the polarity of the poles of a permanent magnet in their vicinity. Therefore the body in which these currents have been induced will be pushed out of the field of the permanent magnet. Qualitatively this accounts for the phenomenon of diamagnetism. Quantitatively we have to take into account that the electro-magnetic force acting upon a moving electron will be proportional to the radius of curvature of its orbit while the magnetic moment due to its changed orbital velocity is also proportional to its radius of curvature. In the end therefore the momentum caused by the change in orbital velocity will be proportional to the square of the radius. This is the important result for the chemist. It tells us immediately, that diamagnetism is mainly due to the outermost electrons of atoms or to the electrons circulating around the largest areas of molecules.

On the basis of plausible assumptions about the mean radii of electron orbitals in atoms or ions it was possible to add up the diamagnetic moments of all electrons within them and thus to calculate the diamagnetic susceptibilities of these species. Calculated and measured atomic susceptibilities (magnetic moment induced by unit field strength per gramme-atom) are in fair agreement. For example we can compare them for the rare gases:

Gas	He	Ne	Ar	Kr	Xe
Measured value	1·9	6·75	19·54	28·0	42·4
Calculated value	1·68	5·67	16·65	29·33	44·78

Taking krypton for example, the detailed summation shows that the moments due to the two innermost electrons are negligible, that those caused by the eight electrons in the second shell are only 0·5 per cent while the eighteen in the third shell are responsible for 18 per cent. Hence 81·5 per cent of the total diamagnetic susceptibility is due to the eight outermost electrons of the krypton atom, because these move in orbitals with the largest radii.

Among organic compounds it was possible to assign diamagnetic increments to most of the bond types as for example to C—C, C—H, C—Cl, C=O, etc., and the molecular susceptibility was found to be the sum of these more or less empirical increments (*P. Pascal*). Sometimes such calculations allow us to distinguish between two conceivable structural formulae of an unknown compound.

In special cases some electrons are moving more or less freely along a number of neighbouring bonds, as for instance the six electrons of the "double bonds" in benzene move around the whole hexagonal ring. Diamagnetic measurements confirm this assumption in a beautiful way: the benzene crystal is much more diamagnetic in the direction perpendicular to the plane of the rings (which are parallel in the crystal). In this direction the susceptibility is 91·2 units, as against 37·3 in the other two directions. The magnetic field of these six delocalized electrons, moving in much larger circles than those which remain in bonds between atom pairs, is much stronger, because it is proportional to the square of the average radius.

In naphthalene and anthracene the delocalization extends over two and three rings respectively, yielding respectively 187·2 and 272·5 units of diamagnetic susceptibility in the direction perpendicular to the ring plane whereas in the other two directions naphthalene has 39·4 and 43·0 and anthracene 49·9 and 52·7 units. These two directions are equivalent for benzene but unequal for naphthalene and anthracene, which of course possess a longer and a shorter axis:

The diamagnetic susceptibility of graphite is sixty times as large perpendicular to its layers as in its other two axis directions. Of course, graphite is the last member of the series of condensed carbon rings of which the three afore-mentioned compounds constitute the beginning.

Paramagnetism is due to the magnetic moments which exist in atoms or molecules. These are aligned by an external magnetic field so that north pole comes to lie beside south, and thus the whole chain of elementary magnets becomes more or less aligned and strengthens the already existing magnetic field. Of course such an aligned set of magnets is drawn into the external magnetic field.

What can these elementary magnets be? The spin of an electron represents probably the simplest elementary magnet. Whenever there is more than one electron in an atomic or molecular system, they can either form pairs and thereby compensate, annihilate, their magnetic moments or stay with their spins parallel thus reinforcing each other. This however, is not the whole story. Electrons within atoms and molecules are distributed among orbitals, according to the wave functions of the Schrödinger equation. Except for the spherically symmetrical *s* orbitals all orbitals are more or less eccentric, and they have an oriented angular momentum around the nucleus, or the nuclei, as the case may be. This oriented orbital motion involves rotation of an electron in a—somewhat hazy—orbital and is therefore equivalent to a closed electric current. And closed electric currents have magnetic moments. These again may compensate each other if two electrons are swirling around in the same orbital in opposite directions. Therefore the sum total of the electric moment caused by an electron is the vectorial sum of the moment it owes to its spin and that which it owes to its non-spherical orbital. Both spin and orbital moments can be annihilated by another electron having the vectorially opposite values; only those which are not compensated remain effective.

It should be kept in mind, that whereas not all atoms or molecules contain uncompensated spins or orbitals, all atoms and molecules do contain revolving electrons. Hence all of them are diamagnetic, even those which actually show paramagnetism. Therefore the measured paramagnetic moment has always to be corrected for the amount of diamagnetism that works against it. This diamagnetic correction can be taken from similar atoms or molecules which are not paramagnetic, or by calculating the diamagnetism from electronic models.

Paramagnetic and diamagnetic susceptibility are in the same relationship to each other as permanent and induced electric moments. Permanent paramagnetic and dipole moments are oriented by the external field whereas the thermal motion tends to bring disorder into this orientation. As long as only a small fraction of the maximal possible orientation has been achieved the attained order, the electric or magnetic moment of a given volume, is inversely proportional to the absolute temperature because temperature is disorderly motion. For paramagnetic materials this law was first found by P. Curie. Induced electric and induced diamagnetic moments lie always in the direction of the exciting field and are therefore not disturbed by different orientations of the molecule as a result of its thermal motion. They are independent of temperature.

As long as the picturesque Bohr model of atoms was in vogue it was quite easy to calculate magnetic moments of spins and orbitals. However, quantum mechanical principles are much more difficult to visualize and it is impossible

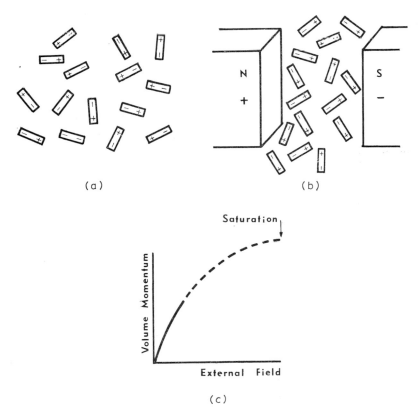

Fig. 102. **Paramagnetism**

    (*a*) Magnetic molecules are oriented at random because of their thermal motion if there is no external field.

    (*b*) An external field brings some orientation among the disordered molecules but thermal motion keeps working against this.

    (*c*) Orientation by a field induces a magnetic moment into the whole volume of the paramagnetic material. Orientation is at first proportional to the field but approaches a limiting value when most of the small magnets have already become oriented, that is, at strong field strength and low temperature.

for the author to give even an idea of the calculations leading to numerical values of these moments and to the laws describing their addition. Suffice it to say that the results differ somewhat from those predicted by the simple Bohr theory and that the experimental values agree with those calculated by quantum mechanics, and not with those derived from the Bohr model. For instance, an electron in a planar "Bohr" orbit ought to have an orbital magnetic moment while an electron which surrounds the nucleus spherically ought to have none. The hydrogen atom has no orbital moment. In an external magnetic field or in each other's fields the elementary magnets (revolving charges) cannot settle in all possible positions. Only a few are "permitted," just as only a few of all possible electron distributions around nuclei are compatible with solutions of the Schrödinger equation. Hence transitions from one magnetic orientation to another are just as discontinuous as transitions from one orbital or one vibration or one rotation into another, and the energy involved in such a magnetic transition is just as well defined.

Coming back to chemical entities, it is immediately clear that atoms or molecules with an uneven number of electrons must of necessity be paramagnetic: the moment of the unpaired electron cannot be compensated for. This is true of all atoms of uneven nuclear charge, that is with uneven number of orbital electrons. It is true too, for those rather rare molecules with an uneven number of electrons, as for instance NO, NO_2, ClO_2 and for organic "radicals" such as triphenylmethyl, $(C_6H_5)_3C—$, which contains trivalent carbon and is formed by the splitting of its dimer, hexaphenylethane: $(C_6H_5)_3C—C(C_6H_5)_3$. The chloride of monovalent mercury, calomel, would contain an uneven number of electrons if it had the monomer formula $HgCl$. However it is not paramagnetic, which speaks for the dimer formula Hg_2Cl_2 in accordance with measurements of its vapour pressure.

There are paramagnetic molecules—though not many—for instance oxygen (O_2) which contain an even number of electrons. We may logically call them "biradicals"; for some reason of quantum mechanical stability they contain two electrons which are not arranged in a manner to compensate their magnetic moments.

Uncompensated electrons occur as a rule when a subshell of neighbouring atoms is being filled in, as in case of the transition metals, the lanthanides (rare earths) and the actinides (*see* pp. 241–4). On page 272 we have already mentioned the rule announced by Hund which says that these electrons are being built into the successive elements so that the next one always occupies a new suborbital as long as it finds one empty, and only proceeds to enter an occupied suborbital with spin compensation if no empty orbital is available. It looks as if electrostatic repulsion would be more important than magnetic attraction of compensating spins. Fig. 103 shows the magnetic moments of ions of the elements between potassium and gallium. Two regularities show up in this graph. Firstly, all ions with the same number of electrons have the same momentum: K^+, Ca^{2+}, Sc^{3+}, Ti^{4+} are neighbours and all have lost all of their electrons beyond the complete shell of argon. None of these ions is paramagnetic, because they have the same number of electrons as argon which is highly symmetrical and has all its electrons paired. Neighbours in pairs of adjacent elements differ by one electron only. Whenever the higher neighbour forms an ion, with one positive valency higher than its left-hand neighbours, it has retained the same number of electrons as the latter and has the same paramagnetic moment. For instance: Ti^{3+}—V^{4+}, Mn^{2+}—Fe^{3+}.

The second interesting phenomenon is the form of the curve in Fig. 103. Paramagnetism is absent at its two ends, the first containing ions with the structure of argon and the other with the structure of the nickel atom with ten electrons more than argon. These are the ten *d* electrons which are being fitted into the subshell. We see how the magnetic moment increases for the first five of the incoming *d* electrons because, according to Hund's rule they

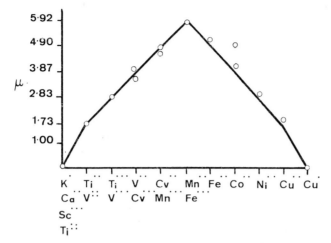

Fig. 103. **Ionic magnetic susceptibilities in the first period of transition metals**

Paramagnetism increases while the first five *d*-orbitals are un-occupied or singly occupied and decreases while full occupation of the orbitals takes place (*see* Fig. 104). Ordinates in "Bohr magnetons" comprise spin and orbital momenta.

do not compensate, but occupy vacant orbitals. The next five, however, have no choice other than to enter half-filled orbitals with spin-compensation and thus gradually detract from the paramagnetic moment until, when the whole subshell is full, no paramagnetism remains.* Fig. 104 is a schematic representation of how the subshell 3*d* is being filled.

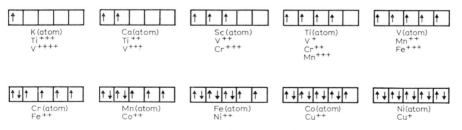

Fig. 104. **Filling up of the 3*d* subshell with electrons according to Hund's rule**

Note the pairing-off of spins and compare with magnetic moments of Fig. 103.

---

* The ordinates of the graph are "Bohr magnetons" that is units of the spin momentum of an electron. The numbers are not integers because they represent the quantum mechanical sum of spin and orbital momenta due to 1, 2, 4 or 5 unpaired electrons.

These same inner incomplete subshells play a part in the formation of complex compounds. As a matter of fact Cr, Fe, Co and Ni are among the most characteristic complex-forming elements. They are also the metals that form carbonyls and related compounds (*see* 339) because these incomplete subshells are capable of absorbing electrons, either from carbon monoxide molecules or from other complexing molecules or ions. In some cases inter-penetration of the electron systems of the central ion and of the complexing species is so strong that it manifests itself in the paramagnetic properties of the complex. For example iron stands eight places after argon in the system and hence must have eight electrons more than argon. These are generally thought to be arranged so that two are in the N shell (main quantum number = 4) while six occupy the still incomplete *d* subshell of its M ($n = 3$) electrons. The paramagnetic moment of divalent iron compounds, the so-called "ferro" compounds amounts to $5\cdot2 \sim 5\cdot4$ units, as for example in anhydrous ferrous chloride ($FeCl_2$) or ferrous sulphate ($FeSO_4$). The molecular paramagnetic moment remains unchanged in the complexes $[Fe(H_2O)_6]$ Cl_2 and $[Fe(NH_3)_6]Cl_2$ or $[Fe(NH_3)_2]Cl_2$. Thus there seems to be no interaction between the complex formation and the magnetically incomplete set of *d* electrons. The following three complex salts, however, have no paramagnetic moment at all: $[Fe(dipyridyl)_3]Cl_2$, $K_4[Fe(CN)_6]$ and $K_3[Fe(CN)_5 CO]$. The dipyridyl molecule contains two complexing N atoms, thus six in all, just as there are six complexing groups in the two other compounds. Let us assume with Sidgwick that in these cases more intimate complexing is due to two electrons of each complexing group being donated to the central atom in complex dative bonds (*see* p. 332). This makes 12 donated electrons in all while the divalent ferro ion has lost two from its eight electrons outside its argon shell, as electrons of valency. $12 + 8 - 2$ gives 18 electrons outside the argon shell: exactly the number possessed by the very stable configuration of the krypton atom. Neither krypton, nor these ferro-complexes are paramagnetic. This ideal symmetry is, however, lost as soon as we extract an electron from the structure. This happens in the ferri-analogue of $K_4[Fe^{2+}(CN)_6]$, i.e. in potassium ferricyanide, $K_3[Fe^{3+}(CN)_6]$. One electron which was contributed by the fourth potassium atom is missing now, the whole structure contains one electron less than krypton. It is consequently not compensated magnetically, it has a paramagnetic moment of two units.

The same difference holds for the non-ionic complex $[Co^{3+}(NH_3)_3F_3]^0$ which is not paramagnetic and for $K_3[Co^{3+}F_6]^{4-}$ which has the moment $5\cdot3$. The trivalent cobalt ion has six electrons in its incomplete *d* shell, the same number as we have assigned to the divalent ferrous ion. For some reason that we do not yet understand, the first complex is dative, with 12 electrons entering the sphere of the cobalt ion while the second seems to be based on simple electrostatic attraction of the fluoride and trivalent cobalt ions, without covalent interaction between them. Maybe the compact, symmetrical F^- ion holds its electrons too tightly to allow of their partaking in a dative bond.

Ferromagnetism is a special case of paramagnetism which only occurs in the crystalline state. Examples of paramagnetic materials are the element iron, the compound Fe_3O_4 which as a mineral is called magnetite and was probably the first magnetic material ever observed, and some alloys which contain at least one metal with an incomplete subshell. In these crystals the magnetic fields of neighbouring paramagnetic atoms or ions somehow manage to orient their neighbours so that whole regions are built up where all paramagnetic moments are parallel, even without the influence of an external field. An external field can only bring adjacent regions parallel to each other and when this has happened the material is as magnetic as it can be: all its moments are parallel, we say it is magnetically saturated. Its magnetization is not going to increase any further with an increasing field and it has attained some million times the magnetization of normal paramagnetic materials at normal temperature. This is so because only a small orientation is achieved by available field strengths in normal paramagnetic materials, the disorienting effect of the heat movement is too intense. Only in the neighbourhood of the absolute zero is it possible to magnetize paramagnetic materials near to saturation. At high temperatures, however, ferromagnetism is destroyed also by increasing disorder.

Why some materials are able to become ferromagnetic and others not, is one of the most difficult problems of physics. It seems that the elementary magnets must be strong but shielded from each other to a certain extent. This happens when they are localized in incomplete subshells below non-magnetic valency electrons. The reader is advised to turn towards more competent authorities than the author to obtain some insight into this complex question.

Nuclear magnetism. The nuclei of atoms are built from protons and from neutrons (*see* 449) both of which possess angular momentum. Hence the proton and other nuclei with uncompensated nuclear spin also have magnetic moments: they are rotating electrical charges. But they are some ten thousand times smaller than the electron orbitals, and magnetic moments being proportional to the square of the "radii of currents", nuclear moments are negligibly small in comparison to moments due to electron spin and orbitals.

However, these very small nuclear moments have been turned into an invaluable tool for the chemist in a method called "nuclear magnetic resonance" (n.m.r.). The material to be examined is placed in a strong magnetic field which makes the axes of the spinning nuclei precess around the field, like a spinning top in the field of gravity. Calculations by quantum mechanics have shown that this precession can change its direction discontinuously, in the simplest case by 180°, against the field. Each discrete direction of precession involves a definite magnetic energy relative to the field, which can change its value only in discontinuous "jumps". If a rotating magnetic field is superimposed on the steady field, the spinning nucleus will be able to absorb energy and change its state of precession whenever its spinning frequency is equal to the frequency of the rotating field; then both "resonate" like two tuning forks. This loss of energy in the external magnetizing circuit can be observed and measured as a function of the rotating field frequency. The resonating frequencies of the nuclei in the probe can thus be ascertained. More often, the rotation frequency of the field is kept constant and the direct magnetic field is continuously changed, thereby causing change in the frequency of precession. In this case resonances occur at given values of the steady direct field.

In organic compounds neither the carbon (^{12}C), the oxygen (^{16}O), or the nitrogen (^{14}N) nuclei have magnetic moments, all three being internally compensated. Other isotopes (p. 416) of these elements as for instance ^{13}C or ^{17}O *do* have magnetic moments but occur in extremely low concentration in Nature. Thus we mostly encounter only the protons, the hydrogen nuclei, as nuclear magnetic entities. These would have a precessional frequency determined by their moment and the direct field, if no other influence were acting upon them. But two influences do act on them, and just those two which are important for the chemist.

The first external action is due to the magnetic shielding effect of the electrons moving around the nuclei. The more electrons encircle the proton, and the nearer they are to it, the greater this shielding effect. Small as it is, with modern instruments it can be measured very well as a shift in the resonance frequency: it is as if the external field were weaker than it actually is. This so-called "chemical shift" is *characteristic of groups* adjacent to the proton. Fig. 105 shows some n.m.r. curves of simple organic molecules obtained by slightly changing the strength of the direct field around the value of 10,000 gauss, and observing how large the change must be to induce energy absorption by one of the many kinds of protons in the compound. Comparison of such shifts in different compounds allows us to ascribe each shift to a special grouping in the neighbourhood of the protons, and thus to resolve the structure of the molecule into known groups. Note how small the shift is compared to the absólute value of the field: some 10 to 100 milligauss as against 10,000! But it can be resolved very well.

The second external influence shows up in the fine multiplet structure of the shifted resonance lines, which is due to the magnetic fields exerted by neighbouring protons on each other. In acetaldehyde,

$$(H_3C—C—H)$$
$$\overset{\|}{O}$$

for instance, the three protons in the methyl group are shielded by a different electron environment from that of the proton of the aldehyde group, and therefore their resonance frequency underlies different chemical shifts. But within these shifted frequencies there is a splitting due to the action of the methyl protons on the aldehyde proton, and vice versa. The precession of the protons can be aligned with or against the main field. Therefore the aldehyde proton may have two orientations and each of these has a slightly different influence on the methyl protons: their resonance frequencies are split up into doublets, one component of which is caused by aldehyde protons precessing in one direction and the other component by those precessing oppositely. On the other hand, there are four statistically independent orientations for the three methyl protons as a group, and therefore there are four different ways in which they can influence the resonance frequency of the

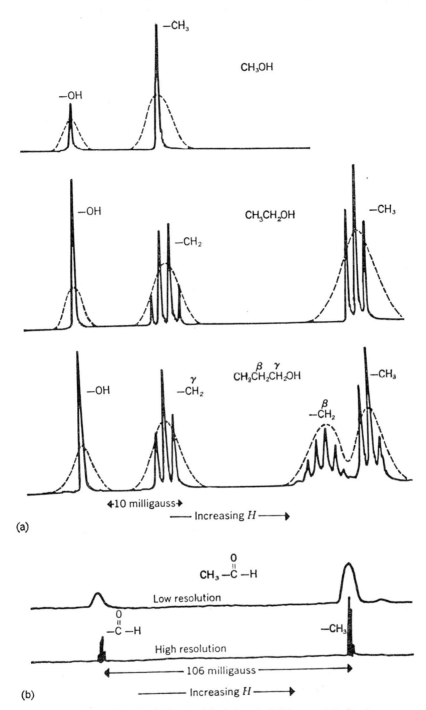

Fig. 105. **The n.m.r. spectra of several simple compounds at a frequency of 40 megacycles per second and a magnetic field of 10,000 gauss**

In (a) the *solid lines* give the high resolution spectra and the *dashed lines* show the appearance of the spectra at low resolution where the splitting arising from the interaction of the nuclei would not be observed.

(From G. M. Barrow, *Physical Chemistry*, p. 316, Fig. 10–41 (New York, McGraw-Hill, 1961). By courtesy of the McGraw-Hill Book Co.)

aldehydic proton. This latter resonance is therefore split into four fine components, near, but distinct from, each other (Fig. 106).

Unpaired *electrons* in a molecule can also be forced into spin transitions by a magnetic field of appropriate frequency. The difference from n.m.r. lies in the magnetic moments, that of the electron being about a thousand

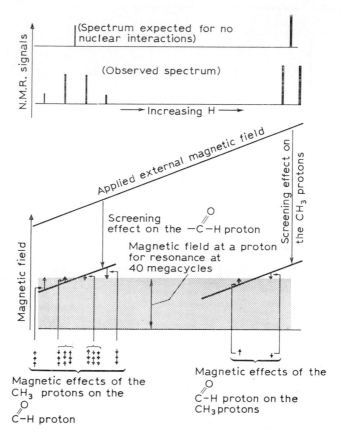

Fig. 106. **A schematic representation of the n.m.r. spectrum of acetaldehyde, CH$_3$—C—H, and its interpretation in terms of**

O

the screening effects and nuclear interactions

The frequency of precession of the nuclei is gradually increased by increasing the external field, and one after the other the nuclei come to resonance with the rotating field of 40 million cycles per second. Observe: (1) The screening effects on the two kinds of proton which subtract from the external fields; (2) The magnetic effect of the four possible combinations of CH$_3$ protons on the CHO proton and of the two possible positions of the CHO proton on the CH$_3$ protons. Note that intensity of a resonance line is proportional to the statistical probability of combining proton directions.

(From G. M. Barrow, *Physical Chemistry*, p. 319, Fig. 10–42 (New York, McGraw-Hill, 1961). By courtesy of the McGraw-Hill Book Co.)

times greater than that of the proton. Absorption of energy from a rotating magnetic field at a frequency determined by an applied constant magnetic field is a sign of "radicals" with unpaired electrons. Fine modulations of this frequency are again dependent on the presence of nuclei with magnetic moments in the neighbourhood of the unpaired electron. The method is named "electron spin resonance" (e.s.r.).

62. DIPOLE MOMENTS

Since molecules are composed of atoms, and atoms of positively charged nuclei and negative electrons, the question arises whether the "centre of gravity" of all positive charges necessarily always coincides with that of all negative charges. Whenever these centres do not coincide, the molecule or ion will behave as an electric dipole.

A macroscopic electric dipole can be built by connecting two spheres by an insulating rod and giving opposite electric charges to the spheres. Such an assembly would orient itself in an electric field, as a magnet with its two poles orients itself in a magnetic field. Both dipole and magnet are acted upon the more strongly by an external field, the greater their respective charges and the farther these are from each other, because the force acting on the "poles" is proportional to their charge, or magnetic strength, while the moment of rotation increases in proportion to the distance that separates them. Evidently there is no orienting force at all if the charges or poles coincide, that is if their separation is zero. Hence the relevant quantitative measure of dipole or magnet strength is the moment, the product of charge or pole strength and the distance between the charges or poles. It is called the dipole moment for electrical charges, and the magnetic moment for magnets.

The macroscopic, measurable effect of molecular dipole moments manifests itself in the dielectric constant of materials. This constant shows how many times the Coulomb force between two charges is smaller, if they act on each other surrounded by the material in question, than it is in a vacuum. Or how many times more electricity is necessary to charge a given capacitor to the same voltage if the electrodes are separated by this material instead of by a vacuum. Qualitatively it is clear what is going on in both cases. In the first, the two charges surround themselves (in the liquid or gaseous phase) with the dipolar molecules which orient themselves with their oppositely charged ends towards the charges while their other ends point in a more or less disordered way outwards. A part of the original charge is thus neutralized towards the medium, and weakens the Coulomb force of the charge. In the case of the capacitor, the elementary dipoles set up counter-directed electrical double layers adjoining the electrodes (Fig. 107) so that a given charge on the electrodes generates a smaller field and thus a lower voltage between them than it would if unpolarizable vacuum were used instead.

Thus electric moments of molecules and the dielectric constant of the medium containing these molecules are causally interconnected. It is, however, important to note that electric dipole moments in molecules are not

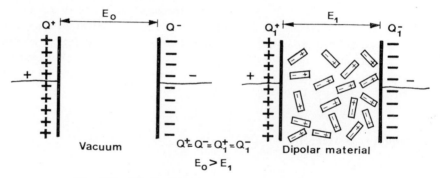

Fig. 107. Dipole molecules as dielectric of a capacitor

The same amount of electric charge on the plates of a capacitor gives rise to a stronger field between the plates if they are separated by vacuum than if they are separated by a dipolar (or induced dipolar) material. Molecules with dipole moments (permanent or induced) orient themselves in the field and accumulate negative charges near the positive plate and vice versa, which weakens the field of the original charges.

necessarily their permanent attributes (*see* Section 75). They may be—and to some extent always are—also due to the polarizing effect of the external electric field in which the molecules find themselves. Electrons are displaced in one direction in a field, and positive nuclei in the other, so that such "induced" moments necessarily come into being. They do this the more so the weaker the force between electrons and nuclei, that is the larger the atoms

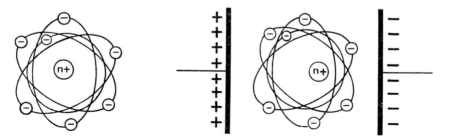

Fig. 108. Dipole moment induced in an atom

Orbital electrons of an atom are attracted in one direction in an electrostatic field while nuclei are attracted in the other. Thus dipoles are induced by polarization of the atom. (Note that electron-orbitals are not really as well defined as in the figure.)

in question, because the outermost electrons are then far from the nuclei and more weakly bound. Detailed calculation shows that polarizability is proportional to the volume of the atom. This induced polarizability is, however, independent of the heat movement, of the temperature, because it is always induced in the direction of the external field no matter how the molecule is oriented (at least, as long as the polarizability of the atom or molecule is independent of direction).

That part of the electric moment of a medium which is due to fixed structural dipole moments becoming oriented in an external field and thereby somewhat disturbing their random thermal orientation, depends necessarily on the temperature. The field and the dipole moment interact to give ordered orientation whereas the unordered molecular motion, which is what we call the temperature of the medium, tends to counteract this ordering effect. As long as the orientation is not too large it is therefore proportional to the field, and to the dipole moment, and inversely proportional to the absolute temperature, in complete analogy with paramagnetism (p. 282 and Fig. 102). This is how one differentiates between the part of the electric moment which is due to induced dipoles and the part due to the orientation of fixed dipoles. The part which would remain at very high temperatures is due to induction, the rest to dipole orientation. One generally measures the capacitance of a capacitor filled with the material at different temperatures and determines the two components of the dielectric constant. At great dilutions, as for instance in the gas phase, the dielectric constant is proportional to the dipole moment. In other cases the effect of neighbouring dipoles on the field strength, around each dipole molecule has to be taken into account.

For the chemist both the fixed and induced moments of a molecule are important. The fixed dipole moment adds further insight into the structure of the molecule, showing towards which parts of it the electrons are preferentially attracted. It gives us an idea about the distribution of electric charges within a molecule. The induced moment or polarizability shows us how easily these electric charges can be made to depart from their positions in the equilibrium state, how "soft" their electric structure is against the influence of external—or even internal—fields (*see* pp. 346, 351). Both fixed and induced moments give rise to forces of attraction between neighbouring molecules thereby increasing, for example, the boiling-point of compounds.

Symmetrical molecules should possess no dipole moment; even if some moments are built into them, these compensate each other. Therefore a molecule which could be imagined either in a symmetrical or in some unsymmetrical structure can be assigned to one of these classes by determining whether it does, or does not, possess a dipole moment. Water, for instance, has a dipole moment of $1 \cdot 8 \times 10^{-10}$ electrostatic units. The charge of the electron being $4 \cdot 8 \times 10^{-10}$ e.s.u. and atomic distances a few Ångström units (10^{-8} cm), one would expect dipole moments to be of the order of magnitude of 10^{-18}–10^{-17} units, as we see is the case. For water this rules out the linear configuration H—O—H and only allows for a bent structure

which also has been confirmed by calculation of its moment of inertia from its infra-red spectrum. The same holds good for ammonia with a moment of $1 \cdot 5 \times 10^{-18}$ units: the molecule cannot be planar. It has to be a deformed tetrahedron, with its lone pair of electrons at one apex. This lone pair is responsible, by the way, for ammonia being able to add a fourth proton and to transform itself into the symmetrical, positively charged, ammonium ion, which has no dipole moment. Carbon dioxide has no dipole moment and its atoms must therefore be arranged symmetrically in a straight line: $O{=}C{=}O$ as confirmed by its spectrum.

A single molecule of an ionic compound is necessarily a dipole, because it is composed of one or more positive and negative ions alongside each other. Only symmetrical ionic arrangements, say two negative ions on two sides of a cation or similar combinations, would lack an overall dipole moment, even if the bonds in themselves were electrically polarized.

Dipoles attract each other head to tail and even more efficiently by lying inverted one beside the other and forming two loci of head-to-tail attraction (Fig. 109). This is the way ions aggregate into ionic crystals as, for instance,

Fig. 109. **Attraction between dipoles**

Dipoles can attract each other "head to tail" in series, increasing their moment, or can form quadrupoles, octapoles, etc., which underlie shorter-range forces than dipoles because of mutual compensation.

the sodium chloride molecule which only exists as such in the gas phase at very high temperatures. The higher the dipole moment the stronger the attraction between the ions and this is the reason why salts which crystallize in alternate series of positive and negative ions only evaporate at high temperature.

There is a single positive ion that is an exception: the proton, which is a naked hydrogen nucleus having lost its electron. It is the only bare nucleus in chemistry, without electrons surrounding and shielding it from outside influences. Therefore it can be drawn into the electron system of any anion it combines with and hence it can come much nearer to the middle of this negative ion than can any other cation. The resulting dipole moment is therefore much smaller and the dipole-dipole association much weaker too. This is the reason why hydrogen chloride (HCl) is a gas while all alkali halides are hard crystals and difficult to evaporate. (The dipole moment of NaCl is around 13×10^{-18}, that of HCl around 1×10^{-18}.)

We shall see in Part 6 that chemical bonds range from those we call extremely ionic to those that are highly "covalent". In the first case, one or more of the outer electrons of an atom is split off and incorporated into the originally incomplete electron system of another atom of a different element. These transferred electrons contribute to the symmetry of the anion which is thus formed and they stabilize it.

In covalent bonding (*see* p. 322), the electrons from both atoms are located between the atoms which they hold together. But they are not necessarily exactly half-way between the two atoms. According to the electronegativity of the participating atoms, their centre of gravity will be nearer to one of

them, for instance nearer to chlorine than to carbon in a C—Cl linkage.
This fact shows up in the dipole moments of molecules.

On comparing the dipole moments of many molecules, we see that the
moments are to a first approximation characteristic of the bonds, and add up
vectorially, when many are present in the same molecule.

For instance, let us compare the same substitutions on the symmetrical,
non-polar benzene and methane molecules. Chlorobenzene has a moment of
1·56 debye (1 debye = D = 10^{-18} e.s.u.), and iodobenzene 1·25 D. On the
other hand, methyl chloride has a moment of 1·9 D and methyl iodide

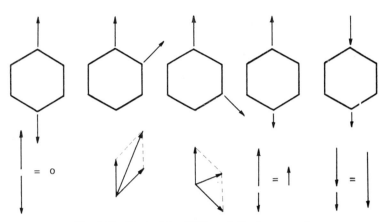

Fig. 110. **Vectorial addition of dipole moments**

1·65 D. The differences between the chloro-and iodo-derivatives are 0·31 and
0·25 D, respectively, in the two cases. Ten different aliphatic ketones

$$\overset{\displaystyle O}{\underset{\displaystyle (R\text{—}C\text{—}R')}{\|}}$$

from acetone to hexamethyl-acetone to methyl nonyl ketone, all have their
dipole moments between 2·69 D and 2·79 D. Evidently the moment belongs
to the carbonyl (C=O) groups and is not disturbed by the nature of the two
alkyl groups attached to it.

Substitutions on the benzene ring show some interesting regularities too.
Chlorobenzene, as mentioned above, has a moment of 1·56 D. From
among the three dichlorobenzenes, the para-isomer is symmetrical in the
sense that the directions of the two C—Cl bonds are exactly opposite;
accordingly this isomer has no measurable dipole moment. From the other
two isomers, the ortho- has a moment of 2·25 D and the meta- one of 1·48
D. Qualitatively this is quite clear if we draw the vector diagrams: the resultant of the two vectors enclosing 60° must be larger than that of those which
enclose 120°. Quantitatively the agreement is not very good. We should
expect 2·70 D for the ortho- and 1·56 D for the meta-isomer. It is very probable that the two large, somewhat negative chlorine atoms repel each other,
thus increasing the angle to greater than 60° in the ortho-isomer. Inductive

effects between the charges may also play their part in accounting for the difference. Such deviations from the exact vectorial sum are the rule, but the qualitative trend is always clear.

Para-substituents on the benzene ring are sufficiently far away to be regarded as sterically or even inductively independent. They teach us another very interesting lesson. Chlorobenzene has a moment of 1·56 D, nitrobenzene of 3·78 D. The compound in which these two substituents are opposite to each other, *p*-chloronitrobenzene, has a moment of 2·36 D which is nearly exactly the difference of the first two, 2·22 D. It seems probable that both moments have the same sign, negative relative to the benzene ring, and would partially compensate each other according to their individual strengths.

On the other hand aniline (aminobenzene) has a moment of 1·6 D and methyl benzoate one of 1·8 D. The compound methyl *p*-amino-benzoate has a moment of 3·3 D which is very near to the sum of the first two, 3·4 D. This can be easily understood by remembering that the amino-group is positive in organic chemistry, so that the direction of its dipole moment ought to be, and evidently is, opposite to that of the negative benzoic ester group. One moment points away from the benzene ring, the other towards it, so that they add up as shown by the experimental values.

Agreement between simple vector calculation and experiment is not always as good as in these examples, but even the deviations can be of use to the chemist because they point to steric and inductive interactions which may be of value in understanding the structure of molecules.

63. THE ORIGIN OF SPECTRA

No matter in what way we introduce energy into matter it can always happen that light is emitted. We may bombard matter with electrons or irradiate it with electromagnetic waves, we may heat it or sometimes perform with it some chemical reaction: it will emit X-rays or regular light, it will fluoresce, glow or show chemiluminescence. The emission spectra of gases are characterized by the fact that they are composed of distinct wavelengths. In the spectroscope we observe sharp lines and between these lines no light at all. In the X-ray spectra of anticathodes bombarded by electrons we also observe sharp lines but in between them there is always a background of continuous radiation of lower intensity. But incandescent solids and liquids generally emit continuous spectra: these include light of all wavelengths, with a maximum which is mainly determined by the temperature.

Let us start with those cases where the emitted wavelength is sharp, and well-determined. We shall need to use an apparatus which may be easily controlled and varied, to invoke this light emission.

Typical is the method of *J. Franck* and *G. Hertz* (1912) which we mentioned quite briefly in the last paragraph of Section 48. We remember: the experiment was performed with free electrons, emitted from the incandescent cathode of a vacuum tube, such as the cathode of a radio valve (Fig. 111*a*).

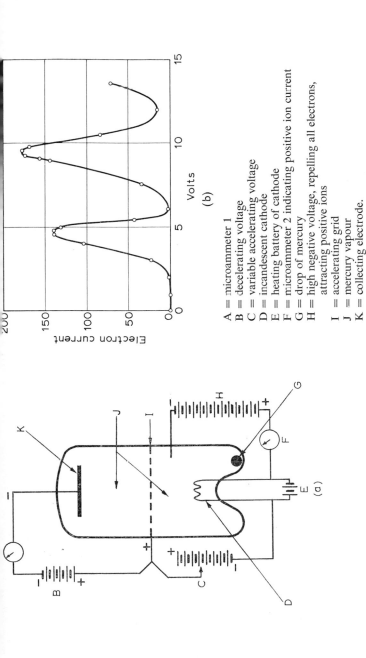

A = microammeter 1
B = decelerating voltage
C = variable accelerating voltage
D = incandescent cathode
E = heating battery of cathode
F = microammeter 2 indicating positive ion current
G = drop of mercury
H = high negative voltage, repelling all electrons,
    attracting positive ions
I = accelerating grid
J = mercury vapour
K = collecting electrode.

Fig. 111. **The Franck–Hertz experiment**

(*a*) Electrons from the incandescent cathode are accelerated to different velocities by a variable voltage on the grid. Between grid and collecting electrode they are decelerated. The current on microammeter 1 is due to the difference of acceleration and deceleration. At certain velocities characteristic of the gas atoms in the tube the current falls off because the electrons loose energy to the atoms.

(*b*) Current on microammeter 2 as function of the accelerating voltage. The current decreases each time the energy of the electrons is sufficient to excite the Hg atoms from their ground state into a higher level of energy. This energy is then emitted as ultraviolet light by the Hg atoms.

(From A. Eucken, *Lehrbuch der chemischen Physik*, p. 742, Fig. 163 (Leipzig, Akademische Verlagsges., 1930). By courtesy of Akademische Verlagsges.)

However the vacuum within the tube is not very good in this case. There is present a rare gas or mercury vapour at a pressure of some tenths of a millimetre, to a few millimetres of mercury. The electrons are accelerated by a given voltage drop to a determined final velocity. With this velocity they collide with the gas atoms and we observe the phenomena of light emission which correspond to specific terminal electron velocities.

In order to achieve this end the tube is fitted with a fine metal grid opposite the cathode and this grid is made positive with respect to the cathode by a variable, known, voltage which accelerates the negative electrons. The electrons are accelerated in this electric field and occasionally collide with the atoms. In spite of the fact that they are deflected by each collision, they are always under the accelerating influence of the field and therefore move in parabolic paths between two collisions, like tennis balls in the field of gravity. Hence they move on the average towards the grid although not in a straight line. On arrival at the grid, a part of the electron beam strikes its surface and enters it, flowing back within the circuit as an ordinary current. But another part passes through the empty spaces of the grid and reaches the space beyond the latter. The future of the electrons in this part of the beam will be determined by the electric field which exists in this region of the tube. Fig. 111 shows that there is a third electrode in the tube, after the grid, and this is made negative with respect to the grid in order to repel the oncoming electrons. If the potential between the grid and this electrode is exactly equal to the potential between cathode and grid, this second field is just sufficient to decelerate the electrons until only their irregular thermal motion remains. If the decelerating potential between grid and third electrode is smaller than the accelerating potential between cathode and grid, the electrons will retain a part of their velocity and will enter the last electrode, to pass into its circuit where they can be measured as a current by a sensitive ammeter.

Now a simple experiment begins. The decelerating voltage between grid and third electrode is kept constant but the accelerating voltage between cathode and grid is slowly increased from zero. As long as this voltage remains below the other, the third electrode is going to repel all electrons from its surface and there will be no current in its circuit. As soon as the force of acceleration is greater than that of deceleration, the electrons in the neighbourhood of the collecting electrode will have sufficient energy to impinge upon it and the current on this collector will gradually increase with the accelerating voltage. The interesting fact in this experiment is that the electrons have not lost any energy in spite of their having suffered many collisions with the atoms that they have encountered during their flight from grid to collector. The collisions only change their direction but not their velocity. They arrive at the collector with the same kinetic energy they acquired at the grid: this is the reason why the decelerating voltage must be equal to and cannot be less than, the accelerating voltage in order to give zero current at the collector. The electrons must have collided with the atoms just as elastic balls would collide with other elastic balls some ten thousand times their mass, that is without appreciable loss of velocity.

However, at a well-defined velocity of the accelerated electrons, the whole phenomenon changes. The current at the collector drops suddenly: something must have happened to rob the electrons of their kinetic energy within the gas. If the tube contains mercury vapour, this drop in the current sets in at an accelerating voltage of 4·9 V (Fig. 111 (b)). At exactly the same time the vapour in the tube begins to emit radiation, the well-known ultraviolet spectral line of mercury at a wavelength of 2,536·7 Å appears. The kinetic energy of an electron that has "fallen" through 4·9 V can be easily calculated (p. 202) and it turns out to be exactly equal to the energy of the 2,536·7 Å light quantum, calculated according to the Einstein relation $E = h\nu$. The energy of the electron has served to make the mercury atom emit this characteristic light quantum. As long as the electron has not sufficient energy to do this it has no effect on the mercury atom. It is just reflected.

On increasing the accelerating voltage further, between cathode and grid, the phenomenon is practically repeated. The electrons begin to reach the collector again, because though they have lost energy equivalent to 4·9 V acceleration, they have now received more energy than that, and as soon as the deceleration is overcome again, there will be a current on the collector. However, at exactly twice 4·9 V, i.e. at 9·8 V, the same phenomenon will occur again. The electrons have acquired once more the amount of energy which the mercury atoms seem to be able to accept, and now the mercury atoms once more take this energy away from the electrons. The ammeter falls back a second time while the intensity of the emitted ultraviolet radiation doubles.

The whole phenomenon is relatively independent of the gas pressure in the tube. It only changes if the pressure has become so low, that is, the mean free path of the electrons has become so long, that a considerable number of the electrons are able to acquire more than 4·9 V energy between two subsequent collisions. It now becomes apparent that there is more than one characteristic amount of energy which the atoms are able to accept. The mercury atom is also able to accept well-defined doses of energy above 4·9 V, and as soon as the electrons are able to offer this energy to them they will take it: the current will fall again. This time light of different wavelength from 2,536·7 Å will appear. The importance of this experiment lies in the fact that the acceptable doses of energy are absolutely well-defined and characteristic for each sort of atom. Atoms can be excited with certain amounts of energy and are then able to emit light of certain wavelengths. It is evident that these characteristic energies must give us an insight into the structure of the atoms.

Let us further increase the energy which an electron may acquire between two consecutive collisions, by increasing the accelerating potential and also increasing the mean free path of the electron by decreasing the pressure. At a certain point we arrive at a state of affairs when the current at the collector increases like an avalanche. At the same time the current is not confined to the collector but also begins to flow through a second collecting electrode which has a considerably higher negative potential—a potential so high that it would surely repel all electrons of the given velocity. Whatever is reaching this electrode must surely be positively charged. The electrons must have become energetic enough to remove an electron from the atoms, to

"ionize" them, and leave positively charged species, i.e. positive gas ions
behind. These of course are readily attracted by the high-potential negative
electrode. This is the reason why the whole process has changed into an
avalanche. An original electron freed a second one from an atom; this other
was able to free a third one and so on. The number of electrons partaking
in the process is no longer limited by the emission of the cathode, and the
current flowing is limited only by the resistance of the external circuit. The
voltage at which the above occurs is the characteristic ionization voltage or
ionization energy of atoms which we have spoken about in the preceding
section.

However, an atom has in general more than one electron and accordingly
more than one ionization potential. The second, third, etc., electron can also
be knocked out of its shell, but only with ever-increasing energy, because the
net positive charge on the resulting ion which holds these electrons increases
after each loss of an electron.

The photoelectric effect was nothing but the same process of ionization
with the difference that the energy necessary to detach an electron was provided by a light quantum instead of by an impinging electron. But the amount
of energy required, as given by the Einstein equation, remained the same.

Not only ionization, but excitation to light emission can be performed by
light quanta as well. If the mercury vapour is irradiated within a silica tube
(which allows ultraviolet light to pass through it) with light of shorter wavelength, another interesting phenomenon occurs. Visible and near ultraviolet light pass through the tube without anything happening. But the moment we arrive at the characteristic wavelength of the mercury atom which it
emitted on being attacked by 4·9 V electrons, the picture changes. This
ultraviolet light of 2,536·2 Å is dispersed by the mercury vapour in all directions: the tube begins to reradiate. The mercury atoms must have absorbed
these light quanta which seem to be to their liking and, having been brought
temporarily into a state of higher energy, have lost this again by radiation.
But the excited mercury atom does not "remember" from which direction it
has received the original quantum. The re-emitted, so-called "resonance,"
fluorescence radiation proceeds in all directions of space according to the
laws of chance: the whole tube radiates in all directions. The phenomenon
can be observed with the naked eye in sodium vapour. Its resonance radiation is yellow.

Condensed matter, solids, liquids and to some extent even high-pressure
gases, do not radiate such well-defined wavelengths as do gases at lower
pressures. This means that the doses of energy they are able to handle are
not well defined either. There is some difference in the colour of light emitted
by different materials at "red heat" of the same temperature but the difference
is not very characteristic and the spectra are all continuous. To every emitted
light quantum there are neighbouring light quanta of slightly smaller or
greater energy which may also be emitted—or absorbed as the case may be.

It can be shown that the electrons of adjoining atoms or molecules in
condensed matter are so near to each other that they are always under each
other's influence. No single electron of an atom can be "excited" into a

higher energy level without affecting adjacent atoms: the whole system is excited simultaneously. It can be shown that the number of permitted energy states increases in the course of this interaction. Fundamentally this is a consequence of the Pauli principle. As long as two electrons belong to two atoms far away from each other they are independent. But as they begin to interact they are not permitted any more to have identical quantum numbers or energies. The more complicated system allows more quantum states, more and more energy levels become available and at last practically any state of energy can be acquired. The spectrum becomes continuous.

64. ATOMIC SPECTRA

Atomic spectra may be observed in monatomic gases or in electric discharges of polyatomic molecules where the energy of the electrons is sufficient to break molecules up into individual atoms. Thus the spectrum of atomic hydrogen may be observed in molecular hydrogen in energetic discharges at low pressures.

We are now able to form a picture of the origin of spectra. Every atom has a fundamental state of its electron shells wherein all of its electrons reside in the states prescribed by the lowest energy solutions of the corresponding Schrödinger equation and by the Pauli exclusion principle. Whenever this stable arrangement absorbs sufficient energy in one manner or other to allow one or more of its electrons to leave the lowest state of energy and change into a possible state of higher energy, then this latter event will occur. The atom attains a so-called excited state or, if the energy is sufficient, one or more of its electrons may leave it permanently so that it becomes ionized. If the energy is taken from light waves, the atom chooses from the multitude of wavelengths those which would just enable one of its electrons to reach one of its possible excited states. The difference of energy between the original state and the excited one, divided by Planck's constant (h) gives the frequency of this light quantum. The differences are sharp and well-determined, so that the absorbed frequencies (wavelengths) will be sharp and well-determined too. The spectrum will be composed of sharp absorption lines.

The frequencies of these lines, or the characteristic voltages in the Franck–Hertz experiment, inform us about these energy differences within the atoms. Each dose of energy which can be absorbed shows that the atom possesses an excited state with just this amount of energy above the lowest, the "ground" state. These possible states of energy are called the energy "terms" of the atom and we are confronted with the problem of combining these terms in order to arrive at all the possible frequencies of light (more exactly of electro-magnetic radiation) which the excited atom is able to reradiate in the course of its falling back into the ground state.

We shall only discuss the most fundamental points of this process. The spectrum of any element has a shortest wavelength among its spectral lines. This corresponds to the event when the most strongly bound electron is forced by some means to leave the atom; it then returns into it again. In the spectra of the lightest elements this wavelength is still in the ultraviolet part of the spectrum, but as the atomic number increases, it tends towards, and soon comes to lie in, the X-ray region.

It is worth while to glance at these shortest wavelengths along the whole set of atomic spectra, with increasing atomic numbers. The X-ray spectra of the elements are drawn above each other in Fig. 112. They not only contain

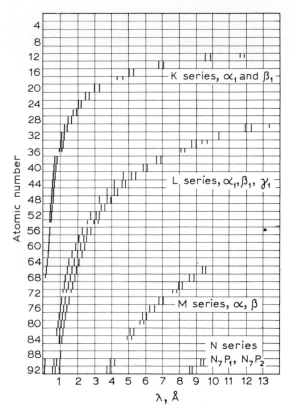

Fig. 112. **X-ray spectra of elements; their atomic number increasing downwards**

(From A. Eucken, *Lehrbuch der chemischen Physik*, p. 759, Fig. 170 (Leipzig, Akademische Verlagsges., 1930). By courtesy of Akademische Verlagsges.)

the shortest wavelength spectral line but those of their immediate neighbours too which seem to form a group beside them. Farther away we find similar groups of spectral lines as the atomic number increases. The analogous lines of the spectra fit well into a smooth curve and they seem to be governed by a rather simple rule. It was discovered by the young British physicist *Moseley* (1913) who unfortunately lost his life in the first world war. The dependence of the frequency of the shortest X-ray lines on the atomic number of the element in the periodic system is given by

$$v = AZ^2 \quad \text{or} \quad \sqrt{v} \sim Z \qquad . \qquad . \qquad . \quad (64.1)$$

The frequency is proportional to the square of the atomic number, A being a constant throughout the system.

This equation gives some insight into the structure of the atom, especially into the bonding energy of its most intimately bound electron—the most intimately bound because it must be evident by now that the shortest wavelength, that is the highest frequency and energy, are mobilized if the most energetically bound electron is concerned. Without even trying to go through the exact wave-mechanical calculation of the energy of the internal electron it is still possible to understand the Moseley equation in a qualitative way. As the atomic number and hence the positive charge upon the nucleus increases two things happen which work together in binding the innermost electrons more intensely. The primary cause is that the increased Coulomb attraction pulls the electron with a force proportional to the charge on the nucleus. But as a consequence thereof the electrons are really drawn in, nearer the nucleus. The original equilibrium between zero-point energy (indeterminacy energy) and electrostatic attraction becomes unbalanced because the latter increases with the charge. They will balance at a smaller average radius of the "smeared out" electron. It can be shown—relatively easily in the Bohr model, somewhat less easily by the Schrödinger equation—that this radius will decrease in proportion to the increasing nuclear charge. Now electrostatic energy is proportional to the charges and inversely proportional to the distance between them.* Therefore the binding energy increases once in proportion to the charge and once again in proportion to the charge, because of the decreased distance: we now have the Moseley relation with its quadratic dependence.

What about the groups of lines in the X-ray spectra of Fig. 112? There is a first group of lines near each other and there are other groups farther away from the first. The explanation is as follows. Whenever an electron has been ejected by some means from its original state, another electron will "fall" back into this state, sooner rather than later. The lines within a group represent the energy liberated by these falling electrons as they descend into the atom. The most energy is liberated when the electron is free, ionized, beside the atom (now a positive ion). According to the Einstein relation this is equivalent to the largest energy difference and hence the greatest frequency or shortest wavelength. (It is not shown in Fig. 112.) Conversely this is the minimum frequency which absorbed X-rays must possess to enable them to ionize the innermost electron of an atom. The subsequent lines within the first group represent the falling back of electrons from lower and lower energy states into the lowest state of all. The last line of the group on the right-hand side belongs to the electron which falls back from the next-to-lowest shell into the lowest one. (It is designated by the subscript α.)

Each group of X-ray lines represents a shell of electrons from which the original electron was knocked out, and each line within a group represents

---

* Electrostatic force is inversely proportional to the square of the distance between the charges. But energy is force times distance, thus electrostatic energy is inversely proportional to distance.

one of the higher shells whence it returns. (We neglect the question of different energy states within a shell itself.) In the course of building up the periodic system from its successive electron shells we have already learned that these are designated by the letters K, L, M, N . . . etc. These symbols are also applied to the groups of X-ray lines in the spectra. The group to the extreme left, caused by replacement of a vacancy in the innermost shell, is designated the K-line group of X-rays, and so on. The farther away the shell, the less the energy and hence the longer the wavelength.

We have already discussed the fact that the inner shells almost completely screen the amount of their own charge from the charge of the nucleus. Only nearly, because their distribution is not quite spherical and because the electrons in the outer shells are not all confined to spherical probability distributions either. (The "p" distributions were double-cigar-shaped, etc.) Owing to this shielding effect the shells which are described by higher principal quantum numbers (n) are attracted less towards the nucleus, therefore they are farther away, and for both reasons their binding energy decreases with their principal quantum number. Therefore the energy which is liberated when an electron from one of these shells leaves it and is replaced by one of the outer electrons, decreases also. The wavelength increases accordingly from the K group to the L, M, etc., groups of X-ray lines. But within such a group, as long as nothing fundamental is changed, the Moseley relation prevails. Only it is not the actual atomic number which now stands for the nuclear charge but a net quantity which is less by the amount of shielding. Let us say $(Z - B)$ instead of Z (Z = atomic number, B = shielding).

It is best to visualize the Moseley relation by plotting the frequency instead of the wavelength, as a function of the atomic number, because frequency and not wavelength is proportional to the energy differences. It is even better to plot the square root of the frequency against the atomic number because in this case the Moseley relation becomes linear. (Fig. 113.) The linearity of these square-root frequencies is very remarkable indeed! But only as long as no fundamental changes occur. At well-specified points of the higher X-ray groups there is a break. And this break occurs exactly at the elements of the system where from purely chemical reasoning we decided in Section 60 that a subgroup of electrons is beginning to be built below an already existing shell of electrons. It is of course clear, that in such a case the new subgroup electrons help to screen a part of the nuclear charge which up to then was free to reach the higher electrons. The grip of the nucleus loosens, the energy, and thus the frequency, increases more slowly than it would if the subgroup had not appeared. The straight line goes on increasing, but its slope has decreased.

The periodic system and the spectra of the elements thus agree in the interpretation of the properties of the transition group elements. Chemical behaviour made us assume that within these groups the new electrons entered a deeper region of the electronic system—there was much less difference between the chemistry of neighbouring elements than was usual within the "normal" part of the system. Chemistry is dependent on the outer electrons because these are utilized in forming chemical bonds; hence it was quite logical to assume that these outer electrons remained the same throughout the transition groups. Now, the X-ray diagrams of Moseley confirm our assumption. They show convincingly that electrons are built into deeper

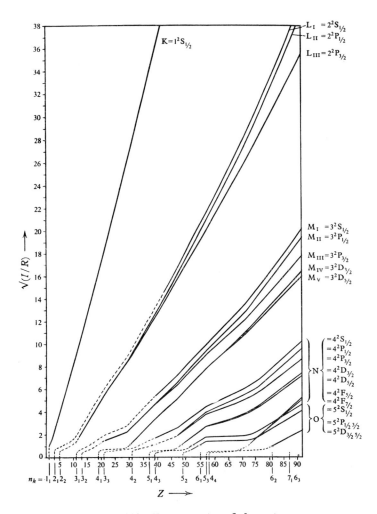

Fig. 113. **X-ray spectra of elements**

The square root of the frequencies as a function of the atomic number (Z). Note that the slope of the straight lines changes each time a subgroup of electrons is being built in below the electrons giving rise to the spectral lines.

(From H. G. Kuhn, *Atomic Spectra*, p. 237, Fig. IV. 4 (London, Longmans, 1962). By courtesy of Longmans, Green & Co.)

shells of the atom whenever transition groups of elements appear. The wave-mechanical theory assures us that this is what ought to happen in these regions of the system: we previously by-passed certain possible solutions of the wave equation because for a time they demanded too much energy but now the time has come when the energy of other possible solutions is even higher so that the by-passed solutions may now be considered.

The electrons on the surface of the atom are responsible for its chemistry, as we have stated. But they are responsible also for the longest wavelengths in the atomic spectra because, as they are farthest away from the nucleus, their binding energy is smallest. It so happens that the energy necessary to excite them into the next highest possible quantum state generally lies within or near to the visible part of the spectrum. On page 249 we have noted that these visible spectra have similar character for elements in the same column of the periodic system: the spectra of alkalis, alkaline-earth metals, etc. are group-wise similar in character. Thus there is a fundamental difference between the X-ray spectra which are due to the deep-seated electrons and the visible spectrum which is due to those in the surface shell. Whereas the inner electrons are strongly under the influence of the nucleus and feel its increased pull from element to element, the surface electrons are well shielded from its influence and are rather susceptible to the repulsion of their neighbours within the same shell and therefore to the number of neighbours they have in the same shell. This is the reason why the X-ray spectra are not periodic—a new series appears with each period (K, L, M, etc.)—but underly the regular increase of Moseley's rule. On the other hand, the surface electrons show periodically recurring possible energy changes—that is spectra—according to whether they are alone in the surface shell of the atom or have a partner or several partners there.

The hypothesis that each emitted spectral line corresponds to the *difference* of energy between two permitted electronic states has a very important consequence. As a matter of historical fact it was this consequence which made it possible for *Balmer* in 1885 to arrange the spectra of simple elements into sequence, and for *Bohr* to explain these sequences by his ingenious atomic model (1913).

We shall discuss only the simplest case. Take two spectral lines of one element, both belonging to the K series, that is, to the innermost electrons of its shells. We said that these two lines are emitted when an electron falls back into its original deepest state from two different excited states within the atom. If this is true, it is easy to see how great the energy difference between these two excited states must be, from which the electrons were able to fall back. The frequency of the first line multiplied by Planck's constant (h), gives the energy of the first excited states above ground level, the frequency of the second line times h, that of the second. The difference between these two energy differences shows how much more energy there should be in one excited state than in the other. If this is the case, we are able to predict the energy which would be set free if the electron could pass from the higher excited state into the lower—not into the ground state this time! If such a transition is possible, this means the emission of a new spectral line: this energy difference has only to be divided by h to obtain its frequency (Fig. 114).

Let ν_{1-0} be the frequency of the first original line and ν_{2-0} that of the second. The energy difference between the two excited states is thus $h(\nu_{2-0} - \nu_{1-0})$

and this in its turn must be the energy difference between the two excited states hv_{2-1}

$$h(v_{2-0} - v_{1-0}) = hv_{2-1} \qquad . \qquad . \qquad . \qquad . \quad (64.2)$$

or simply
$$v_{2-0} - v_{1-0} = v_{2-1} \qquad . \qquad . \qquad . \qquad . \quad (64.3)$$

The differences between frequencies must be spectral frequencies themselves. This combination principle, due to *Ritz* (1908), is valid throughout the realm

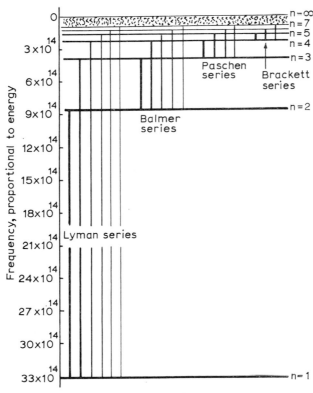

Fig. 114

The energies can be calculated by multiplying the frequencies by Planck's constant, h, equal to $6·62 \times 10^{-27}$. The length of the vertical lines represents the energy difference between possible states of the electron and hence the frequency of the emitted wavelength.

(From G. M. Barrow, *Physical Chemistry*, p. 64, Fig. 3–8 (New York, McGraw-Hill, 1961). By courtesy of the McGraw-Hill Book Co.)

of spectroscopy and made it possible to find the permitted energy levels, so-called energy "terms" of atoms and of molecules. If in our example the second K line was emitted by an electron falling back from the L shell into the K shell and the first from the M shell into the K shell (the so called K_{α} and K_{β} lines) then the difference frequency must be due to the falling back of an M electron into the L shell and must hence represent the least energetic of the existing L lines of the atom (L_{α}). This is really the case.

It is clear that any system of electrons must have fewer energy terms than spectral lines, because the spectral lines arise from the possible combinations of a given number of terms. The situation lends itself easily to graphic representation. The terms are shown as horizontal lines above the lowest, the ground term, and the spectral lines are the transitions between two different terms within this system. They can be well represented by vertical lines. Light absorption is possible along a vertical line pointing upwards if there is an electron in the lower state and if there is a place in the state above. Light emission is possible if the reverse is true. (Not all such transitions are equally probable and some of them are actually forbidden in normal cases, but this would lead us too far into theoretical spectroscopy, that is into the quantum mechanical theory of electron shells.)

Disregarding the exclusion rules there are

$$\frac{n(n-1)}{2}$$

possible combinations of only n terms. It is clear what an immense degree of clarification we have achieved if it has become possible to reduce all measured spectral lines to combinations of their respective energy terms. These terms are the characteristic energy states of the atom or molecule and it is these terms which must be explained by the quantum theory of the electron shells.

The principal quantum number n was responsible for the distance of the most probable electron density from the nucleus. The other quantum numbers determined the shape of this distribution, whether spherical or cigar-shaped and so on, and they represented a momentum of rotation around the nucleus whenever they were not of the spherical type. Whereas the principal quantum number explained the important macro features of spectra according to Moseley's law, the other quantum numbers and the spin explain the finer and finest details.

The picture becomes more complicated if we leave the monatomic gases and begin to analyse the spectra of di- or polyatomic molecules. The energy of such a system is not exclusively determined by the relative states of electrons around a nucleus but is also affected by the relative energies of the nuclei with respect to each other. Nuclei may vibrate along the straight lines which join them or along other well-defined directions, and the energy or frequency of these vibrations is also subject to quantum-mechanical limitations. Just as the wave equation of an electron has only discrete sensible solutions with given energy terms so also has the wave equation of nuclei. Besides this the molecule as a whole may rotate about its axes and it can be shown that these rotations too are subject to quantum limitations: only certain discrete rotational momenta are solutions of the wave equation. It can be shown by calculation and proved by measuring the spectra that the energy necessary to excite the intramolecular vibrations is smaller by one or two orders of magnitude than the energy necessary to excite electrons. Therefore the corresponding spectra lie in the much less energetic near infra-red region, around 5μ. The energy necessary to excite the rotations is again smaller by one order of magnitude and pure rotational spectra accordingly lie in the far infra-red around 50 μ (Fig. 115). Often, however, vibrations and rotations change their states simultaneously: the small energy differences of rotation are superimposed upon the greater ones of vibration and one obtains spectra

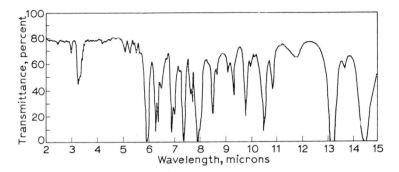

Fig. 115 (a). **Vibrational spectrum of acetophenone, CH₃—CO—C₆H₅**

100 per cent means total transmission, absorption peaks point downwards. Some peaks are characteristic to fingerprint given chemical groups and are used to this end in organic chemistry. The peak at 3·3 μ is characteristic of the CH₃ group, that at 5·9 μ of CO, and those at 13·2 and 14·4 of C₆H₅.
(From *Encyclopedia of Science and Technology*, v. 12, p. 584, Fig. 5 (New York, McGraw-Hill, 1960). By courtesy of the McGraw-Hill Book Co.)

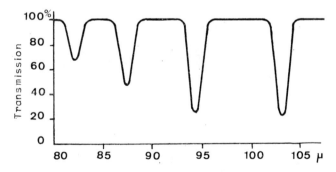

Fig. 115 (b). **Rotational spectrum of the PH₃ molecule in the far infra-red**

Absorption peaks point downward.

which show fine structure in the vibrational lines. Even all three: electronic excitation, vibrations and rotations, may change at the same time and this effect brings both vibrational and rotational differences into the visible or ultraviolet region of the spectrum. Each spectrum line which would otherwise have corresponded to a single electronic transition now appears as a series of bands with substructure. Each band corresponds to a vibrational transition and each fine line within a band to a transition between different

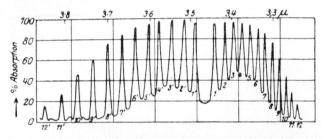

Fig. 116 (a). **Rotational fine structure of a vibrational band in the spectrum of HCl gas**

(From A. Eucken, *Lehrbuch der chemischen Physik*, p. 742, Fig. 163 (Leipzig, Akademische Verlagsges., 1930). By courtesy of Akademische Verlagsges.)

Fig. 116 (b). **Bands in the electronic spectrum of N_2**

Each band corresponds to a given difference in vibrational energy, the substructure in the band to different transitions of rotational energy.

From H. Sponer, *Molekülspektren*, v. 2, p. 58, Fig. 21 (Berlin, Springer-Verlag, 1936). By courtesy of Springer-Verlag).

permitted states of rotation (Fig. 116). These spectra have been used by spectroscopists to determine important molecular data. Vibration is evidently determined by the forces acting between the nuclei and by their masses, whereas rotation is determined by the moments of inertia of a molecule in its three directions of rotation. Thus intermolecular forces and molecular distances emerge from these molecular spectra.

This is the point where spectra and the kinetic theory of gases meet. The founder of the statistical theory of gases, L. Boltzmann, observed that the molecular specific heats of all monatomic gases are equal. It is quite natural that this should be so, because we know that temperature represents nothing but the average kinetic energy of the molecules and this is the same for all molecules at the same temperature. Heating simply increases the kinetic energy by the same amount for different substances and thus raises the temperature to the same extent. A monatomic gas molecule cannot acquire any other kinetic energy than that in the three directions of space required for translational motion. Not so the di- or polyatomic molecules. These may add rotation or rotations to their translational motion and the nuclei within them may vibrate relative to each other. This means that kinetic energy is distributed over more than three translational "degrees of freedom." There are more independent forms of rotation and vibration as the number of atoms in a molecule increases. Classically at least, any energy received by a molecule has an equal chance of becoming translation of the molecule as a whole, or energy of rotation or of vibration within its structure.* On the whole, more energy is necessary to increase the average translation of a di- or polyatomic molecule than that of a monatomic one. The heat necessary to raise its temperature by one degree, the molecular specific heat, must therefore be higher. Experiment shows that this is so. The specific heats of diatomic molecules are higher than those of monatomic gases; but they are, to a first approximation, equal among themselves. The molecular specific heats of more complicated gas molecules increase further because the energy has more choice of where to go, and they gradually differ more and more from each other according to the different sizes of possible rotational and vibrational quanta (Table VIII).

Summing up, we can now understand how it comes about that atoms are not well represented by solid impermeable balls. They are built in a rather loose manner from the nucleus and its surrounding, not very precisely localized, electrons. Hence they are permeable to any small fast particles we happen to shoot at them. We also understand now how the marvellous regularities in the periodic system of elements come about and we have even gained a slight insight into the origin of spectra. At any rate we have become

---

* In quantum mechanics there are well-defined energy quanta of vibration and rotation, less energy cannot be accepted. Sometimes this minimum of acceptable energy calculated from the wave equation, is so high that the kinetic energy at normal temperature is too small to offer it. Then the particular rotation or vibration is "frozen in"—is not going to be "excited" at the given temperature. This is why simple atoms do not rotate at normal temperatures and why a series of vibrations become excited one by one as the temperature increases. But this has only a secondary effect on the present argument.

TABLE VIII. Molar Heat (Specific Heat for One Mole) of Gases at One Atmosphere Pressure and 20°C and Constant Volume

			Calories/ degree C
Monatomic gases:	Ar	Argon	2·98
	He	Helium	2·98
Diatomic gases:	H_2	Hydrogen	4·86
	O_2	Oxygen	5·03
	N_2	Nitrogen	4·93
	Cl_2	Chlorine	6·00
	CO	Carbon monoxide	4·98
Triatomic gases:	CO_2	Carbon dioxide	6·80
	H_2S	Hydrogen sulphide	6·20
	SO_2	Sulphur dioxide	7·50
Tetraatomic gas:	NH_3	Ammonia	6·65

Monatomic gases can only increase their kinetic energy of translation while gases containing more than one atom in their molecule also increase their energy of rotation and sometimes also that of intramolecular vibration.

certain that the name "atom" (indivisible), was given somewhat too rashly to the smallest units of the chemical elements. They are not ultimate particles, they can be broken up into nuclei and electrons and have a very characteristic internal structure of their own. In Part 7 we shall learn that not even nuclei are indivisible. But before descending as far as that we now have to discuss the forces which act between atoms when they combine with each other. This is the problem of the chemical bond which was quite beyond our reach as long as we were not acquainted with the parts and structure of the single atoms.

Part 6

The Chemical Bond

65. SATURATION OF BONDS?

Atoms combine and separate and each structure of a combination corresponds to a definite compound with properties of its own, distinct from all others. So far so good. However this is not enough in itself. Why do atoms combine at all? Why are there certain limits to their combination? We could well imagine that all combinations represented by the formulae HO, H_2O, H_3O, . . . HO_2, HO_3 . . . H_2O_2, H_2O_3, . . . H_3O_2, H_3O_3 . . . and so on existed! Instead we find only H_2O and H_2O_2 realized as stable entities and OH, HO_2 as transient products just capable of being identified in the course of some reactions. The same applies to combinations of other elements. There must be some governing principle in Nature which allows the binding only between a very limited and well-defined number of atoms. Up to now we have simplified the answer by introducing the experimental concept of valency. But this was more a word than an explanation. The question of the nature of the chemical bond thus emerges immediately at the beginnings of chemistry.

The nature of the chemical bond seems to be definitely discontinuous. After it has combined with a number of atoms the binding capacity of a given atom seems to be exhausted, saturated. The second law of stoichiometry (p. 21) tells us that compounds and elements react with each other in different constant proportions, but mostly in proportions of *small* integers. The force that attracts atoms is not like gravitation which cannot be saturated. It resembles more the force between electric charges or magnets: a charge or a pole becomes saturated by its partner.

But even this is not quite true. As long as charges or poles are not allowed actually to coalesce, they only move nearer to each other, until they stop at some small distance. Two such neighbouring electric charges form a dipole, four a quadrupole, eight an octapole, and so on (*see* Fig. 109). Although the electric field around such multipoles decreases much more quickly than the field around a single charge, it still exists. There remains an electric field around them, a field which is no longer spherical because the multipole is not spherical either. This field can be easily constructed by adding up the individual fields of the differently charged particles of the packed multipole. There thus remains a force acting on other charged particles at its surface and there is no reason why it should not grow further by attracting them one by one. Even for electric or magnetic forces there seems to be no saturation.

However, this kind of non-saturation is no stranger in chemistry. Leaving aside for the moment the question why a sodium atom loses its electron to a chlorine atom yielding sodium and chloride ions, respectively, we may inquire into the manner in which these two kinds of ions are built together into a rock-salt crystal. A Na^+ ion attracts a Cl^- ion to form an electric dipole. Two such dipoles attract each other head-to-tail and yield a quadrupole. Two quadrupoles yield an octapole. The process may go on indefinitely (Fig. 109b) in all three directions of space. Today we know that it really does go on, the result being an actual crystal of sodium chloride. There is no way of telling which of the six neighbouring Cl^- ions belongs to a specified Na^+ and vice versa. The originally simple concept of a NaCl molecule in the solid crystal fades away and gives place to the much more indefinite concept of a macromolecule $(Na^+Cl^-)_n$ where n may be an integral number. The crystal is the molecule itself, and while the difference between the dipole and the quadrupole may have been significant, that between the higher multipoles vanishes very quickly. It becomes quasi-continuous and tends rapidly towards the properties of the macroscopic crystal where it does not matter whether it contains a million or a million and one "molecules" built into the same lattice.

Even the ratio $1:1$ between Na^+ and Cl^- ions may be changed to a slight extent. If the crystal is immersed in a solution containing Cl^- ions but no Na^+ ions—say hydrochloric acid—the Na^+ ions at the surface will tend to become covered by excess Cl^- and the whole crystal will acquire a slight negative charge from this surface layer. As a matter of fact such a crystal will move towards the anode in suspension if it is small enough. What about the formula NaCl now? It is the limiting case for no surface adsorption and for no irregular sites within the crystal. An ionic crystal is never really saturated, but the quickly increasing electrostatic repulsion does not allow it to acquire any considerable charge.

This is our way of approach today. But in the beginnings of exact chemistry, NaCl was a unit molecule just as was H_2 or CO or CO_2. We were lucky enough to find the simple, fundamental concepts of valency saturation before the premature knowledge of secondary complications—as illustrated by infinite ionic crystals—was able to confuse us. Now that we have become acquainted with its building stones, the whole complicated edifice reveals its unity of structure.

The gas molecules of CO and CO_2 just mentioned are well defined indeed. They are brilliant examples of saturation and of discontinuous proportions. How do they come about? For more than a century we managed to work our way with atoms which joined hands with each other, which were equipped with a given set of hands even if they chose to keep some of them in their pocket for a while. Now, after learning something about the structure of the atoms themselves, about their electrons and the *discontinuous* solutions of wave-mechanical equations, we have gained much more insight into the chemical bond. But difficult as the quantum mechanical concepts are to many chemists, the *phenomenon of discontinuity* of saturation had long been well known before physicists ever thought about it. Maybe *we* should have guessed at the existence of quanta long ago.

66. TYPES OF CHEMICAL BOND

We know that atoms contain electrons in different, smeared-out shells around their nuclei and that the electrons of the shell on the surface and sometimes those somewhat beneath it are able to interact with other atoms. It is up to us now to answer the question how and why they interact as they actually do.

It is necessary to repeat once more what we have learned about the laws governing the electronic structure of an atom. The structure of molecules is governed by them too, it is only necessary to apply them intelligently to the problem.

An atom presented the problem of arranging a set of electrons around a nucleus. It turned out that the quantum mechanical wave equations of this problem only possessed stable solutions for quite definite arrangements, where the probability of finding the electrons in a definite portion of space was given and where each such solution corresponded to a state of exactly given energy. From all such possible solutions the states of lowest energy content were occupied, provided that external energy did not excite the atom temporarily into a higher state.

A molecule presents the problem of arranging more than one nucleus with its electrons into a similarly permitted, stable structure. The wave equation which is already formidable for the heavier *atoms* with many electrons, becomes even more complicated for this case. It is clear that at present no exact solution can be even attempted, with the exception of the simplest systems such as the H_2^+ ion with two nuclei and one electron, or the H_2 molecule with two nuclei and two electrons.

But very good and sensible approximations offer themselves readily. The deeper-situated electron shells are far from the surface of the atom and are shielded from foreign atoms by the shells nearest the surface. Besides, they are in general saturated as far as concerns the number of electrons which are permitted to belong to them. Therefore it seems a very good idea to treat the inside shells as if they belong only to their parent atom and are not substantially influenced by any neighbour. Thus they do not take part in forming chemical bonds. Using this simplification the problem shrinks to a consideration of the permitted states of electrons within the surface shell or within an incomplete shell beneath the surface. This is a problem which lends itself more easily to mathematical treatment, as least as long as the atoms interacting within the molecule are not very numerous. It has been solved for simple problems and it is being solved for new, more complicated problems day after day, with some necessary approximations, of course. The "orbitals" of these electrons no longer belong to any single atom but to the whole molecule. They are hence called "molecular orbitals". In ideal cases these molecular orbitals belong only to a single bond between two neighbouring atoms and are not so difficult to treat, but they do sometimes extend across "vast" regions of large molecules, when they are said to be delocalized.

In this section we shall only review the main types of the chemical bond without occupying ourselves with any of them in detail. We shall discuss them again, one by one, in the subsequent sections. However, we are going to begin by repeating very much what we have learned already about zero-point motion and zero-point energy because this concept cannot be overemphasized as the foundation of the chemical bond energy. It is not an easily understandable concept and it is probably worth while to explain it once more instead of referring simply to Section 59.

Let us assume that a strong field of (electrostatic) force delineates the position of an electron within a small space. The uncertainty relation of Heisenberg (p. 210) demands then, that we only know its momentum inaccurately. Momentum is mass multiplied by velocity. Now, if we do not know exactly how great the momentum of a given particle is, but are able to assign a probable limit to this inaccuracy we can equally well say that the electron certainly does not remain stationary. If it did, its momentum would be exactly zero and we know it cannot be *exactly* zero because it cannot have *exactly* any value. Therefore the electron most probably moves. If it moves, it must have kinetic energy, and the more of this it has, the greater is its momentum. Because of this uncertainty its average momentum is the greater, the smaller the space in which the electron is enclosed. Thus the more precisely an electron (or any other particle, of course) is localized, the higher will be its kinetic energy. The original concept of absolute zero temperature where all thermal motion comes to a standstill is no longer valid: a new kind of motion persists and gives rise to a new kind of kinetic energy. We call it, therefore, zero-point energy.

Let the average, probable, diameter of the space where we have confined our electron be Δx. Heisenberg's principle demands that its average, probable momentum must then be

$$\Delta(mv) = \frac{h}{\Delta x} = m\Delta v \qquad . \qquad . \qquad . \qquad (66.1)$$

where $\Delta(mv)$ represents the mean uncertainty in momentum and Δv that in velocity. (As long as the velocity is appreciably less than c, the velocity of light, the mass may be regarded as constant: *see* the section on relativistic motion of the electron, p. 203). Knowing that kinetic energy is $mv^2/2$ we arrive at the most probable kinetic energy of this zero point motion as:

$$\Delta E_0 = \tfrac{1}{2}m(\Delta v)^2 \qquad . \qquad . \qquad . \qquad (66.2)$$

or, substituting for Δv,

$$\Delta E_0 = \frac{h^2}{2m(\Delta x)^2} \qquad . \qquad . \qquad . \qquad (66.3)$$

For chemists this is a very important discovery. It means that whenever an electron is free to move about in a space which is larger than that in which it was originally enclosed, its kinetic zero-point energy decreases. We have gained energy. On the other hand we have to expend energy if we wish to restrict the electrons probable sphere of action to a more limited space.

This difference in zero-point energy is sometimes called by an extremely unfortunate nomenclature, "energy of resonance." Even the mechanical picture which gave rise to the term "resonance," is extremely unclear in this context and is scarcely worth entering into. Nothing whatever resonates here: resonance means a coincidence of frequencies and only in the case of very highly symmetrical electron distributions around, say, two identical nuclei can it be postulated that the two frequencies of the electron, one in respect of each nucleus, are identical. Nothing, nothing whatever resonates. The concept was attacked violently by Soviet scientists because they feared that something metaphysical had been introduced into the "materialistic" world of science. It is certainly true that a most unsuitable name has been introduced and it was right to attack it irrespective of metaphysical implications. Unfortunately the term "resonance" is now used very widely and we cannot safely neglect its existence; we had to mention it. But we shall not use this term again and strongly advise against its being used at all. Until a better word is coined for this difference in zero-point energy on expanding or contracting the zone of an electron, we shall use the term "delocalization energy". The limiting structures for the position of electrons in a molecule, which however are not really preferred within their indeterminate domains are called "mesomeric" ($\mu\acute{\epsilon}\sigma\sigma\varsigma$ = the middle) because the average real structure is in between them. For example, the two Kekulé structures for benzene:

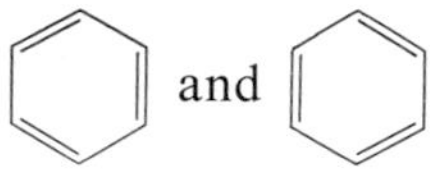

and

or the three for the nitrate ion:

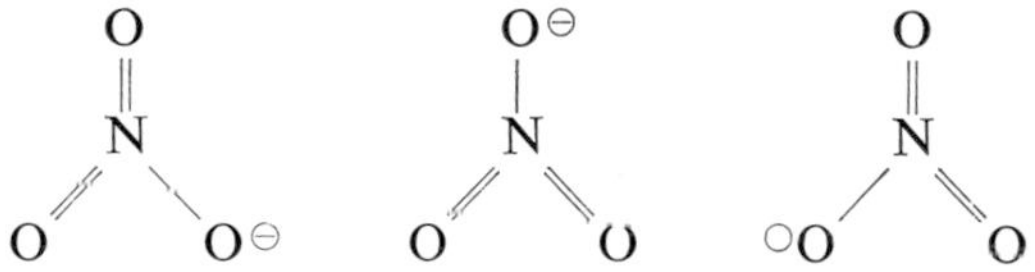

It often happens that one or more electrons of an atom extend their spheres of action into a neighbouring atom or across a series of consecutive atoms in their vicinity. This has become possible because the attractive force of the other nucleus or nuclei has made it easier for them and because often they are able to neutralize their spin-momenta by overlapping "orbitals" with electrons in these other atoms. As a result of this encroachment into a larger volume of space the average zero-point energy of the electrons has decreased, because their uncertainty in space has become greater and therefore their momentum has become better defined. We have won energy of delocalization. This is of course in addition to having gained simple electrostatic interaction energy through the attraction by the other nuclei, and also in addition to spin-pairing energy, if such a pairing has come about. This kind of bond between atoms is called the "*covalent*" (or sometimes, *homopolar*) bond. The combination of electrostatics and wave mechanics decides when and which of these bonding orbitals become possible.

In contradistinction to this covalent bond, another event may occur between neighbouring atoms. The electronic structure of an atom may contain one or more electrons at an average distance from the nucleus where the Coulomb attraction is relatively small. On the other hand, a neighbouring atom may have such a structure that it is able to incorporate these electrons into its own electron shells, either nearer to the nucleus than they were in the first atom, or at any rate so that they are given more space in the new system of "standing waves" than they originally possessed, hence decreasing their delocalization energy. Once more it is the sum of electrostatic, delocalization and sometimes spin-coupling energy which makes it worth while for the electron or electrons to go over from their parent atom into another. An electric ion pair has been built up in this way and this always demands electrostatic energy, but the three other energies mentioned more than compensate for it in this case. Using another jargon: the electron affinity of the second atom is greater than the ionization energy of the first. Two ions of opposite charge are formed and these attract each other with bonds which we call *ionic, polar* or *heteropolar.*

However, the two bonds mentioned are only limiting cases. It is not necessary that the electrons of a covalent bond should be just midway between the adjoining nuclei. According to the electronic structure and nuclear charge it may happen that the state of minimum total energy is reached if the electrons are on the average nearer to one nucleus than to the other. In this case the covalent bond is *"polarized"*, and the charge distribution between the partners is not quite even. The bond becomes the more asymmetric electrically, the more the "electronegativity", that is the electron-attracting capacity of the partners, differs. In the limiting case, the electrons belong to one partner only: we have here reached the state of the ionic bond, whereas we have the ideal covalent bond before us if the electro-negativities are equal and the electrons are symmetrically disposed with respect to the two nuclei. The state, the "orbitals" of minimum energy depend on the solution of the relevant wave equation between the effective positive ions and the valency electrons.

Though it is the surface electrons which interact with each other in most cases, the Pauli principle sometimes allows foreign electrons to be built into a subsurface shell of an atom if this shell happens not to be quite filled with electrons. In this case a delocalization within this shell becomes possible, with the corresponding lowering of the total energy. New "standing waves" will now be formed which were not possible before. Interesting classes of compounds possess this kind of bonding, the existence of which could not be explained by older concepts of chemistry. Complex compounds and metal carbonyls belong to this group.

Electrons which are far from the nucleus and from the core of the rest of the atom may leave this core without becoming attached to any individual partner, provided that many atoms of this sort are sufficiently near to each other. The combined system of atoms may have more permitted states for electrons than there are electrons and thus the latter are able to move freely across the whole system in the lattice of the atomic cores which have been left with a net positive charge. Substances showing this behaviour are the *metals*, their alloys and the intermetallic compounds. We shall deal with them later on.

In many molecules, for example, water, ammonia and hydrogen chloride, the "centres of gravity" of the positive and negative electrical charges do not coincide. The resultant dipole or multipole forces lead to attraction of neighbouring molecules if they are suitably oriented (Fig. 109), giving rise to liquids, solids, hydrates or other solvates. Under special circumstances a hydrogen atom within a molecule—being a bare proton within the cloud of bonding electrons—may attract the electrons of an atom in a neighbouring molecule and so act as a bond between the two. We call it the *hydrogen bond* and it becomes manifest by unduly strong cohesional forces (as shown by high boiling-points, for instance) as compared with those of the existing dipole moments (*see* Section 62). Sometimes we find hydrogen bonding between two atoms of the same molecule.

An ion or a dipole has an electric field. Atoms or molecules without electric charge or electric asymmetry may become polarized in such a field and this induced dipole moment serves to bind the otherwise non-polar molecules to their polarizing agents.

There remain the forces which act between chemically saturated atoms or molecules. These are effective over relatively small distances and are responsible for the fact that even rare gas atoms condense to liquids at low temperatures. At very low temperatures these forces cause the rare gases to form simple crystal lattices. They do the same with saturated molecules such as methane, that is with molecules which have no dipole moments to account for their attracting each other. They combine rare gas atoms with water molecules into solid hydrates. In the gaseous state they cause molecules to attract each other and thus cause deviations from the simple gas laws which were based on the assumption that no such interaction exists—and on the assumption that the molecules are negligibly small. More precise equations of state for gases take account of these forces and of the volume occupied by molecules. One of the most known, though by no means perfect, equations of this type is due to *van der Waals* (1837–1923) and these particular forces between molecules are named after him: *van der Waals' forces*.

This force increases with an increasing number of electrons within the attracting units: it is greatest between emanation (radon) atoms and smallest between helium atoms in the row of rare gases, and it increases with the molecular weight of organic compounds in homologous series. For example, in the alkane series, methane (CH_4) is a gas, pentane (C_5H_{12}) a liquid and octadecane ($C_{18}H_{38}$) a solid. We might feel tempted to ascribe this force to the increase of mass, but this would involve gravitation and it is easy to show that gravitation is much too weak a force to hold together molecules against thermal agitation. It can be shown that the force is due to irregular oscillation of electron shells within atoms or molecules giving rise to transient electric moments and thus to attraction.

It may happen that any, or the sum total, of these intermolecular forces exceeds the weakest bond strength within a molecule. In such a case it is not possible to distil the compound without decomposition because its molecule will prefer to break into pieces rather than to loosen its bonds with neighbouring molecules. Sometimes it may not even melt without decomposition.

The common characteristic of forces arising from polarization is that they do not become saturated—they add up in the same way as gravitation. This is the reason why we generally call them physical forces, to distinguish them from the so characteristically discontinuous chemical forces which *can* be saturated when they have formed chemical bonds, and which do not then act any further.

Very special arrangements of atoms in space may result in the formation of holes of very critical size and shape in the solid structure. If atoms or molecules happen to exist which fit in well with the geometry of these holes, they might become captured and built into the structure by simple van der Waals' forces, without real chemical binding taking place. They are called "clathrate" compounds from the Latin, *clathratus*, encaged. Stoichiometrical relations which are found in such cases are caused by geometrical and not by energetic factors.

Solutions also originate from the action of intermolecular forces. The solute of course tends to spread over as great a space as is at its disposal, driven by the thermal motion of disorder, of increasing entropy (p. 147). But a solute molecule would dissolve no more readily than it evaporates into free space, were not the attraction of the solvent molecules there to help it break the bonds which hold it fast into its own crystal or liquid phase.

In the following sections we shall have a brief look at the types of bonding and cohesive forces we have enumerated here.

67. VAN DER WAALS' FORCES

The name of this force commemorates the scientist who gave us a relatively good approximation of the volume–temperature–pressure relations ("state equation") of non-ideal gases. He attached two corrections to the equation of state for ideal gases: he considered the attractive force between molecules when they begin to approach each other and he took into account that the actual disposable space for a gas molecule is less than the total volume in which it is enclosed, because it cannot overlap other molecules. During discussion of the Loschmidt derivation of the Loschmidt number we dealt with this volume effect. Now we have to deal with the forces.

Atoms and molecules always interact, always attract each other because they have electron shells. These may become polarized and the resulting transient dipoles attract each other. As we know, even helium with its two electrons and extremely symmetrical structure can be liquefied: a sign that even helium atoms attract each other to a slight extent. The polarizability of an atom or a molecule also determines its reaction to an electromagnetic field and thus towards light, as discussed for instance on p. 346. This is the reason why those atoms or molecules which easily react with incident light in refracting and dispersing it to a larger extent, are the same which are easily polarized by each other and therefore attract each other with relatively strong van der Waals' forces. Hence these are often termed "dispersion forces." Their theory is due to Debye as far as polarization is concerned and to London in its quantum mechanical development.

We cannot follow here the exact calculation. But it can be made clear how the forces come about. Suppose two molecules are near to each other and, owing to the continuous and indeterminate motion of their electrons which persists even at the absolute zero of temperature, they are in general not in a state of perfect symmetry as far as distribution of the electron cloud around the nucleus is concerned. This asymmetry is eliminated in the time average because no one direction is preferred against another, but during most of the time it exists at every moment. Thus every molecule can be represented during most of its life by a small electric dipole or multipole (Fig. 117).

Fig. 117. **Mutual polarization of transient electric moments of molecules results in attraction (Van der Waals' forces)**

Owing to the asymmetrical electric charge distribution, there is an electric field around such a multipole and if a neighbouring molecule comes into its sphere of influence this will become polarized by it to a small extent. The direction of polarization will be, of course, such that the positive end of a dipole on the first molecule will create a small accumulation of negative charge pointing towards it on the second molecule and vice versa. Thus inducing and induced dipoles (or multipoles) will always be situated so that an attraction results. In reality there is no difference between polarizing and polarized molecule, both act in both ways at the same time. It can be proved that this situation does not lead to a breaking up of the atoms or molecules by increasing the mutual polarization until the molecule becomes labile. At least in general this is not the case, though one could argue that just this sort of thing happens when an alkali atom yields its electron to a halogen atom. In the great majority of cases no such mutual ionization occurs because the polarizing forces are weak and the resistance to polarization increases at first with the extent to which the molecule has been polarized.

Whenever two molecules (or segments of them) meet in a way that $+-$ comes head-to-tail with $+-$ the dipoles will induce an additional polarization which increases the original moment. Whenever they meet $+-$ against $-+$ or $-+$ against $+-$, that is head-to-head or tail-to-tail the polarization tends to diminish the original moments. This means that the head-to-tail situation which gives rise to attraction attracts with more force than the repulsive force of the symmetrical situation: the overall force on the average between many molecules will be a force of attraction. These are the van der Waals' or dispersion forces.

It is clear that the larger the molecule, the more parts, the more atoms it contains, and each of them contributes to the sum total of van der Waals' forces exerted and suffered by it. The more so, the more electrons the atom contains and the more loosely they are bound: the larger its polarizability. Hence the total force between molecule and its neighbours increases with the size of the molecule and with the number of electrons in an atom. Radon (Em) condenses more easily than helium because it has 86 electrons as against 2 in helium, and because the external electrons of radon are at more than twice the distance from the nucleus of those of helium.

Methane is a gas, pentane a liquid, eicosane a solid in the paraffin series, only because the atoms along the hydrocarbon chains of neighbouring molecules all attract each other and it is the sum total of this attraction that holds the molecules together. Small as a single van der Waals' interaction may be compared with the energy of a real chemical bond, many of them in large molecules may add up to such an extent that a valency bond in a molecule is broken rather than allow neighbouring molecules to separate: the compound will not boil, or sometimes will not even melt without decomposition, that is without rupture of its molecules.

68. THE COVALENT BOND, I

Hydrogen is the simplest element and has only one electron in its system. Its positive ion, the proton, has no electron around its nucleus at all. Compounds of hydrogen may belong to quite different classes of chemical bonding: covalent, ionic, metallic and naturally "hydrogen bonding" also. In the course of making a review of different bond types it will be expedient always to cast a glance at the respective compounds of hydrogen, because of their relative simplicity. We shall begin by considering the covalent bond between two hydrogen atoms.

Two protons may be bound by one or two electrons into a positive molecule ion H_2^+ or the hydrogen molecule H_2 respectively. The former system is simpler, having only one binding electron, but it is not stable because it tends to neutralize its net positive charge by capturing a second electron from its surroundings. However, under the impact of ions and electrons in electrical discharges hydrogen molecules often lose an electron for an instant and the spectrum of the resulting species has been identified and investigated.

The stable probability distribution of this single electron in the field of the two nuclei was amenable to exact quantum mechanical treatment. The two protons which would otherwise repel each other are bound electrostatically by the common electron which is distributed so as to spend nearly all its time between the nuclei (Fig. 118). At the same time, the electron has acquired extra space wherein to move, compared with its positions around a single proton, so that there is also a net gain in zero-point energy. These two factors stabilize the species. The electron has a ground state of lowest energy and other excited states which represent possible solutions of its wave equation, say higher "electronic Chladny figures" of the problem. The energy differences between possible states are absorbed, or emitted, in its spectrum in accordance with the Einstein relation $E = h\nu$. Calculated and experimental frequencies of the spectrum coincide.

This bond of the hydrogen molecule-ion is very interesting because it represents the very rare type of "one-electron" bond between two atoms.

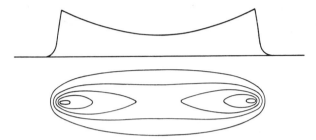

Fig. 118. **The electron distribution function for the hydrogen molecule-ion, H_2^+**

The upper curve shows the value of the function along the line through the two nuclei. The lower figure shows contour lines, increasing from 0·1 for the outermost to 1 at the nuclei.

Generally, at least two electrons combine to form a bond (Fig. 119), increasing thereby the electrostatic attraction they exert on the two atoms they combine and adding the energy of spin compensation to the balance. The energy of the bond H_2^+ is quite high, amounting to 61 kilocalories per mole, that is *more* than half of the energy which holds a hydrogen molecule together. The protons are 1·06 Å apart, which is exactly twice the length of the most probable distance of the electron from the proton in the hydrogen atom. The reason why two protons may be bound together by a single electron while the great majority of other atoms cannot, lies in the fact that there are no electrons shielding the protons from the binding electron in this case and it can therefore attract them near to itself. Owing to the fact that the two nuclei are identical the electron will be, on the average, halfway between them, ensuring a maximal attraction.

The next step in building more complicated systems involves the hydrogen molecule with its two binding electrons between the protons. Just as there were "orbitals" of different symmetry around a nucleus (spherical s orbitals, cigar-shaped p orbitals, etc.), so there are bonding orbitals of different symmetry for electrons between and around positive atomic cores or rump-atoms,* between two protons in the present case.

---

* We use the term "atomic core" or "rump-atom" for whatever remains from an atom on disregarding those of its electrons which serve for bonding.

The simplest orbital in an atom is the spherical *s* orbital. Mathematical solution of the wave equation for two nuclei shows that the simplest orbital around an axis joining two nuclei or two positive atom cores is an elongated figure of rotational symmetry around this axis (i.e. an ellipsoid). The laws of quantum mechanics and the Pauli principle only allow one *s* orbital in each shell of an atom, containing a maximum of two electrons of opposite spin.

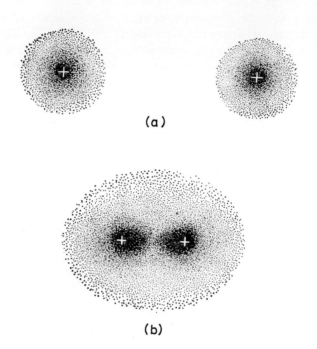

(a)

(b)

Fig. 119. **The electron density in the H$_2$ molecule**

(a) separated atoms; (b) atoms covalently bound in a molecule.

(From G. M. Barrow, *Physical Chemistry*, p. 224, Fig. 9–2 (New York, McGraw-Hill, 1961). By courtesy of the McGraw-Hill Book Co.)

These same laws applied to a molecule only allow a single such "sigma" (σ) orbital in the ground state of a bond, again with a maximum of two electrons of opposite spin occupying it. Thus there is still room for a second electron, between two protons, in this type of orbital in the H$_2^+$ ion: the result of course is the H$_2$ molecule. The bond energy between the two hydrogen atoms in the molecule, that is, its energy of dissociation into two atoms is 102·6 kcal/mole. It is again the sum total of the electrostatic attraction between nuclei and the electrons which are mostly between them (Fig. 119), the spin-coupling energy, and the delocalization energy of the electrons which are now able to extend their sphere of action around the two protons into an increased volume. However, the two electrons also *repel* one another. Two electrons attract the two protons more strongly than the single one in the

H_2^+ ion: the distance between the protons decreases from 1·06 Å to only 0·74 Å in the H_2 molecule.*

In the course of the (indeterminate) motion of the bonding electrons around the nuclei, it may sometimes happen that both electrons are very close to one nucleus, leaving the other practically bare. If this state of affairs were to last any length of time we should be dealing with a positive and a negative hydrogen ion, one beside the other. The electrostatic force of this arrangement also contributes to the sum total of binding energy but only to the extent of a few per cent because the configuration is improbable. It will occur only seldom, the two electrons thus near to each other exerting mutual repulsion.

This H—H bond in the covalent hydrogen molecule may well be regarded as the prototype of the simple covalent chemical bond between any two atoms. For example the covalent bond acts between two neighbouring carbon atoms, and between carbon and hydrogen atoms in organic compounds. The reason why carbon is mostly tetravalent, is that it possesses four electrons in its outer (L) shell, and by accepting, e.g., four further electrons from four hydrogen atoms it is able to form four single, covalent bonds of two electrons each. The eight electrons complete a "rare gas" shell (*see* Fig. 101).

The fundamental difference between compounds which are held together by covalent bonds and those which owe their existence to attraction of oppositely charged ions lies in the fact that once the possible number of covalent bonds is saturated, no further force remains to add other parts to the molecule. If the hydrogen molecule consisted to any considerable extent of the ionic type we have just considered, H^+H^-, it should be able to add an opposite dipole to form the quadrupole

$$H^+H^-$$
$$H^-H^+$$

and so on in the same manner as on p. 294 for NaCl.

---

* In most "popular" or quasi-popular writings on the H—H bond this situation is presented in an unduly complicated form. Many readers may have seen this presentation already and will feel insecure about it. Therefore it seems worthwhile to deal with it further.

It is generally stated, that two hydrogen atoms repel each other if they approach with parallel spinning electrons. This is all right and is only a special case of the Pauli exclusion principle: they would be unable to enter the same deepest state ("orbital") with their spins being parallel and the energy to reach the next higher state would already be excessive. After this, an approach with anti-parallel electrons is tried and we are told to our deepest regret that even this does not help: they are going to repel each other. We know however, that actually they do not. Then comes the veritable *deus ex machina* and we learn to our great joy that a mathematical miracle is in store, and though no single solution of the wave equation is adequate to explain attraction, a permitted linear combination of the two solutions happens to have lower energy and so saves the situation.

What actually happened was that the mathematical approximation towards the solution with lowest energy content was rather complicated. Bringing two H atoms nearer and nearer to each other without deforming the distribution of the electrons around the nuclei is evidently no good as a physical approximation of the facts. The electrons gradually enter the sphere of attraction of the opposite nucleus, become electrostatically attracted to it too and at the same time expand their sphere of action and win energy of delocalization. All this is only taken account of when the two primitive mathematical solutions are combined. But this is no physical complication. Surely no resonance!

But the hydrogen molecule does not do this, neither do the molecules O_2, N_2, Cl_2 and the other halogens or CO, CO_2, NO, etc., nor the organic compounds methane (CH_4), methyl chloride (CH_3Cl), etc. Covalently bound molecules are firmly built but neighbouring molecules only attract each other with much weaker forces. In ionic compounds it is often impossible to tell which pair of ions belong together. The electrostatic force may well be equal between any adjacent pair and holds the whole crystal together with undiminished force. Ionic crystals are mostly very hard and do not vaporize easily. They only dissolve if they are brought into contact with a solvent of high dielectric constant so that the electrostatic force between dissolved ions decreases. (In other words: the dissolved charged ions become surrounded

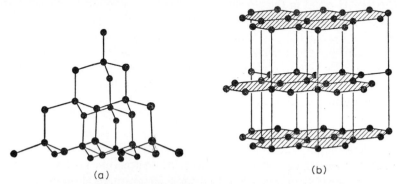

(a)   (b)

Fig. 120. **The crystal structure of diamond and graphite**

(*a*) Diamond; (*b*) graphite.

(From W. Hume-Rothery and G. V. Raynor, *The Structure of Metals and Alloys*, p. 53, Figs. 27, 28 (London, Institute of Metals, 1962). By courtesy of the Institute of Metals.)

by dipolar solvent molecules.) Covalent solid molecules however, vaporize more easily and dissolve in solvents which are relatively similar to them, especially in polar nature (or the lack of it), but which need not possess a high dielectric constant. No electric charges are separated during their dissolution.

The situation changes entirely if the series of covalent bonds occur in chains, in planes or in an infinite three-dimensional structure. Thus carbon atoms in diamond may interlock with each other indefinitely along their four tetrahedral directions, forming an infinite number of covalent bonds (Fig. 120) with each other. Alternating atoms of C and Si do the same in carborundum. In both cases the resultant structure may be regarded as a single molecule, and since the forces, the covalent forces, which hold it together are strong, they are very hard molecule-crystals indeed. In the alternative structure of elementary carbon, in graphite, an infinite series of aromatic rings is built, one beside the other but there remains no covalent bond to bind these molecular layers into stacks. The result is a much longer C—C distance in the "stack" direction, and the constituent planes slip very easily over each other. Because of this loose bonding in one direction graphite breaks up easily into lamellae and is therefore soft (Fig. 120). Colloidal graphite (also MoS_2 with an analogous structure) is a well-known lubricant.

Those plastic materials which possess a covalently bound, three-dimensional structure such as bakelite or "cross-linked" polystyrene, are also insoluble, they cannot be melted without decomposition and are relatively hard because each piece of plastic is a covalently linked macromolecule and covalent bonds have to be broken if it is mechanically damaged.

69. THE COVALENT BOND, II: MULTIPLE BONDS

So much for single bonds. But what if two atoms are bound by multiple bonds, as we have learned they may be (p. 82). In ethylene,

we have twice four electrons of the carbon atoms and four single electrons of the hydrogen atoms to form the bonds. Eight of these 12 are built into the four C—H bonds. At least two others must bind the two carbon atoms to each other. What about the remaining two? They could either remain unused on their parent carbon atoms or combine into a second bond between them. Generally in such cases they combine. Though, as we have mentioned already, no more than two spin-coupled electrons can be placed into a "sigma" bond, there are other, less symmetrical solutions of the wave equation for electrons between two atomic cores. The next type of solution—as far as energy is concerned—is termed "pi" (π) bonding and is not axially symmetrical; it has only a plane of symmetry which contains the bond axis (Fig. 121). This is the probability distribution which is available for the next pair of electrons after the first pair has occupied the axially symmetric sigma states. This means that there is a maximum of probability of finding these "pi" electrons somewhere above or below a plane but not in the plane itself, just as the p electrons were to be found in two directions around a straight line across the nucleus but never in the nucleus itself (Fig. 100). Quite inadvertently we have found a situation which seems to account for the mechanical rigidity of double bonds. A sigma bond with its rotational symmetry remains the same whether we twist the two parts of a molecule relative to each other or not. The second electron pair of a double bond, however, is fixed with respect to a plane. It resists distortion by rotation of the two halves relative each other and hence introduces rigidity into the double bond.

Simple bonds between two carbon atoms are found by X-ray analysis to be 1·57 Å long, with very small deviations from this value. Double bonds are formed by two pairs of electrons and their average length is only 1·33 Å. That of triple bonds is only 1·20 Å. The bond energies, as measured by the amounts of heat liberated by total combustion, increase from single to triple bond just as the bonding energy of the hydrogen molecule was greater than that of the H_2^+ molecule-ion.

In benzene, however, we encounter a structure in which three out of six bonds should be double according to the Kekulé formula, but where actually the six C—C distances are equal, 1·39 Å each. After what has been said on p. 86 about the complete absence of ortho-isomers among substituted

π— bond overlap

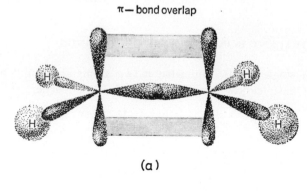

(a)

π bond between the two p_z orbitals

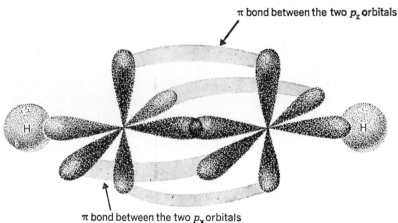

π bond between the two p_x orbitals

(b)

Fig. 121. **Electronic structure of double and triple bonds:**
ethylene and acetylene

The bonds in the middle, in axial contact with each other are the sigma (σ) bonds, those parallel to each other, connected in the— distorted—figure by grey strips are the pi (π) bonds. Actually the parallel p-orbitals over-lap in the pi bond; the figure is distorted in the horizontal direction to make it clearer.

(From G. M. Barrow, *Physical Chemistry*, p. 234, Fig. 9–8; p. 235, Fig. 9–9 (New York, McGraw-Hill, 1961). By courtesy of the McGraw-Hill Book Co.)

benzenes, we are not very much astonished that the Kekulé formula with fixed double bonds forsakes us once more. Purely chemical reasoning has brought us to accept a structure of hexagonal symmetry; now we are able to understand how this comes about. Each of the six carbon atoms in benzene is bound by a sigma bond to its two neighbours and by another sigma bond to a hydrogen atom. This disposes of three electrons from each carbon atom; a fourth one remains. Following the idea of *Heitler* we imagine these six electrons spin-paired into π states, occupying the space on both sides of the ring, like the π electrons in ordinary double bonds. However, unlike the case of ordinary double bonds we do not imagine the electrons linked between particular pairs of carbon atoms but allow them free, indeterminate motion above and below all six carbon atoms of the ring. This involves more delocalization than we have encountered up to now and therefore a very considerable decrease in momentum, that is in kinetic zero-point energy. The structure of benzene and of similar "aromatic" structures such as naphthalene and pyridine becomes very stable. As the jargon has it there is "stabilization by resonance."

A similar delocalization occurs if double bonds are near to each other in a chain of atoms. It has long been known to organic chemists that two or more double bonds, each separated by a single bond, behave in a remarkable way. On addition of, say, a molecule of bromine (Br_2) to such an unsaturated chain it adds to the two ends of the chain instead of adding to any specified double bond:

$$CH_2\!\!=\!\!CH\!-\!CH\!\!=\!\!CH\!-\ \cdots\ =\!\!CH\!-\!CH\!\!=\!\!CH_2 + Br_2$$

$$\rightarrow CH_2Br\!-\!CH\!\!=\!\!CH\!-\!CH\!\!=\ \cdots\ -\!CH\!\!=\!\!CH\!-\!CH_2Br \qquad (69.1)$$

The whole set of double bonds changes position during this simple addition so that they finally occur between pairs of carbon atoms which were originally bound by single bonds. *Thiele* (1899) recommended for this reason that one should represent the whole system of such "conjugated" double bonds by a set of "partial" valencies, abolishing the distinction between single and double bonds and substituting them by something like $1\frac{1}{2}$ bonds:

$$CH_2\!\cdots\!CH\!\cdots\!CH\!\cdots\!CH\!\cdots\ \cdots\ \cdots\!CH\!\cdots\!CH\!\cdots\!CH_2 \qquad . \quad (69.2)$$

Later on, when bond distances were measured, it was found that in fact the single-bond distance in conjugated systems was shorter and the double-bond distance longer than the "pure" values. We now assume that all this is due to the fact that the π electrons of adjacent carbon atoms in C—C "π" bonds are easily delocalized along the chain. Owing to the fact that the six carbon atoms of a C_6 conjugated chain may be arranged with very little straining of the tetrahedral angle, into the very symmetrical hexagon, this delocalization becomes complete for benzene.

The stabilization through delocalization ("resonance") is, of course, not confined to organic compounds. On p. 56 we have become acquainted with the fact that the four oxygen atoms in the sulphate group, SO_4^{2-}, are equivalent, in spite of the fact that we are tempted to write a structural formula for Na_2SO_4 with valency bonds:

$$
\begin{array}{ccc}
Na\!-\!O & & O \\
\diagdown & & \diagup \\
 & S & \\
\diagup & & \diagdown \\
Na\!-\!O & & O
\end{array}
$$

It is the same temptation as to write benzene according to Kekulé's formula. We must not succumb to it in either case because there is no chemical means of discriminating between the four oxygen or the six carbon atoms respectively. We therefore write instead the "complex" formula,

$$Na^+ \begin{bmatrix} O & & O \\ & S & \\ O & & O \end{bmatrix}^{--} \quad Na^+$$

The two electrons which belonged to the sodium atom are in the complex anion but they cannot be distinguished from the others. They are part of a delocalized electron system between the central sulphur and the four oxygen atoms. Energy, that is stability, has been acquired by this process:

$$\begin{array}{ccc} \overset{\cdot\cdot}{\underset{\cdot\cdot}{O}} & & \overset{\cdot\cdot}{\underset{\cdot\cdot}{O}} \\ & S & \\ \overset{\cdot\cdot}{\underset{\cdot\cdot}{O}} & & \overset{\cdot\cdot}{\underset{\cdot\cdot}{O}} \end{array}$$

The same holds for CO_3^{--}, NO_3^-, ClO_4^-, etc., ions.

70. THE COVALENT BOND, III: ELECTRONIC FORMULAE

We stated in the preceding section that surplus electrons generally combine along a bond. This is not always the case. First of all there is a general limiting condition concerning the number of electrons which may be arranged into the valency shell of each atom. This number may not exceed the number which is the limit in the atom-shell itself.

This means that hydrogen cannot enter into true covalent bonding with more than two electrons around its nucleus, because this is the limiting number of electrons for the K-shell (He atom). Accordingly hydrogen is always mono-covalent with one pair of bonding electrons around it. (The hydrogen bond is not accepted as a covalent bond, though it seems to be near to one.)

Nitrogen may add three hydrogen atoms, each with an electron, to its existing complement of five electrons in the L shell in order to arrive at the L shell maximum of eight. Therefore in ammonia the L shell around the nitrogen atom is occupied by 3×2 electrons forming the N—H simple bonds, and two of its original electrons remain unbonded around it. Following *Lewis* we often write covalent formulae by representing each electron in the valency shells by a dot:

$$H:H \qquad \overset{\displaystyle H}{\underset{\displaystyle H}{:N:H}}$$

Sometimes two spin-coupled electrons are drawn together in the form of one line instead of two dots:

$$\text{H—H} \qquad \begin{array}{c} \text{H} \\ | \\ |\,\text{N—H} \\ | \\ \text{H} \end{array}$$

resembling very much the original valency-bond formulae of the chemists, but differing in that each unbonded pair in the valency shell receives a separate line also, whereas a single electron in the valency shell remains represented by a dot:

$$|\,\overset{.}{\text{N}}\!\!=\!\!\text{O}\diagdown$$

Molecules of this type, with electrons the spin of which has not been compensated by a partner, are accordingly paramagnetic, are generally very reactive and easily combine with similar partners, e.g. to form

$$|\,\overline{\text{Cl}}\text{—}\overline{\text{N}}\!\!=\!\!\text{O}\diagdown$$

where, in a happy ending, each atom receives its maximum complement of eight electrons around it.

The nitrogen molecule retains its triple-bond character in its formula with three bonding pairs, and one single pair (the lone pair) on each nitrogen atom

$$:\text{N}\vdots\text{N}: \qquad \text{or} \qquad |\,\overline{\text{N—N}}\,|$$

There is something anomalous, however, about the oxygen molecule. It is paramagnetic, that is its molecule is itself a small elementary magnet. This could not be the case if it had the regular double-bond structure with eight electrons around each atom:

$$\overset{..}{\underset{..}{\text{O}}}\vdots\overset{..}{\underset{..}{\text{O}}}$$

which would well account for all of its L electrons. It must have electrons in its molecule without their spins being paired, in so-called "three-electron bonds":

$$:\text{O}\overset{..}{\underset{..}{:}}\text{O}:$$

Spectroscopic evidence corroborates this assumption. But why this state of uncoupled spins and incomplete valency shells is more stable than the expected regular structure would have been, cannot be explained in simple terms. It is a very special consequence of the structure of the oxygen atom,

Halogens form molecules according to the normal octet scheme

$$:\overset{\cdot\cdot}{\underset{\cdot\cdot}{F}}:\overset{\cdot\cdot}{\underset{\cdot\cdot}{F}}:$$

Written with lines instead of dots this is again very near to the old valency-bond formula:

$$|\,\overline{F}\!\!-\!\!\overline{F}\,|$$

The higher halogens would be capable of arranging more than eight electrons in their M, N, etc., shells but we have already seen during the build-up of the periodic system that after completion of an octet shell of a rare gas the next electron always arranges itself in the next higher shell while the remaining subshells are filled up only later on. The same applies here: simple covalent compounds are not built with more than eight electrons in the valency shell. The Cl_2 and other halogen molecules are built like F_2.

71. THE COVALENT BOND, IV: THE DATIVE BOND

Up to now we have only considered covalent bonds of which the bonding electrons belonged originally to both partners. This is not always so. The free, unbonded electron pairs of atoms in a molecule may very well be used to form a new covalent bond by themselves. Recall as one of the simplest cases the ammonium compounds (Section 16). An ammonia molecule is able to add a hydrogen ion—a proton—and yield the very stable ammonium ion, NH_4^+. What is this process like when written in Lewis formulae?

$$\begin{matrix} H \\ :\overset{\cdot\cdot}{N}:H \\ H \end{matrix} + H:\overset{\cdot\cdot}{\underset{\cdot\cdot}{Cl}}: = \left[\begin{matrix} H \\ H:\overset{\cdot\cdot}{N}:H \\ H \end{matrix} \right]^+ + \left[:\overset{\cdot\cdot}{\underset{\cdot\cdot}{Cl}}: \right]^- \qquad (71.1)$$

The resulting structures are more symmetrical, the nitrogen atom being surrounded only with N—H bonds and the chlorine atom only with electrons.

But the newly-attached proton did not bring its electron into this new household: it left it behind with the chlorine ion. The electron pair which was able to bind this proton was "donated" entirely by the nitrogen atom. Hence the term "dative" for this kind of bonding.

The free proton does not exist in aqueous solutions. It is attached to a free electron pair of a water molecule in the same manner:

$$\begin{matrix} H \\ H:\overset{\cdot\cdot}{\underset{\cdot\cdot}{O}}: \end{matrix} + H:\overset{\cdot\cdot}{\underset{\cdot\cdot}{Cl}}: = \left[\begin{matrix} H \\ H:\overset{\cdot\cdot}{O}:H \end{matrix} \right]^+ + \left[:\overset{\cdot\cdot}{\underset{\cdot\cdot}{Cl}}: \right]^- \qquad (71.2)$$

The crystal structure of ammonium perchlorate and of the monohydrate of perchloric acid are nearly the same, which shows that the monohydrate is practically the perchlorate of the OH_3^+ "oxonium" ion:

$$(OH_3)^+ \,(ClO_4)^- \text{ and } (NH_4)^+ \,(ClO_4)^-$$

The energy with which ammonium and oxonium ions are formed is large and this accounts for the enormous heats of dilution when concentrated perchloric or sulphuric acids are diluted with their first molecules of water.

But the dative bond is by no means restricted to compounds of the proton. Any compound which lacks an electron pair to make up the number of electrons of its rare-gas shell, may combine with another compound which has a free pair at its disposal. For instance the "Lewis acid" boron trifluoride (BF_3) readily combines with ammonia

$$
\begin{array}{ccc}
:\overset{\cdot\cdot}{F}: & H & :\overset{\cdot\cdot}{F}:H \\
:\overset{\cdot\cdot}{F}:\overset{\cdot\cdot}{B} + :\overset{\cdot\cdot}{N}:H = :\overset{\cdot\cdot}{F}:\overset{\cdot\cdot}{B}:\overset{\cdot\cdot}{N}:H \\
:\overset{\cdot\cdot}{F}: & H & :\overset{\cdot\cdot}{F}:H
\end{array}
\qquad (71.3)
$$

or with water, or with ether, where the oxygen atom also carries two free electron pairs

$$
\begin{array}{c}
R:\overset{\cdot\cdot}{\underset{\cdot\cdot}{O}}: \\
R
\end{array}
$$

(R represents a monovalent organic group).

The dative bond is always electrically polarized to some extent, because the electron pair which originally belonged wholly to its own parent atom is now shared, that is displaced, towards the receptor atom, making the latter somewhat negative. On the other hand the donating atom has lost a part of its negative surroundings and becomes positively polarized. This polarizing effect plays a great role in making reactions possible which would not take place without it and this is the reason why electron-donating compounds or groups, as well as electron-accepting ones, very often act as catalysts in chemical reactions.

Which is the simplest electron acceptor in chemistry? Evidently the bare proton itself, the H^+ ion. Every chemist knows how many reactions are brought about or speeded up by the mere presence of acids, of hydrogen ions. This is what we term acid catalysis and, beginning with the hydrolysis of esters and compounds of the sugars, there is no end of the reactions it helps to accomplish.

Another long series of mostly organic reactions is catalysed by the mere presence of electron-accepter Lewis acids, BF_3, $AlCl_3$ etc.:

$$
\begin{array}{cc}
:\overset{\cdot\cdot}{F}: & :\overset{\cdot\cdot}{Cl}: \\
:\overset{\cdot\cdot}{F}:\overset{\cdot\cdot}{B} & :\overset{\cdot\cdot}{Cl}:\overset{\cdot\cdot}{Al} \\
:\overset{\cdot\cdot}{F}: & :\overset{\cdot\cdot}{Cl}:
\end{array}
$$

These attach themselves to free electron pairs and render the molecule electrically vulnerable by polarizing it.

$$
C_6H_5-CO-Cl + C_6H_6 \xrightarrow{AlCl_3} C_6H_5-CO-C_6H_5 + HCl \qquad (71.4)
$$

The simplest assumption concerning the intermediate is:

$$(C_6H_5\!-\!CO)^+ \left(Cl \mid \overset{\displaystyle Cl}{\underset{\displaystyle Cl}{Al}} \mid Cl \right)^-$$

Also the presence of groups or compounds with free electron pairs has often strong catalytic action. Many reactions are known to take place only in the presence of at least traces of water, as for example the oxidation of carbon monoxide to carbon dioxide by oxygen gas or the oxidation of sodium metal by oxygen gas.

Ammonia and even more so the tertiary organic bases, e.g. $(CH_3)_3N$ and

 are well known catalysts for a series of reactions, e.g.:

$$C_6H_6 + SO_2Cl_2 \xrightarrow[\text{amine}]{\text{tertiary}} C_6H_5\!-\!SO_2\!-\!Cl + HCl \qquad . \quad (70.5)$$

Reagents which have free electron pairs at their disposal are capable of adding a proton, among other electron-lacking species. The proton being "the bare nucleus" these reagents are often termed "nucleophilic". They comprise H_2O, NH_3, OH^-, Cl^-, etc. On the other hand groups of compounds with fewer electrons than they are capable of binding in their sphere of valency are termed "electrophilic". For instance, H^+, BF_3, $AlCl_3$, etc. It is open to discussion whether it was a good idea to extend these group names to reactions. However, it seems to be too late to worry about it now because they have gained international acceptance. A nucleophilic attack on a molecule means, for example, an attack by a hydroxyl ion or a water molecule, while the attack by a proton is called electrophilic. There is no choice at present other than to remember their meanings.

72. THE COVALENT BOND, V: BORON HYDRIDES

On extrapolating the hydride series HF, H_2O, H_3N, H_4C to include the boron compound one would expect a compound with the composition H_5B. However, no such compound exists, the simplest "borane"—as a boron hydride is called—being the compound B_2H_6. Some even more complicated compounds, corresponding to the two series B_nH_{n+4} and $B_{n+4}H_{n+10}$ respectively are known, all of them being very unstable and prone to decomposition.

The fact that BH_5 does not exist shows that these hydrides are not ionic compounds, built by the electron of hydrogen going over to the central boron atom. Such a structure, with a complement of eight electrons in the L shell of a (-5)-charged boron ion, would be conceivable. However, the accumulation of such a high charge upon a single atom does not seem to be possible owing to the mutual repulsion of electrons.

Regular σ bonds with five hydrogen atoms cannot be arranged around a boron atom, first because there are only eight $(3 + 5)$ available electrons instead of ten for five bonds, and secondly, because even if there were ten

electrons, there is no place for five such bonds around the boron atom in its
L shell. This seems to be the reason for its very extraordinary set of hydrides.
They would not exist at all, were the boron atom not so small that they can be
held together by an average number of electrons per bond which is less than
two. The elements in the periodic table to the left and below boron, namely
beryllium and aluminium, combine with hydrogen to form the simple hydrides
BeH_2 and AlH_3. These compounds are salt-like and probably contain the
negative hydride ion (H^-). We learned to know it in lithium hydride (p.
270).

It is extremely interesting, that no compound BH_3 is formed, analogous to
the halides BF_3, BCl_3, etc. The number of electrons would be exactly sufficient
for a σ bond between the boron and each hydrogen atom. But in this very
special case there seems to remain an unsaturation, a force which is capable
of binding two such BH_3 units into B_2H_6. The only solution we can postulate
is a sort of delocalization of the whole electron system. By spreading out the
12 available bond-electrons over the 7 bonds of B_2H_6, the gain in delocalization energy seems to outweigh the fact that there is only 6/7ths of an electron
pair per bond. One-electron bonds are not present because the molecule is not
paramagnetic: all electron spins must be compensated. The problem is
not yet definitely solved. Lewis postulates that the six electron pairs occupy
alternately the seven bonds in quick succession ("resonate" between the seven
so-called mesomeric structures) with an atom-pair, (or rather ion-pair),
unbonded for a short instant each time:

$$
\begin{array}{ccc}
\text{H H} & & \text{H H} \\
\cdot\cdot\ \cdot\cdot & & \cdot\cdot\ \cdot\cdot \\
\text{H : B   B : H} & & \text{H B : B : H} \qquad \text{etc.} \\
\cdot\cdot\ \cdot\cdot & & \cdot\cdot\ \cdot\cdot \\
\text{H H} & & \text{H H}
\end{array}
$$

Others postulate a double bond between the boron atoms with inserted
protons in between. This is quite unprecedented also. Nuclear magnetic
resonance (p. 287) has proved that two hydrogen atoms are bound otherwise
than the other four.

The whole situation is saved, however, the moment a donor molecule
appears to present the missing pair of electrons in a dative bond. The compounds $H_3B:N(CH_3)_3$ or $H_3B:CO$ are quite stable. Two alkali atoms may
contribute these two electrons also, giving rise to an ionic compound, e.g.
$Na_2^{2+}(B_2H_6)_2^-$ which is also stable.

The hypothetical structures of the higher boranes would lead us too far
into a very specialized chapter of inorganic chemistry. Diffractional structure
determinations have revealed the existence of B_4 rings and BH_3 groups in
these compounds, but once more there are fewer electrons available than are
necessary to form sigma bonds all round. It is the same problem as with
B_2H_6 but on a more complicated scale. All these compounds are labile,
probably because of the lack of bonding electron pairs. We have only discussed this group of compounds because they represent the weakest point in
the electronic theory of the chemical bond, but they still leave a way out of
the impasse which does not seem to be improbable.

73. THE COVALENT BOND, VI: COMPLEX COMPOUNDS

In Sections 16 and 20 we have spoken about such relatively simple groups as the NH_4^+ or SO_4^{--} ion as complexes, conforming to the definition of Werner. In the preceding few sections we have gained insight into their electronic, covalent structure and the forces holding them together.

However, another group of compounds is termed complex in a more specialized way: these are compounds in which ions and/or molecules are bonded around elements of the transition series. We shall become acquainted with them now.

When the central metal ion in question is coloured, its colour often changes to a great extent according to the groups which surround it. Its magnetic properties also often change considerably. All this points to a strong interaction of the complexing groups with the electronic structure of the central ion. This was no doubt one of the main reasons why chemists occupied themselves intensively with the syntheses of such complex compounds long before A. Werner managed to arrange them into a clearcut system, and the modern theory of the chemical bond gave a satisfactory explanation of their inner structures.

Many compounds crystallize by including a number of solvent molecules in their crystal structure. Sodium carbonate crystallizes with ten, seven or one molecules of water, sodium sulphate with ten or seven or without any at all. These water molecules are partly bonded around the ions, partly into the structure between them. We have good reason to suppose that even in solution there is a solvation envelope around ions, especially if they are small and highly charged so that the electric field around them is strong. (The Li^+ ion which is small moves more slowly in aqueous solution in an electric field than does the K^+ ion which is much larger. Probably this is because Li^+ is able to attract a greater number of water molecules around itself.) But these molecules of "water of crystallization" or "solvent of crystallization" are loosely bound and are exchanged with other solvent molecules in solution, a fact that is easily demonstrated today by using isotopically or radioactively labelled solvent molecules (p. 416).

There is another kind of water of crystallization, however, which belongs much more firmly to its central ion and represents only a special case of some group attached to the ion. Trivalent chromium for example has six water molecules in each of its chloride molecules. By heating a chromi-chloride solution, its original violet colour turns green and another compound can be isolated which contains one water molecule less. But this is not the whole story. Whereas all three Cl^- ions of the violet salt are easily precipitated with the classical chloride reagent, the silver ion, as insoluble silver chloride, only two of them precipitate from the green salt. Some very strong bond must have been formed which has "immobilized" the third ion. The missing ion can certainly not have remained as a free chloride ion, and in fact does not. This can be shown by ascertaining the number of independent particles in the solution by Raoult's method of freezing-point depression (p. 47). The freezing-point depression of the green solution is only three-quarters of the original depression of the violet solution: from the four separate entities containing the chromium and the three chloride ions, one has disappeared. At the same time the electrolytic conductance of the solution decreases by rather more than the same factor: the number of ions has therefore decreased. Werner explained the phenomenon by ascribing the six water molecules to a firmly-bound complex around the central Cr^{+++} ion and postulated that the violet-to-green transition meant the exchange of one of these molecules of water for a chloride ion. Writing the complex part of the molecule in brackets we have

$$[Cr(H_2O)_6]^{+++} + 3Cl^- = [Cr(H_2O)_5Cl]^{++} + 2Cl^- + H_2O \qquad (73.1)$$

One chloride ion has lost its freedom. The number of separate entities has decreased from four to three, and not only has the total number of ions decreased but the charge upon the central ion has decreased too: a reason for its moving more slowly in an electric field.

The process goes further on continued heating and another green compound appears with only a single *free* chloride per molecule. The depression of the freezing-point and the lowering of the conductivity decreases further and only four molecules of H_2O remain in the crystal. In the series of the $[Pt(NH_3)_6]Cl_4$ to $K_2[PtCl_6]$ complex compounds all seven possible combinations have been prepared from the hexa-ammino to the hexachloro compound. The tetrachloro-diammino compound $[Pt(NH_3)_2Cl_4]$ is remarkable for being electrically neutral: the four monovalent anions serve to neutralize the charge on the tetravalent Pt^{++++} ion *within* the complexing first sphere around it. They are bound firmly and there are no charged particles left to move in an electric field. From here on, beginning with the ammino-pentachloro compound there is more negative than positive charge within the complex. It has become an anion and moves towards the positive pole, the anode. Of course the odd anion within the complex must have come from somewhere. It might come for instance from its alkali salt according to the equation:

$$[Pt(NH_3)_2Cl_4] + KCl = K^+[Pt(NH_3)Cl_5]^- + NH_3 \qquad . \qquad (73.2)$$

and the alkali ion remains as cationic partner to the now anionic complex. It may be exchanged for a hydrogen ion to yield the complex acid, $H_2[PtCl_6]$ from $K_2[PtCl_6]$. Fig. 122 shows this series together with a similar series from among the Co^{+++} complex compounds.

Throughout the series of transformations the number of ligands around the Cr^{+++} ion has remained six. There is a fixed "coordination number" for every central ion, nearly independent of the nature of the coordinatively complex-bound ligands. It is not necessary that *water* is exchanged for an ion. Molecules such as NH_3, H_2O_2, organic amines, etc., and anions may take its place. A very well-known example of an ammine complex is the deep blue cuprammonium $[Cu(NH_3)_4]^{++}$ ion which results from the addition of ammonia to a blue aqueous solution of a divalent copper salt. Ammonia has here taken the place of the four water molecules which are responsible for the blue colour of cupric salts: cupric sulphate crystallizes with five

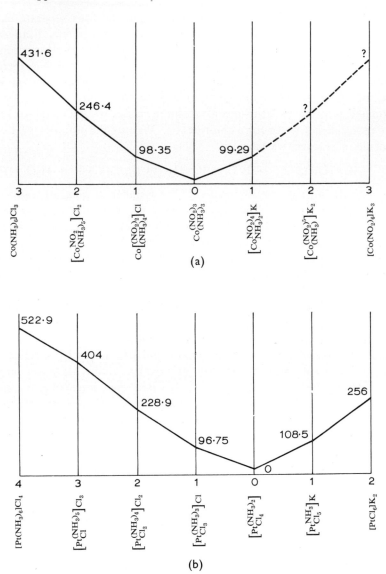

Fig. 122. **Molecular conductivities of complex solutions**

(*a*) A cobalti-complex series.
(*b*) A platini-complex series.

The ordinate represents the electrical conductivity of equimolar solutions; it is roughly proportional to the number of separate ions in the solution.

(From A. Werner and P. Pfeiffer, *Neuere Auschauungen auf dem Gebiete der anorganischen Chemie*, Sammlung "Wissenschaft", Vol. 8, pp. 226, 227 (Brunswick, Vieweg, 1923). By courtesy of F. Vieweg and Sohn Verlag.)

molecules of water of crystallization of which four may be readily removed. The resulting monohydrate is white! (It attracts four water molecules vigorously and may therefore by used to dehydrate, for example, 96 per cent alcohol.)

The coordination number around a central ion is nearly always one which corresponds to the number of corners on a regular polyhedron: 2, or 3 (representing a line and a triangle respectively, as embryonic polyhedra); 4, 6, 8 for tetrahedron, hexahedron and octahedron respectively. One has the impression that ligands must be placed in equivalent positions around the central ion and that only corners of regular bodies are geometrically equivalent. Which of these numbers actually occurs seems to be determined by the relative magnitudes of central ion and ligand and by the charge upon the former. More ligands will be bound if there is more space around an ion, or if its electrostatic field is stronger.

The force which is responsible for this complex formation is twofold. First, it is the strong electrostatic field around these central ions with small ionic radii. Molecules of water, and ammonia, which are known to be built asymmetrically and to possess electric dipoles (Section 62) are attracted through their negatively charged ends. Negative ions are attracted because they *are* negative. All ligands also become polarized to some extent under the influence of the central field and thus either acquire a dipole moment even if they had none initially or else increase it if they did possess one.

But this is not the whole story. There are complex compounds with the same molecular magnetic moment as that of the central ion. There are others where the moment decreases after complex formation. Knowing that the magnetic moment of the transition-metal ions is due to the incomplete shell of d or f electrons (Section 61) where the magnetic moments of the electrons are not yet compensated, it seems probable that the ligands contribute electrons to this incomplete system. There is dative bond formation with ligand electron-pairs entering the incomplete subshell of the central ion. There are cases when the subshell reaches completion by virtue of these ligand electron pairs: then no paramagnetic moment remains. In other cases the subshell is not quite completed and a diminished paramagnetic moment results. Sometimes the repulsion of ligands forces d-electrons so near to each other that their spins become coupled and the resultant magnetic moment decreases (*see* p. 285, Section 61).

These magnetic measurements give us a very deep insight indeed into the bonding within the complexes. As a matter of fact it is even deeper than we have had the opportunity of learning here; there are more detailed characteristics of the types of covalent orbitals involved which may to some extent be read from the experimental results.

74. THE COVALENT BOND, VII: CARBONYLS AND SIMILAR COMPOUNDS

The bonds in these compounds, which belong to one of the strangest groups in inorganic chemistry, are somewhat similar to the covalent, dative complexes we have just encountered. They are well-defined compounds, the best-known among them being nickel tetracarbonyl and iron pentacarbonyl, $Ni(CO)_4$ and $Fe(CO)_5$. Both are formed by reacting the metal with carbon monoxide gas at a suitable elevated temperature and pressure. They are gases or low-boiling liquids and decompose into their components if the temperature is raised further. This is the way in which nickel is "distilled" and purified in the Mond process and also, in which the valuable, extremely fine carbonyl-iron powder for transformer cores is obtained. Iron pentacarbonyl has also been added to motor fuel. It decomposes during the explosion and helps to destroy peroxides in the fuel which could cause "knocking" in the cylinder.

Only metals from the transition groups form carbonyls, metals from the sixth to eighth column of the periodic system. Only the simple nickel and iron carbonyls are formed at normal pressure; for this reason the reaction serves to separate nickel not only from copper but from cobalt too. The influence of high pressure in favour of the general reaction is easily understood if one considers the enormous decrease in volume during its course: four and five molecules, that is volumes, of carbon monoxide for nickel and iron compounds respectively, are reduced to one molecule (volume) of metal carbonyl vapour. Carbonyls of chromium, molybdenum, tungsten and manganese can only be prepared in a complex reaction between an organic magnesium halide reagent (e.g. CH_3MgBr, one of the so-called "Grignard" compounds) and the anhydrous metal halide under CO pressure in ether: the Grignard reagent abstracts the halogen from the metal halide and the nascent atom reacts with carbon monoxide. On the other hand RuI_2 reacts with carbon monoxide, first to yield $Ru(CO)_2I_2$ from which silver metal is able to extract the iodine, and under carbon monoxide pressure the pentacarbonyl is formed.

Before going into the theory of the metal carbonyl bond it is expedient to survey the formulae. Chromium, molybdenum and tungsten from the sixth column form hexacarbonyls, $Cr(CO)_6$, etc. Iron, ruthenium and osmium two places after them in the periodic system form pentacarbonyls, nickel which stands two places after iron has a tetracarbonyl. Besides these simplest members of the group Fe and Ru have carbonyls of the type $M_2(CO)_9$ (M standing for the metals) and there exists a $Fe_3(CO)_{12}$ also. Manganese and rhodium only exist as the type $M_2(CO)_{10}$, cobalt and iridium as $M_2(CO)_8$ and $M_4(CO)_{12}$. It is worth while to note the formulae of some mixed compounds where a part of the carbon monoxide is substituted by another unsaturated gas, nitric oxide (NO), e.g.

$$Co_2(CO)_8 + 2NO = 2Co(CO)_3(NO) + 2CO \qquad . \quad (74.1)$$

$$Fe(CO)_5 + 2NO = Fe(CO)_2(NO)_2 + 3CO \qquad . \quad (74.2)$$

These formulae suggest some correlation between the ratio M:CO and the position of the metal in the periodic system. Metals in the same column have the same type of carbonyl and carbonyl-nitrosyl. Once the fundamental solution of the riddle has been found, it is not so terribly complicated at all.

Following *Sidgwick*, let us assume that each carbon monoxide molecule contributes two of its electrons to the bonds of these compounds, in the same way as did those complexing groups which interfered with the magnetic moments of complex salts, described in the preceding section. Nitrogen has one electron more than carbon, being its neighbour to the right. Hence nitric oxide (NO) has also one electron more than carbon monoxide (CO); it belongs to the rare class of compounds with an odd number of electrons. These are sometimes classified as stable radicals. Nitric oxide is, of course, paramagnetic because the spin of the odd electron has none to compensate it. Let us assume that nitric oxide contributes also its lone electron in addition to its pair, that is three electrons in all. On this assumption it is easy to see that the number of electrons around the metal atom in the monometal carbonyls and carbonyl-nitrosyls is always equal to the number of electrons in the next highest rare gas atom: krypton, xenon or radon. This means for instance, that the eight 3d electrons of nickel are supplemented with two to fill the 3d shell completely and the two electrons in its 4s shell (nickel is divalent!) are supplemented with a further six, giving the complete sub-shell and the external octet of krypton. Altogether we have used eight donated electrons. They have been donated by four carbon monoxide molecules: we have the tetracarbonyl.*

Now it is evident why only the metals with an even number of electrons can form monometal carbonyls: they need an even number of surplus electrons and carbon monoxide is able to supply them with this even number. Atoms with an odd number of electrons, those that belong to the odd columns of the periodic system, cannot solve their problem so easily. But they are able to solve it the moment they associate themselves with nitric oxide molecules which we have supposed contribute their lone electron also, as well as one of their pairs. In $Co(CO)_3(NO)$ the number of electrons around the cobalt atom is again equal to the number in a krypton atom.

The principle should be applicable beyond the limits that we find accomplished *de facto* and we must look for some reason why transition elements below the sixth group do not make use of this reaction. Evidently they would have to combine with more than six ligand molecules; titanium, for instance, needs 14 electrons to become like krypton and this necessitates seven carbon monoxide molecules. Very probably this would already exceed the coordinative capacity, the free, available space around the central atom. On the other hand, the elements to the right of the eighth group, beginning with copper, silver and gold could go on adding carbon monoxide too; cadmium, for example, would only need three molecules to simulate krypton. However, no pure carbonyl has even been prepared from these metals. What is the difference? All these elements, after the eighth column, possess a complete 3d subshell of electrons beneath their external shell which is just being built. This seems to indicate that pure carbonyls are only formed if the d subshell of the metal is incomplete; dative bonding to the external shell alone does

---

* Let us not complicate matters by going into the question of whether the 10 d electrons and the octet are in their original atomic orbitals or in some sort of combined, hybridized orbitals which might be more stable solutions of the extremely complicated wave equation.

not seem to furnish sufficient energy. It is interesting however, that Cu^+ very readily absorbs carbon monoxide molecules giving rise to complex ions. In gas analysis cuprous solutions are used to determine carbon monoxide in gas mixtures by selective absorption.

The more complicated $Fe_2(CO)_9$ is formed from $Fe(CO)_5$ in ultraviolet light and is a yellow solid. Its structure has been determined by electron diffraction experiments and revealed a threefold axis of symmetry which can be best represented by the structure in which each iron atom receives $6 + 3 = 9$ electrons from the carbon monoxide molecules.

The latter are arranged in triangles perpendicular to the Fe—Fe axis. However, Sidgwick's rule demands 10 electrons for an iron atom to become like krypton. One is missing. Now, the distance between the two iron atoms across the triangle of carbon monoxide molecules is only 2·47 Å which is compatible with the idea that an Fe—Fe covalent bond exists in the axis of the molecule. The two iron atoms mutually assist each other to reach the structure of krypton. (The covalent diameter of iron is 2·33 Å.)

What about the carbonyls of the odd-electron elements? Their structure allows a similar explanation, Each cobalt atom in $Co_2(CO)_8$ would need nine additional electrons to arrive at the configuration of krypton. Although the spatial configuration of this molecule is not yet completely resolved, we may suppose that a direct covalent Co—Co bond is at work here too, holding together the two parts of the molecule. However, in this case this would be the only bond between the two halves, each cobalt atom completely binding four molecules of carbon monoxide to itself.

Carbon monoxide may be replaced not only by nitric oxide but also by amines, e.g. ethylenediamine ($H_2N.CH_2CH_2.NH_2$) which has two amino groups with one lone pair of electrons each, like ammonia. Accordingly it may replace in normal cases two carbon monoxide molecules. More complicated compounds probably carry metal–metal covalent bonds. The tertiary, benzene-ring-like amine, pyridine:

has one free electron pair on its nitrogen atom and thus replaces one carbon monoxide molecule.

Even in this very complicated group of inorganic compounds it is worth while to mention those which contain hydrogen. They are the so-called carbonyl hydrides. For instance in alkaline alcoholic solution iron penta-carbonyl reacts thus:

$$Fe(CO)_5 + 2OH^- = Fe(CO)_4H_2 + CO_3^{--} \qquad . \quad (74.3)$$
carbonate ion

A solution of this curious yellow compound may be boiled without decomposition as long as air is absent but it immediately absorbs oxygen and even reduces vat-dyes. It can be oxidized to $Fe_3(CO)_{12}$ by simply losing its hydrogen atoms and linking iron atoms instead.

As the carbonyl hydride is an acid, its hydrogen atoms may be exchanged for metals and it thus yields salts, for instance, $NaH(CO)_4Fe$ and $Cd(CO)_4Fe$. The pure hydride is a volatile yellow liquid, solidifying at $-70°C$ and decomposing above $-10°C$. Its salts are more stable than the acid, probably because the acid may decompose with evolution of the very stable hydrogen molecule.

It is not clear, how the hydrogen atoms are bound in these compounds, though the rule of Sidgwick is mostly valid as for example, in $Fe(CO)_4H_2$, already mentioned. Two hydrogen atoms with one electron each replace a carbon monoxide molecule which contributed two electrons. The proton, which has thus lost its electron is of course acidic, as always. This compound has six ligands around the central atom and six can always be arranged at the corners of an octahedron which is a highly symmetrical regular polyhedron, while the five ligands of the original pentacarbonyl could be at best arranged in the less symmetrical trigonal bipyramid (three ligands at the apices of a triangle, one above, and one below). It may be that the increase in symmetry favours a more stable wave equation.

Halogens, especially iodine, may also enter carbonyl compounds and there is a transition between the more carbonyl-like compounds and those which are rather complex salts with free halogen ions, as was the case with Cu^+. However, we shall not go into detailed inorganic chemistry. The reason why we have concentrated so much attention on this very special group of inorganic compounds is that they represent a very interesting type of covalent bonding. With this group we leave the covalent link altogether to become better acquainted with the other extreme, the ionic bond.

75. THE IONIC OR HETEROPOLAR BOND, I

In purely ionic bonding one kind of atom has lost one or more electrons and these have been built into the surface electron shells of the other atoms. Positive and negative ions result and these are then held together by electrostatic forces. The negative surface shells of electrons keep the ions apart. They would exert repulsion if the ions were to approach to a smaller distance than they actually do in ionic crystals. From the point of view of pure electrostatics the whole system of charges should of course collapse, because this would be the best way to acquire (by the way, an "infinite" amount of) energy. However, we have already seen that the Heisenberg relation keeps the electrons expanded because zero-point energy is gained when they occupy an increased volume of space. The whole problem of why an atom may transfer electrons entirely to another one is determined by the wave characteristics of the atoms in question. In this section we are not inquiring again into this cause, but we accept the fact that in certain cases the phenomenon does occur. We are now only interested in its consequences.

The rebirth of the electrostatic theory of valency some forty years ago at the hands of Kossel was dealt with sufficiently in Section 43. It might well be worth while to reread it at this stage because we have learned a good deal in the meantime. The very interesting argument concerning the acidic and basic properties of hydroxy compounds, as well as the increasing acidity in the row of chlorine oxyacids from hypochlorous to perchloric acid, remains now as impressive as it was originally, though today probably most chemists will tend to consider the bond between oxygen and chlorine as a polarized covalent bond.

Maybe it should be pointed out here that the real difference between the electrostatic and the covalent description of the facts is not so immense as it would seem at first sight. The electrostatic theory supposes that to a first approximation ionization takes place between two atoms of different electron-attracting power ("electronegativity"). The ions which are thus formed by electrons having passed over from atom to atom now proceed to attract each other. Cations are relatively small because (1) they have lost electrons, (2) the remainder of the electrons are attracted more strongly to the nucleus. For opposite reasons, anions are relatively large. During interaction of small cations with a strong electric field at their surface (caused by their decreased radii) and between larger anions, the electronic structure of the anion becomes polarized or deformed, the outer shell being drawn towards the cations at the cost of giving up its spherical symmetry. Thus the electron surplus which the anion received in the course of ionization tends to some extent to be drawn back towards the cation which has originally lost it. It will thus come about that the bond between the small and highly charged cations and the large anions becomes very much deformed compared with bonds between larger cations and small anions. The bond between K^+ and F^- will be much less deformed, much more purely ionic than the bond between, say, Al^{+++} and I^-, or between Cl^{7+} and O^{2-} in perchloric acid.

The covalent view begins at the opposite end. Electrons are primarily arranged in pairs to form covalent bonds between atoms, say between Al^{3+} and I^-, or Cl^{7+} and O^{2-}, but owing to the special nature of the wave functions around these respective atoms they are not symmetrical, with their maximum probability in the middle between the two atoms, but are drawn more or less closer to the negative atom. The result is a cloud of electrons, with its centre of gravity near the negative atom, pointing towards the positive neighbour. This is very much the same situation we arrived at from the electrostatic point of view.

There should be no doubt about it that the quantum mechanical picture is on a higher level. It explains why the whole electrostatic structure does not collapse and why covalent bonds become saturated. It explains the maximum number of covalent bonds which may occur with elements in different horizontal rows in the periodic system. But it is extremely interesting to note how far one can go from the other, from the electrostatic extreme of the picture.

The relevant data needed for the electrostatic theory are the charge and the radii of the ions and their polarizability in an electric field. All three are accessible from independent experiments and fit very well into the picture.

The charge upon an ion may be derived formally, as for example, the charge $+7$ on the chlorine atom in perchloric acid, taking into account that, of the eight negative charges of the four oxygen ions only one is neutralized by the hydrogen ion. But the fine structure of X-ray spectra tells the same story. To a first approximation it is true that electrons in an external shell are farther from the nucleus than inner electrons. But on refining the picture, one sees that there is a finite probability of finding an electron of an external shell near the nucleus—relatively small as this probability may be (Fig. 99). As a consequence of this state of affairs the external electrons repel the inner ones to a certain extent, and these will be bound more loosely than if fewer external electrons were present. Being bound more loosely means that less energy is needed to remove an electron from the influence of the nucleus: it means an X-ray edge of K-shell ionization (Section 64) of somewhat longer wavelength. This is exactly what experiment reveals. The edge of K-shell ionization in the X-ray spectrum of the chlorine atom is more towards the longer wavelengths than in hypochlorites, and goes on moving towards somewhat shorter wavelengths through the series hypochlorites, chlorites, chlorates to perchlorates. On the other hand, the negative chloride ion has one more electron than the atom and this contributes somewhat towards loosening the bond between the innermost electrons and the nucleus: the K-edge of the chloride ion is displaced towards longer wavelengths! Thus the removal of electrons from the atom surface can be observed in the X-ray spectra, it is no mere hypothesis. To what extent an electron has been removed, how far it has been transmitted to the partner atom, is much more difficult to ascertain. But the series from Cl^- to formal Cl^{7+} is continuous.

The radius of ions in crystals can be determined today with great accuracy by X-ray diffraction methods. Not only the position of the centre of atoms can be determined but a whole map of electron densities within the lattice can be constructed. Because the diffraction picture is the sum of diffractions from all space elements within the lattice, and the more light there is diffracted from a space element the higher the electron density and the probability of finding electrons within it. Such an electron map resembles a map of mountain ranges with contour lines around the peaks as in Fig. 61. The boundaries of the individual atoms are clearly visible from these distribution diagrams as the minima of electron densities. The radii of some ions are given in Table IX.

TABLE IX. Radii of Some Ions in Crystals (Å units)

Li$^+$	0·60	Be^{2+}	0·31	B^{3+}	0·20	O^{2-}	1·36	F$^-$	1·36
Na$^+$	0·95	Mg^{2+}	0·65	Al^{3+}	0·50	S^{2-}	1·70	Cl$^-$	1·81
K$^+$	1·33	Ca^{2+}	0·99	Sc^{3+}	0·81	Se^{2-}	1·98	Br$^-$	1·95
Rb$^+$	1·48	Sr^{2+}	1·13	Y^{3+}	0·93	Te^{2-}	2·21	I$^-$	2·16
Cs$^+$	1·69	Ba^{2+}	1·35	La^{3+}	1·15				

Ag$^+$	1·26	Mn^{2+}	0·80	Zn^{2+}	0·74
NH$_4^+$	1·48	Fe^{2+}	0·75	Cd^{2+}	0·94
		Co^{2+}	0·72	Hg^{2+}	1·10

76. THE IONIC BOND, II: POLARIZABILITY

Let us now drop the assumption that ions are electrically and mechanically rigid spheres. Being built from a positive nucleus and a negative cloud of electrons they may become deformed in an electric field, the electrons being attracted and the nucleus repelled by a positive charge, or vice versa. Thus an induced dipole moment arises (*see* Section 62). Cations are smaller and more strongly knit than anions; they are much less deformed than the latter. K$^+$ and Cl$^-$, for instance, both have 18 electrons, but the nucleus of the potassium atom has 19 positive charges while that of chlorine has only 17. Accordingly the former ion has a radius of only 1·3 Å compared with 1·8 Å of the second. Divalent Ca^{2+} has a radius of 1·0 Å; divalent S^{2-} with the same number of electrons, a radius of 1·8 Å. The polarization of cations may therefore be neglected to a first approximation.

The polarizability of ions or atoms is revealed not only in the interaction between adjacent atoms but also in the interaction between the particle and an electric field. If the field is static, the molecules or groups which possess electrical asymmetry also become oriented to some extent and this tends to obscure the deformation of electron shells. But very rapidly alternating electric fields are unable to act upon the heavy, relatively inert, groups while the light electrons are able to catch up with them. Such an alternating electric field is the electromagnetic wave field of light; it alternates with the frequency c/λ, where c is the velocity of light and λ the wavelength. For visible light the frequency is say, 10^{14}–10^{15} vibrations per second, whereas from infra-red spectra we know that vibrational and rotational frequencies of molecules are some orders of magnitude slower. Thus only the electrons resonate to the tune of incident visible light. The heavier nuclei have no time to change their position between the passage of the top and bottom peaks of a wave. The electrons which are brought into forced vibration by the incident light become themselves emitters of electromagnetic waves which interfere with the original incident light and produce resultant waves which

travel at lower speed than the velocity of light *in vacuo.** This is the reason why light changes it velocity in matter and thus is refracted from its original direction if it enters a medium at an angle different from the perpendicular.†
The magnitude of the refractive index is a measure of the interaction with the electrons in matter. The more weakly the electrons are bound, the easier it will be for them to follow the forced vibration, and so the refractive index of the

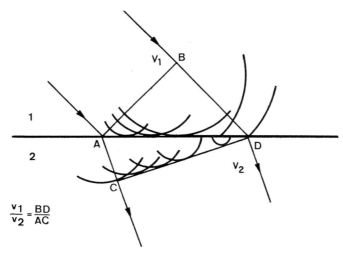

$$\frac{v_1}{v_2} = \frac{BD}{AC}$$

Fig. 123. **Refraction of waves between two media in which their velocity is different; according to Huygens**

material will be higher. Table X shows the refractive index of some atoms and molecules: it increases with their actual dimension. It can further be shown that to a good approximation it increases in proportion to the volume occupied by the particle, by its external electrons. Thus the ease with which the electron shell can be deformed by the electric field of a neighbouring ion can be read from the table of refractive indices and we arrive quantitatively at the result we obtained in a more qualitative manner, that the heavier, larger particles with higher negative charges are more easy to deform.

We have mentioned the determination of polarizability from the refractive index. This is satisfactory for atoms which are equal to each other. But an ionic compound contains cations and anions. How should we apportion the experimentally measured refractive index between them? It is the same problem as occurs with atomic or ionic volumes, susceptibilities, etc. Only very careful comparison of refraction data for ions of elements from both sides of

---

* The difference between "wave" and "group" velocity may be observed on the surface of a pond. Any periodic disturbance generates waves and their velocity may be observed directly. But where two sets of waves interfere areas of piled up and depressed water are formed and the movement of these areas, wave groups, occurs with a velocity which is by no means the velocity of the original waves.

†The principle of Huygens states that the resultant wave-front is the superposition of elementary waves within a bundle. Fig. 123 shows how this results in changing the direction of an incident beam of light in a medium where its velocity has decreased.

TABLE X. The Increment by which the Refractive Index of a Gas is greater than that of a Vacuum (1·000000 by definition) increases with the Size of its Molecule

Gas	Refractive index, n_D, for yellow sodium light	Difference against vacuum: $n_D - 1$	Average radius of the gas molecule
			Å
He Helium	1·000035	0·000035	0·95
H_2 Hydrogen	1·000143	0·000143	1·19
H_2O Water	1·000249	0·000249	1·36
O_2 Oxygen	1·000271	0·000271	1·50
Ar Argon	1·000281	0·000281	1·43
CO_2 Carbon dioxide	1·000448	0·000448	1·66
Xe Xenon	1·000703	0·000703	1·75
Cl_2 Chlorine	1·000773	0·000773	1·85
Br_2 Bromine	1·001132	0·001132	2·02

a rare gas, and refraction data of the rare gas atom itself, allows us to ascribe individual refraction values, i.e. polarizabilities, to individual ions. For example K^+ and Cl^- have the same number (18) of electrons as the rare gas argon between them, Ca^{2+} and S^{2-} are also isoelectronic with argon. The refractive index per KCl or CaS pair is, however greater than that of two argon atoms, because it increases more in the anions than it decreases in the cations. As a matter of fact, these values are not quite constant: the polarizability decreases with increasing polarization so that the small cations with their strong fields polarize their neighbours to a relatively smaller extent than does a larger, less highly charged cation.

If the anion is surrounded by cations, most of the polarizations cancel. This is not so in an individual molecule. The two protons within the electron shell of the oxygen atom in water polarize this shell. In effect: two dipoles are formed opposite the protons within the molecule, the electrons are drawn somewhat towards the protons. *Hund* has shown that the two protons are in labile equilibrium when they are located symmetrically, opposite to each other across a diameter of the molecule. Because any small movement of a proton in a direction perpendicular to the diameter yields two dipole moments at an angle to each other, which add up to a resultant dipole within the molecule according to the law of the parallelogram of forces, Fig. 124. This resultant dipole is perpendicular to the line connecting the protons, with its negative end nearer to the protons, attracting them more strongly than they are repelled by the positive, further, end of the dipole. The result is a force pulling the protons more together around the molecule, until the mutual repulsion of the positive protons just compensates this attraction by the induced dipole. This will be the equilibrium form of the molecule: triangular with a dipole moment. Analysis of rotational and vibrational spectra has shown that the water molecule is really triangular; the H—O—H angle is 105°. The high dielectric constant of water is due to a strong dipole moment. The same argument leads to the postulate that ammonia must be a triangular pyramid with the nitrogen atom at its apex. This again is borne out by experiment.

However, the quantitative results of this electrostatic model are not very good, especially as far as the angles of these molecules are concerned. But it has to be borne in mind that the polarizabilities at such enormous field strengths as occur around a naked proton are certainly very different from those taken from refraction data. On the other hand, the covalent-bond theory explains the same triangular and pyramidal molecules with higher accuracy,

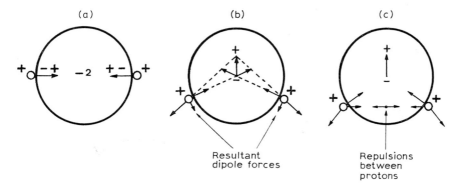

Fig. 124. **Hund's explanation of the non-linear water molecule**

Only Coulomb forces and polarization are made use of, no quantum mechanical effects.

(a) The two small protons polarize the oxygen atom. The linear configuration is in labile equilibrium.

(b) This becomes clear if we allow a small deviation from the straight line arrangement and consider the forces. The two induced dipoles of the oxygen atom add up to a vertical resultant which reacts in turn on the two protons. The resultant force of the positive and negative end of the dipole tends to move the two protons further away from the linear configuration.

(c) These will move around the oxygen atom until the repulsion between the two protons equals the resultant dipole force due to polarization (*see* (b)). This, now, is the equilibrium configuration.

the angles between *p*-electron orbitals being 90° according to theory. A tetrahedral arrangement of *s-p* hybrid bonds as for tetravalent carbon (p. 273) would yield an angle of 109° between substituents; two "lone pairs" of electrons in water and one such lone pair in ammonia would occupy the empty corners of the tetrahedron. All in all, however, the electrostatic theory is not so far behind the theory of covalent bonds as it is generally supposed to be and there is probably a deep common root of truth underlying both ways of approximation.

These, then, are the data from which to build an electrostatic theory of compounds: charge, radius and deformability of ions.

77. THE IONIC BOND, III: MELTING POINTS, BOILING POINTS AND CRYSTAL STRUCTURE ACCORDING TO THE ELECTROSTATIC HYPOTHESIS

Take the compounds corresponding to the general formula AB, that is those having two ions of the same charge—of course with opposite sign. Such compounds are, for instance, NaCl, CaO, AlN, SiC, where of course the difference in electronegativity decreases along the series. But suppose for argument's sake that all of them were ionic. The ions will be attracted, as we said, until the repulsion between their external electrons produces equilibrium separation. (*Born*, after whom this repulsive force is named, approximated it before the advent of quantum mechanics by an r^{-10} power law). As stated earlier (p. 313), the field of force of such a dipolar molecule is by no means saturated but can act further in all directions. Thus the well-known three-dimensional crystal lattices are formed. Many symmetrical structures may be derived for the AB ionic compounds and the most decisive factor determining their special structure will be the ratio of the ionic radii within them. The total electrostatic energy of such a lattice may be calculated to a good approximation by adding up the attractive and repulsive Coulomb energy between all ions and the repulsive Born energy between the neighbours. (Owing to the fact that Coulomb forces only fall off inversely with the square of the distance they may act between second or third neighbours, but the Born repulsion falls off with the inverse 10th power of the distance and is for all purposes already zero between second neighbours.) This so called "lattice energy" is thus the energy set free if the lattice is imagined to condense from a low-pressure vapour of its constituent ions.

The lattice energy depends mainly on the Coulomb forces because they act even at relatively large distances. The Born repulsion force decreases so rapidly that it only appears as a correction in the total energy content. Suppose we only add up the Coulomb energies between all constituent ions within a lattice. This energy will be the greater the nearer the ions are to each other, and the greater the number of ions that are in immediate contact. The smallest distance between neighbours, the "lattice constant" is the sum of the radii of positive and negative ions. The lattice energy, like all Coulomb energy, is inversely proportional to it. The number of ions in contact depends on the specific geometry of the lattice: a positive ion may be amidst eight negative ions and vice versa, as in caesium chloride (CsCl; the so-called space-centred cubic lattice) or amidst six as in sodium chloride (NaCl; face-centred cubic lattice) or four as in zinc sulphide (ZnS; *see* Fig. 58). In the course of summing up all kinds of Coulombic energies between ions of a lattice this geometry expresses itself as a numerical factor, the result of a summation of all relevant sites. It is not very different from unity, being 1·763 for the caesium chloride lattice, 1·747 for the sodium chloride lattice and 1·641 for the tetrahedral structure. Clearly this so-called "Madelung constant" must decrease if there are fewer first neighbours because the attraction between first neighbours dominates the whole calculation. The rest, that is the term which is multiplied by this constant, is the usual Coulomb energy: the product of the charges divided by their smallest distance.

One might suppose that if most energy is gained by arranging ions into an octacoordinated caesium chloride lattice, then this is the lattice that is going to be formed by all AB-type compounds. However, its occurrence is relatively rare. Caesium chloride and bromide crystallize in this structure, sodium and potassium chlorides form hexacoordinated lattices, and zinc sulphide the tetracoordinated structure.

There must be then, and in fact there is, another factor which determines the stability of lattices apart from the Madelung constant. Whenever the ions A and B are not of equal size it may happen that the bigger ions already touch each other before their smaller partners of opposite charge reach them. In this case it is of no help that further energy could be evolved if they were able to touch: they simply cannot. This evidently happens the more readily the more of the ions B surround the ion A, or vice versa, and thus it happens most readily with the octacoordinated caesium chloride structure. As long as the radii of the two ions are approximately equal, as in the two caesium salts mentioned, this structure with the greatest Madelung constant may and does exist. Potassium and even more so sodium ions, however, are much smaller than the caesium ion and the halide ions around them come into contact with each other before reaching them. Therefore the structure with fewer halide ions around the alkali ion comes into existence: six chloride ions may still be packed around a sodium ion without touching each other, so that they are all in contact with the sodium ion. The zinc ion is even smaller in proportion to the sulphide ion than the sodium ion is to the chloride ion; here even six sulphide ions would contact each other before touching the zinc ion and therefore only the tetrahedral coordination may exist.

Pure geometry shows us at which ratio of ionic radii ($r_c : r_a$) the CsCl type of structure must become labile because of anionic contacts. It turns out that the ratio must fall below 0·73 and that if it falls further, below 0·41, even the NaCl hexacoordinated structure must give way to a tetrahedral arrangement. Quite similar considerations hold, of course, for the structures of compounds of the type AB_2 and A_2B and so on. The order of the experimentally realized structures is always according to theoretical expectation. The quantitative agreement, the limiting $r_c : r_a$ values, may be in slight error because of polarizability of the anions and/or small variations of the ionic radii, if they are under the influence of a strong field of neighbouring ions.

Polarization of anions is the cause of some special lattice structures in crystals. Cadmium iodide (CdI_2), a yellow compound, is built in layers. A layer of cadmium atoms is flanked from above and below (Fig. 125) by layers of iodide ions. This is a structure intermediate between a molecular lattice and an ionic lattice: one cannot tell which iodide ion belongs to a given cadmium ion from among its neighbours, but one can always tell to which composite sandwich layer any ion belongs. This comes about by the strong polarizing effect of the small, divalent cadmium ions upon the easily deformed iodide ions, which *are* deformed to such an extent that their distal end, away from the cadmium ion practically loses its electrostatic attraction.

Crystals which are held together in all directions by electrostatic forces are hard. Their hardness increases for two reasons when the charge of the ions increases: (1) because these charges determine the Coulomb attraction, and (2) because the radii of the ions themselves decrease with increasing charge, and thus again contribute to raising the Coulomb attraction. For the same reason the melting-points and boiling-points increase also; the latter especially are a measure of the energy inherent in the structure. For example, in the series of AB compounds: LiF melts at 842°C, BeO at 2,500 BN at 2,730° and diamond above 3,500°. The extreme hardness of diamond as well as of CSi (silicon carbide, carborundum), would fit well into the picture, but here we are surely beyond the domain of heteropolar crystals.

Fig. 125. **The layer lattice of cadmium iodide**

Black spheres: cadmium; *white spheres:* iodine.

(From J. M. Bijvoet, N. H. Kolkmeyer and C. H. Macgillavry, *X-ray Analysis of Crystals,* p. 164, Fig. 126(c), English ed. (London, Butterworths, 1951). By courtesy of Butterworth and Co. (Publishers) Ltd.)

The C—C bond in diamond is as covalent as the bond between two carbon atoms in a chain: the electrons are on the average in the middle between the atoms. They are nearer the middle the more similar in electronegative character are the atoms in question. Carbon and silicon are electronegative in the ratio 2·5 : 1·8 according to a very empirical scale due to Pauling.

Compounds of hydrogen are a class apart. The proton is the only naked positive ion, that is without surrounding electrons. Therefore nothing prevents it from invading the electron cloud of neighbouring negative ions and settling there at a certain distance from the repulsive, positive nucleus. The radius of the water molecule is, for instance, 1·35 Å but the protons are 0·35 Å within its surface. Thus a specific sort of compound arises. The electrostatic forces between proton and anion are to a great extent already saturated *within* a molecule. Only a residual dipole moment remains, owing to the fact that the proton and the nucleus of the other atom do not coincide. Hence no ionic lattice can be formed, since the force is not sufficiently great. We have compounds which are either gases at room temperature or liquids which easily volatize: the hydrogen halides, water, hydrogen sulphide, ammonia, etc. On condensation only dipole forces are at work except in the hydrides of the first row (HF, H_2O, H_3N) where, as we shall see soon, a special effect of the hydrogen ion, the so-called hydrogen bond, is also active.

There are other compounds which, though held together by electrostatic forces, do not form ionic lattices either. Take for example the chlorides of the alkaline-earth metals. Going from the bottom of the appropriate column of the periodic system upwards we have the radium, barium, strontium, calcium and magnesium chlorides: all of them salts with ionic crystals. As is generally the case with the melts of ionic crystals, the melts of these chlorides are tolerable ionic conductors. They do not boil easily. But the chloride of beryllium, $BeCl_2$ is different. It melts at 440°C, sublimes easily and its melt is a very poor conductor of electricity having a resistance about 400 times that of molten magnesium chloride ($MgCl_2$). What has happened here? The small central beryllium ion lies between the two large chloride ions which are of six times its diameter (Fig. 126) and there is no place for chloride ions of other beryllium chloride molecules to approach it. We say that the beryllium ion is shielded by its neighbours. In such cases naturally no ionic

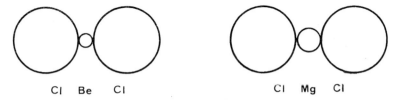

Cl   Be   Cl   Cl   Mg   Cl

Fig. 126. **The beryllium chloride and the magnesium chloride molecule**

crystal can exist—it is always possible to tell which chlorine ions belong to a certain beryllium ion, unlike the case of the sodium chloride crystal for instance. We are dealing with well-defined $BeCl_2$ molecules within the lattice and this crystal is held together by higher multipole and van der Waals' forces which are much weaker than forces acting between ions. Hence the softness, and low melting- and sublimation-points of the compound. And hence also the poor electrical conductivity: the molecule holds together as a discrete unit—its separate ions are not free to change their partners.

We may draw a slanting line across the periodic system, under HCl, $BeCl_2$, $AlCl_3$, $TiCl_4$. The chlorides above the line melt easily or are already liquid at ordinary temperature, like $TiCl_4$, and these same compounds are poor conductors of electricity. The chlorides below the line are hard, melt and evaporate at higher temperatures and their melts are ionic conductors of electricity. They are ionic crystals—salts as we know them. But this is not because the atoms of the former group are not "metallic" enough to yield salts. Just consider their fluorides. AlF_3 for example is a hard solid melting above 1,000°C: a salt in all respects with nothing of covalent bonding about it. But the corresponding chloride melts at 181°C, the bromide at 255°C and the iodide at 381°C; they are non-conducting, soft and volatile. The small fluoride ions were simply not able to shield the aluminium ion from neighbouring fluoride ions of other molecules whereas the larger halogen ions were able to do so. The melting-points and other cohesive properties rise again from chloride to iodide only because they have more and more easily deformable electron shells. Thus their polarizability is greater and this contributes to the van der Waals' force between adjacent molecules (*see* Section 67). But even aluminium iodide (AlI_3) is far behind aluminium fluoride as far as cohesion is concerned: the large van der Waals' forces are still much behind the ionic attraction of the latter. The break between fluoride and chloride marks the beginning of complete shielding. Other groups of the periodic system behave in a similar manner.

Not only the geometry but the energetic relations of ionic lattices may also be attacked from the electrostatic point of view. *Van Arkel* and *de Boer* showed that very many interesting features of experimental data can be explained in this way with plausible assumptions. However, there certainly is a limit where the electrostatic interpretation fails, not only in the molecules of elements such as H_2 and O_2 but also in the bonds between two atoms of the same element, for example between two carbon atoms in organic compounds. Compounds like carbon monoxide (CO), nitric oxide (NO), carbon dioxide (CO_2), and so on, should form ionic lattices if they were built on the same principles as in Na^+Cl^-. X-ray and electron-diffraction measurements of bond length, of interatomic distances, reveal that the distance between covalent-bonded atoms differs from the sum of radii taken from the ionic compounds of the same elements. Thus we may safely conclude that though polarization makes an ionic bond less polar, and though differing electron affinity makes a covalent bond polar to some extent, the two types of bond in their extreme forms are definitely different.

78. THE IONIC BOND, IV: THE HYDROGEN BOND

If we glance at the curves representing melting-points and boiling-points of hydrogen compounds (Fig. 127) we observe a remarkable irregularity of these properties in the first period, from nitrogen onwards. While the boiling-points, etc. generally decrease with decreasing molecular weight because of the smaller van der Waals' forces between the molecules, ammonia, water and hydrogen fluoride (NH_3, H_2O, HF) are exceptions. Were it not for this, water should be a gas down to about $-100°C$! All organic and inorganic compounds containing amino (NH_2) or hydroxyl (OH) groups show similar deviations from the behaviour of normal related compounds in which the NH_2 or OH group is replaced by, say, a halogen atom of similar size and dipole moment. Ethanol boils at a much higher temperature than does ethyl chloride, not to mention the fluoride!

The common cause of this state of affairs is seen to lie in the electrostatic force exerted by a proton attached to one molecule, upon the electron shell of another atom in a neighbouring molecule. This other atom should possess external electrons which are not all tied up in chemical bonds, for example the lone pairs in oxygen, nitrogen and the halogen atoms. Hydrogen atoms bound to carbon do not show such aberrations since they are bound more tightly than in an OH or NH_2 group.

This outwardly-directed force can only act if the proton has not penetrated deeply into the electron shell of its negative partner and this is naturally the case if the whole electron shell of this partner is not deep enough in

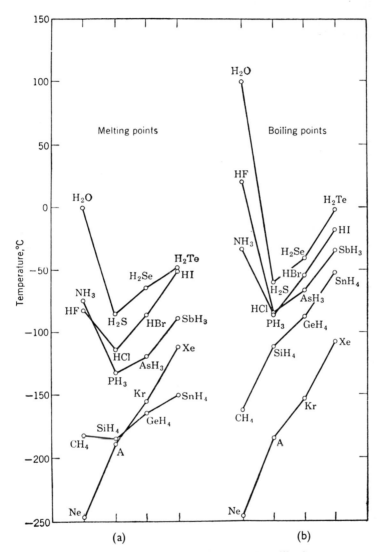

Fig. 127. **Melting points (*a*) and boiling points (*b*) of rare gases and hydrides of elements of the 4th to 7th column of the Periodic system**

(From 11. Moeller, *Inorganic Chemistry*, p. 190, Fig. 6.3 (New York, Wiley, 1952). By courtesy of John Wiley and Sons).

itself: if it belongs to the first period of the periodic system. Of course, dipole–dipole interaction must be taken into account too, and maybe it is not even quite safe to try to distinguish between two names relating to very nearly identical phenomena. Methane (CH_4) cannot become hydrogen-bonded because it is already shielded: the carbon atom in the centre is surrounded by its own hydrogen atoms and there is no free electron pair left around it. It is not necessary that the two atoms which are connected by a hydrogen bond should belong to different molecules: an atom may interact just as fully with another active atom within its own molecule. This may express itself in one conformation of a molecule being stabilized from among the many which are possible because of the free rotation which can take place around single bonds. A hydrogen atom on a hydroxyl group becomes more strongly attached to its parent molecule if a carbonyl $\left(\begin{array}{c}-C-\\ \| \\ O\end{array}\right)$ group lies beside it.

Therefore it loses some of its reactivity in a grouping like

$$-C-\!\!\!-\!\!\!-C-$$
$$\||$$
$$OO$$
$$\diagdown\diagup$$
$$H$$

In this conformation* the hydrogen atom remains bonded within its own molecule, but is drawn a little away from its oxygen atom partner by the neighbouring oxygen atom. The conformation

$$-C-C-$$
$$\||$$
$$OO$$
$$\diagdown$$
$$H$$

would be more reactive but less stable.

In view of the overwhelming importance of water for life and for chemistry in general, it is worth while to attack the problem using water as an example. The arrangement of the two hydrogen atoms in water is not linear (p. 293). The molecule has a large dipole moment and the angle in

$$\text{H}\text{H}$$
$$\diagdown\diagup$$
$$\text{O}$$

is slightly smaller (105°) than the angle in a tetrahedron (109°). The arrangement of covalent bonds tends to be tetrahedral as in methane (CH_4) but the electrostatic repulsion of the two hydrogen nuclei drives them further apart. According to the covalent model of the water molecule, two corners of this "nearly tetrahedral" structure are occupied by the hydrogen atoms while its other two corners contain the remaining two electron pairs of the L shell around the oxygen atom. The hydrogen bonds originating in the electrostatic attraction of the hydrogen nuclei, will naturally be directed towards these lone electron pairs of neighbouring water molecules. Owing

---

* Possible geometrical arrangements of a molecule resulting in rotation of one of its parts against another around a single bond are called its conformations. *See also* pp. 69–70.

to the fact that the hydrogen nucleus is the bare proton itself, it is virtually a point mass compared with the two oxygen atoms which it joins and it is completely screened, so that no third atom of any kind is able to approach it (Fig. 128). The hydrogen atom has the maximal coordination number of two in hydrogen bonds.

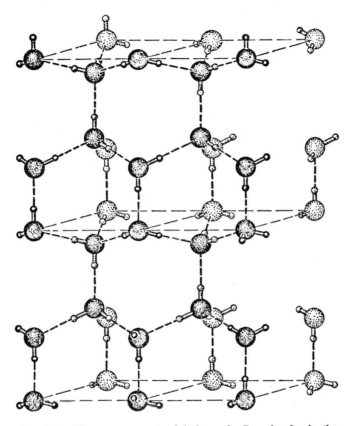

Fig. 128. **The arrangement of (schematized) molecules in the ice crystals**

There is one proton along each oxygen–oxygen axis, closer to one of the oxygen atoms. Their position could be changed arbitrarily as long as the above condition prevails and each oxygen atom has two neighbouring hydrogen atoms.

(From L. Pauling, *The Nature of the Chemical Bond*, p. 465, Fig. 12.6, 3rd ed. © 1960 by Cornell University. Used by permission of Cornell University Press.)

X-ray and electron diffraction measurements as well as infrared spectra assure us that the O—H distance in free molecules of water vapour is 0·96 Å. In ice crystals, however, each molecule is surrounded by four others—a very loose structure, by the way. Between these five molecules four hydrogen bonds are active, so that each hydrogen atom still belongs to its parent molecule but its bond length has increased to some extent, being 0·99 Å. Its distance from the oxygen atom of the other water molecule is 1·77 Å, that is, very considerably greater. Still, a certain amount of Coulombic attraction is at work already and this is the reason why the structure of the ice lattice is so extremely loose with only four molecules around each other instead of the twelve which would be found in closely packed lattices of spheres. The structure of ice is determined by the four hydrogen bonds around each molecule and not by the general van der Waals' forces. The consequence is that ice is less dense than it would be in the absence of any hydrogen bonding. Its heat of sublimation is 12·2 kcal per mole. Out of this, 3 kcal can be ascribed to the van der Waals energy (as estimated from the heats of sublimation of molecules with the same number of electrons as water but without hydrogen bonds, e.g. methane, CH_4). Thus there remain 9 kcal for the two hydrogen bonds: there are four hydrogen bonds around each molecule, but each bond only half belongs to one of the partners. This gives 4·5 kcal per mole for an OH.O hydrogen bond and very similar values are found in other cases.

The heat of fusion of water is very small, only 1·44 kcal per mole. This shows that very few hydrogen bonds can have been destroyed during the melting of ice. Hence they must persist in water, and water near its freezing-point must have a loose, tetrahedral grouping of its molecules, at least on the average. This is the reason why water has an anomalous density and specific heat in the neighbourhood of 0°C. The structure remains loose until it is broken up by thermal motion and then water contracts on heating, in contradistinction to all other well-behaved compounds. The heat necessary to raise its temperature in this region is of course abnormally high, because the hydrogen bonds are just being broken. The low heat of fusion is compensated for now.

The same bond energy can be derived from the heat of sublimation of hydrogen peroxide (H_2O_2). It amounts to 14·1 kcal per mole against 5·0 for the isoelectronic ethane, $H_3C.CH_3$. The difference is again 9 kcal which has to be halved. The hydrogen bond energy for alcohols

$$\text{RO} \overset{\displaystyle H}{\underset{\displaystyle H}{\diamond}} \text{OR}$$

is somewhat greater, 6 kcal per mole, and the distance between the proton-bound neighbouring oxygen atoms is accordingly less than in water, only 2·70 Å against 2·77 Å. This is due to the fact that the carbon atom in alkyl radicals is slightly negative relative to the hydrogen atom and thus the total attraction is increased.

Dielectric constants are related to dipole moments (p. 291) by the fact that an electric field is able to some extent to orient dipoles against their thermal disorder. Thus a relatively smooth curve can be drawn between dielectric constants of liquids and the dipole moments of their molecules, measured as free molecules in the vapour state (Fig. 129). It is immediately evident that the dielectric-constant values of hydrogen-bonded molecules do not fit into the normal curve: the dielectric constant of the liquids is appreciably higher than that of the corresponding normal compounds. This difference is easily explained by the assumption that a head-to-tail association of these dipoles by hydrogen bonds effectively increases the dipole moments in the liquid state.

Hydrogen fluoride (HF), the first in the series of the halogen hydrides is again abnormal with respect to its cohesive properties and its dielectric

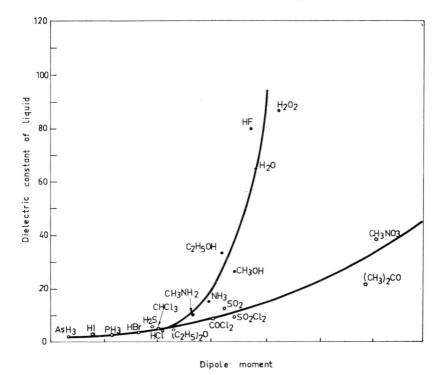

Fig. 129. **Dielectric constants of some liquids as a function of their dipole moments**

They clearly separate into two distinct curves; points of the steep curve belong to molecules in which H atoms can be attracted to lone electron pairs of neighbouring molecules.

constant. It is a liquid at normal temperatures whereas hydrogen chloride, hydrogen bromide and hydrogen iodide are gases. Its vapour density is higher than it would be if it contained only simple HF molecules; a plot of the vapour density as a function of total pressure shows that the average molecular weight tends to approach the normal value of hydrogen fluoride as the pressure decreases. This has to be ascribed primarily to the dissociation equilibrium

$$(HF)_2 \rightleftharpoons 2HF \qquad . \qquad . \qquad . \qquad . \qquad (78.1)$$

because naturally the bimolecular state occurs more often at higher pressures where collisions are more frequent. But detailed consideration of the experimental data reveals that higher polymers, $(HF)_3$, $(HF)_4$, etc., are present also.

The dimeric structure prevails even in solid salts, and acid salts such as KHF_2 and $NH_4.HF_2$ are known to exist. Such acid salts are entirely absent in the rest of the halogen hydride series and their structural unit is undoubtedly the hydrogen-bonded anion $F.HF^-$.

The dielectric constant of hydrogen fluoride is again higher than that of normal compounds of similar dipole moment. But in the gas phase the dipole moment begins to decrease on the average, with the formation of trimer and higher polymer molecules, because the valency angle around the hydrogen bonds is about 140°. The head-to-tail association soon degenerates into bent structures and ring-like configurations with intramolecular compensation of the dipole moments (Fig. 130). The fact that each fluoride ion is

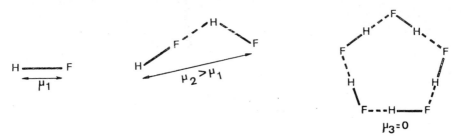

Fig. 130. **Hydrogen-bonded oligomers of HF and their dipole moments**

only able to participate in two bonds with hydrogen, $HF.HF.HF...$, is the reason why only linear chains or rings can be formed, and no tetrahedral structures as in the ice lattice. It seems that the recurrent unit structure in solid hydrogen fluoride is the ring $(HF)_4$. The calculated diameter of the closed octet of the fluoride ion is 0·91 Å, in excellent agreement with the experimental H . . . F distance, 0·92 Å.

79. METALS, I

The metallic state is confined to solids and liquids—a gas cannot be a metal. (It may be that highly ionized gases which occur in discharges and consist mainly of positive ions and free electrons, a state called "plasma," are quite near to what could be termed a metallic gas.) But the metallic state is not confined to elements. Solid and liquid solutions of metals in each other, and intermetallic compounds, are metals also. They conduct electricity without transport of matter, they are good conductors of heat and reflect light very strongly. Their electrical conductivity and their optical properties immediately suggest that they must contain many mobile electrons. However, this hypothesis was rejected for a long time because independently moving electrons should possess a temperature-dependent kinetic energy of their own and this would show up as an additional component of the specific heat. The law of Dulong and Petit (p. 44) stated that the atomic heats of elements were equal, irrespective of whether they were metals or not. This is so because temperature means average energy of independent particles (p. 156) and thus the same number of independent atoms contains the same amount of energy. If there were independent electrons in metals, the rule of equipartition of energy would give them the same amount of average energy as the atomic cores (ions) and accordingly more heat would be spent in raising their temperature by one degree than in raising the temperature of non-metallic elements. The atomic heat should then be at least doubled—doubled in those cases where each atom has split into only one electron and one core (rump atom). One "independent particle" has then split into two.

The advent of quantum statistics solved this problem. We now know that the Pauli principle is valid for all electrons in a piece of metal. They all belong to the same system and hence they cannot be in identical quantum states, two electrons of opposite spin being sufficient to fill any of the discrete states allotted to them by quantum mechanics. It follows that the states of low momentum, that is small kinetic energy, are filled quickly and the more electrons we have in a piece of metal, the higher the momentum, that is energy, we have to assign to those which come later. This results in an entirely different statistical distribution from that of pure chance with which we became acquainted in the Maxwell–Boltzmann distribution (p. 146). The total energy does not vanish on approaching absolute zero. This is not new to us, for it is the same zero-point energy that we have already encountered (p. 260). But this zero-point energy is enormous for a macroscopic piece of metal. In a molecule we had only to deal with a few, or at most a few dozen electrons, whereas here we have quandrillions of them. Accordingly they must be in states of high momentum, i.e. energy, to satisfy the Pauli principle, and the average zero-point energy of the electrons in macroscopic pieces of metals corresponds to a temperature about 3,000°K in a normal Maxwellian distribution. This new type of statistics includes the Pauli principle and has a name of its own: we call it Fermi-Dirac statistics in honour of the two great physicists who clarified its essence. Even at the absolute zero the average electronic energy in a metal lattice must be as high as this, and so it is clear that a rise in temperature of the lattice itself by a few hundred degrees scarcely affects the energy distribution of the electrons, which are already much "hotter." Only very few electrons will receive sufficient energy from atomic lattice vibrations to be raised into states of somewhat higher energy than they already possess. Therefore they scarcely absorb energy for themselves when the metal is heated: they do not contribute to its specific heat, as is found by experiment to be the case (Fig. 131.)

This state of affairs has luckily been confirmed by direct measurements. The wavelength of X-rays emitted from excited atoms depends, as we know, on the difference of energy between the original higher state from which the electron "fell," and the energy of the lower state into which it "falls" during this process. The wavelength is inversely proportional to the energy difference according to the Einstein relation $E = h\nu$, since $\nu = c/\lambda$ where c is the velocity of light. Suppose we create free spaces, vacancies, for electrons to fall back into, by ejecting electrons from the K or L shells of metals by X-ray irradiation or electron bombardment. These inner shells are practically undeformed and belong to the individual atoms, only the surface electrons being free in the metal. The vacancies thus created will shortly be filled by one of these free, metallic, conductivity-electrons. If, however, the energies of these are not equal but are determined by the Fermi-Dirac distribution, the energy of the electron falling back into the vacancy will be greater or smaller, according to the energy it happened to possess within the limits of this distribution. Thus one and the same electronic state within an atom will be replenished with the liberation of different amounts of energy, according to chance, according to the energy of the conducting electron. Therefore the

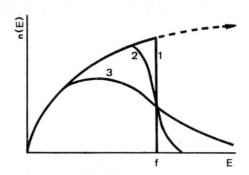

Fig. 131. **The "Fermi distribution" of electrons in metals at three different temperatures**

The Pauli principle prohibits two electrons from occupying the same state in a piece of metal. The states of low energy (on the left) are soon filled and higher states of energy (*E*) are occupied one after the other. The number of states with the same energy but different impulses *n(E)* slowly increases along the upper curve. At the absolute zero point (*curve* 1) the electrons fill the highest levels corresponding to their number and there are no electrons above this level. At a somewhat higher temperature a few of the high energy electrons are moved into higher energy levels by thermal motion (*curve* 2) so that fewer remain below the "Fermi level" (*F*), which is the limit of *curve* 1. Those raised into higher energy levels are distributed with decreasing probability beyond *f*. *Curve* 3 represents the situation at a temperature much beyond room temperature, where the thermal effect is appreciable.

X-rays emitted in this process will not all have been caused by the same energy difference and in consequence will not have exactly the same wavelength. Instead of the sharp X-ray *lines* that we were used to when dealing with individual atoms (say atoms in a metal vapour) we observe a *band* having a more or less sharp edge on the short wave-length side, that is towards higher energies. These very important investigations are due to *Skinner* who was able to calculate from the energy-density distribution of these spectra the distribution of electronic energies in the conductive state in the metal. He confirmed that the edge of the emission band towards high energies became sharper the lower the absolute temperature. This happens because at low temperature the distribution of electrons towards states of higher energy ends abruptly when all electrons are placed in their respective state and there

are simply no more at hand. At higher temperatures some energy is transmitted from the lattice-atom vibrations to these electrons and therefore the
abrupt end of the energy distribution is slightly degraded, the more so, the
higher the temperature. This distribution curve of electrons among increasing

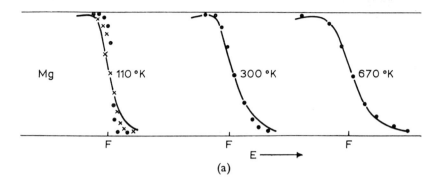

(a)

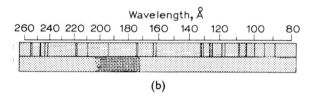

(b)

Fig. 132

(a) The curves represent the energy distribution of the high-
frequency (= high energy = shorter wavelength) side of an
edge in the X-ray spectrum of Mg. A deep-seated electron has
been knocked out of its original level and is being replaced by
another electron from among those in the higher levels. The
energy distribution of the latter does not end abruptly, however,
because of thermal motion (*see* Fig. 131). Therefore electrons
fall back from slightly different levels of energy, the more
different, the higher the temperature. Hence the energy of
the emitted X-ray quanta is also distributed around a value
decreasing less or more steeply at high energy values.

The curves are actually those in Fig. 131, their ordinate
being the intensity of the emission and their abscissa the energy
corresponding to the frequencies. The Fermi level (*F*) is indicated as reference point at all three temperatures.

(From W. Hume-Rothery, *Atomic Theory for students of Metallurgy*, p. 107, Fig. 30 (London, Institute of Metals, 1955). By
courtesy of the Institute of Metals and the Royal Society.)

(b) The X-rays emitted by transitions between sharp energy states
of electrons in free atoms are compared with those emitted by
transitions from the "diffuse conduction band" of (a).

successive states of energy is called their Fermi distribution and can be calculated by a fairly simple application of quantum mechanics to statistics,
taking care not to violate the exclusion principle of Pauli, that is never placing
more than two electrons in the same state. The experimental curves of Skinner
are fair representations of this Fermi distribution (Fig. 132).

However, in a number of cases an interesting phenomenon appeared. The observed emission band had two parts and in between them there was only a much weaker emission or sometimes no emission at all. On subsequent translation of these emission curves into the curves of electron numbers belonging to specified energies Skinner had to conclude that

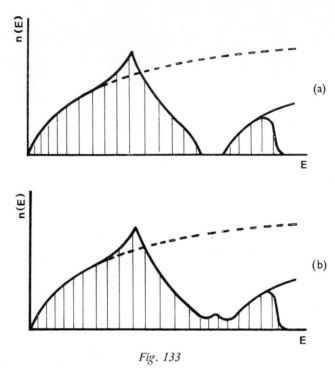

Fig. 133

The broken line is the energy distribution curve of a great number of electrons, moving unhindered by crystal geometry (Fig. 131). It can be called an "electron gas". Total reflections on internal crystal planes exclude some electron-wavelengths (= velocities, that is energies) while directing some into energies which would already been full in the electron gas. The gap between two "Brillouin zones" can be absolute (*a*) or may contain only a small number of electrons (*b*), according to the geometry of the lattice. The shaded area represents the part of the permitted states actually filled with electrons at a given temperature.

electrons of certain energies were either altogether absent or at any rate that their probability showed distinct and characteristic minima at certain energy values.

We must return to the fact that the movement of electrons is associated with waves. In the experiments of Davisson and Germer, electrons of certain specific velocities, that is electron waves of specific wavelength were reflected from crystal planes at certain angles of incidence. This is analogous to the reflection of electrons from the "Bragg" planes in crystals (p. 123). The same happens within a crystal with its own mobile electrons, if it has any. There will be some electron wavelengths which simply cannot propagate within a crystal because they are immediately reflected. These wavelengths correspond to given momenta and therefore to given energies: from all the energies within the Fermi distribution those which correspond to such a total inner reflection will be excluded. Or, if they are only reflected in some directions, their number will decrease in comparison with the situation where these reflection do not occur. This is exactly what Skinner observed in his X-ray wavelength distributions: some values of electron energy within the Fermi distribution were either lacking or at any rate occurred much less often. What we actually observe is an interaction of the Fermi distribution of electrons with the movement of these electrons within a lattice. The minima which are superimposed upon the original Fermi distribution are due to the lattice geometry (Fig. 133).

It can be shown that according to the nature of the atoms and lattices in question these minima in the distribution of electronic energies may either be only minima without actually forbidding electrons to move with this energy at all, or may completely exclude a definite range of energies from their movement. The calculations are extremely involved and the qualitative picture we are trying to form for ourselves is certainly inadequate. But up to now there seems to exist no simpler treatment of the phenomenon. Suffice it to say that the energies which are permitted to electrons within a given lattice fall into more or less distinct zones. They are named *Brillouin* zones after the physicist who first postulated their existence.

Now, very different situations may arise according to whether these zones are actually in contact or are separated from each other by a zone of entirely forbidden energies. As long as they are in contact through depressions of the Fermi curve there is always a possibility of raising the energy of the whole electron system by almost infinitely small steps. The permitted states within the curve are so near to each other as to form practically a continuum. If an electron from among the most energetic ones (from those at the right-hand end of the curve) receives some energy from, say, an external electric field it can always pass over into a state of somewhat higher energy while others may enter the state that it previously occupied. In short, such a material will conduct electricity because an applied field is able to accelerate its electrons by any small amount.

The situation changes if the zones of permitted energies are separated by gaps (Fig. 134) and it so happens that the zone before such a gap is full of electrons. Now it is no longer possible to add energy to these electrons below a certain amount. This is because the top electron of the filled zone must obtain at least the energy necessary to jump over the gap into an allowed state of the next Brillouin zone. It is the same case as with electrons within an atom: they are not able to acquire any given amount of energy, but the energy offered to them must be at least large enough to bring them into the first empty permitted state above the one that they are in (Franck–Herz experiment on p. 297). If the gap between the two zones is so large that it cannot be bridged by normal amounts of thermal or photo-energies, we are dealing with an electric insulator. If, however, the gap is relatively small, of the order of say 0·1–1 electron-volt (eV), it becomes possible to concentrate by chance so much energy in one electron that it is able to jump over into the next, the conductivity band. Thus a small but definite conductivity ensues and this conductivity increases with increasing temperature because the more energy they have at their disposal, the more of the low-sited electrons will manage to bridge the gap of forbidden energy. We are dealing with so-called semiconductors, a class of compound with rapidly increasing technical importance.

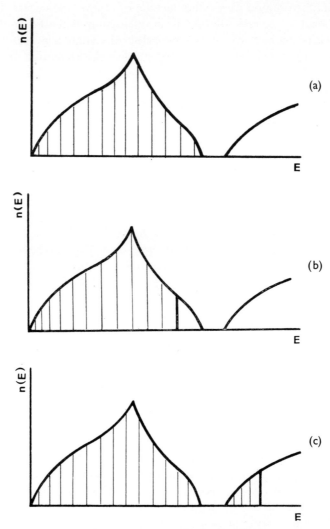

Fig. 134. **Electron energy-distribution curves**

(*a*) Insulator. Thermal energy is insufficient to allow electrons to jump over the energy gap of forbidden states.

(*b*) *P*-type semi-conductor. Admixture of traces of an element with a smaller number of electrons creates some vacant places in the nearly filled zone ("holes" of quasi-positive charge).

(*c*) *N*-type semi-conductor. Admixture of an element with more electrons brings some of them into the empty, high-energy, zone.

As we just have learned and understood, the electrical conductivity of semiconductors increases with temperature. This is the crucial difference between them and the true metals whose conductivity always decreases on heating. The reason for this decrease is the growing disorder within the lattice, due to the thermal vibrations of atoms. Electron waves may accommodate themselves to regular lattices but must of necessity become disturbed if the spacings of the lattice in which they move vary from one place to another. The electrons "collide" with the atoms in the disordered lattice and are unable to arrange themselves in an orderly, secure pattern. Thus they transfer energy to the atoms, the lattice absorbs energy and becomes warmer: ohmic resistance and ohmic heat have ensued. All this is of course true for the semiconductors also, but in their case the thermally increased number of electrons which are at all free to partake in conduction, by far outweighs the effect of ohmic resistance and thus, in spite of this existing resistance, their conductivity increases. Graphite (in the planes perpendicular to the benzene hexagons of p. 326), selenium, silicon, boron and the now-famous germanium of the first transistors, belong to this class of materials.

The energy necessary to bridge the gap between two distinct Brillouin zones need not necessarily be of thermal origin. It may be derived by absorption of light. In such cases, as for example in grey selenium, the material conducts electricity if it receives light: we experience photoconductivity.

We said that there will be no conductivity whenever a Brillouin zone is filled with electrons and a gap of forbidden energy ranges follows. But if it is not filled, there is no reason why electrons should not be able to increase their energy within the boundaries of the unfilled zone. A certain amount of electrical conductivity becomes possible, the greater the more free space there is in the zone. Now, it is possible to change the number of electrons within a zone by changing its chemical composition, or let us say rather by alloying the element in question with one of its neighbours. Take for instance germanium, from the fourth column of the periodic system with four external (N-shell) electrons which in the given lattice structure happen to fill a zone. A small quantity of gallium with only three external electrons may be alloyed with it without changing its lattice. The number of atoms, and with this the structure of the Fermi curve, has not changed, but some electrons are missing. The zone is no longer full. Accordingly the electrons within the zone become able to acquire some additional energy, and the material begins to conduct electricity. The more electron "holes" have been introduced by increasing amounts of gallium the greater the conductivity. Of course this is only true within certain limits since the general lattice structure must not become deranged.

The opposite way of changing the number of electrons is to alloy germanium with an element which contains more electrons than it does itself, for example with arsenic. This element of the fifth group contributes one electron more to the household of the germanium lattice than does a germanium atom and thus the number of electrons exceeds the number of places available in the valency band (valency zone) of the Fermi distribution. These surplus electrons have to occupy places in the higher conductivity zone because no other place is available. In this way the conductivity zone begins to be, though sparingly, populated and the conductivity increases with the arsenic concentration. As a matter of fact it is one of the most difficult processes of purification to eliminate all impurity atoms from the lattice of a semi-conductor such as germanium or silicon and thus arrive at its maximal, intrinsic electrical resistance which is so important in the manufacture of good photocells and transistors.

The two kinds of conductivity are opposed to each other: in the first case we have defect-conduction due to missing electrons in a filled band, and in the second case electron conduction due to some surplus electrons in a zone which has previously been empty. The defect-conductance may be crudely approximated by the picture of bubbles moving within a fluid, while the electron conductivity resembles drops of fluid flying about in space. The "bubbles" behave like positive electricity (lacking negative charges) moving around, and the effect of an external magnetic field upon such a conductor (Hall effect)* has the opposite sign for defect-conductors as for electron-conductors. They are often designated by *p* or *n* for positive and negative, as far as carriers are concerned.

80. METALS, II: METALLIC ELEMENTS

The place of metals in the periodic system of elements is very neatly defined. They are to be found on the left-hand side and in the lower part of the system; the right-hand upper corner contains the non-metals. The reason is easy to find. The left-hand side begins with the most electropositive elements, the alkali metals which very readily give up their single valency electron in the course of chemical reactions. No wonder that this same electron is also given up as easily within the elementary lattice itself. The alternative would be to bind, say, two sodium atoms together with a covalent bond, in analogy to the H_2 molecule, but for this to happen the distance of these outside electrons is much too large. They prefer to leave their parent atoms behind as positive ions and to expand their domain throughout the whole sodium crystal. In this way they evidently gain all the zero-point energy that is to be won in such an expansion of range. The radii of the alkali ions are known from the structure of their salts. The volume of sodium metal which contains one sodium atom is, however, much larger than this ionic volume. The ions are not in contact with each other but are embedded in the electron gas of their lost valency electrons. Accordingly these metals are all very soft and they all melt easily. They crystallize in the cubic body-centred lattice with eight nearest neighbours to each ion. But four other ions are at relatively close distances around them so that all in all they are surrounded practically by 12 neighbours. Metallic elements nearly always crystallize in relatively

---

* If a conductor is placed in a magnetic field which is perpendicular to its current a voltage is induced perpendicular to both. This is the Hall effect.

simple structures because they are spherical, without any complications of molecular structure, and pack together like balls.

The elements of the second column are the alkaline-earth metals. Their electrons are bound somewhat more strongly but they still lose them readily in their metallic lattices as they do in their compounds. They are somewhat more dense because divalent ions are embedded in an electron gas, which is twice as concentrated, and accordingly they are harder and melt at higher temperatures than the alkali metals. They crystallize in the hexagonal and cubic close-packed arrangement of spheres, each ion having the maximal number of 12 similar neighbours (*see* Fig. 43).

Boron, at the top of the third column of the periodic table is not metallic. It has a rather complicated structure with six nearest neighbours to each atom and is a semiconductor of electricity. It is extremely hard, similar to diamond, silicon and carborundum. How its three external electrons manage to hold this structure together with such an intense force is not clear, as is the case with practically all of the valency problems connected with the boron atom. That it is not a metal is probably due to the fact that its valency electrons are near to a nucleus with an effective unshielded charge of three units.

The element below it, aluminium is already a metal. The charge of the nucleus acting on its valency electrons is three also, but they are farther away by one shell and thus less strongly bound. It crystallizes in the face-centred cubic lattice which is one of the densest arrangements of spheres. All trivalent elements below aluminium are naturally metallic—the valency electrons are even more remote from the nuclear field than those of aluminium.

In the fourth main group, carbon occurs as a perfect insulator of extreme hardness in diamond, and as a two-dimensional metal with very weak bonds between the layers, in graphite. Practically, graphite is a semiconductor because the individual layers are not in metallic contact. Silicon and germanium are semiconductors and the metallic state only begins in this column with tin. Tin has a low-temperature modification which is on the verge of being metallic. Lead, below tin, is a metal but its ion occupies much more space in the lattice than that of gold, for example: $3 \cdot 5$ Å as against $2 \cdot 9$ Å, in spite of the fact that the nuclear charge is greater in lead. This suggests that maybe only two of its external electrons are released to the conductivity level while the ion ("rump atom") retains the other two, just as in the divalent plumbous compounds. A similar situation prevails with thallium (3 valency electrons), where again the monovalent ion is known to occur in many of its compounds. So it is not necessary that all external electrons should be given up, and fine differences in lattice dimensions help us to find out where some remain attached to the atom. It means, in other words, that the metals thallium and lead contain Tl^+ and Pb^{++} ions embedded in conductivity electrons instead of Tl^{+++} and Pb^{++++} ions which would be expected of elements of the third and fourth columns.

Going to the right of the periodic system the effective, unshielded nuclear charge increases further and keeps the electron system of the atoms together: covalent bonds are formed between atoms of the same element in its lattice but valency electrons are only given up into the conductivity level in the lowest rows of the system, where they are farthest away from the nucleus. Antimony in the fifth row of the fifth column is a metal of medium conductivity while in the sixth column its neighbour tellurium is a good semiconductor rather than a bad metal. In the group of the halogens, the seventh column, only the radioactive astatine begins to show some conductivity. The rare gases all have strongly bound, highly symmetrical electron systems and the energy needed to disturb this high symmetry, where electronic states are spread out spherically around the nucleus, is much too great to allow them to become metallic. As a matter of fact they do not form covalent bonds with each other: they remain gases and only condense at low temperatures because of the van der Waals' forces between their atoms.

The electronic structure of the transition elements is very complicated because their outermost *s*, *p*, and *d* subgroups of electrons are in overlapping ("hybridized") bands. They include metals of the highest melting points such as tungsten (W) and they also include the ferromagnetic metals with incomplete, magnetically uncompensated subshells. They often possess different lattices at different temperatures: manganese has four such modifications, two of them of a complicated structure. So have tungsten and molybdenum, suggesting that in these elements some critical, labile stage has been reached. The minima of atomic radii occur around the metals of the eighth group. Since compressibilities and coefficients of expansion go parallel with atomic volumes it is understandable that these metals are hard. The bonding forces between the ions in the bulk metal are at their highest in this group and ferromagnetism is an interaction which correlates electronic spins of neighbouring ions across considerable ranges of the lattice.

It is possible to draw some conclusions about the state of valency of the "core" ions in the metallic lattice from their magnetic behaviour. Copper is not paramagnetic whereas the ion Cu^{++} should be so. Hence copper metal must be built from cuprous (Cu^+) ions and each atom only gives up one of its electrons to the conduction band. Nickel has just sufficient electrons to fill its *d*-subshell but two of them are in a level with the next principal quantum number. However, magnetic measurements show that in the metallic state the majority of these higher-level valency electrons must have fallen back into the *d*- subshell because the paramagnetism corresponds only to an average of 0·6 missing electrons per *d*-shell of an atom. Thus the electronic structure of the metal is not necessarily one of the structures shown by the atom in its compounds. It is interesting to note that on absorbing hydrogen gas to the extent of 0·6 atom per metal atom, the paramagnetism of nickel and of the analogous palladium and platinum disappears: the electrons which have been missing from the *d*-subshell have been furnished by the hydrogen atoms. This means, of course, that the proton must have remained naked within the lattice: a true alloy of hydrogen with these metals. This surely explains why hydrogen diffuses so readily through them in the solid state and why these metals are excellent catalysts of hydrogenation reactions: not only is the strongly bound molecule H_2 dissociated into its atoms, but even the atoms have already given up their electrons to the lattice and remain in a state of highest reactivity.

Copper, silver and gold (Cu, Ag, Au) in the first subgroup of the periodic system resemble the alkalis in so far as they possess one external electron above a rather complete shell: a shell of 18 electrons with complete *d*-sub-shells in this case, and not a shell of eight as in the alkalis with only their *s* and *p* electrons. The external electron is bound loosely here too: the atomic conductivity of these renowned excellent conductors is high indeed; though it remains below that of the alkali metals. But the structure of the ions is more compact: the 18-electron shell exerts a stronger van der Waals attraction towards its similar neighbours than did the eight-electron shell. The ions approach nearer to each other within the "electron gas". The structure itself is changed from the less compact space-centred cubic to the extremely compact face-centred type and the density is higher. The ions are practically adjoining, the high compressibility and low melting-point of the alkalis has disappeared. The three metals are also much harder than the alkalis.

81. METALS, III: ALLOYS AND INTERMETALLIC COMPOUNDS

Solid solutions are not uncommon in chemistry, but they are most striking among the metals and the greater part of metallurgy depends on them. Tech-nically important alloys may consist of solid solutions and intermingled phases of intermetallic compounds. The stoichiometry of these compounds is often unusual and even among the solid solutions phase transitions are known to occur which obey quite special rules of a novel stoichiometry.

In order to gain insight into this mass of *prima vista* incoherent data we have to make clear that the bonding forces between metals are of more than one type and they may be at work simultaneously. First of all quasi regular compounds may form if the electrochemical characters of the reacting metals are sufficiently different from each other: one metal behaves as an anion against the other which plays the role of a cation. Further on, a solution of a metal in another may be random or may form a more or less regular dis-tribution with a given ratio of the different atoms to each other. These ordered solutions are more symmetrical than the random distribution and are called superlattices. The forces which cause the stability of the superlattice may be due to small differences in radii and/or in electrochemical character. Evi-dently, if no such differences existed, there would be no reason to arrange the atoms into any pattern of higher symmetry because they would practically not differ from each other.

This process may go so far that the whole structure including the stoichio-metry is primarily determined by spatial relationships between the atoms, if one sort is so small that it may be fitted into the free spaces between adjacent atoms of the other sort. Finally, there is a special sort of metallic bond between different metal atoms which may be best designed as the formation of electronic compounds; here the ratio of the number of conducting elec-trons to the number of atoms (that is, rather, to the number of lattice sites) within the basic cells of the crystal lattice is the determining factor. Very often more than one of all these factors operate together and thus we must not be astonished that it has been very difficult to gain insight into the problems of alloys, and that even today some of them wait for solution.

Compounds between different metals might be in harmony with their normal valency as in Mg_2Sn, Mg_2Pb, Mg_2Si, Mg_2Ge, Be_2C and Cu_2S. All these, for instance, crystallize in the same structure as the ionic compound CaF_2 (Fig. 135) and the difference in electronegativity between the partners is sufficient to warrant a quasi-ionic, or at any rate a highly polar structure. We may imagine that the atoms are bound by highly polarized ionic, or very asymmetrically situated, covalent bonds, which amounts to much the same thing. The four pairs of bonding electrons fill the first Brillouin zone of these lattices: they are insulators or semiconductors as crystals. But their melts are metallic because there are no defined Bragg planes (pp. 123 and 364) in the melts to reflect the electrons.

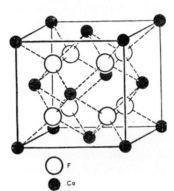

Fig. 135. **Calcium fluoride-type crystal lattice**

(From W. Hume-Rothery, *Electrons, Atoms, Metals and Alloys*, p. 301, Fig. 135 (London, Iliffe, 1948). By courtesy of Iliffe Books Ltd. and the Institute of Metals.)

Other metals, for example sodium, not only form the regular valency compounds with elements of the fourth, fifth and sixth columns as in Na_3As, Na_3Bi, Na_2Te, but also yield a series of more complicated compounds which resemble the well-known polysulphides and polyiodides, for example Na_2S_5 where the sulphur atoms are coordinated to each other; or NaI_3 where the I^- ion coordinates an iodine molecule (I_2). Such compounds are Na_3As_3, Na_3As_5, Na_3As_7, Na_2Te_2 . . ., Na_2Te_4, etc. The first, normal-valency series of these compounds is insoluble in liquid ammonia, but these complex anionic compounds dissolve in it to yield deep-coloured solutions and can even be electrolysed. This increased solubility may be due to the increased radius of the complex anion: the Coulombic force which binds it alongside the cation in the solid lattice has decreased. Na_4Pb_9, for example, yields sodium at the cathode and lead at the anode. The compounds seem to be ionic although only one of the metal atoms in the anion carries its normal charge. The others attach themselves to it, building a miniature negatively charged metallic complex. The charge is "spread out" across the complex anion. Cationic and anionic lead may react in liquid ammonia solution to yield the metal:

$$Na_4Pb_7^{4-} + 2Pb^{2+}I_2 = 9Pb + 4NaI \qquad . \qquad . \quad (80.1)$$

These polyanionic complexes become labile if the polarizing electric field of the cation is strong. This is the reason why they are most stable with the heavy alkalis and the ammoniates. This is analogous with the polyiodides: $NaI_3.H_2O$ decomposes on dehydration but CsI_4 is stable without water. The anions seem to be held together by polarization of the uncharged metal atoms by the original anion; polarization from a cation disturbs them.

Ordered solid solutions of one metal in the other occur under rather special circumstances. The two kinds of metal atoms must have in general similar radii, within a limit of some 15 per cent, to make it possible for one kind to intrude into the lattice of the other at all, without distorting it to an extent which makes its existence impossible. Lithium, with its small atom, is immiscible, even in the liquid phase, with the other alkali metals!

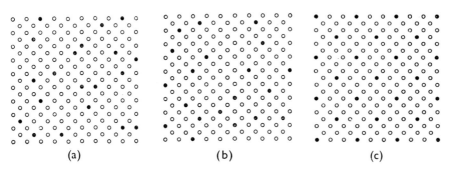

(a)   (b)   (c)

Fig. 136. **Structures of copper-gold alloys**

(*a*) Random distribution after quenching.
(*b*) Annealing begins to establish order, at first at a short range. Copper–gold pairs are favoured as against gold–gold pairs.
(*c*) Long annealing establishes long-range order.

(From W. Hume-Rothery, *Electrons, Atoms, Metals and Alloys*, pp. 276–7, Fig. 122 (London, Iliffe, 1948). By courtesy of Iliffe Books Ltd. and the Institute of Metals.)

The radii being within this limit, a certain amount of difference in electrochemical character (positive-negative, polarizing-polarizable) seems to be necessary to stabilize a particular structure within the solid solution: a structure in which different atoms are more likely to be beside each other than they would be, if the distribution were strictly along the lines of probability (as if white and red balls were distributed by chance over the sites of a simple lattice).

Such a case seems to hold for the pair, gold–copper. Their radii are 1·28 and 1·46 Å respectively, and copper has 29 electrons as against the 79 electrons of gold, which increases the polarizability of the latter. If alloys of this pair of metals are quenched by cooling them quickly from their melts, their structure is random (Fig. 136). On cooling them slowly or on annealing the quenched alloys something very interesting happens. At the beginning a so-called short-range order is established. Copper atoms are more likely to be surrounded by gold atoms and vice versa, but many deviations still occur and it is impossible to predict the occurrence of a gold atom at some spacings distant from a copper atom. When, however, the ratio of gold to copper is 1:1 or 1:3 and annealing is prolonged, the far-range order becomes practically established also, and copper and gold atoms alternate in the first case, while in the second gold atoms arrange themselves into the corners of the face-centred cubic lattice leaving the face-centre sites to the copper atoms (Fig. 137). It is evident that these ordered superlattices are more stable, more free from geometrical and electrochemical stress than the random arrangements. But the energy differences involved (mainly from polarization of the gold atoms) are not great and order may be established step by step without the abrupt appearance of a new phase.

Experimentally the existence of these superlattices shows up in the X-ray diffraction patterns. Certain planes within the crystal are filled with more

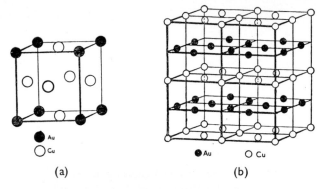

● Au
○ Cu
(a)

● Au　　○ Cu
(b)

Fig. 137. **Structures of copper-gold alloys**

(a) Space lattice of annealed superstructure of an alloy containing three times as many copper as gold atoms.
(b) Space lattice of annealed superstructure of an alloy containing equal numbers of copper and gold atoms.

(From W. Hume-Rothery, *Electrons, Atoms, Metals and Alloys*, p. 331, Figs. 154, 155 (London, Iliffe, 1948). By courtesy of Iliffe Books Ltd.)

strongly reflecting gold atoms (more electrons!) instead of copper atoms and the intensities of the diffraction lines change accordingly. The electric conductivity of the ordered superstructures exceeds that of the random orientation because the electron waves fit more easily into an arrangement with strict periodicity.

It is clear that the stability of such a regular arrangement of different metal atoms increases with the difference of electrochemical character between the components. We arrive at a structure that resembles faintly the structure of heteropolar compounds where order is nearly absolute and dislocations very rare. Thus, beginning with the face-centred cubic structure of Cu–Au, we see the same stabilized structure in Cu–Be, Ag–Mg, Fe–Al, Ag–Zn, Au–Zn, Ni–Al, etc. As in the case of ionic structures (p. 351), an appreciable difference in radii of the components makes it impossible to attain a highly coordinated structure and other structures with fewer neighbouring atoms are resorted to: Na–Tl, Li–Zn, Li–Cd, Li–Ga, Na–In, and Li–Al have structures where each atom has four neighbours of the same sort and four others of the opposite partner. At quite specific ratios of the radii some new, closely-fitting structures may become possible as specific solutions of this atomic jig-saw puzzle. In Al_4Ba, for instance, each barium atom has 16 aluminium atoms and 4 barium atoms in its neighbourhood.

82. METALS, IV: ELECTRONIC COMPOUND ALLOYS

Although the structural sequence we are going to deal with occurs quite often, we shall consider it with respect to the different lattices of the copper–zinc system which includes the common and important alloys we call brass.

Beginning on the copper side of the alloys we first encounter the solid solution of dispersed zinc atoms within the face-centred cubic copper lattice, called α-brass (Fig. 138). At around 30 per cent (weight) of zinc this structure is discontinued and a new, body-centred lattice, so called β-brass is formed instead. At somewhere about 50 per cent a new, very complicated structure (γ-phase) appears, consisting of 52 sites within the repetitive unit of the lattice. As the zinc concentration continues to increase this parent structure is retained, but one after the other atoms drop out of lattice sites leaving vacancies behind them. Of course this cannot continue indefinitely. On further increasing the zinc percentage this loosened structure collapses and a new phase, the dense, so-called ε-brass appears with a closely-knit, hexagonal structure. This is followed finally by the lattice of metallic zinc, a nearly hexagonal lattice with intruding copper atoms at some of its lattice sites (η-brass). For some reason, no phase was designed by the Greek letter δ.

Now, identical series of intermediate phases are known to occur in a number of systems, e.g. in Cu–Si, Cu–Al, Cu–Zn–Al, Cu–Sn, etc. *Hume–Rothery* was the first to remark, that the limits of stability in these seemingly very different systems obey a new kind of stoichiometry: the stoichiometry of the electrons versus lattice sites. Thus the upper limit of stability of the primary, so called α-phase (zinc dissolved in the copper lattice) sets in when the ratio of valency electrons which are free in the metallic lattice to the number of atoms exceeds $4:3$, and the composition of the new, β-phase begins with the ratio $3:2$. Compositions in between separate into both these structures giving rise to a microscopically heterogeneous alloy. It is quite irrelevant how this ratio came into existence. The ratio $3:2$ for instance, is built up in Cu–Zn (with monovalent copper, *see* p. 370) from $1 + 2 = 3$ electrons for $1 + 1 = 2$ atoms; in Cu_5Si, $(5 + 4):(5 + 1)$; in Ag_3Al, $(3 + 3):(3 + 1)$, etc. It is very interesting to note here that the transition metals iron, cobalt, nickel and palladium also partake in β- and γ-phase alloy structures, but the Hume–Rothery rule only applies to them if we assign them no conduction electrons at all. This is not quite as absurd as it seems to be at first sight. The d-subgroup which is below the external group of two electrons in these atoms is gradually being filled up along the series of transition metals (p. 277) and the energy difference between an electron in the external shell or in the subshell below it, is very slight. For instance, monovalent cuprous (Cu^+) compounds have the d-shell filled whereas the blue, better known, cupric series of Cu^{++} ions has lost one electron from this

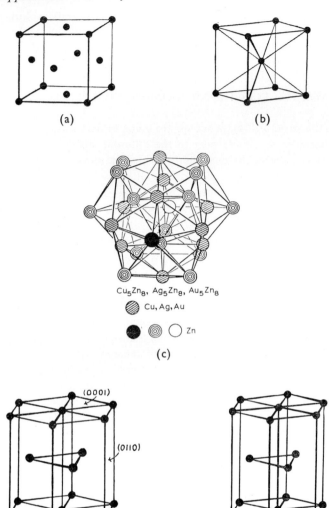

(a)

(b)

Cu_5Zn_8, Ag_5Zn_8, Au_5Zn_8

◙ Cu, Ag, Au

● ◎ ○ Zn

(c)

(0001)

(0110)

(d)

(e)

Fig. 138. **Structure of copper and zinc and of their alloys, the brasses**

(a) Lattice of copper, cubic face-centred. May contain some zinc atoms in solid solution: α-brass.
(b) Body-centred cubic lattice of β-brass, contains more Zn.
(c) Complicated lattice of γ-brass, contains more Zn.
(d) Hexagonal lattice of ε-brass, contains more Zn.
(e) Nearly hexagonal lattice of pure zinc; may contain some copper atoms in solid solution. (Vertical axis longer than for hexagonal).

(a, b, d, e from W. Hume-Rothery, *Atomic Theory for Students of Metallurgy*, p. 114 Fig. 36; p. 115, Fig. 37 (London, Institute of Metals, 1955). By courtesy of the Institute of Metals. (c) from W. Hume-Rothery, *Electrons, Atoms, Metals and Alloys*, p. 315, Fig. 143 (London, Iliffe, 1948). By courtesy of Iliffe Books Ltd. and the Institute of Metals.)

subshell. It is easily conceivable that the energy set free on forming the alloy lattice is sufficient to overcome this slight energy difference and that these metals really exist in the lattice with their external electrons drawn back into their d-subshell, without contributing electrons of their own to the common lattice structure.

The electron-to-atom ratio in the complicated γ-phase is $21:13$ and this ratio may be maintained not only by appropriate mixing of metals of different valency but also by dropping atoms out of the elementary geometrical lattice unit so that the number of electrons within this elementary cell remains constant. This was why we defined the Hume–Rothery rule at the beginning of this chapter as the ratio of electrons to "*lattice sites*", though until now, all lattice sites having been occupied by atoms, this seems to be the same as if we had stated "ratio of electrons to atoms". Here, however, empty lattice sites become possible and we see that it is the number of sites and not the number of atoms that really matters. This ratio $21:13$ is present in some ternary alloys also, e.g. in Cu_6Zn_6Al $((6 + 12 + 3):(6 + 6 + 1))$ or $Cu_8Zn_2Al_3$ $((8 + 4 + 9):(8 + 2 + 3))$ and $Na_{31}Pb_8$ $((31 + 32):(31 + 8))$ which is near, but not identical with the simpler ratio of Na_4Pb!

The Hume–Rothery ratio for "ε-phases" is $7:4$, corresponding to $CuZn_3$ $((1 + 6):(1 + 3))$, Ag_5Al_3 $((5 + 9):(5 + 3))$.

Thus alloys of quite different "stoichiometric formulae" come under the same heading and show the same structure because of the Hume–Rothery rule. In order to understand these quite unexpected ratios to some extent it is necessary to go back to the theory of the Brillouin zones (p. 365). We remember that the curves relating electron number to the sum of the zero-point energy (Fermi energy) of electrons in a lattice had maxima and minima because of the limits of these Brillouin zones for the electrons in a lattice where total reflection of electron waves excluded some values of electron energy. In such cases, as the number of electrons increases the new ones per lattice unit have to be accommodated in higher-energy states, the intermediate states being excluded by total internal reflection. Thus they soon begin to fill up the adjacent Brillouin zone and necessarily the sum of the energy of all electrons is now higher than it would have been if all lower states had been open. However, it is closed, nothing can be done about it. There is only a radically different alternative: the whole lattice structure may change and with it the set of internal total-reflections which are of course bound up with the geometry of the lattice. It may happen, and does happen that another lattice structure is able to accommodate the same number of electrons at a smaller sum-total of electronic energy because a zone of forbidden energies has been displaced. If we plot the curves of total energy against number of conducting electrons for different lattice structures (Fig. 139) it happens that, at the beginning the curve of one lattice is lower than that of another but at a given number of electrons per lattice unit they intersect and the curve of the other structure lies lower. At this point, then, the second structure has become poorer in energy and thus more stable than the former. We have reached a point where one structure ends and another begins. It is one of the "magical" electron-per-site numbers of Hume-Rothery.

All this is open to theoretical calculation and at least for the first one or two Brillouin zones the calculations are accurate enough. Later on, on reaching higher zone-limits, other structures may take over which have been calculated also but are more difficult to evaluate. The first peak on the curve of the number of electrons versus energy in a face-centred cubic lattice lies at 1·36

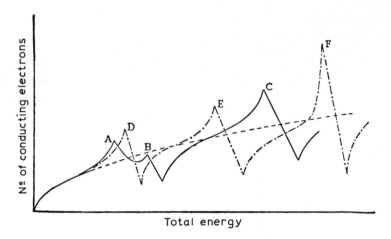

Fig. 139. **Electron distribution curves for face-centred and body-centred lattices**

The shape of the curves is determined by the crystal structure. The number of electrons which have to be accommodated under the curves is the total number of valency electrons: it increases by one each time a Cu atom is replaced by a zinc atom. Hence areas which have been shaded in Figs. 133 and 134, advance towards the right with increasing concentration of zinc in the brass. Whenever the total energy of the system (shaded areas) under one curve exceeds the total energy below the other curve, the lattice with higher energy becomes labile and changes into the one with lower total energy. In the case of this figure, the change occurs at the atom ratio 1·36:1, so far as other energy differences may be disregarded.

———— Electron distribution curve for face-centred lattice.

– · – · –   Electron distribution curve for body-centred lattice.

– – – –   Fermi distribution curve for an "electron gas" without crystal lattice.

(From W. Hume-Rothery, *Atomic Theory for Students of Metallurgy*, p. 281, Fig. 116 (London, Institute of Metals, 1955). By courtesy of the Institute of Metals and the Royal Society.)

electrons per atom, while that of the body-centred cubic lattice is at 1·48. Thus, as the concentration of free electrons per cell rises from 1·36 to 1·48 the curve of the face-centred lattice falls off rapidly while that of the body-centred lattice continues to rise. Therefore the average energy of the electrons is less in this region for the body-centred lattice and (if other kinds of energy differences, e.g. electronegativity, remain unchanged) energy is set free if the face-centred lattice changes over to the body-centred lattice. This is what actually happens: the Hume–Rothery limits for the end of the α-phase and for the beginning of the β-phase were 4:3 = 1·33 and 3:2 = 1·50, respectively, that is very near to the theoretical values 1·36 and 1·48.

The same calculation shows that at the upper limit of the β-phase the γ-phase becomes stable at an electron concentration of 1·54 electrons per lattice site, whereas the Hume–Rothery ratio 21:13 is equal to 1·61. The ratio 7:4 for the hexagonal phase is also confirmed.

It should be kept in mind that the number of electrons can only be increased if we substitute a metal atom with more valency electrons instead of one that possesses fewer. We referred to increasing the concentration of electrons but actually increased the zinc concentration relative to copper. In the transition stages both phases are present, for example some of the crystallites in a certain brass may belong to the α-phase while other crystallites alongside it belong to the β-phase. The relative amount of β:α increases with increasing zinc concentration. Naturally the mechanical properties of an alloy are determined to a large extent by the simultaneous presence, and the relative amount, of such different phases. The old magic art of metallurgy begins to open up for scientific understanding.

The Hume–Rothery ratio 3:2 corresponds, as we have seen, to the "compound" formula $CuZn$. But it would be a grave mistake to regard it as a compound in the regular sense of chemistry. First of all, the theoretical ratio was not even 1·50 but only 1·48 which happens to be near. Secondly, it is in no way necessary that the copper and zinc atoms should be ordered as in a superlattice. A random arrangement satisfies the requirements of the zone theory just as well, as long as the concentration of electrons over the lattice sites is the same and as long as the dimensions of the lattice, the distances between adjacent internal planes, are not changed either. And lastly, we have only defined the *limits* of compositions, solid solutions of one metal in a mixed stable phase being allowed as long as another lattice structure with less total energy does not take over. It makes no sense at all to force these limits into strict chemical formulae.

83. METALS, V: INTERSTITIAL ADMIXTURE

Up to now we have considered alloys in which atoms were exchanged for other atoms of approximately similar radii. There exists, however, another type of solid solution. Here the difference in radius is so great that the original structure of the metal need not be disturbed on introducing into the lattice new atoms of a smaller kind: they fit into the holes between the atoms of the original structure. The structure of the parent metal may be changed in some cases, but whenever smaller atoms of another kind are built into holes between larger ones in a metal, we speak of an interstitial structure.

Simple measurements of density show already whether the added atoms are built into the holes, or replace atoms of the other metal. For interstitial compounds the volume does not increase, or hardly increases, on admixture of the smaller element (Fig. 140).

Naturally the smallest atoms are most easily fitted into such interstices; for example those of hydrogen, lithium, boron, carbon, nitrogen and, to a small extent oxygen, that is, the first-row elements of the periodic system. It should suffice to remember that steel belongs to this group of compounds,

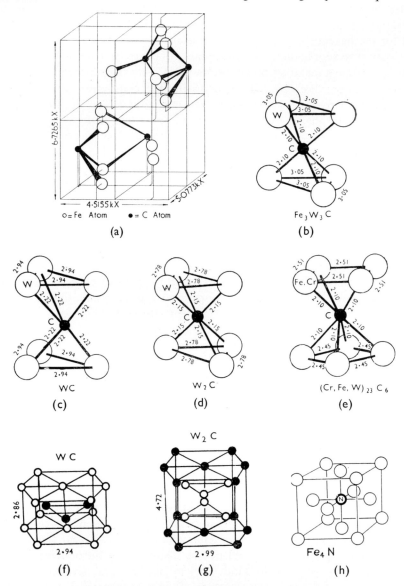

Fig. 140. **Lattices of interstitial metal compounds**

in order to appreciate their importance. Their hardness and cohesive force increases to a great extent; naturally so, because holes are replaced with atoms which introduce additional bonds into the lattice. And bonds mean cohesion. The carbides of tungsten (WC, W_2C) constitute the hardest admixture to special steels which serve as knives on the turner's lathe.

It is not necessary that the interstitial compound be a metal. It might perchance be a compound of regular valency. For instance, the gamma form of aluminium oxide (alumina) has nearly the density it would have if oxygen atoms were placed within the aluminium lattice. Maybe this explains why the surface oxide layer adheres so firmly to aluminium as to protect it from further oxidation by the air or moisture, even though it is sufficiently electropositive to react with water if the oxide layer is removed (for example, by amalgamating its surface!) Corundum, another modification of Al_2O_3 has an even more compressed structure with higher density; no wonder that its hardness ranks immediately after that of diamond.

The carbides, nitrides, and borides are metallic in their electrical conductivity, decreasing in the order: metal, carbide, nitride, boride. Some of them are even capable of becoming superconductors in the neighbourhood of the absolute zero of temperature: a typical metallic behaviour.* Interstitial compounds represent the metals with the greatest cohesion of their lattices. HfC and TaC melt around 4,150°K, and the highest melting-point ever encountered belongs to the ternary carbide, Ta_4ZrC_2, which melts at 4,215°K. They are nearly as hard as diamond and very resistant, e.g. against acids.

The limiting radius ratio for interstitial addition in between closely packed spheres is—from reasons of pure geometry—1·7:1. (Spheres of unit radius just find space for themselves in holes between closely built spheres of 1·7 units radius). The technically so-important pair of atoms Fe:C is slightly below this limit. Therefore the iron carbide "cementite" which is contained in steels, Fe_3C, must be built from the loosest structure which is known for iron, the so-called gamma structure, stable at high temperature. Only thus can the carbon atoms be accommodated between the iron atoms. Ni_3C is much less stable than cementite and Co_3C cannot be isolated in the pure state.

Cr_3C_2 has a very remarkable structure. Its carbon atoms are in a zig-zag row, so X-ray analysis reveals, 1·64 Å apart from each other, embedded in a lattice of chromium. It resembles a slightly dilated paraffin chain with chromium atoms instead of hydrogen atoms. And in fact, dissolution in acids yields paraffins.

In Fe_4N the iron atoms are in the cubic closest packing and the nitrogen atoms in the centre between six adjoining iron atoms. AlB_2 is different. It is built from layers of boron hexagons, like the carbon hexagons of graphite, with aluminium atoms between these layers.

Throughout the whole discussion it must be borne in mind that we are not dealing with continua. In building a house we may choose between bricks of any size, or make our iron structures and concrete parts to any size we desire. In the domain of chemistry we have a limited number of atoms at our disposal, with atomic and ionic radii of their own. These must be fitted into structures. We simply do not have any other building material at our

---

* The conductivity of all metals increases with decreasing temperature. But there are some metals and alloys which suddenly lose practically all their resistance at a critical temperature, so that a current generated by induction in a ring of such a superconductive metal is capable of flowing undisturbed for hours, or in some cases even much longer.

disposal. Radii of different atoms fit more or less into a geometrically optimal arrangement: we cannot alter it. The great variety of structures is the result of a series of compromises between geometry and physical reality. The radius ratio is the most important factor, and it primarily determines the jig-saw problem. On second approximation, differences in electron affinity, quasi-covalent bonding and electron-lattice site ratios become important also. All of them together determine structures and properties.

84. CLATHRATE COMPOUNDS

Somewhat similar to interstitial atomic lattices are crystals where molecules of one kind are built into lattices of other molecules just because they happen to fit into the free spacings.

If nickel cyanide is crystallized in the presence of benzene by shaking its aqueous ammoniacal solution with the latter, a precipitate is formed in which a benzene molecule is trapped in each hole of the cubic lattice. The dimensions happen to agree, and van der Waals energy can be gained by inserting anything alongside any other set of atoms because the induced moments of the electron shells attract each other. The energy is not great, and could not change the structure. But if the structure happens to agree with the dimensions of the enclosed molecule even this small amount of energy is sufficient to stabilize the adduct.

The molecules of hydroquinone (quinol) are arranged in such a manner in the crystals that they are able to accommodate molecules like HCl, HBr, H_2S, CH_3OH, SO_2, CO_2, HCN, etc., in between them. At a pressure around 40 atmospheres, even the rare gases argon, krypton and xenon can be trapped in such structures. Three organic molecules are able to enclose one molecule of the clathrate (caged) compound.

A very interesting and even technically important clathrate system is the one between urea

$$O{=}C\underset{\textstyle NH_2}{\overset{\textstyle NH_2}{\big<}}$$

and the straight-chain paraffins. The crystals of urea are built in such a manner that pairs of its molecule join together leaving an empty channel between them which runs through the lattice (Fig. 141). The dimensions of this lattice happen to accommodate exactly a straight paraffin chain: $CH_3{-}CH_2{-}$ . . . ${-}CH_2{-}CH_2{-}CH_3$ along these channels. If the chain of a paraffin is branched:

$$CH_3{-}CH_2{-}CH\underset{\textstyle CH_2{-}\cdots{-}CH_2{-}CH_3}{\overset{\textstyle CH_2{-}CH_2{-}\cdots{-}CH_2{-}CH_3}{\big<}}$$

there is no room for it in the urea crystal. Thus if one has to separate straight-chain and branched-chain paraffins, which is by no means an agreeable task, one has only to prepare a saturated solution of urea in methanol, add the mixture of paraffins and induce crystallization. The mass of urea crystals occludes all straight-chain paraffins while the others remain in the methanol. The crystals are filtered off and dissolved in water, whereupon a layer of straight-chain paraffins floats on top and can be easily recovered. This is a very interesting and important example of a naturally occurring jig-saw arrangement.

Fig. 141. **Normal hydrocarbon chain built into the long channels of the urea lattice**

It is held there by van der Waals' forces.

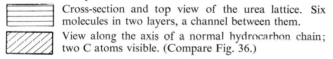

Cross-section and top view of the urea lattice. Six molecules in two layers, a channel between them.

View along the axis of a normal hydrocarbon chain; two C atoms visible. (Compare Fig. 36.)

Evidently the crystal structure suitable for enclosing clathrate molecules or atoms must be a loose, open one. But it must be held together with sufficient force to guarantee its existence. The cavities within its structure must either have some access from without to enable the clathrate component to enter it, or the clathrate component must be available in sufficient concentration at the time the enclosing lattice is being formed to be ready on the spot when the cavity is about to be closed.

85. SOLUTIONS

Two entirely different forces are at work whenever two or more compounds dissolve in each other.

The first is not a real force at all, but is the tendency to acquire a configuration of great probability. Given a certain volume, there is a higher probability of finding molecules evenly dispersed throughout it than beside each other in a corner (p. 143). If there were no forces acting between them, molecules would disperse; their thermal, kinetic, motion would not allow them to cluster together. The smaller these forces, the more the molecules will disperse. They will form gases and vapours and fill all the space they are able to enter. In the same way they will intermingle with molecules of a solvent if the forces which kept the solvent molecules together are not great in turn.

The second force which makes one compound dissolve in another is a real attraction between the molecules of these two compounds. In this case even considerable cohesion of solute molecules between each other may be overcome because the force between solute molecules A and solvent molecules B may be greater than the force acting between A and A or greater than the force acting between B and B. By the way, the distinction between solute and solvent is sometimes not so clear-cut as one would suppose from everyday life. We say that sugar, salt, and carbon dioxide gas dissolve in water as a solvent because the reverse process, the dissolution of water in solid sugar, salt, or carbon dioxide gas is very much less extensive. Therefore we consider these three as being solutes. But what if water and alcohol, or benzene and acetone are mixed with each other? The distinction becomes meaningless and surely unimportant for these cases of complete miscibility.

It is easy to say whether the first or the second cause of dissolution prevails. Whenever the sum total of forces which hold the two pure phases A–A and B–B together are greater than the forces which tend to mix the two into A–B, energy has to be expended to bring about dissolution. That it is brought about at all is due only to the cause we have mentioned first, to the increase of probability (of "entropy", *see* p. 147) during this process. Therefore dissolution proceeds, but the difference in energy must be covered by the heat content of the system, which will cool during the process of dissolution. This happens if, for instance, common salt, ammonium chloride, or fixing salt (sodium thiosulphate) is dissolved in water. On the other hand, heat is evolved on dissolving concentrated sulphuric acid in water; one has to take care of one's eyes and hands in doing this in the laboratory. The same applies when sodium or potassium hydroxide dissolves in water. In these cases the attraction between A–B is greater than the sum of attractions A–A and B–B and the difference in energy appears as the positive heat of solution.

The sign of the heat of solution only indicates, however, whether the energy or the entropy of mixing is more important. To some extent both play their part during dissolution: in one case working hand in hand, in another energy working against entropy, against the geometrically most probable distribution.

Fundamentally a solution may be solid, liquid or gaseous. Generally, however, we consider gas molecules to be so far apart that their interaction is of secondary importance and thus a gaseous solution simply implies a mixture of different gases. Within this "solution" the energy term is not important and the tendency to permeate space dominates. As pressure increases, the gas molecules come nearer to each other, the forces of attraction begin to increase (e.g. the laws of ideal gases are followed even less than normally) and the energy term may not be neglected any more. Liquid solutions are normal in the ordinary sense of the word. Solid solutions are, for example, some of the alloys we have dealt with, and mixed crystals, where an ion or a molecule of similar shape and magnitude occupies some lattice sites of a parent crystal. For instance, the purple permanganate ion, MnO_4^-, has similar dimensions to those of the sulphate ion, SO_4^{2-} and a long series of solid solutions between $KMnO_4$ and $BaSO_4$ can be prepared. This would not be possible if the two respective positive ions, K^+ and Ba^{2+} were of very different size; in fact their radii are practically the same (1·33 and 1·35 Å). In this case, the water-soluble $KMnO_4$ is built into the lattice of the insoluble $BaSO_4$, and as long as it is only present in small amounts it becomes enclosed and thus insoluble itself.

Liquid solutions may be ionic or not according to whether the solute dissociates into electrically charged ions or not. Dissociation and thus dissolution of ionic compounds itself is facilitated in solvents of high dielectric constant, because this constant tells us by which factor the force between two electric charges is diminished in the given medium compared with the force *in vacuo*. Liquids having a high dielectric constant, such as water, liquid ammonia, liquid hydrocyanic acid, and to some extent methanol, are good solvents for electrolytes, for ionic crystals, because the force which holds the ions together is diminished in these solutions. Solvents such as benzene, petrol and paraffin oil on the other hand have very low dielectric constants and thus any ions which happened to be dispersed in them would attract each other with nearly undiminished force and so quickly recombine into solid crystallites.

The term "dielectric constant", however, is a term of macroscopic physics. What happens between the ions and the molecules of the solvent which has a high dielectric constant? It is not difficult to answer this question. High dielectric constant means easy polarization of the solvent by electric fields, either because the solvent consists of electric dipoles as molecules and/or because its molecules become asymmetric, are polarized themselves (p. 292) in an electric field. In both cases the electrically charged ion becomes surrounded by solvent molecules in such a way that the oppositely charged— or polarized—ends of them come into contact with the surface of the ion. Electrostatic energy is thus set free and stabilizes the new, dissolved and "solvated" configuration. Water with its triangular molecule having the negatively charged oxygen atom at its apex, is the prototype of such a solvent: water molecules will attach themselves by their oxygen ends to cations, the more strongly the smaller the latter are and the more positive the charge they carry. Anions are generally larger because they have built electrons into their structure instead of having lost them (p. 346) and therefore they solvate or form hydrates less than the cations. But when they do so, they attract the two hydrogen atoms of the water molecule triangle: the hull of hydration around a cation is of opposite structure to that around an anion.

If the solute is not ionic, the dielectric constant, that is the dipole moment and the polarizability of the solvent are unimportant. Instead of these, the general van der Waals attraction between molecules or atoms of any kind becomes more pertinent. It is less than the electrostatic force we have just been discussing, but the force which holds together a non-ionic crystal is also much weaker than the force between adjacent ions. Thus there is no reason why benzene should not easily dissolve paraffin wax, and it does dissolve it. But the heat effects accompanying this dissolution are of course very small.

An old chemical proverb, probably dating back to the alchemists, states that *"similia similibus solvuntur"*: similar bodies dissolve each other. This proverb is true. Compounds containing hydroxyl groups are likely to dissolve each other and compounds with C—Cl, C—Br, etc., bonds do the same as a general rule. This leads us into the finer details of dissolution. A bond between two atoms has a geometry and an electrical structure of its own. Two such bonds beside each other (Fig. 142) may easily form a labile but still to some

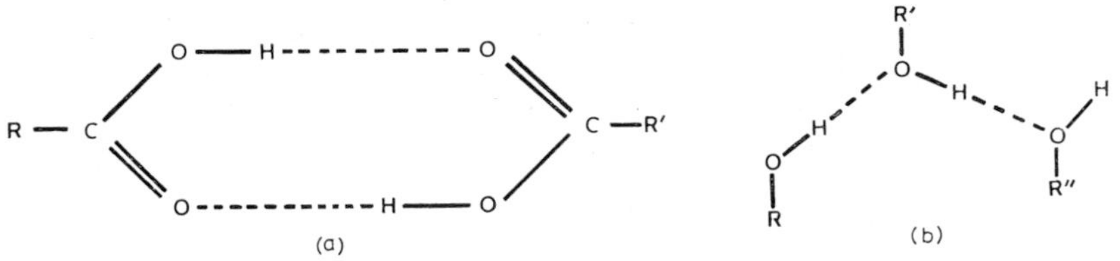

Fig. 142. **Hydrogen-bonded carboxylic acid** (*a*) **and alcohol** (*b*) **molecules**

extent coherent structure which liberates some energy, e.g. by hydrogen bond (p. 354) or quadrupole formation (pp. 294–313). They "fit" better than other bonds with different dimensions and electrical disposition. Therefore the structure A—B between the two sorts of molecules will be favoured and the tendency to dissolve increases.

It is important in this connexion to distinguish between macroscopic and microscopic polarity or apolarity. A molecule may contain two highly polar bonds exactly opposite each other as, for example, dioxan or para-dichlorobenzene:

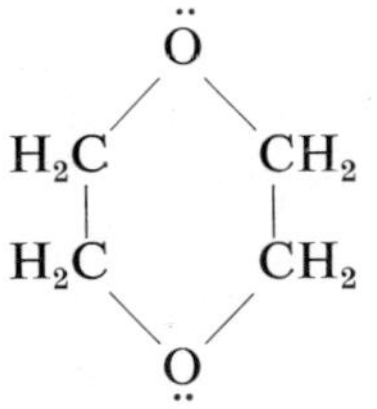

Dioxan

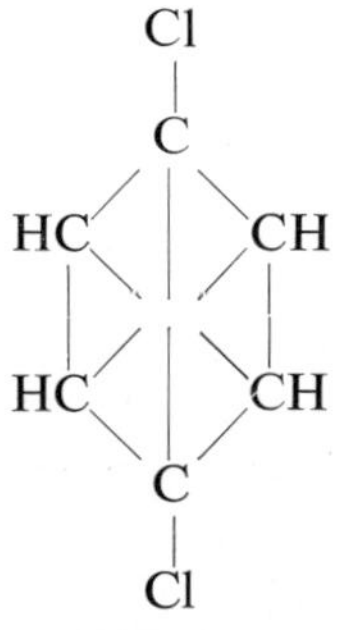

p-Dichlorobenzene

The two dipole moments compensate each other, and the molecule as a whole will not be turned round, oriented, in an electric field. Its dielectric constant will be low, as long as polarization is not important. But seen from a neighbouring molecule which is as small as the solvent molecule itself this compensation is no longer valid: the individual moment of the C—Cl bond or the triangularly bound oxygen atom may and do come into action. This is the reason why dioxan is such an extraordinarily versatile solvent: it may act as a microscopic polar molecule towards polar groups or as a compensated dipole-free system with dipole-free molecules and even as partner of a hydrogen bond. It is extremely suitable for the simultaneous dissolution of polar and apolar compounds.

An exact theory of solutions, especially of liquid solutions is still outstanding and will probably be extremely complicated. This is not surprising in view of the fact that the theory of the liquid state itself is much less advanced than either the theory of gases or of crystals. The former can be approximated by the picture of randomly moving molecules with negligible interaction during their free path of flight between collisions. The latter is a very regular arrangement of atoms or molecules with a periodically recurring pattern. A liquid is much less regular and the interaction between neighbours is certainly not negligible.

Some beginnings have been made for such simple cases as those where the forces between A—A, B—B and A—B are not very different in apolar solutions and for dilute solutions of electrolytes (p. 182). Maybe some similarities for very concentrated electrolyte solutions will emerge in the not very distant future, but we cannot yet be certain about it. At all events, solutions will remain one of the domains where the nearly limitless variability of distinct chemical entities will dominate, making exact theoretical advance very difficult and leaving much room for the intuition of good chemists.

Part 7

Atomic Nuclei

86. RADIOACTIVITY

We have learned by now that the overwhelmingly largest portion of an atom consists of the space where the more or less delocalized electrons are to be found, while practically all the atomic mass is concentrated into the tiny nucleus of some 10^{-13} cm diameter. Everything we denote by "chemistry" is determined by the structure of the successive electron shells around this nucleus especially of the outer shell; we dealt with them in Part 5 of this book. As far as the chemical structure of matter is concerned we have seen the fundamentals and we might well have concluded our studies at the end of Part 6. However, the structure of atomic nuclei has reached such exteme importance in the last two decades that it seems worth while to inquire a little into the fundamentals of this structure also, and obtain at least a glimpse of that sub- or super-chemistry, that is being unravelled before our eyes. We shall see that a certain amount of similarity exists between the structure and reactions of atomic nuclei and that of their electron shells. There is the enormous difference, however, that the spatial dimensions are about a hundred thousand times smaller, and the energies at least as many times greater, than for atoms in the realm of normal chemistry. In the course of our discussions during previous chapters we have already alluded to atoms of the same element with different atomic weights (isotopes, *see* pp. 39, 55, 231, 288), and to radioactivity (pp. 17, 250, 279). Therefore it is virtually imperative for us to go on and explain these facts to some extent, that is, to dive into the subatomic dimensions of the atomic nuclei.

In Section 53 we noted but could not explain the fact that the atomic weights of the light elements in the periodic system are very often near to integral numbers, much more often than should occur by pure chance. If we assume that it is really not by chance that this comes to be, we have no choice other than to ascribe some sort of structure to these nuclei and explain this interesting fact in a way which is a consequence of their structure.

In Section 58 we mentioned alpha particles which are emitted from radioactive elements in the course of their disintegration. We used them to show how tiny the nuclei must be which scatter them if they happen to by-pass them at a small distance. Now we must learn how these alpha particles were first discovered: we shall have to deal with the phenomenon of radioactivity. And in dealing with it we shall gain some insight into the structure of nuclei.

Radioactivity was discovered by *H. Becquerel* in 1896, in Paris. He observed that a salt of the element uranium (the heaviest element known at that time) gave a photographic effect upon photographic plates or paper, even through a black envelope. A similar effect was produced by X-rays.

He also found, again similarly to X-rays, that this active radiation was absorbed more strongly the higher the atomic weight of the material through which it had to pass. Very soon it also became apparent that a series of materials became fluorescent in its neighbourhood and that the air around it acquired electrical conductivity—just as with X-rays. We know, of course (p. 250), that this conductivity is due to the fact that the molecules of air are knocked apart into electrons and positive ions by some concentrated attack of energy. The whole series of effects proved to be due to the presence of uranium: the metal itself and all its compounds showed them in a degree that was proportional to their content of uranium.

More exact chemical analysis, however, soon revealed the fact that this new phenomenon, termed radioactivity, was not confined to uranium itself. Other chemical fractions which were isolated from pitch-blende, an uranium mineral—showed the same effect. The renowned French scientist *P. Curie* (1859–1906) and his Polish wife, *M. Sklodowska* (1867–1934) undertook this chemical analysis under extremely difficult circumstances and succeeded eventually in isolating two new, strongly radioactive elements: polonium and, later on, radium (1898). In order to appreciate the enormous difficulties of this work it is worth while noting that one ton of mineral yielded only 0·1 g of radium metal! Only the strong ionization of air by radioactive material allowed the Curies to follow step by step the accumulation or depletion of it in different fractions of their preparative analysis. Radium at last accumulated in the fraction which contained barium and had to be separated from it by a long series of fractional crystallizations. All the chemical and physical properties of the new metal, including its spectrum, placed it among the alkaline-earth metals. Its atomic weight was 226 and this value fitted well into the periodic system, below barium.

Radium is approximately a million times as radioactive as uranium. Its salts fluoresce by themselves in the dark; water or other compounds in its neighbourhood are easily smashed into their constituent atoms. It generates a steady flow of heat: 25·2 calories per gramme of radium in an hour. (This means that the temperature of 1 g of water would rise by 25·2°C.) And it was not used up in this process—or at least so it seemed. Was it the *perpetuum mobile* at last? Is energy generated from nothing? For a long time there was no explanation forthcoming.

Radium and its compounds are solids at normal temperature. When freshly prepared, so that the air or gas which previously surrounded it is removed, the fresh air around it is free from radioactivity. But already in the course of a few hours the situation changes and any gas which surrounds the radium or its compounds manifests an increasing radioactivity until, after a few days, a limit is reached. The surrounding gas may now be removed from the vicinity of the radium. It remains radioactive, but not for ever. In less than four days the gaseous radioactivity has fallen to half its original value and it goes on halving and halving during successive periods of the same length. At the same time the vessel which contained the gas becomes radioactive on its internal wall, without any visible material having been deposited there. Still, it is safe to assume that some sort of condensed material must have accumulated on the walls, only in so small an amount that it cannot be seen.

It was possible to isolate the radioactive gas and to determine its density and thus its molecular weight. One gramme of radium accumulated 0·65 mm³ of this new gas if it was allowed to stand long enough. This was the limiting value. Of course this amount is very small, but not too small to be handled. Its molecular weight was found to be 222. Owing to the fact that radioactivity is an extremely sensitive indicator allowing one to determine the whereabouts of any radioactive element, it was even possible to examine the chemistry of this new element which was called radium emanation, or emanation or radon for short. (RaEm or Em or Rn). That is, it would have been possible to examine its chemistry if it had had any. But it had none. It behaved as a rare gas, and with its molecular weight of 222 it fitted well into the column of rare gases in the periodic system. Today, no doubt, it would be possible to prepare its fluorides, etc., as was the case with xenon and krypton (p. 278).

All this was extremely remarkable and did not fit in with the behaviour that one expected of chemical elements. Radium *seemed* to be a chemical element. No chemical method of attack could disintegrate it and it had all the characteristics of an alkaline-earth metal. Heating or the action of electricity would not decompose it. But if we simply wait without doing anything to it at all, it goes on evolving a rare gas, seemingly for ever and without any appreciable loss of weight. No external influence is able to change the velocity of this gas evolution: it goes on unchanged in the vicinity of the absolute zero point or at the highest temperatures which can be achieved in the laboratory. It does not matter which compound of radium we deal with, nor is it influenced by the electric or magnetic fields at our disposal. Whatever we do with it, this new gas in its turn decreases to one half of its initial amount in exactly 3·85 days while a deposit of some unmeasurably minute amount of radioactive material is generated by it. The evolution of heat during these radioactive processes is also independent of all external influence. What does all this mean?

E. Rutherford (1871–1937), the veteran master of radioactive research, and his associate, the chemist *F. Soddy* (1877–1956) were the men who found the answer to this question in 1902 and it has remained true up to the present day. They supposed that radioactive elements were not stable entities but went on changing, one into the other. Now we have arrived at the difficulty in definition which we mentioned in the beginning (p. 17). Why do we call these substances elements if they change? The direction of this change is quite obvious: radon (emanation) is lighter than radium, so that it needs must be—and is—the product of disintegration. Why do we not say then, that radium is a compound of radon with something we still have to find out about? Why call it an element? But radon is not more stable than radium itself since it decomposes even more quickly. And something new, some new, solid, radioactive material, ensues. Is radon a compound too?

Our answer to this question in Section 5 was, that it is much more sensible to decompose things step by step than to progress by huge leaps to what at the moment seems to be an ultimate building stone. The energy of decomposition processes in Nature is either within the limits that we are used to for chemical reactions or is some million times as great. In between there are no decompositions at all. We agree, sensibly, to call those reactions chemical reactions which proceed by absorption or evolution of amounts of energy which belong to the lower orders of magnitude. We have seen that not only is the energy of radioactive disintegrations a million times greater than these chemical energies: the process is also invulnerable to energies of normal, chemical intensity. It belongs to another order of magnitude, to another realm of occurrences in Nature. Therefore it is practical to agree that radium *is* a chemical element as long as it remains radium. It is certainly an alkaline-earth metal until something very special happens and it transforms spontaneously into another chemical element, radon.

The theory of Rutherford and Soddy also explains the source of the seemingly infinite energy and unceasing emanation from radium. Both come from the decomposing radium itself and they are neither infinite nor unceasing. According to this theory radium is being consumed in the process, but so slowly that it could not be observed by ordinary measurements. If we could live to see all the radium from a piece before us completely disintegrated, the evolution of emanation would stop and if no other energy-producing process went on instead, the production of energy would stop also. Radon itself is a good example of this, because it decomposes so much more quickly than radium. It is halved every 3·85 days. And parallel with the amount left behind, the amount of solid radioactive products formed by it decreases, just as the amount of heat evolved in the gas phase by the decomposition of the radon decreases also.

Let us try to calculate how long it should take for a piece of radium to be halved by natural disintegration. We have mentioned that $0.65 \, \text{mm}^3$ of radon is always found alongside one gramme of radium if the latter is enclosed long enough in an evacuated vessel to attain this limiting value. This evidently means that radon is now decomposing at the same rate as that at which it is formed by decomposition of radium; this is the reason that its "equilibrium" amount does not change. This situation is logically called radioactive equilibrium and it is evident that the amount of a radioactive element in equilibrium is the greater the slower its decomposition, the longer its "half-life".

The percentage of atoms of a given radioactive element decomposed in a given time is constant and we have seen that it cannot be influenced from outside. Hence the time necessary to lose half the atoms is a characteristic constant also: it is a measure of the intrinsic lability which we do not yet understand. The absolute amount of material disintegrated during a short time interval is thus proportional to the amount initially present (M) and inversely proportional to the half-life of the element, t. In equilibrium, the number of disintegrating parent atoms must be equal to the number of disintegrating daughter atoms. In the case of the pair Ra and Rn for example

$$M_{\text{Ra}}/t_{\text{Ra}} = M_{\text{Rn}}/t_{\text{Rn}}$$

From these four quantities only t_{Ra} is unknown to us and thus we may calculate it from this equation. If M_{Ra} was 1 g and M_{Rn} was 0·65 mm³ or $6·47 \times 10^{-6}$ g, t_{Rn} being 3·85 days, we arrive at the half-life of radium, $t_{Ra} = 3·85/(6·47 \times 10^{-6}) = 595,000$ days, that is 1,630 years.

It takes 1,630 years to lose half of any given initial amount of radium by spontaneous disintegration! We begin to see why it was impossible to observe that it was losing weight or ceasing to evolve heat and radon.

But let us think about the amount of energy evolved during the disintegration of one gramme of radium. It evolves $4·53 \times 10^{-8}$ g of radon in an hour, at the beginning. The ratio of atomic weights Rn : Ra is 222:226 and thus the amount of radium lost in this time must have been $4·62 \times 10^{-8}$ g per hour. This much radium has set free 25·2 cal of heat. This was the heat we could not previously ascribe to any visible process. Hence the disintegration of a whole gramme of radium will evolve $5·52 \times 10^8$ cal, half of this heat in the first 1,630 years. This is equivalent to the heat produced by the combustion of 70 kg of coal! 1,570 cal are set free when 1 g of nitroglycerine explodes. Now we see the difference in the order or magnitude between chemical and radioactive energies. It is the difference between a decomposing compound and a disintegrating element.

We have seen up to now that radium produces radon, and radon in its turn produces something we may call "radioactive precipitate" on the walls of the vessel. But is it probable that nothing else is being produced? The atomic weight of radium is 226 while that of radon is only 222. What happens to the difference? This difference amounts to 4 units of atomic weight and we encounter it not only in the transformation of radium into radon but in all cases of spontaneous radioactive decomposition where the atomic weights of parent and daughter element are found to differ. As we shall see, many of these radioactive decompositions give rise to daughter elements the atomic weights of which are found—by ordinary means—to be unchanged. Looking up the element with atomic weight 4 in the periodic system we find that it is helium. It was long known that natural gases which are found in the neighbourhood of radioactive mineral deposits are rich in this rare gas. The suspicion arose that there might be some connexion between these two facts and that it was really helium which was being formed in the course of some radioactive disintegrations. This suspicion was, however, only proved to be true in the course of detailed examination of the rays emitted by radioactive elements.

87. RADIOACTIVE RAYS, I: ALPHA RAYS

Many of the phenomena connected with radioactive decay proceed along straight lines. We are able to stop the blackening of photographic material alongside radioactive preparations, or the fluorescence of the ionization of gases, by putting a piece of lead in the way. In the "shadow" of the lead we do not observe anything. All this is very similar to what happens with X-rays. For a time it was supposed that the two radiations were of the same sort,

But it soon became apparent, that there is a very important difference between them. X-rays go on undisturbed through electric and magnetic fields but the rays from radioactive elements behave differently. They do not even behave in a homogeneous manner: a part of them goes on undisturbed too, but another part is deflected into a right-handed circle in a magnetic field while a third part is deflected the other way. According to the laws of electrodynamics positively charged moving particles are deflected to the right in a magnetic field (Fig. 143), while negative particles, for example, electrons, to the left. The positive components were called alpha rays, the negative ones beta rays, while those which proceeded without deviation were called gamma rays. Let us become acquainted with them in sequence.

Alpha rays only traverse a few centimetres in air of normal pressure and then disappear. If the pressure is reduced, the range of these rays increases

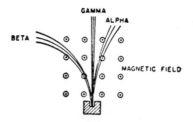

Fig. 143. **Deflexion of radioactive rays in a magnetic field**
(From S. Glasstone, *Sourcebook on Atomic Energy*, p. 57, Fig. 2.11 Princeton, N.J., van Nostrand, 1952). By courtesy of the D. van Nostrand Co., Inc.)

in inverse proportion. They have similar definite ranges in other gases too, the heavier the gas the shorter the range. In comparing these ranges it proves to be a good approximation to consider the total number of orbital electrons within the gas molecules traversed by the rays. The alpha rays are deflected not only by a magnet but also by an electric field and it is confirmed that they possess a positive charge. Using a method which is based upon the same principle as the one described on p. 201 and in Fig. 87 for electrons, it was possible to determine the ratio of their charge to their mass and to determine their velocity. It was only necessary to place a radioactive material in place of the incandescent cathode of the vacuum tube and to choose appropriate field strengths. Alpha and beta rays deviate in opposite directions and can be examined separately. These experiments showed that the velocity of alpha rays was around $1/10$–$1/20$ of the velocity of light and their e/m ratio was 1.44×10^{14} electrostatic units per gramme, irrespective of the particular radioactive element from which they were derived. The velocities differed from element to element but were constant for each.

All this seems to prove that alpha particles are the same no matter from which element they originate, but they start with different characteristic velocities. It was important to determine both their charge and mass, not only the ratio of these. This was done in an ingeniously simple manner by *Regener* in Germany and soon afterwards by *Rutherford* and *Geiger*. They

actually counted the number of alpha particles emitted into a given angle of
space and then determined the charge transported through this angle on to a
very sensitive electrometer. The fundamental set-up is sketched in Fig. 144.
Lead slits determine the aperture for the alpha-ray beam coming from a
radioactive preparation in a vacuum tube. Behind the second set of slits one
can arrange either a fluorescent microscope slide with microscope, or a
sensitive electrometer. If the preparation is weak enough it is possible to
observe the individual arrival of each alpha particle on the slide and to count

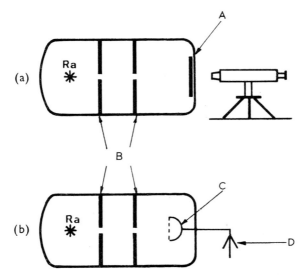

Fig. 144. **Determining the charge of alpha particles**

    (*a*) The number of alpha particles impinging in a given time is
        counted as scintillations on the screen.
    (*b*) The charge transferred to the Faraday cage during an identical
        period of time is measured on an electrometer.

        A = fluorescent screen   C = Faraday cage
        B = lead slits   D = electrometer.

them throughout a time interval. These individual light spots on the fluorescent screen are called scintillations. We have once more arrived at the limit
where the fate of individual elementary particles can be observed as in the
Millikan experiment on the charge of the electron (p. 194). Though we
are never able to foretell when and where a scintillation will take place on the
screen, the average frequency in the course of time, and the average density
of small flashes over the whole area is constant. Now we change over to the
apparatus where the electrons impinge upon the collector of the electrometer instead of upon the screen, and determine how great is the positive
electric charge that has been transmitted to it in a given time. On dividing
this charge by the number of alpha particles we know have impinged upon
the collector during this time, we arrive directly at the positive charge of an
individual particle. It proved to be 9.56×10^{-10} electrostatic units. Exactly
twice the charge of an electron!

The mass of the particle follows immediately from the charge and the e/m ratio $1\cdot44 \times 10^{14}$ found above: it is $6\cdot6 \times 10^{-24}$ g. The mass of the electron is $9\cdot0 \times 10^{-28}$ g and that of the hydrogen atom $1\cdot66 \times 10^{-24}$ g. We see that the mass of the alpha particle is of the order of magnitude of that of an atom, not of an electron. Knowing that atomic mass is concentrated in the nucleus we might as well say that it is of the order of magnitude characteristic of light nuclei. Actually it is just four times the mass of the proton, thus being equal to the mass of the helium nucleus! Helium is the second element in the periodic system and its nucleus has a positive charge of 2. These values are

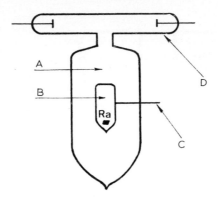

Fig. 145. **Helium from radium**

Initially both bulbs are evacuated. Alpha particles from Ra pass out of the thin inner bulb, accumulate in the thick-walled outer bulb and prove to have become He gas.

A = evacuated bulb
B = evacuated inner bulb containing Ra
C = Pt holder, which also serves to equalize charges in the two
    bulbs
D = Crookes spectral tube.

exactly as for the alpha particle: charge and mass agree. As we know, helium was found in the neighbourhood of radioactive minerals in natural gases. Are alpha particles and helium atoms really the same in the end?

They really are. The alpha particle is a helium nucleus emitted from the exploding nucleus of a radioactive atom which therefore loses four units of atomic weight, as we have seen. This can be proved directly. Rutherford and Soddy (1903) enclosed a piece of radium in an evacuated glass tube and found after a while that the evolved gases showed not only the new spectrum of radon but also the well-known spectrum lines of helium. On condensing the radon with liquid air, only helium remained in the gas phase.

Eventually, in 1909, the same scientists showed beyond doubt that helium was associated with the alpha rays themselves, by repeating this same experiment in a very thin-walled glass tube which was itself built into a larger evacuated tube. The alpha particles pierced the thin glass wall and after a few weeks helium was detected by its spectrum in the external vessel, though it had been perfectly evacuated and sealed at the beginning of the experiment. The hypothesis of disintegrating nuclei was completely verified (Fig. 145).

It is only now that we know what projectiles we have used in Section 58 to probe into the depths of the atoms. The projectile has been the free nucleus of a light atom, of helium, and the deviation of these projectiles from their original directions had enabled Rutherford and Chadwick to determine the dimensions of heavier nuclei. Or rather, to ascertain the smallest distance to which nucleus and alpha particle could approach each other without another force becoming active between them beyond the Coulombic repulsion (p. 253).

We have seen that alpha particles only proceed a distance of some centimetres in air of normal pressure. During this short path they lose their enormous kinetic energy, practically by collision with molecules and, by ionizing and knocking out one or two of their electrons. The mass of an alpha particle is some 7,000 times the mass of an electron and hence, according to the laws of elementary mechanics, it is the light electron that is going to be knocked out of its path with its newly acquired energy while the alpha particle will proceed in its original direction at a slightly reduced speed. Reduced, because it has lost energy to the electron, first by ionizing it and then by ejecting it with a certain amount of kinetic energy. On the average it loses a given amount of energy with each ionization. At the end, when it has lost all that has kept it in straightforward motion it goes on with the normal thermal velocity of a helium atom and is knocked about itself. On meeting electrons at this stage it is able to attach them to itself in its K shell, without having them detached again by impacts. It attaches them and continues its life as an ordinary atom of helium with its two electrons.

Alpha particles which have been emitted by radium traverse 3·3 cm of normal air, some hundredths of a millimetre of an aluminium foil or some tenths of a millimetre of mica. During this journey each one generates some 130,000 ion pairs. Their initial velocity has been determined in the experiments described on p. 394 as 1.5×10^9 cm/sec. Knowing now the mass of an alpha particle we are able to calculate its kinetic energy from $mv^2/2$, and arrive at the value 7.4×10^{-6} erg. Let us now compare this energy with that of an average gas molecule at room temperature, as given in Section 30. There we found the value 5.6×10^{-14} erg: the kinetic energy of a radium alpha particle is more than one hundred million times greater than the thermal energy of a molecule at normal temperature!

We may also calculate the average energy transmitted to an ion pair during the act of ionization, by dividing the total initial energy of the particle by the number of ion pairs it creates, i.e. by 130,000. The result is 5×10^{-11} erg or some 50 electron-volts if we translate it into the language used in discussing electron energies. (This energy would be that acquired by an electron after falling through a potential drop of 50 volts.) Comparing our results with the ionization energies in Section 63 we see that they are of the same order of magnitude.

We should not take leave of our alpha particles without having learned about an ingenious method, introduced by *C. T. R. Wilson* in 1899 which makes their paths directly visible and allows us to take photographs of them. The method permits the observation of the paths of beta rays and also of other charged, rapidly moving particles and opens up a wealth of information in the study of spontaneous and artificial radioactivity, cosmic rays and elementary particles in general. The instrument is called the Wilson cloud chamber after its inventor, or the cloud chamber in general.

As discussed, particles with great kinetic energy carrying electric charges are able to interact with gas molecules, to ionize them. If the same gas contains some supersaturated vapour which should be about to condense but is unable to find the necessary condensation nuclei to do so, these ions

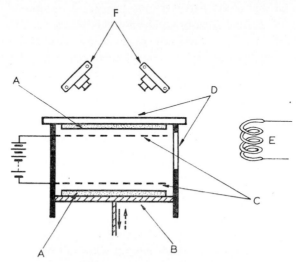

Fig. 146. **Wilson condensation chamber**

As the piston moves down the gas expands and cools, and the water vapour in it becomes supersaturated. It condenses on gas molecules ionized along the track of charged particles. These condensation tracks are photographed by flashlight. The charged grids pull out the ions, the piston goes up, and the chamber is again ready for exposure.

A = moist gelatine	D = glass windows
B = piston	E = flash
C = metal grids	F = cameras

in the air will serve as centres for initiating condensation. The first few molecules of, say, water vapour attach themselves to an ion and from then on, the submicroscopic droplet will progressively increase in size until the supersaturation is terminated, until there is no more vapour in the atmosphere than should be in equilibrium with water at the given temperature. These droplets lie in a row along the path of the particle where the ions have been generated. If appropriately illuminated these lines of droplets can be easily photographed. Fig. 146 shows a sketch of such an apparatus. The radioactive material or other source of rays is at the left of the chamber. On the top inside, and on the upper surface of the piston, there is a layer of moist gelatine to keep the atmosphere saturated with vapour. Oversaturation is produced by sudden downward movement of the piston: the air expands adiabatically (*see* p. 151) and cools to some extent, while the vapour which was saturated at the beginning becomes supersaturated at this lower temperature. At the same moment a flash of light illuminates the chamber and a photograph is taken. If two photographs are taken in directions perpendicular to each other, the two pictures allow a stereo reconstruction of the paths (Fig. 147). Between two exposures the piston compresses the air again and by thus heating it, causes the droplets to evaporate. An electric field pulls out the ions from the chamber and the stage is set for the next exposure. Very many exposures can be obtained in this way and there is some probability that a few of them will contain an interesting phenomenon. If cosmic

Fig. 147. **Stereophotograph made in a Wilson chamber**

Alpha tracks in nitrogen. Only three tracks meet in the chamber: an alpha particle hits an N nucleus which disintegrates into an H and O nucleus.

(From E. Rutherford, J. Chadwick and C. D. Ellis, *Radiations from Radioactive Substances*, p. 302, Plate X, Fig. 1 (Cambridge Univ., Press, 1951). By courtesy of the Cambridge University Press.)

rays or showers are photographed it is possible to combine the Wilson chamber with a set of monitoring devices around it which only allow exposures when they have been penetrated by a ray or particle, thus avoiding unnecessary photographs.

A few years ago the same principle was adapted to the local evaporation of slightly overheated liquid hydrogen in a chamber, yielding bubble-lines. Like supersaturation of a vapour, overheating of a liquid can also be overcome by ions acting as nucleating agents. The new method has some technical advantages over the original one.

Fig. 148 shows how alpha particles emanate from a speck of radium. They traverse a certain length of path and then disappear. This is their path, which we have already mentioned, before they lose their kinetic energy. It ends at the limit of their ionizing capacity. On careful inspection of the picture we also find a path which changes its direction at a certain point. A particle has come so near to an atomic nucleus that it has been deflected in a hyperbola. We *see* the photograph of a track that was calculated on theoretical principles for Fig. 97.

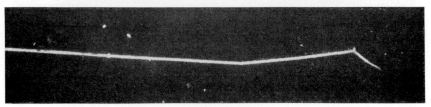

Fig. 148. **Equal path length of alpha rays from identical dis-integrations**

One path changes its direction, it has come near a nucleus. (Compare Fig. 97.)

((*a*) from J. Chadwick, *Radioactivity and Radioactive Substances*, p. 24, Plate I (London, Pitman, 1961). (*b*) from E. Rutherford, J. Chadwick and C. D. Ellis, *Radiations from Radioactive Substances*, p. 58, Plate III, Fig. 2 (Cambridge Univ. Press, 1951). By courtesy of Cambridge University Press.)

88. RADIOACTIVE RAYS, II: BETA AND GAMMA RAYS

The rays which have been shown deflected to the left in Fig. 143 are deflected just as well by the field in the apparatus of Fig. 87 as the alpha rays, but in the opposite direction. Their velocity and e/m value can be calculated from their deflexion in the same manner. The e/m value is not absolutely constant this time but shows some decrease as the velocities increase. However, the e/m of the slower particles is exactly the same as that of the electrons in a cathode-ray tube. In order to determine e and m separately a similar set of experiments is necessary as in the determination of the charge and mass of alpha particles. The only important difference lies in the fact that beta

particles are not energetic enough to evoke scintillations and they have to be
counted by another method.

The instrument to do this is seen in Fig. 149(*a*) and it is the prototype of the
now familiar Geiger-Müller counter tube. It was invented by Rutherford
and Geiger for this specific purpose. The particles fly from the source
through a set of slits, which determine the aperture, between two metal

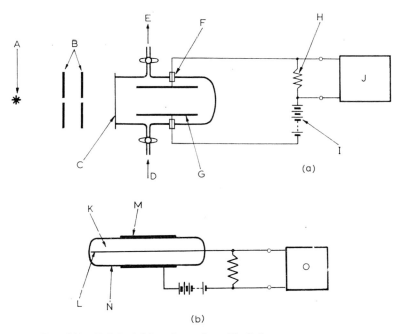

Fig. 149. **Original (*a*) and, modern (*b*) Geiger counters**

(*a*) A = preparation
  B = slits
  C = thin window
  D = gas inlet
  E = gas outlet to vacuum

  F = insulator
  G = metal plate
  H = resistance
  I = high voltage source
  J = instrument registering
    pulses.

(*b*) K = low-pressure filling
    gas (argon +
    alcohol)
  L = insulated wire
    electrode

  M = metal tube
  N = glass tube
  O = instrument registering
    pulses.

plates and into the electrometer. During the part of the experiment when the
number of particles has to be counted over a certain period of time, gas at
low pressure is allowed to enter the tube containing the metal plates. The
plates are connected to a high-tension battery; as long as the gas in between
them remains an insulator, there is no current in this circuit. But the moment
a beta particle enters the tube it ionizes the gas molecules and thus renders the
gas conductive. A short current pulse goes across the tube until all ions have
been extracted by the plates. (Modern Geiger counters are metal tubes filled
with low pressure gas, with an axial, insulated wire as second electrode, Fig.
149(*b*)). These pulses are registered and counted, and they give us the number

of beta particles. The sum of their charges is then determined in the electrometer in vacuum during a known period of time, in the same way as was
done with the alpha particles. The individual particles follow each other
in a completely random and irregular manner, but their average frequency
over a longer period remains constant. Dividing the total charge by the
number of particles gives again the charge of a single particle. The result is
$4{\cdot}77 \times 10^{-10}$ negative electrostatic units; the well-known charge of the
electron. As the e/m ratio also is as great as for electrons, the particle mass
must also be equal to that of the electron ($0{\cdot}9 \times 10^{-28}$ g). Beta-ray particles
are therefore electrons. Now we understand why the value of e/m decreased

Fig. 150. **Wilson chamber photograph of beta particles**

(From E. Rutherford, J. Chadwick and C. D. Ellis, *Radiations
from Radioactive Substances*, p. 66, Plate IV, Fig. 2 (Cambridge
Univ. Press, 1951). By courtesy of Cambridge University Press.)

slowly with increasing velocity of the beta particles: they had approached the
velocity of light and their mass began to increase because of the relativity
effect (p. 203).

The Wilson chamber photograph of beta particles is shown in Fig. 150,
and their track is appreciably thinner than the track of alpha particles, and
less continuous; they move very quickly, spend less time in the neighbourhood of a gas atom and the probability of ionizing one out of the many is
therefore smaller. They are light enough to be knocked about somewhat in
collisions. Fig. 151 shows the circular track of a beta ray which flew through
a magnetic field and was accordingly forced into this circle. The chamber of
the apparatus was between the poles of a strong electromagnet in a magnetic
field of, say, H gauss. Equation (46.2) on p. 202 connects the mass, the charge
of the moving body, its velocity and the strength of the field in which it moves,
with the radius of its circular track: the centrifugal force mv^2/r must be
exactly balanced by the electro-magnetic force Hev so that we have

$$r = \frac{mv}{He} \qquad . \qquad . \qquad . \qquad . \quad (88.1)$$

e is constant, H is kept constant during the experiment, and r can be measured, so that we arrive directly at the momentum mv of the charged particle. As long as the velocity is not near the velocity of light and relativistic corrections are unimportant the mass m may be taken as constant, so that we calculate the velocity of the particle from the radius of curvature of its track.

Fig. 151. **Wilson chamber photograph of beta particles in a strong magnetic field**

(From W. Gentner, H. Maier-Leinitz and W. Bothe, *An Atlas of Typical Expansion Chamber Photographs*, p. 10, Fig. 13 (Oxford, Pergamon, 1954). By courtesy of the Pergamon Press Ltd.)

All alpha particles from a given radioactive element have the same initial velocity and they travel the same distance in a given medium (Fig. 148). Not so the beta particles. If all the beta particles which leave a radioactive element in the same direction are bent into circles in a magnetic field and photographed or observed in a Wilson chamber, it is seen that they possess practically all velocities up to a given maximum. This maximal velocity of the "beta-ray spectrum" is sharp, and well-defined. There are some more frequently occurring velocities, "beta-spectrum lines" below this maximum, but on the whole any velocity may and does occur. The *maximal velocity* is just as characteristic for a beta-emitting element as is the alpha velocity for, say, radium. One should imagine that it represents the energy of the nuclear explosion. Assuming that the energy contents of the parent nucleus and the daughter nucleus are constant and characteristic of the species, it is only natural that the difference between them, the energy of the decomposition reaction, should be constant also. This holds for alpha decompositions but seemingly not for beta ones. Where is the energy of the explosion from an atom which has emitted a very slow beta ray? It has been assumed that it is carried away by an electrically neutral particle of extremely small or zero mass which could hardly interact with matter and would therefore escape detection. The value of such an *ad hoc* assumption to explain something is not very great while it is not corroborated by some other evidence. In this instance it was decades before such independent proof of this strange particle, the so-called neutrino was forthcoming, but now its existence seems to have been ascertained, as we shall see in Section 105.

Beta particles traverse matter more easily than do alpha particles: we have seen that they interact more rarely with matter during their flight and therefore they loose their initial energy only after a greater distance. But their mass is equal to that of the electrons they act upon during their ionizing action and therefore they change their direction to some extent at each collision. The beta tracks in the Wilson chamber are not quite so straight as the alpha tracks and the longer a beta-ray "bundle" travels through matter the less parallel it remains. It is slowed down more effectively by heavy elements with many electrons than by light ones, just as were the alpha rays. 0·2 cm of lead absorbs the strongest beta rays and 0·1 mm of aluminium absorbs about half of them. Absorption and deviation are of course very dependent on velocity, as the meticulous investigations of Lenard on electrons have shown.

We shall soon learn that natural radioactive elements disintegrate either by alpha or beta emission. Whenever both types of rays are observed to emerge at the same time, it is fairly certain that they are emitted by different disintegrating elements, one beside the other. For instance, by parent and daughter or "grandchild" elements of different types. There is, however, the third sort of radiation that we have spoken about, the *gamma radiation*. This is absorbed by matter to a much smaller extent than either of the first two, and is found to accompany all beta radiations but seldom the alpha particles. It could be, for all we know, a very energetic corpuscular radiation or a light quantum of very high energy, like the X-rays or even higher. It is a very "hard" that is, a penetrating radiation and we have seen that it carries no charge. We shall soon become acquainted with the neutrons, particles with mass but without charge and, *a priori*, gamma rays might be something of this sort.

This question, too, was resolved at last by *Rutherford*, and *A. N. da C. Andrade* in 1914, who determined the wavelength of gamma rays by diffracting them with crystals at a grazing angle of incidence, in a similar manner to the diffraction of X-rays (p. 122) by grazing incidence on artificial gratings. The wavelength of gamma rays is so short that even at grazing incidence they needed crystal lattices as diffraction gratings. The same value for the wavelength was obtained in an entirely different set of experiments by liberating photoelectrons from matter by impinging gamma rays and measuring their velocity, as with X-rays (Section 49). From the kinetic energy of these photoelectrons we arrive at the frequency of the generating photons through the Einstein equation $E = h\nu$ and the frequency that was obtained agreed with the one obtained by the diffraction experiments using the relationship $c/\lambda = \nu$. It also became evident that the sharp lines within the "beta spectrum" were groups of photoelectrons liberated by the gamma rays within their own material, because the beta particles in these lines possessed the calculated kinetic energy.

From among the three kinds of radiation with which we are now familiar, the gamma is by far the "hardest" to absorb. A few millimetres of aluminium suffice to absorb practically all alpha and beta particles but hardly diminish the intensity of a gamma ray. The hardest gamma rays lose only half their intensity in 6 mm of aluminium or 1·5 mm of lead. But it should be kept in mind that hardness is not a necessary attribute of gamma rays. Very soft gamma quanta are known also, and if one desires to separate them from the accompanying corpuscular radiations it is necessary to deflect these by electric or magnetic fields so as to leave the soft gamma quanta undeflected.

The methods we have mentioned allow us to examine the wavelengths, the spectra, emitted from any radioactive element. It has proved that these wavelengths are exactly defined and characteristic of the element: veritable gamma-ray spectra. We know today that they carry with them the energy of an activated nucleus that failed to reach its final steady state after beta emission.

89. DETERMINATION OF THE MASS OF ATOMS FROM RADIOACTIVE DATA

We have learned of independent methods which all lead to the same atomic dimensions in Parts 2–4 of this book. The thickness of monomolecular fatty acid layers (Section 24), the structure of crystals (Section 27), the diffusion and viscosity of gases (Section 38), the Brownian movements of ultramicroscopic particles (Section 35) and the charge of the electron combined with data on electrolysis (Section 45) all gave the same result. But in view of the fact that this agreement is of supreme importance and is one of the main objects of this book, it seems worth while to stop for the length of this section and see how radioactive methods yield the same dimensions in their turn.

The first method has been described already, without it having been pointed out that it means a determination of atomic dimensions. This is the absolute determination of the charge of alpha particles and of beta particles, that is of electrons by counting their number and measuring the sum total of their charge. This is as absolute a method as was used in Millikan's experiment, and yielded the same values for the elementary charge (naturally twice the elementary charge for the alpha particle, the helium nucleus.) By dividing the Faraday number (p. 187), the charge associated with a gramme-equivalent in electrolysis (96,400 coulombs) by this elementary charge we arrive at the Loschmidt (Avogadro) number. The original determinations gave $6\cdot2 \times 10^{23}$, in rather good agreement with the best value of $6\cdot05 \times 10^{23}$.

The second, independent method is based on the fact that it is possible to count the number of scintillations from alpha particles emitted by a speck of radium within a known space angle. Thus the number of alpha particles emitted all around the speck can be easily calculated, for a given period of time. Knowing the weight of the speck we now know how many alpha particles are emitted by one gramme of radium in a second. It is an enormous number: $3\cdot4 \times 10^{10}$, 34 milliards (or "billions" in American counting). Now, Rutherford and Boltwood managed to determine in very tricky experiments that one gramme of radium gives off 39 mm³ of helium gas in a year, that is $1\cdot29 \times 10^{-6}$ mm³ in a second. This weighs $2\cdot30 \times 10^{-13}$ g and its 34 milliardth part, the weight of a single helium atom is $6\cdot77 \times 10^{-24}$ g according to this experiment. One mole of helium weighs $4\cdot00$ g, hence the Loschmidt number, the number of atoms in a mole, turns out to be $5\cdot9 \times 10^{23}$, not so bad a value after all!

The mass of an alpha particle was already determined however in our deflexion experiments, from e/m and from the charge e (p. 396), and there the result was $6\cdot6 \times 10^{-24}$ giving the more accurate number of $N = 6\cdot06 \times 10^{23}$. This is another, independent value.

A further method is based on the amount of radon forthcoming from one gramme of radium in, say, one hour. By dividing this mass by the number of scintillations, that is disintegrations, in the same time one arrives at the absolute mass of the radon atom and finally again at the Loschmidt number.

A last method makes use of the energy emitted by a gramme of radium. This amounts to $25\cdot2$ cal or $1\cdot05 \times 10^9$ ergs per hour, while $34\cdot10^9$ disintegrations are going on per second. The energy of one alpha particle is thus $8\cdot6 \times 10^{-6}$ erg. Knowing its velocity from the deviation experiments to be $1\cdot5 \times 10^9$ cm/sec the formula for kinetic energy yields the mass $7\cdot1 \times 10^{-24}$ g for a helium atom or $N_L = 5\cdot6 \times 10^{23}$ atoms/mole. This time the deviation is great, but this can be explained. We have committed a fault in ascribing the whole kinetic energy to the alpha particle. The law of action and reaction in mechanics determines that the parent nucleus, the radon, also receives impulse and therefore kinetic energy, though of course much less than that of the alpha particle because it is 55 times heavier. And a small part of the energy escapes in the form of gamma rays.

We were able to observe elementary processes one by one in the course of radioactive investigations. We have been able to count the number of individual particles taking part in the processes. Therefore it is no wonder that these observations yielded the masses and energies of these individual particles. All such data can be easily transformed for comparison into values of the Loschmidt (Avogadro) number which connects the macroscopic world with the world of the atoms. We have used the original, historical values of the experimental results instead of more refined modern ones in order to give an idea of the agreements arrived at in those days when every agreement between independent determinations was a new victory for science.

90. RADIOACTIVE SERIES

We have only mentioned a few radioactive elements up to now, but have always spoken of them as if there were many. Now we are about to make their acquaintance. They occur at the end of the periodic system: all elements beyond lead are radioactive. The idea seemed natural that the system is really limited by the phenomenon of radioactivity: the heaviest nuclei seem to be held together only weakly and hence they are liable to disintegrate. Above a certain limit they may not have been able to come into existence at all, or they must have had ample time for disintegration—since the ancient chaos when, in the course of some grand cosmic happening, they were born out of something. Only their lighter descendants have endured to meet us. There are only a few radioactive elements among those of medium atomic weight: potassium and rubidium are beta-active; samarium, a rare-earth metal, is alpha-active, technetium was only prepared by an artificial nuclear reaction and is beta-active.

We have seen that one element gives rise to another in the course of its radioactive transmutation. It had to be ascertained whether all known radioactive elements may be ordered into a single genetical series, whether there were more than one such series among them and how they followed one another. It was not at all easy to answer these questions and it took some twenty years to clear them up. Elements with a medium half-life, like radon, were the easiest to work with, because they lived long enough to be handled in the laboratory but disappeared sufficiently soon to yield their descendants without contaminating them themselves.

We have understood on p. 392 why an element accumulates in radioactive equilibrium if it is long-lived. The chemistry of such an element may be examined at leisure. The short-lived elements must be rare and they must be isolated and examined quickly. Sometimes they occur in such minute quantities that even their sparingly soluble compounds remain in solution, because there is not enough substance present to give saturation. If this occurs it is sometimes helpful to add an analogous compound of a similar non-radioactive element to the solution and precipitate it: the traces of the radioactive sister element will often be absorbed in the precipitate. This is the case with $RaSO_4$–$BaSO_4$ for instance. Conversely, from experiments which show which other element is capable of dragging with it compounds of a radioactive element one is able to draw conclusions as to the chemical nature of this trace element and even find its place within the system. The fact that radioactive elements ionize the air in their surroundings makes it easy to determine their whereabouts after a reaction.

Emanation (radon) for instance is a gas. It may be transferred to a vessel and pumped off soon after: a radioactive deposit will have formed on the walls of the vessel within seconds. Of course, it cannot be seen, but it can be easily shown that the walls emit alpha rays at the beginning. The situation changes, however, minute by minute. The intensity of alpha radiation decreases, with a half-life of about three minutes, while the beta radiation keeps on increasing. After a lapse of some hours, another alpha radiation begins to manifest itself. First and second alpha radiations can be kept apart easily by their different ranges in air: 4·75 and 6·94 cm respectively. They *must* be distinct. Chemical examination shows that the first alpha-active element is similar to tellurium, since it follows it in the course of tracer reactions. Let us suppose that its place in the periodic system is really below that of tellurium. From what we have observed it is clear that its daughter element is beta-active. Working very quickly it is possible to show that this element may be precipitated with lead. However, another beta-active element may be isolated too, being similar to bismuth. It is found that the beta-active element which is similar to lead changes into the analogue of bismuth with a half-life of 20·8 minutes. This second beta-emitter disintegrates with a half-life of 19·5 minutes and yields the second alpha-active element that we have mentioned. The latter is again similar to tellurium, though different from the first one, and from another alpha-emitter, polonium (p. 360) which has the same chemical properties also. But their half-lives and alpha ranges are quite specific and distinct.

The elements we have just been discussing are designated by the letters RaEm, RaA, RaB, RaC, RaC′. It would really lead us too far into details if we wished to follow all radioactive series in the same manner. But we have to mention some fundamental rules. Table XI (p. 410) shows us that *whenever an element emits alpha rays, its daughter element will lie two places to its left in the periodic table whereas beta emission leads to an element immediately on the right-hand side of its precursor.* This is the law announced by Rutherford and Soddy. As we know what the number of an element in the periodic system means in the context of its nuclear charge (p. 256) and what alpha and beta rays are (pp. 395, 402) this rule seems quite natural. But at the time when these strange elements were first collected and examined it did not seem so natural, and *K. Fajans* and *F. Soddy* independently had to find the rule from a mass of experimental results. It is now easily explained. Each alpha particle reduces the positive charge of its parent nucleus by two, hence the daughter element must be two places before it in the system. Each beta particle on the other hand, removes a negative unit of charge and so increases the positive charge by one unit; the element one place to the right of the original must appear.

Of course, not only the change in nuclear charge may be ascertained thus.

The mass of the daughter atom is fixed also by the fact that an alpha particle has an atomic weight of four while the weight of an electron may be neglected to a first approximation. Hence atomic weights decrease by four units after alpha-emission and remain unchanged after beta emission. We have already seen that the difference of atomic weights of Ra and RaEm is actually four. Uranium, the heaviest natural element, has an atomic weight near to 238, while that of radium is 226. If we assume that radium is derived from uranium and know uranium to be alpha-active, it follows that two further alpha-active elements must lie between the two to fill the gap. Our assumption that radium is a great-grandchild of uranium with possibly a series of beta-emitters in between them, is corroborated by the fact that the ratio U:Ra is constant in most of the uranium minerals in Nature. It looks like a radioactive equilibrium of elements with different half-lives. There is mostly 2·7 million times as much uranium as radium in a mineral and according to what we have said about radioactive equilibria this could mean that the half-life of uranium is this multiple of radium's half-life. This yields (p. 392) 4·5 thousand million years for the half-life of uranium! After prolonged searching, the two intermediate alpha-emitters were found at last: they are called uranium II and ionium, the parent element of radium itself. Thus the three alpha particles which are consecutively emitted by U, UII and Io account for the difference $3 \times 4 = 12$ between the atomic weights of uranium and radium.

In between them, however two-beta emitters were found also. Beta-emission only affects the position in the periodic system, not the atomic weight, because it means a change of charge but not of mass. (The mass of the electron only amounts to 1/1838 atomic weight unit, *see* p. 203.) As often as it was possible to determine the atomic weight of radioactive elements one after the other in a series, the Rutherford–Soddy–Fajans rule was always observed. However, often enough the short life of a radioactive element made such a determination impossible and in these cases the atomic weight has been conversely calculated by the rule.

For a long time three radioactive families were known: the uranium, thorium and actinium families. But on reflection, we miss a fourth one, because within each family atomic weights may only differ by 4 units (the weight of the alpha particle) and there seems to be no *a priori* reason why all kinds of atomic weights should not occur. The descendants of uranium must have atomic weights according to the general formula $(238 - 4n)$, those of thorium $(232 - 4n)$ (or $236 - 4n$) and those of actinium $(227 - 4n)$. Later on, it became apparent that the actinium series began with an uranium atom with the atomic weight 235, a very important kind of nucleus in our atomic age! Thus the series was $(235 - 4n)$. The difference in the atomic weights of ^{235}U and Ac is $2 \times 4 = 8$. But the fourth mathematically possible series beginning with an element 237 was not found for a long time. A fifth series would necessarily coincide with the first one because its parent atom, 234 is the daughter of U $(= 238 - 4)$:UII. In the course of the tremendous development of nuclear science this 237-based series was found at last. Its ancestor element does not occur in Nature, it is synthetic, man-made,

TABLE XI

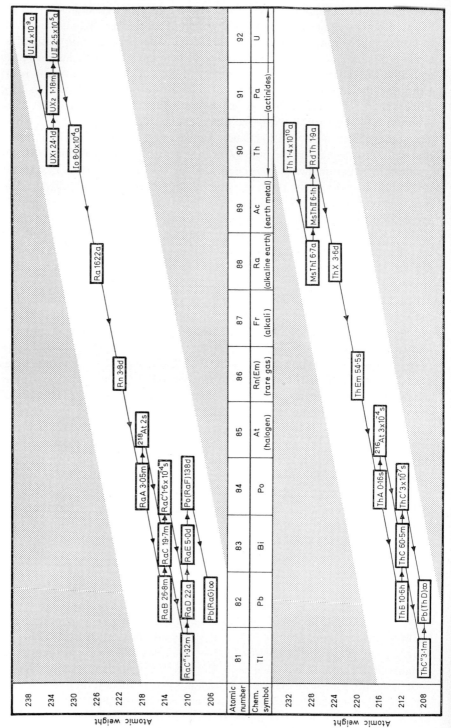

Atomic number	81	82	83	84	85	86	87	88	89	90	91	92
Chem. symbol	Tl	Pb	Bi	Po	At (halogen)	Rn(Em) (rare gas)	Fr (alkali)	Ra (alkaline earth)	Ac (earth metal)	Th	Pa (actinides)	U

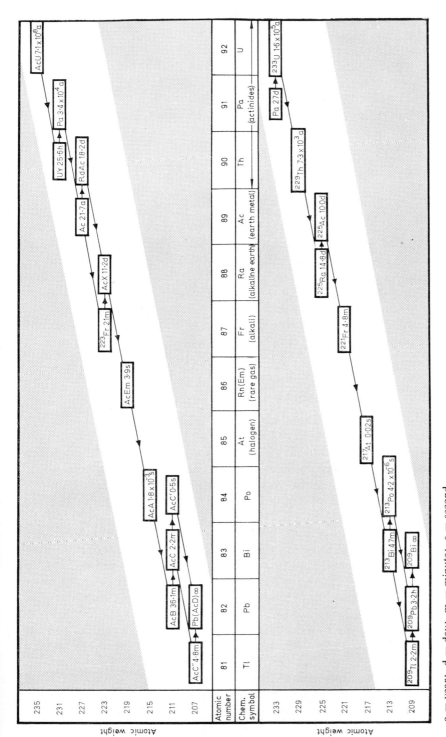

Atomic number	81	82	83	84	85	86	87	88	89	90	91	92
Chem. symbol	Tl	Pb	Bi	Po	At	Rn (Em)	Fr	Ra	Ac	Th	Pa	U
					(halogen)	(rare gas)	(alkali)	(alkaline earth)	(earth metal)	(actinides)		

a = year; d = day; m = minute; s = second.
α–decay: the atomic weight decreases by four units, the atomic number by two.
β–decay: the atomic weight is unchanged, the atomic number increases by one.

neptunium with the atomic weight 237, the right-hand neighbour of uranium in the periodic system of elements (p. 442).

Table XI shows the family relationships of the four radioactive series. Fundamentally it is simply the last two rows of the periodic system. Each family is connected by a series of lines and the limit of natural radio-activity is set by the vertical line after thallium.

The families show a somewhat similar genesis and this similarity has been emphasized in the nomenclature. All families contain a rare gas, an emanation, preceded by two alpha-emitters and followed by an alpha- ("A") and a beta- ("B") emitter. Then each family line splits into two alternative directions: the "C" element disintegrates by beta-emission into a very short-lived, long-ranged alpha-emitter or by alpha-emission into a beta-emitter (C' and C" respectively). We receive the impression that these alternative disintegrations only differ in sequence: a very labile (short-lived) nucleus wants to get rid of an alpha particle and a beta particle as well, and at one time it may be the alpha that is first emitted, at another time the beta. The resulting "D" element is always the same. All these "D" elements are similar to lead but whereas AcD and ThD are stable, are no longer radioactive, RaD emits a beta particle to yield the other beta-emitter RaE which in turn gives rise to the alpha-active polonium. The disintegration of this element, which was the first to be discovered by the Curies, yields RaG, stable, ordinary lead.

91. ISOTOPES

We must pause for a moment to consider some implications of what has been demonstrated (Table XI). Up to now we have considered the periodic system number of an element as its chemical characteristic. It meant the number of positive charges in its nucleus and hence the sum total of its orbital electrons which in their turn (at least the outer ones) have been responsible for its chemistry (Section 59). But we generally assumed (except for some allusions on pp. 39, 55, 231, 288) that a well-defined atomic weight belongs to each position in the periodic system, even if the magnitude of this atomic number did not seem to be so important as far as chemical properties were concerned. (Even the sequence of atomic weight against periodic system number differed with the pairs K-Ar, I-Te and Co-Ni!). In Table XI, however, we have written a whole series of elements with different atomic weights into the place of the eighty-second element, lead (Pb). Is it reasonable to do this? And are we free to write three different emanations in place of element 80? And elsewhere too? The concept of a chemical element has become muddled to some extent and it must now be cleared up.

Let us begin by experiments. If it is really true that RaG, AcD, ThD and RaD are all lead as far as their chemistry is concerned, they must behave as such. Do they? But first where can we find enough RaG or ThD, for example, when it takes milliards of years until they are formed from their ancestor atoms? Evidently we have to go back to our ancient minerals where radioactive equilibrium has had time to become established. We should assume that lead found in a uranium mineral and lead found in a thorium mineral must have been formed at least partially from these ancestor elements. They should be RaG and ThD. Therefore it is necessary to analyse lead samples, prepared from the said minerals and to determine the atomic weight of their lead by analysis. *T. W. Richards* and *O. Hönig-schmidt* (1914), the most famous analysts of America and Europe respectively, undertook this examination on a series of uranium and thorium minerals. They actually found that the atomic weight of their lead samples varied according to their origins! Whereas lead normally has the atomic weight 207·2, Table XI shows that the atomic weight of RaG must be 206 if each alpha particle emission involves a decrease of 4·00 atomic weight units.

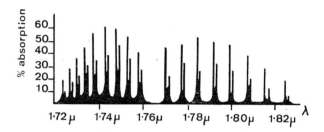

Fig. 152. **The rotation–vibration spectrum of hydrogen enriched in deuterium**

The lines are split because of the slower rotation of molecules containing heavy hydrogen. The lower peaks belong to these molecules.

On the same basis ThD must be 208. It was found that the atomic weight of lead from a uranium mineral called broggerite is 206·06 while that of lead from thorite is 207·77! Both kinds of lead were identical from the chemist's point of view, nor could they be distinguished chemically from ordinary lead. But the one needed 123·4 per cent of iodine based on the weight of uranium to form PbI_2, the other 122·3 per cent. This is what we call a difference in atomic weight.

It becomes necessary therefore to distinguish between the concept of a chemical element and a physical atomic species. As often mentioned, chemical properties are dependent on the electronic structure of the atom and this is determined by the number of the element within the periodic system. The effective weight of an atom on the other hand is surely one of its important physical properties. So is also the question whether it is radioactive or not, and the type of its radioactive decay, if any. In consequence of the difference in atomic weight there is a difference in density too, and owing to the different inertial moments within a molecule there is a slight difference in the spectra also. Even vapour pressures are very slightly different.

As a matter of fact, all these differences increase if the percentage difference between the atomic weights within a single chemical element increases. In the case of lead it only amounts to 1 per cent. But the doubling of certain spectral lines in small residues of evaporated liquid hydrogen (Fig. 152) revealed (H. C. Urey, F. G. Brickwedde, 1931) that there must be a heavy hydrogen atom also, and owing to differences in vapour pressure and electrolytic decomposition it became possible to isolate this heavy hydrogen, or deuterium, as it is now called (Urey, 1932). Its atomic weight is 2, an increase of 100 per cent compared with ordinary hydrogen. This involves quite substantial differences in physical properties: the freezing-points, boiling-points, etc., differ to a considerable extent as shown in Table XII.

TABLE XII. Physical Properties of Normal and Heavy Hydrogen and of Normal and Heavy Water

Physical property	H_2	D_2	H_2O	D_2O
Boiling point (at 760 mm) .	20·4°K	23·5°K	100°C	101·4°C
Melting point (at triple point*)	13·96°K	18·72°K	0°C	3·8°C
Molar heat of vaporization .	225 cal/mole	304 cal/mole	9·719 Kcal/mole	9·960 Kcal/mole
Molar heat of fusion . .	28 cal/mole	47 cal/mole	1·435 Kcal/mole	1·522 Kcal/mole
Molar volume (liquid) . .	26·15 ml	23·17 ml	18·0 ml	18·09 ml
Density (at triple point) . .	0·0767	0·1725	—	—
Density (at 20°C) . . .	—	—	0·9982	1·1059

* Triple point means the melting point of a pure compound under its own vapour pressure.

Owing to the fact that the energy content of molecules with this heavy hydrogen must differ from that of the ordinary hydrogen compounds there is some difference in their chemical equilibria also. Hence it is not quite correct to maintain that the chemical properties are identical, though these always vary in a parallel, well-predictable, manner. Only in the case of heavy atoms, where the percentage difference is slight, is the chemical difference so small as to escape detection.

However, the confusion concerning the fundamental concept of an element was already embarrassing enough before the advent of heavy hydrogen, and it had to be resolved. This was eventually achieved by *F. Soddy* and *F. Paneth* in 1913. They proposed to reserve the term "element" for species of atoms which had the same serial number in the periodic system, that is the same nuclear positive charge. Such substances are, as we have seen, identical or nearly identical in their chemical properties. Within the limits of the definition of an element, however, they proposed to distinguish between different "isotopes" which may have different atomic weights. (The Greek word ἴσος is already known to us from "isomers," while τόπος means place. Isotopes occupy the same place in the periodic system.) Thus RaD, RaG, ThD, etc., are all is otopes of lead, and the three known emanations are istopes of each other. The radioactive properties of isotopes (half-lives, penetrating power of rays) may be entirely different, their physical properties may differ to a small extent, and chemical differences between them may be extremely small.

Apart from the concept of isotopic elements, it has become necessary to define as "isobars" those species which possess the same atomic weight but still are different elements because the positive charge on their nuclei is different. Elements which evolved one from another through beta-decay are necessarily isobars because the charge on their nuclei differs by one unit, whereas only the negligible mass of an electron has been lost in the process. The atoms of UX_1, UX_2 and U_{II} are examples of isobars.

We are only now able to understand the remarkable regularity discussed in Section 53. According to Prout's hypothesis all elements should have been built from hydrogen and hence their atomic weights should be a multiple of the atomic weight of hydrogen. However, this was only found to be the case in the majority of the light elements and became exceptional among the heavier ones. This spoke against the hypothesis. The situation becomes fundamentally different the moment we allow for the existence of isotopic atoms. Any natural element may then well be a mixture of different isotopes and the atomic weight found experimentally may well be only an average value. It is thus possible that the atomic weights of isotopes really do obey Prout's hypothesis. The question had to be, and was, examined anew and it was found that Prout was right in his inspiration. For instance, the atomic weight of chlorine is 35·46 and it was proved that the element is a mixture of atoms of weights 35 and 37, in the ratio of approximately three to one.

92. SEPARATION OF ISOTOPES

In order to decide the question we have just referred to it is necessary to take one chemical element after another and try every sort of refined fractionation technique in an effort to separate them into isotopes. *J. N. Brønsted* and *G. v. Hevesy* were the first to accomplish such a partial separation (1920). Their first experiments were performed with mercury. Its atomic weight in Nature is 200·6. v. Hevesy made use of the fact that the mean kinetic energy of all gas molecules is the same at a given temperature. Therefore the velocity of a lighter mercury atom must be somewhat greater than that of a heavier one in order to keep the value of $mv^2/2$ constant. The number of light mercury atoms which are able to arrive at and break through the surface of liquid mercury will therefore be a little greater than if it were only proportional to their concentration, and it was possible to condense a fraction of mercury in a high vacuum which, according to the very precise measurements of Hönigschmidt possessed an atomic weight slightly below 200·6.

This kinetic effect was made use of by *W. D. Harkins* and later by *G. Hertz*. They allowed chlorine or neon gas to diffuse through porous earthware tubes and found that the lighter was in slight excess after this diffusion. An ingenious system of interconnecting a series of such diffusion tubes and of recirculating parts of the gas stream allowed them to achieve a substantial separation of the respective isotope. Later on, during the war against Hitler, this same method was extended on an immense scale to separate the hexafluoride of the important fissile material ^{235}U from that of normal ^{238}U.

As mentioned above, the most spectacular separation of isotopes was accomplished by Urey in 1932 when he managed to prepare deuterium. Later on a third isotope of hydrogen, tritium, with atomic weight three was isolated too as a product of nuclear reactions.

During the technological struggle to isolate ^{235}U another method was also used to separate the two uranium hexafluoride molecules. It is called thermal diffusion and is based on the fact that if a temperature gradient is established within a mixture of gases or liquids the heavier component of the mixture tends to accumulate at the end where the temperature is higher. *Clusius* invented an arrangement in which the temperature gradient also established circulation within the mixture so that the effect of a single separation could be automatically multiplied.

Early in the beginning of this book we made use of the fact that certain atoms may be "marked" and traced throughout their reactions. This marking was done by the admixture of a specific isotope, the fate of which could then be followed through a series of reactions. We can only now fully understand what this meant. As long as hydrogen is "marked" by addition of deuterium, its presence may be followed simply by measuring the density of the products. This was the case when we followed the reaction of $NH_3 +$ HCl (p. 57) and found that the four hydrogen atoms in the NH_4 ion were equivalent. v. Hevesy introduced this method of isotopic tracers into biochemical reactions and its use has since been extended throughout the wide domain of biochemistry and chemistry in general. The metabolism of calcium or phosphorus in bones, of iron in blood, and so on, can be examined with isotopic tracers. If their detection cannot be accomplished by density measurements or radioactive techniques (that is if the isotopes are stable), it can always be done in the mass spectrograph which will be dealt with in the following section.

Urey and his co-workers succeeded, early in the thirties, in preparing isotopes of the other three fundamental elements of organic chemistry: oxygen, nitrogen and carbon. There is 0·02 per cent deuterium in natural hydrogen, 1 per cent ^{13}C in natural carbon, 0·15 per cent ^{15}N in nitrogen and 0·3 per cent of ^{18}O in oxygen. They could be prepared by repeated reactions owing to the fact that, as mentioned before, there is a slight difference in equilibrium constants of reactions between isotopes. Nitrogen, for instance, was fractionated as ammonia solution in a column where the reaction

$$NH_3 + H_2O = NH_4OH \qquad . \qquad . \qquad . \quad (92.1)$$

took place a number of times in both directions. Radioactive isotopes of most elements are prepared today in atomic piles using a heavy flux of neutrons (*see* Section 100) and are very convenient tracers in all sorts of reactions.

93. THE MASS SPECTROGRAPH

It is often less important to separate isotopic mixtures in weighable quantities than it is to determine the relative percentage of isotopes in natural elements or tracer experiments. The first apparatus which solved this problem was built by F. W. Aston in 1919 and was named a mass spectrograph. It is analogous to an optical spectrograph in so far as it resolves a mixture of different atomic or molecular weight species into its components, just as a spectrograph resolves light in different wavelengths. It may be used with a photographic plate as detector such that the lines which represent the isotopes (on the developed plate) lie besides each other like the lines of a spectrum.

The apparatus itself resembles in its original form the Crookes tube with electric and magnetic deviation, which has been shown in Fig. 87. We have

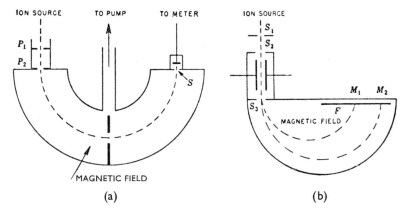

Fig. 153. **Mass spectrometer** (*a*) **and mass spectrograph** (*b*)

(*a*) By changing the strength of the magnetic field, particles of different mass are collimated one after the other on the measuring (counting) instrument behind slit S. P_1 and P_2 are accelerating slits.

(*b*) The accelerated particles are deflected first in an electrostatic and then in a magnetic field. All particles with the same mass are collimated on the same spot on the photographic plate (M_1, M_2).

(From S. Glasstone, *Sourcebook on Atomic Energy*, p. 214, Fig. 8.4, p. 215, Fig. 8.5 (Princeton, N.J., van Nostrand, 1952). By courtesy of the D. van Nostrand Co., Inc.)

seen on p. 200 that an electric field causes a deviation of particles from their original path which is greater the higher their charge, the smaller their mass and the more slowly they move. A magnetic field has a similar effect. According to the equations on pp. 201–2, the deviation is a function of e/m and not of these quantities separately. Whenever the two fields follow each other, the deviations add up, of course. But if they follow each other in the reverse direction, as in Fig. 153 their effects partially cancel. Only partially, because the deviation in the electric field is proportional to the square of the velocity whereas that in the magnetic field is proportional to the velocity itself (Eqns. 46.1, 46.2, pp. 201, 202). It thus becomes possible to choose the two field-strengths in such a way that differences in deviation which are due to different velocities just cancel whereas those due to differences in specific charge (e/m) remain.

All particles of the same specific charge will converge at point *B* of the photographic plate irrespective of whether they moved in their path with the same velocity or not. The detailed theory of the instrument is involved, especially if we take into account several refinements which have been added in the course of the years. These allow for far shorter times of exposure and for far higher precision and definition than in Aston's original instrument. However, all these refinements are no more fundamental for the working *principle* of the instrument than, say, the structure of an apochromatic objective for the principle of the microscope.

Fig. 154 shows the isotopes of the element Cd on a mass spectrogram: the lines at the side of the spectrum show the positions which correspond to integral numbers on the atomic weight scale. The work of Aston and his

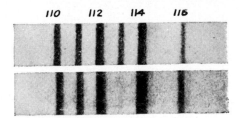

Fig. 154. **Mass spectrum of the cadmium atom**

The lower spectrum belongs to a sample which had been exposed to neutrons (*see* p. 430): the isotope of mass 113 absorbed neutrons and therefore disappeared, increasing the content of isotope 114.

(From S. Glasstone, *Sourcebook on Atomic Energy*, p. 342, Fig. 11.8 (Princeton, N.J., van Nostrand, 1952). By courtesy of the D. van Nostrand Co., Inc.)

followers clarified, in the course of two decades, the whole problem of isotopes. Prout's original, fascinating hypothesis was found to be true: all isotopic atomic weights fell in the direct vicinity of an integral number, if the atom of oxygen (that is, its main isotope: ^{16}O) was taken as 16·000 units. (Currently $^{12}C = 12·000$ is the reference species for chemical atomic weights.) In between the integral atomic weight positions no mass-spectral lines appear. Not only can the individual isotopic atomic weights be measured very exactly, but the intensity of the mass-spectrographic lines also gives us the relative abundance of each isotope. The weighted mean value of the isotopic atomic weights therefore give us the average atomic weight of the sample: this was the "chemical" atomic weight determined by analysis. It happened now and then that this mean value differed from the chemical atomic weight and it was always found on re-examination that the chemical analysis had been incorrect; the mass-spectrographic value was always validated.

Let us glance at the exact isotopic weights of some elements. Table XIII shows the values for the first 18 elements in the periodic system.

It is very important to note that although the isotopic atomic weights are all very near to integers, they still show a small deviation from integral numbers. The deviation is far within the accuracy of the method and it

TABLE XIII. Isotopic Weights of the Isotopes of the first eighteen Elements of the Periodic System

(Based on $^{16}O = 16 \cdot 000000$ according to the old convention)

Element	Atomic weights of its isotopes and relative abundance		
Hydrogen . . .	1·008131 99·984%	2·014725 0·16%	3·017004 10^{-10}%
Helium . . .	3·016988 1·3 × 10^{-4}%	4·003860 ~100%	
Lithium . . .	6·016917 7·39%	7·018163 92·61%	
Beryllium . . .	9·104958 100%		
Boron	10·016169 18·83%	11·012901 81·17%	
Carbon . . .	12·003880 98·9%	13·007561 1·1%	14 . . . 10^{-10}%
Nitrogen . . .	14·007530 99·62%	15·004870 0·38%	
Oxygen . . .	16·000000 99·76%	17·00450 0·04%	18·00485 0·20%
Fluorine . . .	19·00454 100%		
Neon	19·998895 90·51%	21·00002 0·28%	21·99858 9·21%
Sodium (Na) . .	22·99644 100%		
Magnesium . .	23·99300 78·60%	24·99462 10·11%	25·99012 11·29%
Aluminium . . .	26·99069 100%		
Silicon . . .	27·98723 92·28%	28·98651 4·67%	29·98399 3·05%
Phosphorus . . .	30·98441 100%		
Sulphur . . .	31·98252 95·06%	32·9819 0·74%	33·97981 4·18%
Chlorine . . .	34·97884 75·4%	36·97770 24·6%	
Argon	35·97728 0·307%	37·97463 0·060%	39·97549 99·633%

cannot be ascribed to experimental error. This is so throughout the whole periodic system.

Thus the experimenters had rejoiced too soon that they had verified Prout's hypothesis. True, isotopic atomic weights are very near to integral numbers. But mass spectrograms present a new question. Why are they not *exactly* integral numbers? If atoms were built from hydrogen atoms, they should weigh *exactly* some multiple of the weight of a hydrogen atom. And the greatest deviation is with hydrogen itself: we have worked on an $O = 16$ scale and the atomic weight of hydrogen is 1·008 units! The precision of the experiments is beyond any doubt and the deviation reaches nearly 1 per cent of the atomic weight itself. The mass of the electron is less than $1/1000$ of that of the proton; it cannot be the cause. There must be some reason that we have not found as yet.

94. NUCLEAR ENERGY

The explanation we are looking for concerns the energy within the nuclei. The mere existence of radioactivity shows that such an energy must be present because alpha and beta particles are ejected from the decaying nuclei with enormous energies. But analogy with normal chemical thinking makes us assume also that if a body is built from smaller particles and still manages to hold together, it must be held together by energy. If we would know the process by which smaller nuclear units are built into heavier ones we should expect that energy would be liberated in its course.

But what can nuclear energy have to do with the mass of the respective nuclear species? We are accustomed to treat mass and energy as separate concepts. How could we have assumed any interconnexion between nuclear mass and nuclear energy? The answer is given by the theory of relativity.

We have mentioned the so-called theory of special relativity on p. 203, where we learned that the upper limit of physical velocities is the velocity of light. As a consequence of this thesis we have seen that the mass of a body that moves with a velocity near to that of light increases and thus makes it ever more difficult to accelerate it any further, so difficult indeed that no force can accelerate the body beyond the velocity of light. This was in flat contradiction to ordinary mechanics—but ordinary mechanics was evolved from studies on bodies moving at much smaller velocities. In the neighbourhood of c, the velocity of light, the mass tends towards infinity and there is practically no further acceleration. But the energy, $mv^2/2$ of course increases further, with v remaining constant and m increasing, just the opposite to the way it does at low velocities (m constant, increasing v). Thus we might as well say that with increasing kinetic energy there is an increase in mass.

Einstein in his theory of general relativity went a step further and extended this connexion between mass and energy. He assumed that mass not only increases with kinetic energy but does so in all cases when the internal energy of a system has increased. "Energy" is accorded the same two properties which characterize "mass": it has inertia and underlies the law of universal gravitation. Astrophysicists have eventually found that light which passes near to the sun is really attracted towards it.

It is not possible to go through the calculations here which eventually lead to the quantitative relation between mass and the energy increase of a system. It can be done by calculating the relativistic mass increase for a moving body in the vicinity of c and by converting it into kinetic energy. The extension of this same relation to all kinds of energy is a postulate. Let it suffice that the result of all this elaborate calculation is a proportionality between mass and energy, the factor of proportionality being the square of the velocity of light:

$$E = mc^2 \qquad . \qquad . \qquad . \qquad . \qquad (94.1)$$

Experimental physicists failed to recognize the existence of this relation up to the advent of Einstein's theory, because the factor c^2 is so enormous that a system must change its energy by an immense amount in order to make the difference in mass open to observation. In c.g.s. units the factor amounts to 9×10^{20} cm²/sec²; this number of ergs of energy have to be introduced into a system to increase its mass by 1 gramme! In thermal units this amounts to $2 \cdot 15 \times 10^{10}$ kcal.

However, the energy which is set free during a radioactive disintegration is also enormous compared with the mass of the particles which take part in the process. Applying the above equation to these radioactive energies, we find that the accompanying mass differences are of the order of magnitude of the deviation of exact isotopic weights from integral multiples of the mass of the hydrogen atom. Let us consider the first two elements of the periodic table. Hydrogen has an atomic weight of $1 \cdot 008$ and helium $4 \cdot 004$. The difference between the weight of one helium atom and that of four hydrogen atoms is $0 \cdot 028$ atomic weight unit. This difference can be expressed in grammes if we divide by the Loschmidt number, thus changing over from gramme-atoms to individual atoms. The result, the difference between the mass of four individual hydrogen atoms and a single helium atom thus becomes $4 \cdot 63 \times 10^{-26}$ g. According to the equation of Einstein this means that about 5×10^{-5} erg of energy is liberated in the process. It is enormous for a reaction on an atomic scale, but it is of the order of magnitude which we have encountered in radioactive processes. As a matter of fact it is the energy of the hydrogen bomb and of the primary evolution of stars! (3×10^{19} ergs for 4g of He.)

Now, as often as it has been possible to measure exactly the energy set free in the course of a nuclear disintegration and to measure with the mass spectrograph the difference of mass-total between the reacting and the result-ing species, the factor of proportionality between these two has always been exactly c^2, and thus the theory of Einstein has been verified in the most impressive manner.

We have already succeeded in combining hydrogen nuclei to form helium nuclei in the hydrogen bomb, but up to now this reaction has never been tamed down to laboratory dimensions and no measurements were obtained for the elementary process. But the mass deficit, $0 \cdot 028$ g per gramme-atom is known and the energy evolved in an explosion can be estimated. Even before the hydrogen bomb came into existence astro-physicists assumed that the process was going on in "young" stars where the temperature, the mean kinetic energy of the colliding hydrogen nuclei, is sufficiently high to bring about this fusion, to overcome the Coulombic repulsion between the protons (p. 433). These young stars contain much hydrogen, and in the course of stellar evolution its amount decreases, giving way to an increasing amount of helium in stars which for various reasons are thought to be older. The relative abundance of these two elements can be estimated with great accuracy from their emission spectra.

In the next section we shall become acquainted with artificially induced radioactive reactions. We shall see how nuclei are sometimes built up, instead of decaying as in natural radioactivity. We should not be worried about the fact that energy is evolved when hydrogen atoms unite to form helium, as well as when heavy radioactive nuclei disintegrate. The same kind of thing happens in ordinary chemistry: we know of compounds which evolve energy when they decompose and others which evolve heat when they are formed from their constituent parts. Think of dynamite in contrast to the combusion of coal! Evidently the situation in nuclear "chemistry" must be a similar one. The heavy elements at the end of the periodic system would thus be analogous to the explosives, and the light elements to the elementary fuels. The mass spectrograph fulfils the same role in nuclear chemistry as the calorimeter in normal chemical laboratories. It yields the difference of energy between reacting and resulting nuclei as the difference in the sum total of their atomic masses.

Now at last we are able to decide upon the merits of Prout's hypothesis. It seems to have been vindicated at last. Hydrogen really seems to be the primordial element because all isotopes of all other elements have atomic weights which are very nearly exact multiples of that of hydrogen. All large deviations in ordinary atomic weights have been shown to be caused by the respective element being a mixture of isotopes with integral atomic weights. And the small deviations which still remain are the result of energy losses which have occurred in the course of building the complex nucleus from protons. The magnitude of these deviations is defined by the theory of general relativity as the energy set free divided by the square of the velocity of light. How it comes about that the positive charge of composite nuclei is always less, much less, than their atomic weight has still to be discussed. A simple fusion of protons would mean that each mass unit accounts for a unit of positive charge; atomic weight and nuclear charge (that is atomic number) should be the same. Somehow or other positive charge is lost; we shall deal with this problem in Section 97.

95. ARTIFICIAL DISINTEGRATION OF NUCLEI

In spite of what we have just said about nuclei being built from protons we only feel certain about this if we really succeed in knocking protons out of them in some way or another, or in shooting them into a nucleus so that they remain there. But in the course of natural radioactive decay we only observed the departure of alpha particles and electrons from disintegrating nuclei. By the way, the fact that electrons leave the nucleus has led scientists to believe for decades that electrons are one of their building elements and this seemed to account for the fact that their nuclear positive charge was much less than their mass. This seemed a quite satisfactory solution of the question we were forced to ask at the end of the foregoing section, but as we shall see later on it was not quite correct.

The fact that atomic weights of isotopes are not multiples of four shows conclusively that alpha particles—with just this atomic weight—cannot be the sole building stones. A glance at Table XIII (p. 419) convinces us at once that this hypothesis will not work. There must be protons in nuclei.

Once more we are indebted to the ingenious experiments of Rutherford who first observed a proton in the act of being knocked out of a nucleus. Let us remember that the range of alpha particles in a gas of given pressure and composition was found to be constant for each radioactive element. All their alpha particles came to rest after having travelled a certain number of millimetres (Fig. 148). Rutherford examined this range of alpha particles in nitrogen gas when, in 1919, he observed that there were exceptions, very rare but reproducible exceptions, to this rule. A similar phenomenon was already known for gases which contained hydrogen atoms for example H_2, H_2O, NH_3, etc. But it was easy to explain the long-range particles in these cases: they must have been protons ejected from their respective molecules by a head-on collision with the alpha particle. It was as though a billiard ball had collided with a smaller one, that had only one-fourth of its mass. The lengths of these observed paths exactly conformed to such an explanation. (Fig. 155.) The velocities and directions of the striking alpha particle and of the proton that was hit could be read from the Wilson photographs taken in a magnetic field, and the result was the same as would have occurred between billiard balls of the mass ratio $4:1$.

But in the nitrogen which Rutherford used for his experiments there was no hydrogen at all! He took great care to purify it to the utmost extent. And still the long-range particles kept on appearing with the same, though very low, frequency. This is one of the case histories in experimental science where it becomes clear that fundamentally new observations depend upon the utmost precision *and* self-reliance of a scientist. Rutherford believed in his own precision and knew that if such long-range particles still continued to appear they must be explained somehow, that they were real. He went on working with other gases which contain nitrogen but no hydrogen: with the nitrogen oxides. The long-range particles still kept on appearing, with a frequency that was proportional to the total number of nitrogen atoms present in unit volume. Thus the particles must have originated from the nitrogen atom! It was clear from the elementary collision theory of spheres that the nitrogen nucleus itself, with a mass of 14, cannot fly further than the impinging alpha particle projectile of mass 4. On the contrary. At that time the hydrogen nucleus, the proton, was the only nucleus known to be lighter

Fig. 155. **Head-on collision of an alpha particle with** (*a*) **a hydrogen nucleus,** (*b*) **a helium nucleus,** (*c*) **an oxygen nucleus**

The collisions conform exactly with those of a ball colliding with a lighter, an equally heavy and a heavier ball.

(From E. Rutherford, J. Chadwick and C. D. Ellis, *Radiations from Radioactive Substances*, p. 246, Plate IX, Figs. 1, 2, 3 (Cambridge Univ. Press, 1951). By courtesy of the Cambridge University Press.)

than the alpha particle (the deuteron was not then known), so the long range particle was evidently a proton. These particles were subjected to electric and magnetic deviations and actually showed the e/m ratio characteristic of protons (*Peterson* and *Stettner* 1925, in Vienna).

The phenomenon itself was first observed in scintillation measurements, then confirmed by ionization, but it can be most dramatically shown in

Fig. 156. **An alpha particle hits a nitrogen nucleus, expelling a proton and forming an oxygen nucleus**

(From J. Chadwick, *Radioactivity and Radioactive Substances*, p. 57, Plate IV (London, Pitman, 1961).)

Wilson cloud chamber photographs. Fig. 156 shows such a collision. An alpha particle proceeds until its track suddenly bifurcates and one of the new particles has the long range that we have discussed. There is no other assumption open than the one put forward by Rutherford: the nitrogen nucleus was hit by the alpha particle and exploded, emitting a proton! What became of the rest of the nucleus? There is only one other track which starts at the bifurcation, so the alpha particle and the other half of the former nitrogen nucleus must have remained together. The range of this other particle is always very short and it is clear that its mass must be much greater than that of an alpha particle. What can have happened between nitrogen nucleus and alpha particle?

It is time to grow accustomed to the way of writing such nuclear interactions. The logic of the whole process is rather simple in itself. Electric charge and mass must have been conserved during the interaction. Thus, if we had the nitrogen nucleus with mass 14 and charge +7 which was hit by the alpha particle of mass 4 and charge +2, the sum of the masses after the nuclear "reaction" must have remained 18 and the sum of the charges +9. We have successfully identified the proton (mass = 1, charge = +1) as one of the products, so the other must be a particle of mass 17 and charge +8. The charge +8 belongs to oxygen, so the other product of our nuclear reaction must have been the 17-isotope of oxygen an isotope known to exist. The currently accepted convention for nuclear reactions is to write the charge and the mass before the atomic symbol of the nucleus, the mass above and charge below. Remembering that the alpha particle is the helium nucleus, we have the following equation:

$$\ce{^4_2He + ^{14}_7N = ^1_1H + ^{17}_8O} \qquad . \qquad . \qquad . \quad (95.1)$$

This heavy oxygen atom had already been isolated from natural oxygen in Urey's laboratory in the thirties (p. 416). But, what about the balance of energies and masses? Table XIII, of exact isotopic weights, allows us to answer this question. The sum of masses on the left of the equation is 18·012 while that on the right is 18·013. The mass has increased, or in other words, a part of the kinetic energy of the attacking alpha particle has been built into the new nucleus! (These small, relativistic mass differences were disregarded when we said that the sum of masses must remain constant.)

After this fundamental first nuclear reaction it has become possible to bombard a series of higher nuclei with alpha particles and observe long-range particles as products of artificial disintegration up to the atomic number 40. Whenever the particle emitted was a proton, it was possible to make an energy–mass balance for the reaction. And this balance could be confirmed in the Wilson cloud chamber! Because the velocity of the alpha particle and that of the emitted proton could be measured by magnetic deflexion, it was thus possible to calculate whether the "billiard ball" theory of colliding spheres was correct or not. Whenever there was a mass deficiency (measured in the mass spectrograph) during the nuclear reaction it was found that there was a surplus of kinetic energy: the proton's velocity was greater than calculated for simple collisions. And whenever there was a mass surplus, there was a loss in total kinetic energy. The differences in mass and in kinetic energy were in the ratio of the Einstein relation

$$m = \frac{E}{c^2} \qquad . \qquad . \qquad . \qquad . \qquad . \quad (95.2)$$

within the limits of experimental accuracy. No such difference in experimental and calculated velocities occurred whenever the long-range proton came from a hydrogen compound like water, instead of from the interior of a nucleus. The free proton in water behaves as it should—like a billiard ball—and no mass is built in or liberated anywhere when it is hit.

The dream of generations of alchemists had been realized at last. Though only a few atomic nuclei reacted, still the ice was broken and it had become possible to transmute elements: man had made oxygen from nitrogen. Such a transmutation was already occurring in natural radioactivity, but it proceeded with its characteristic speed undisturbed by human interference. Rutherford was the first man to induce a nuclear reaction, and he thus opened a new epoch in the history of science—to say nothing of the history of mankind.

The value of transmutations turned out to be entirely different from what the alchemists were interested in. Transmutations have not become all-important because we have learned to make gold (at least in very small quantities) from other elements, but because we have liberated the sleeping giant of nuclear energy and have become promethean masters—or slaves?—of undreamt of power for good or for evil.

96. NEUTRONS

Artificial nuclear disintegration had been examined for already some fifteen years in the leading laboratories all over the world when a prophecy of Rutherford and W. D. Harkins was at last verified again. They said, in 1920, that an electrically neutral elementary particle was well within the limits of imagination; it was not clear, however, how such a particle could be detected. All particles known up to that date could be detected because they did possess electric charges. How could a tiny particle without this electric field of force around it react with anything? Gravitation is much too weak a force to offer any hope of success, and such a particle would be able to rush past quite near to any nucleus, no matter how highly charged.

The daughter of Marie and Pierre Curie (the veteran heroes of the discovery of radium), *Irène Curie*, and her husband *E. Joliot*, noticed in 1933 that a new kind of penetrating radiation was generated in a mixture of radium and beryllium. The new radiation ionized gases but was not acted upon by electric or magnetic fields and was able to penetrate through lead of appreciable thickness. They thought at first that it was a very hard gamma radiation. But it soon appeared that it behaved differently. Gases of low atomic weight were ionized most by the new radiation, whereas gamma radiation is absorbed and hence acts most intensively on elements of high atomic weight, which have many electrons around the nuclei. Thus it could not be gamma radiation. Photographs in Wilson cloud-chambers often failed to reveal anything, but now and then it could be observed that somewhere within the field of this invisible radiation a track resembling that of an alpha ray originated. The curvature of the track in the magnetic field always showed that it was positive and its velocity was the greater the lighter the gas which filled the chamber (Fig. 157).

It was, however, a pupil of Rutherford, *J. Chadwick*, who managed to show a year later—while his great master was still living—that this strange radiating particle was the neutron itself. It was nearly as heavy as the proton but had no electric charge. The argument ran as follows: Though it is true that neutral particles do not attract electric charges and thus are able to pass near to nuclei, they still do react with the nuclei if they suffer a head-on collision. They react primarily according to the rules which define the collision of billiard balls of different masses. The nucleus that has thus been

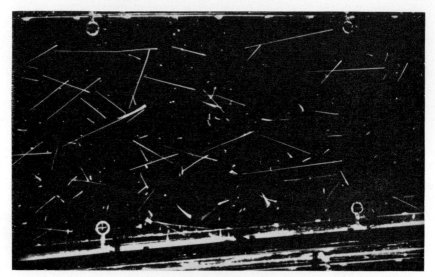

Fig. 157. **Tracks of nuclei knocked out of their atoms by impinging neutrons**

The heavier their mass, the smaller their velocity.

(From W. Gentner, H. Maier-Leibnitz and W. Bothe, *An Atlas of Typical Expansion Chamber Photographs*, p. 109, Fig. 93 (Oxford, Pergamon, 1954). By courtesy of the Pergamon Press Ltd.)

set in motion is of course charged, as are all nuclei, and therefore it acts upon the gas atoms along its track like an alpha particle or a proton or any fragment of a nucleus in an artificial disintegration. This is the reason why such tracks suddenly appeared in the cloud-chamber photographs. What we really observe in such cases is not the neutron itself but the nucleus it has collided with and which has received a part of its kinetic energy. The fact that a neutron has been present only manifests itself by a nucleus going wild, seemingly all by itself.

But this hypothesis is open to quantitative confirmation. The mass of the nucleus that has "gone wild" is known to us. The velocity it has acquired can be determined by magnetic deflexion in the cloud-chamber, and the direction of the incoming neutron is given by the geometry of the arrangement: it must have come from the direction where the radium–beryllium mixture was located. Energy and momentum are conserved in such collisions according to the fundamental laws of mechanics. The nucleus which has been struck would have been practically stationary before the collision (thermal velocities can be neglected in comparison with the velocities of, say, alpha particles) and its mass and acquired velocity are known. The mass and velocity of the impinging neutron, and its velocity after the collision are unknown. The two equations are insufficient for the three unknowns, but Chadwick found a way out of this impasse. Instead of making the experiments in a single gas he worked with two, one after the other. The neutron source remained the same, so there was no reason that the velocity distribution of the impinging neutrons should differ in the two gases. For instance, he worked with hydrogen and nitrogen alternately and thus obtained two further equations: the energy and momentum equation for the second gas as well. The number of unknowns only increased by one because mass and impinging velocity of the neutrons remained the same, only the velocity of the neutron after collision being different. These four unknowns were now calculated from the four equations and yielded as a most interesting revelation, the mass of the new particle. The results were the same no matter which pair of gases Chadwick used in his experiments.

How great, then, was this important mass? The first results seemed to show that it was practically the same as the mass of the proton. Later on, as nuclear reactions became known where neutrons appeared or disappeared and the energy–mass balance allowed more exact calculations, it turned out that there was a slight difference: the neutron was slightly heavier than the proton. Its mass was 1·0091 as against 1·0081 for the latter. Even if we assume that the neutron contains an electron of 0·0055 mass unit as well as the proton, there still remains a surplus of mass which amounts to 0·0045 unit. According to all we have said about mass–energy relations, this must mean that the neutron is richer in energy than the sum of its components. Accordingly we should not be very much astonished if it proved to be labile and to distintegrate with electron emission ("beta emission") into a proton after a while. Subsequent experiments revealed that this is really the case, the half-life of the disintegration is about 20 minutes.

The velocity of a neutron which is emitted in a nuclear reaction is of the order of magnitude of alpha rays or of emitted protons. No wonder, since it carries the same energy of nuclear disintegration as these sister particles. During its journey among atomic nuclei it now and then collides with some head-on and transfers a part of its kinetic energy to them. Its velocity decreases until eventually it is slowed down to thermal velocities, in statistical equilibrium with its surroundings. The time it takes to slow down to this extent depends upon the kind of nuclei that it encounters. A ball bounces back from a solid wall with undiminished velocity but may lose all its velocity at a stroke if perchance it hits centrally another ball, exactly its size. For the neutron this means that it transfers energy only in small amounts to nuclei of heavy atoms but loses it quickly on hitting light ones, especially protons which have virtually the same mass as the neutron itself. Therefore whenever one desires to "cool" neutrons quickly to ambient temperature they are made to pass through a medium where there is an abundance of protons: paraffin or water for instance. Strange as it sounds, they are cooled by water or ice. After travelling a few centimetres in these media they are slowed down to thermal velocity.

We shall see in the coming section that neutrons are able to combine with nearly all kinds of nuclei. The probability that such a union will take place varies from nucleus to nucleus. Here we only need to note, that this property of neutrons enables us to get rid of them if we so desire. It has been found that neutrons of thermal velocity are very readily absorbed by cadmium nuclei and thus a sheet of this metal is an excellent absorber of neutrons that have been previously slowed down (*see* p. 418).

Finally, let us glance at the reaction that helped to discover these wonderful particles. Beryllium has been bombarded by alpha particles—

$$\ce{^{9}_{4}Be} + \ce{^{4}_{2}He} = 3\,\ce{^{4}_{2}He} + \ce{^{1}_{0}}n \qquad . \qquad . \qquad . \quad (96.1)$$

where n stands for the neutron of mass one and zero charge. We might have expected that $^{12}_{6}C$, normal carbon, would be the product of the reaction, but the energy seems to break it to pieces; photographs in the cloud chamber actually show how the three alpha particles radiate apart from the "star" where the reaction took place (*see* Fig. 158 for a similar process). The range

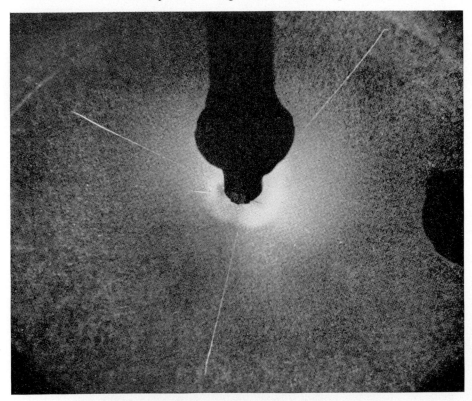

Fig. 158. **A proton hits a boron nucleus and together they disintegrate into three alpha particles**
$$\ce{^{11}_{5}B} + \ce{^{1}_{1}H} = 3\,\ce{^{4}_{2}He}$$

(From R. S. Shankland, *Atomic and Nuclear Physics*, p. 379, Fig. 10 (London, Macmillan, 1955).)

of these three alpha particles is long, a sign of the great kinetic energy of the explosion. Fig. 158 shows a symmetrical threefold star of a similar nuclear reaction in which boron is bombarded with fast protons.

$$^{11}_{5}B + ^{1}_{1}H = 3^{4}_{2}He$$

97. NUCLEAR REACTIONS AND ARTIFICIAL RADIOACTIVITY

The list of "reagents" in our new "nuclear chemistry" gradually increases. At first we smashed atomic nuclei with natural alpha particles, then we were able to induce nuclear reactions with neutrons. The list of reagents has increased further during the last decades as a consequence of the construction of larger and larger electric or electro-magnetic accelerators which are capable of accelerating electrically charged particles to more than hundred million electron-volts of kinetic energy, and are nearing the thousand million electron-volts mark. It would not be impossible to describe the working principle of these accelerators in terms which could be understood by readers who have managed to read this book so far, but it would lead us into an overwhelming complexity of experimental details which are outside our scope. Suffice it to say, that really wonderful ideas, sometimes wonderful in their simplicity but most intricate in their experimental set-up, made it possible to construct something like a dozen fundamentally different machines which either produce the necessary extremely high voltages themselves or induce the particle which is to be accelerated, to traverse a lesser potential drop a series of times until the acceleration adds up to its final value. These latter are the cyclotron, synchro-cyclotron and similar devices. It has also become possible to make a pre-accelerated particle which has nearly reached the velocity of light (and can gain no more *velocity*, only *mass* on acceleration) "surf-ride" an electrical field gradient which travels along an electric "wave-guide" some hundreds of metres long. Such a linear accelerator some kilometres long is about to be built in the United States. The cost of such apparatus is immense, as may be readily imagined. Not only is the final *acceleration* enormous but several types of these machines even permit *currents* of milliampere magnitude to be passed through them: this means that not only are the particles accelerated to fantastic kinetic energies but their number per second is also quite appreciable. We should bear in mind that modern detection methods: scintillation counters, cloud chamber photographs, and ionization tubes enable us to observe single, elementary events of nuclear reactions.

These accelerators can be used to accelerate protons, deuterons or helium nuclei, all of them stripped from their surrounding electrons in a gas discharge. The number of light nuclei which may thus be stripped of all their shell-electrons increases every year, because more energy is available at great concentrations to ionize even their most strongly bound K electrons from around the nuclei. As shown on p. 303, the binding energy increases with the square of the nuclear charge, that is the energy required to strip off the last electron of neon is a hundred times as great as that necessary to ionize the hydrogen atom. These artificially accelerated nuclei "fall" towards the cathode of the apparatus with the enormous kinetic energies just mentioned. If a target containing some element which is to be examined is set in their way they impinge upon its nuclei and bring about much more readily observable nuclear reactions than the weak sources of naturally occurring alpha rays. Weak sources, that is, in comparison with these huge machines because the number of particles available in the latter is far, far greater. Thus the chance of observing a nuclear reaction is far, far greater too. Owing to the fact that the nuclei are very small, only some 10^{-12} cm in diameter, the chance of nuclear collisions occurring is very small indeed and only a few out of very many projectiles will be lucky enough to score a hit.

Let us risk a calculation. How many alpha particles are contained in a current of one milliampere during a second? 1 mA.sec is equal to 1/1000 coulomb and we remember from pp. 197, 257, that the unit charge on a proton is $1·519 \times 10^{-19}$ coulomb. He^{++} carries two such charges, so that 1/1000 coulomb contains $3·15 \times 10^{15}$ such particles. Compare this figure with the number of alpha particles, emitted from one gramme of radium in a second: $3·4 \times 10^{10}$ (p. 406). Thus the accelerator with its current of one milliampere is 100,000 times more efficient, its output equals that of 100 kg of radium metal. Quite a feat!

Now what about the energy of a single particle? On page 202 we calculated that one electron acquires $1·59 \times 10^{-12}$ ergs of energy after having "fallen" through a potential drop of one volt. Hence the doubly charged He^{++} will acquire $3·18 \times 10^{-6}$ erg after having traversed one million volts in an accelerator. According to p. 397, the kinetic energy of an alpha particle emitted by radium is $7·4 \times 10^{-6}$, that is twice as great. With two million volts instead of one we have reached the kinetic energy of radium alpha rays but with an intensity 100,000 times as great as would be obtained from one gramme of radium metal. An enormous success!

Particle sources of such intensity allow us to observe reactions which occur but seldom. The particles at our disposal in such abundance include, however, neutrons as well as the charged protons, deuterons or He^{++} particles. Because there are many nuclear reactions which give birth to neutrons and if we have sufficient "reagents" of high kinetic energy, the product will emerge in great quantities also. If, for instance, a solid target containing deuterium, e.g. heavy ice or (partially) deuterated paraffin, is bombarded by a ray of deuterons:

$$^2_1H + ^2_1H = ^1_0n + ^3_2He, \qquad . \qquad . \qquad . \qquad (97.1)$$

the products are a neutron and a light isotope of helium. (The reaction alternatively yields tritium: $= ^1_1H + ^3_1H$.) One of the best yields of neutrons is given by the reaction

$$^7_3Li + ^2_1H = 2^4_2He + ^1_0n, \qquad . \qquad . \qquad (97.2)$$

a reaction which is probably made use of in the hydrogen bomb. This reaction has been photographed in the cloud chamber and has acquired fundamental importance because it proves that the alpha particle is not an ultimate fundamental building stone of the material world but is itself built from lighter components: neither $^7_3$Li nor $^2_1$H contains sufficient mass *alone* to yield two alpha particles. Yet two were observed emerging from the reaction in the Wilson chamber. Thus at least one of them must have been built up in the course of the reaction.

Particles with an electric charge in nuclear reactions are at a disadvantage compared with neutrons. The positive nuclei electrostatically repel the positive proton, deuteron, or alpha particle and this is the reason why alpha particles fail to disintegrate nuclei with a greater charge than 46 (Pd), that is, nuclei in the latter half of the periodic system. Protons with half the charge of the alpha particle were found to be still effective on nuclei up to nuclear charge 51 (Sb), but not above this at the time the system was systematically scanned for nuclear reactions with neutrons.

Neutrons are of course electrically neutral. Thus they are not repelled from nuclei and may react with them on collision. They indeed do so sometimes. A new sort of force comes into action, evidently the force responsible for holding together nuclei, and this force draws the neutron into the nucleus. It enters, to stay or to induce a nuclear reaction as the case may be. The probability with which the neutron is incorporated into a nucleus varies from nuclear species to species and depends on the velocity of the neutron. That cadmium is an ideal absorber of slow neutrons has been stated already. In general, slow neutrons are more likely to stay within a nucleus, maybe because they spend more time in its vicinity and are more easily captured. But very often there exist sharply defined neutron velocities which are absorbed much better than others. They are similar to the sharp lines of light absorption in spectra of compounds or elements. This so-called "resonance" velocity of the neutron must be in some similar relation to possible energy states within a nucleus as the absorption lines of light are to those within a molecule. In such cases the kinetic energy of the fast neutron is built into the new nucleus to account for the difference in energy which is not quite covered by the mass-deficit.

As mentioned, such nuclear reactions begin by the addition of a neutron, but they do not necessarily end at this stage. A second reaction may follow, and not only after a reaction with neutrons. Reactions which have been started by protons, deuterons or alpha particles may enter a second stage also, after a first, labile product has lived for a certain time. It seems that the ratio of mass to charge in a nucleus cannot exceed certain limits without making the nucleus unstable. Such a labile nucleus corrects this ratio by emitting a negative electron if its charge has been too low (too low positive!) or by emitting a new species, a positive electron or "positron" if it previously had too great a positive charge.

We owe it once again to the Joliot-Curie couple that the first "artificial radioactivity" of this kind was observed in 1933. They bombarded boron with alpha particles and obtained a neutron and a light nitrogen isotope:

$$^{10}_5\text{B} + {}^4_2\text{He} = {}^{13}_7\text{N} + {}^1_0 n \qquad . \qquad . \qquad . \quad (97.3)$$

However, this isotope proved to be unstable and it suffered beta disintegration with a half-life of 10·3 minutes. The beta particle in this instance was not the regular negative electron but its positive counterpart, the positron:

$$\ce{^{13}_7N} = \ce{^{13}_6C} + \oplus \qquad . \qquad . \qquad . \qquad . \qquad (97.4)$$

where the sign $\oplus$ represents the positron and the other species is a heavy isotope of carbon.

Positrons have already been observed in cloud chamber photographs of cosmic rays. *C. D. Anderson* (1932) proved that they were positively charged

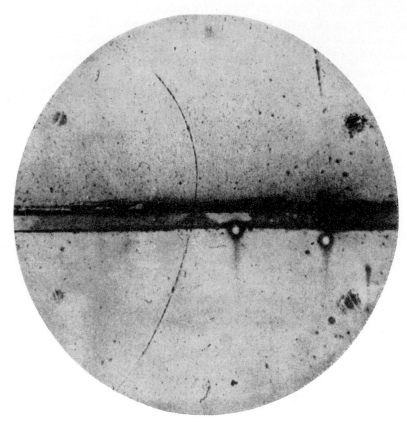

Fig. 159. **Track of a positron in a cloud chamber**

The positron comes from the lower part of the chamber, traverses an Al foil and proceeds after having lost velocity. Velocities are shown by the curvature of the track in a magnetic field: the curvature is stronger in the top part of the picture.

(From C. D. Anderson, *Phys. Rev.*, v. 43, p. 392 (1933). By courtesy of The Physical Review.)

by letting them traverse a thin absorbing foil within the cloud chamber and observing the curvature of their path in a magnetic field (Fig. 159). They must have lost velocity on traversing this sheet and thus it became possible to determine from the change in curvature of the track the direction from which they entered the field. Anderson found that the sign of the curvature was the opposite of the sign an electron would have shown, all other characteristics of the track being identical with those of an electron. The particle must have been a positron. The Joliot-Curies have found this same particle in a terrestrial nuclear reaction!

The positron only lives until it encounters an electron. When they meet they annihilate each other and two gamma rays of opposite direction and same wavelength are emitted. Thus momentum is conserved after this strange conversion of charged particles into electromagnetic radiation. The energy of the gamma rays is 0·511 MeV each, exactly the equivalent of the mass of an electron or a positron according to the Einstein equation $E = mc^2$.

The opposite process also occurs. If a gamma ray has more energy than equivalent to the rest mass of an electron plus a positron it may transform into such a pair in the neighbourhood of a heavy nucleus. The minimal gamma ray energy for this pair formation is twice 0·511 MeV, that is, 1·022 MeV. Charge and energy are conserved in both processes. Knowing that both light quanta and electrons (or positrons) behave sometimes as if they were particles, sometimes as if they were waves, these mutual inter-conversions seem less absurd than they would have seemed fifty years ago.

A nuclear reaction which is the reverse of positron emission had been predicted by *H. Yukawa* and *S. Sakata* in 1936. It is the capture of an electron moving in the innermost, K, orbital of an atom by its nucleus. The charge of the nucleus decreases by one unit after such a capture process as if it had lost a positron. Two years later the effect has been found by *L. W. Alvarez* on a vanadium isotope which spontaneously transformed to titanium, its left-hand neighbour in the periodic system. The process manifests itself by spontaneous emission of an X-ray quantum which belongs to the K-series (p. 302) *of the newly formed element*, in this case of titanium, because the electron which has been lost by nuclear capture from its K-shell is replaced by one from its outer region where it has become superfluous anyway because the charge of the nucleus has decreased by one unit. The same process is known to occur with the $^{40}_{19}K$ isotope which is thereby converted to $^{40}_{18}Ar$. This is probably the reason why argon is found so much more often in Nature than the other rare gases.

It is a quite general rule that whenever heavy particles (proton, deuteron, alpha particle) are generated in a nuclear reaction they leave the nucleus immediately, without any time lag. But light particles: electron, positron (and neutrino, p. 404) take their time, leaving only after some delay with a characteristic half-life of radioactive decay. Maybe this is the most rational proof of the assumption that nuclei are not built from protons and negative electrons, as was held before neutrons had been discovered, but from protons and neutrons instead. These are "ready" to leave the nucleus without notice, whereas the electron and the positron need some time to come into being. Probably by beta decay of a nuclear neutron or positron decay (!) of a proton. The first of these processes occurs spontaneously with a half-life of some 20 minutes if the neutron is free: it yields the proton.

It would serve no purpose to go through the hundreds of nuclear reactions which have been unravelled up to now. We entered the field of special chemical reactions only as far as it was necessary to explain fundamental issues. Here we shall proceed in the same manner, just to show some general types of nuclear reactions.

Let us begin with the reaction of the old alchemists, the making of gold. True, it is made from platinum and with a very poor yield at that, but the conversion has been achieved:

$$^{198}_{78}\text{Pt} + ^1_0n = ^{199}_{78}\text{Pt} + \text{gamma ray} \qquad . \qquad . \qquad (97.5)$$

$$^{199}_{78}\text{Pt} = ^{199}_{79}\text{Au} + \ominus \qquad . \qquad . \qquad . \qquad (97.6)$$

where $\ominus$ stands for the electron. The 198 isotope of platinum captured a neutron and released some of the resulting energy in a gamma quantum. But the new platinum isotope was not in mass–charge equilibrium. It was too heavy for its charge and therefore its positive charge was increased by the emission of a negative electron. This species which is more positive by one unit happens to be gold. The half-life of the beta disintegration is half an hour.

Gamma quanta are not only products of nuclear reactions, they are also able to initiate some. They are like light in photochemistry: there are luminescent reactions where light is emitted and photo-reactions where it is absorbed. For instance, copper of mass 63 emits a neutron (a "photo-neutron") if it is hit by a gamma quantum:

$$^{63}_{29}\text{Cu} + \text{gamma quantum} = ^{62}_{29}\text{Cu} + ^1_0n \qquad . \qquad . \qquad (97.7)$$

followed by positron emission to compensate by a loss of positive charge for the mass loss of positive charge for the mass lost:

$$^{62}_{29}\text{Cu} = ^{62}_{28}\text{Ni} + \oplus \qquad . \qquad . \qquad . \qquad (97.8)$$

Cases have been observed where one and the same nucleus may react in different ways with the same nuclear reagent. It is the analogue of parallel reactions in chemistry:

$$^{31}_{15}\text{P} + ^1_0n = ^{28}_{13}\text{Al} + ^4_2\text{He} \quad \text{and } ^{28}_{13}\text{Al} = ^{28}_{14}\text{Si} + \ominus \qquad . \quad (97.9)$$

$$^{31}_{15}\text{P} + ^1_0n = ^{31}_{14}\text{Si} + ^1_1\text{H} \quad \text{and } ^{31}_{14}\text{Si} = ^{31}_{15}\text{P} + \ominus \qquad . \quad (9.10)$$

$$^{31}_{15}\text{P} + ^1_0n = ^{32}_{15}\text{P} + \text{gamma and } ^{32}_{15}\text{P} = ^{32}_{16}\text{S} + \ominus \qquad . \quad (97.11)$$

$$^{31}_{15}\text{P} + ^1_0n = ^{30}_{15}\text{P} + 2\,^1_0n \quad \text{and } ^{30}_{15}\text{P} = ^{30}_{14}\text{Si} + \oplus \qquad . \quad (97.12)$$

Evidently it is not an easy task to disentangle such a set of parallel reactions. Whenever the new nuclear species is radioactive and lives sufficiently long for chemical separation, this is the method of choice. For instance, the irradiated phosphorus may be dissolved quickly in nitric acid and transformed into phosphate by the addition of alkali. Now some sodium silicate (water-glass solution) is added as carrier of the traces of $^{31}_{14}\text{Si}$ and the phosphate precipitated chemically (e.g. as ammonium magnesium phosphate). A beta activity remains, however, with the silicate and cannot be separated from it; it decomposes with a half-life of 144 minutes. The fact that it again yields phosphorus after its decomposition only follows from the fact that it decayed with emission of a negative electron. The phosphorus which results is not radioactive any more and cannot be detected with radioactive techniques.

Szilárd and *Chalmers* (1934) evolved a very ingenious procedure by which atoms born in nuclear reactions can be followed. For example, it is possible to prove that a phosphorus atom captures a neutron and is transformed to another, radioactive phosphorus atom (equation 97.12). The method is based on the fact that the phosphorus nucleus which has been hit by a neutron (or which goes into any nuclear reaction) receives an amount of kinetic energy by this collision which is many times greater than chemical binding energies and therefore is thrown out of any compound it may be in at the beginning. It would become an individual phosphorus atom and could be extracted, say by carbon disulphide, when necessary after addition of non-radioactive phosphorus to act as "carrier" of the minute quantity of radio-active phosphorus. This radioactive phosphorus now remains indistinguishable from normal phosphorus by chemical means until it eventually disintegrates to silicon.

Neutrons are much more readily detected since artificial radioactivity has been discovered. It is no longer imperative to resort to the cloud chamber in order to find protons which have been struck by neutrons. It is sufficient to expose to neutrons an element which is known to react easily with them and which yields a product that is radioactive. For instance a silver coin will do: leave it for a minute or two where neutrons are around and see whether it discharges an electroscope or gives discharges in a Geiger counter. Ordinary silver does not do these things but its radioactive isotope, which was formed in the nuclear reaction with neutrons, does. It is not difficult to do the same experiment in a quantitative manner.

98. NUCLEAR FISSION

H. Fermi and his school investigated systematically during the thirties in Rome the reaction of all the elements with neutrons. On working with uranium they of course hoped to synthesize a new element, heavier than uranium itself which was the last in the natural system. The idea was obvious: if uranium absorbed a neutron and became beta-active as many other elements did, the emission of a negative electron would increase its nuclear charge to 93: the synthetic element would have been born. And indeed, a number of elements were formed in this reaction and were separated in due course from their parent, uranium. The fact that more than one element was formed did not seem extraordinary: uranium has several isotopes and it was even conceivable that the new element would itself be radioactive and be the parent of the fourth radioactive family that we postulated on p. 409. *Fermi* and his co-workers in Rome and *O. Hahn*, *L. Meitner* and *F. Strassman* in Berlin-Dahlem put in very much hard work in identifying these products in the hope of clarifying the problem of the transuranium elements. But instead of getting clearer, the situation became more complicated. There were more radioactive isotopes than could be accounted for by normal assumptions and they had properties other than they should have had if they were to fit in the place of the transuranium elements in the system. Among others, a beta-active radium isotope was found among the products. By that time L. Meitner had already been forced by the Nazis to leave her famous laboratory but Hahn and Strassman succeeded in identifying it. They decided to concentrate this radium isotope by classical methods: crystallizing and fractionating it after the addition of some barium as a carrier. And it was found that the "radium" isotope remained in the barium fraction! It was not radium at all, it was a beta-active barium isotope. They immediately attempted to identify other remarkable isotopes of this mixture with elements which lie in the neighbourhood of barium and found it was possible to do so. The whole set of new isotopes was shown to belong not to the end of the periodic system, much less to some transuranium element, but to a medium range of the system.

In the course of a few months in 1938–39, electron-emitting isotopes of Se, Br, Kr, Y, Mo, Ma (called today technetium, Tc) Ag, Sb, Te, I, Xe, Cs, Ba and La had been chemically isolated and characterized beyond doubt. Some of them were already known by that time as beta-active isotopes, from direct neutron addition to the respective stable nuclei. Summing up, it could be said that on irradiating uranium with neutrons a whole series of beta-active elements were formed, between the thirty-fourth and fifty-seventh elements of the periodic system. L. Meitner together with O. Frisch drew the conclusion: If this was the case then the uranium nucleus must have split into two after having reacted with a neutron, and the sizes of the two parts thus formed were distributed according to the laws of chance around the middle of the system. It did not always split in exactly the same way but always near the middle. Uranium, being the last natural element must be on the verge of stability and it was not at all improbable that it could not withstand an attack from a neutron.

Bohr immediately calculated that the concentration of positive charge in the $^{235}_{92}U$ nucleus—amounting to 92 units—accounted for such an excessive amount of electrostatic repulsion within the nucleus that it only needed a small amount of additional energy to begin to oscillate around its equilibrium "drop" form, like a drop of water in the wind, and eventually to fall apart. The internal energy of a nucleus is given by its mass defect, that is by the difference between its actual atomic weight and the weight it would have if it had been simply added up from so and so many protons as Prout had postulated. The Einstein equation gives the energy resulting from this mass defect. Whenever this energy of formation, that is the mass defect is small, much energy must have remained in the nucleus after formation. Bohr and Wheeler found that beside the 235-isotope of uranium, a thorium and an actinium isotope also possessed sufficient energy to be liable to fission when a neutron entered them. Experiments confirmed that these three elements did in fact undergo fission on neutron bombardment.

But what sort of nuclei could have been formed in the course of such a fission? The table of atomic weights shows that atomic weight increases more quickly than atomic number. At the beginning of the system atomic weight is about double the atomic number, that is every proton in the nucleus has on the average one neutron as a partner. But the heavy nuclei contain relatively more neutrons. Uranium for example, has the atomic number 92. If it contained the same number of neutrons its atomic weight should be 184, whereas it actually is 238 or 235, according to the isotope considered. If such a nucleus bursts into two medium-sized parts, these fragments will contain relatively more neutrons than they should in proportion to their charge for medium-weight nuclei. A fission of ^{238}U into two exactly equal parts would yield two nuclei of mass 119 and charge $92/2 = 46$. This charge belongs to palladium; its actual atomic weight is however only 107! There are 12 neutrons too many in the hypothetical fission product. So it is with the other fragments in the middle of the system.

There are two ways out of this impasse and both seem to occur during fission. First of all it is conceivable and is actually observed that neutrons are set free during the fission process: one neutron has triggered the process, but more than one can be found among the products. This makes it possible that fragments with a smaller neutron surplus should be formed. But the number of secondary neutrons is relatively small, only two or three in a fission and this is insufficient to balance the difference of 2×12 that we have arrived at tentatively.

The other way to balance the equilibrium between mass and insufficient positive charge in the fragments would be by beta-emission of electrons. This process transforms a neutron within the nucleus into a proton, increasing the charge by one and decreasing by one the number of neutrons in the nucleus. Actually, all fragments of fission are negative beta-emitters. For instance, the first element not to fit into the scheme of transuranium elements, the one which had been mistaken for beta-active radium was beta-active barium, as we have seen. And so on for all of them. One of the most frequently occurring fragments, for instance, $^{140}_{54}$Xe decomposes by electron emission to $^{140}_{55}$Cs, and so through $^{140}_{56}$Ba and $^{140}_{57}$La into $^{140}_{58}$Ce, where the normal ratio for protons to neutrons has been reached at last. This latter nucleus remains the stable end product of this half of this special fission. The other fragment of the same fission is $^{95}_{48}$Sr, the relatively long-lived isotope which makes us so anxious about the effect of atomic explosion fall-out on pastures and eventually on the milk we consume. It too undergoes beta-decay, and its products continue to do the same until, through the beta-active series of Y, Zr, Nb, Mo, and Tc, the nucleus at last becomes stable at $^{95}_{44}$Ru.

Whereas normal fission of a nucleus divides it into only two parts and neutrons there is another process whereby it can be literally smashed to pieces. This process occurs if protons accelerated by some 400 megavolts or neutrons accelerated by about half of this potential drop impinge on nuclei of medium atomic weight. It has been first observed by *G. Seaborg* in 1947 and is called "spallation."

As stated above, Bohr's calculations showed that the really fissile component of natural uranium should not be its main isotope, $^{238}_{92}$U, but the second one which accounts for 0·7 per cent of the natural element, $^{235}_{92}$U. The fact that it is less stable than $^{238}_{92}$U is no chance happening, isotopes of odd atomic weight are generally less stable than their even neighbours, as expressed by their mass defect and by the smaller percentage in which they generally occur in Nature (p. 457). This more labile isotope of uranium is broken into fragments after the addition of a further neutron, even if the neutron is "cold" or "thermal," that is slowed down to thermal velocity. The other two fissile isotopes, those of thorium and actinium respectively, are not quite as labile as uranium. They need high-velocity neutrons to make them explode. Soon after Bohr's theoretical publication, uranium isotopes were isolated on a mass spectrogram and it could be confirmed that the 235-isotope was responsible for the fission.

Thus with the help of $^{235}_{92}$U it seemed hopeful to go on and increase nuclear fission like an avalanche. The number of neutrons is multiplied during each act of fission and the secondary neutrons may go on reacting with more and more $^{235}_{92}$U nuclei, evolving a fantastic amount of energy during this fission, until all $^{235}_{92}$U has reacted. It is easy to compare the mass of $^{235}_{92}$U with the sum total of the masses of its fragments, and the Einstein relation $\Delta E = \Delta mc^2$ gives the energy set free during the process. It proved to be above 200 million electron-volts, that is more than twenty times as much as the energy of an average radioactive disintegration. And this radioactive energy is already some 10^6–10^7 times as great as the energies we have been used to in the course of chemical reactions! The kinetic energy of the fragments was determined in the magnetic field of the cloud chamber: the curvature of their tracks gave their momenta. In other cases the sum total of ion pairs generated in a gas by a fragment was determined, and knowing the energy of ionization of a single pair, the total energy of the fragment was found by simple multiplication. All experimental results confirmed the enormous amount of energy that had been calculated from the mass deficit.

Then, for the first time in the history of mankind it had become possible to sit down and make plans for avalanche-like fissions of atoms. For good and for evil.

The first step seemed to be to separate ^{235}U from natural uranium. The reason for this was that if only every 140th uranium atom is fissile—as in natural uranium—a great part of the available neutrons would be put out of action by absorption in the heavier isotope, and the number of fissions would be insufficient to maintain the avalanche. On dealing with isotopes in Section 92 we have discussed methods for their separation. All these methods were very slow and worked with very great loss of the expended energy. Furthermore most of the methods worked only on gases and the only compound of uranium which vaporizes under normal conditions is its hexafluoride, UF_6. The molecular weight of this compound is 349 or 352, respectively, according to whether it contains the lighter or the heavier isotope. The difference is less than 1 per cent. It was no easy task to use this small difference to separate the uranium isotopes.

As a matter of fact all, or nearly all the methods available were used or at least given a trial during the war against Hitler. The great venture of immigrant, British and American scientists in the United States, the "Manhattan Project," lasted six years before the first atomic bomb exploded over Hiroshima. It came too late to vanquish Hitler, but it shortened the war against Japan. The scientists and the engineers on this project had to start from first principles. The whole idea of nuclear fission was entirely new in all its phases. And it had to be carried out in secret while involving operations on a tremendous scale.

Gas diffusion through batteries of porous tubes, thermal diffusion and even the mass spectrograph were in use to separate the uranium isotopes. This last technique had to be adapted to mass production. Of course a single unit could hardly separate an appreciable quantity, so the number of units was increased until the Federal silver reserve in West Point had to be transformed into silver wire to help build the city of "calutrons," as these electromagnetic separators were called in Oak Ridge.

The exact size of the atomic bomb was of course a strategic secret. Its minimal size was determined by the fact that sufficient neutrons had to be retained within its body to trigger the nuclear avalanche. The problem was this: The chance of a neutron hitting a nucleus is very small because of the tiny effective cross-section of a nucleus. Many of the nuclei are by-passed before one at last is hit. The "free path" of a neutron is long. The nuclei cannot be brought nearer to each other, because their natural distance is determined by the density of the solid. If the piece of uranium is smaller than a certain size, it does not help to introduce neutrons into its body. There will be some fission, but the number of neutrons lost through its external surface will exceed the number created in the same time and there will be no avalanche. However, if the size of the piece is increased, the number of fissions will increase in proportion to its mass, that is to its volume, while the number being lost only increases with its surface. Volume is proportional to the cube of the radius, surface to its square; a certain radius will make it possible for more neutrons to be formed than lost: the process gains momentum autocatalytically in the same way as when an explosive, e.g. TNT, detonates. For chemical explosives it is the heat and/or the shock wave that has to be conserved within their body to cause complete detonation, for atomic-fission bombs it is the number of neutrons. Within seconds, or fractions of seconds, all ^{235}U nuclei are hit by the ever-increasing number of neutrons and all of them explode. Or rather not quite all of them. The enormous energy that is set free volatilizes the whole bomb, unreacted and reacted uranium and all. During its expansion into a cloud the ^{235}U nuclei become ever more distant from each other and from the neutrons, which are also diluted. So a certain fraction of the explosive material is dispersed without reaction.

In practice, the atomic bomb is triggered by allowing two halves of it, which in themselves are each too small for an explosion, to coalesce into a single unit, probably by shooting or dropping one part into the other. There is no need to introduce neutrons, since there are always sufficient at hand to initiate the explosion as soon as the critical size is reached.

99. TRANSURANIUM ELEMENTS

Historically it is not true that the first atomic bomb was made of ^{235}U. In the course of experiments which were designed to gain insight into conditions which make nuclear reactions become critical, the original idea of Fermi to produce transuranium elements was realized incidentally, and it happened that one of these was even more prone to fission than ^{235}U. The circle closed: scientists first looked for transuranium elements and thought they had found them. Then it was shown that what they found was something quite different, fragments of a nuclear process they had not dreamt of, of nuclear fission. This new process promised undreamt of energy concentrations and accordingly everything was done to produce ^{235}U, the raw material for fission, and to determine its properties. But while enormous quantities of both uranium isotopes had been exposed to neutrons it was found that ^{238}U really achieves the trick Fermi originally sought for. It adds a neutron and emits an electron subsequently yielding a new, synthetic element which was named neptunium because the planet beyond Uranus is called Neptune.

$$\mathrm{^{238}_{92}U} + \mathrm{^{1}_{0}}n = \mathrm{^{239}_{92}U} \qquad . \qquad . \qquad . \qquad . \qquad . \qquad . \qquad . \quad (99.1)$$

and

$$\mathrm{^{239}_{92}U} = \mathrm{^{239}_{93}Np} + \beta^- \text{ with a half-life of 23 minutes} \quad . \quad (99.2)$$

Neptunium is an element in its own right, it differs chemically from uranium and thus it can be separated from it by chemical means. No extravagant isotopic separation is necessary. However, as one might have assumed, it is not stable and suffers radioactive disintegration: not fission, but normal disintegration as beta-emitter with the half-life 2·3 days.

$$\mathrm{^{239}_{93}Np} = \mathrm{Pu^{239}_{94}} + \beta^- \qquad . \qquad . \qquad . \qquad . \quad (99.3)$$

This second artificial transuranium element was named after the small planet Pluto, beyond Neptune and thus received the name plutonium. It is of course also radioactive, an alpha-emitter with a half-life of 24,000 years yielding the fissile isotope $\mathrm{^{235}_{92}U}$ as its product. But this was not its main value and the time needed to produce ^{235}U in this way would have been too long anyhow. Plutonium is important, because it is fissile itself, on being hit by a neutron. So it has become possible to manufacture a fissile isotope of a new element by irradiating normal uranium with a strong flux of neutrons and this new element can be separated by purely chemical means from its parent, or rather grandparent element.

Plutonium as well as neptunium have a chemistry of their own, not so very different from that of uranium but sufficiently so to make a chemical separation possible. Remembering the last row of the periodic system we know that the elements from actinium onwards form a subgroup similar to the rare earths, where a subshell of electrons begins to be completed. This is the reason why they do not differ very much in their chemical behaviour because their outermost electronic structure remains the same. Still, they differ sufficiently to allow a technical separation of plutonium and this was accomplished. Not without great difficulties, by the way, because it is a powerful source of alpha rays and can only be handled with the utmost precaution. As a matter of fact huge robot factories had to be built where some 100 chemical operations were carried on automatically, and the supervisors only saw what was being performed behind thick walls of concrete by gazing into mirrors. It was very difficult but it was during the war for human freedom and all means were put at the disposal of the staff involved. Plutonium was isolated and was probably the material of the first bomb.

Plutonium is an element with the positive valencies three, four, five and six and its valency states are characterized by beautiful colours. As with all elements of the actinide group, its compounds have a very characteristic affinity towards ion-exchange resins and may therefore be separated from the others by adsorption on a column of such a cation-adsorbing resin and gradual elution from it by a solution of given acidity. One after another, the actinides are eluted from the column and appear one after another in the solution which comprises the eluate. The radioactivity of the actinides makes their detection easy: radioactivity increases and decreases again in the effluent solution each time a new element is being carried off from the column.

Other isotopes of neptunium and plutonium have been prepared by other nuclear reactions, ^{241}Pu, for instance, by bombarding ^{238}U with highly accelerated helium nuclei (more energetic than natural alpha particles!) in the Berkeley cyclotron. The alpha particle is captured and a neutron is set free:

$$^{238}_{92}U + {}^{4}_{2}He = {}^{241}_{94}Pu + {}^{1}_{0}n \qquad . \qquad . \qquad . \quad (99.4)$$

or using a simplified way to designate nuclear reactions:

$$^{238}_{92}U(\alpha, n)^{241}_{94}Pu \qquad . \qquad . \qquad . \quad (99.5)$$

The first particle in brackets is the one shot at the preceding nucleus, while the second is the one being emitted during its transformation into the product.

This plutonium isotope is beta-active and gives rise to an element of atomic number 95: americium (Am). Alpha bombardment of plutonium in the cyclotron gave the next element: $^{242}_{96}$Cm (curium). Going progressively higher on the scale of atomic weights the stability of the transuranium elements decreases. Their half-life becomes shorter and makes this ultra-microchemistry of microgramme to milligramme quantities a veritable miracle. In spite of this it was possible to irradiate americium and curium in the cyclotron and to identify the two next elements berkelium (Bk) and californium (Cf) (1950).

The elements 99 and 100 were named einsteinium (Es) and fermium (Fm) to honour these famous scientists; they have been discovered in the debris of atomic bomb explosions. Later on they were also prepared from plutonium in a heavy flux of neutrons (1952–53).

Without even isolating a weighable amount of einsteinium (Es) the Californian group of nuclear scientists lead by *S. G. Thompson*, *A. Ghiorso* and *G. T. Seaborg* proceeded to irradiate it again with helium nuclei in the great Berkeley cyclotron.   Dissolving the product and passing it through a separating ion-exchange column only took about ten minutes;   eventually the new element, mendelevium (Md), (after Mendeleev) was only identified by the predicted spontaneous fission of its daughter element, and that only in separately observed individual cases.   Thus this element, $^{256}_{101}$Md was first detected as its individual atoms (1955).   The last synthetic elements prepared up to now are nobelium (No) and lawrencium (Lw), the 103rd element in the periodic system (1961).

   We have concluded on p. 409 that four different alpha-beta desintegration series in all should exist among the radioactive elements, because the alpha particle has the mass four.   If a fifth series existed it should be identical with one of them: an element with the atomic weight $n + 5$ must be the same as one with $(n + 4) + 1$.   But only three such series were found.   The fourth series, predicted by *F. A. Turner* in 1940 was eventually found as originating from ^{241}Pu.   As mentioned above this emits electron-yielding ^{241}Am which in its turn is alpha-active and gives ^{237}Np, the parent (by alpha-emission) of protactinium 233.   This rare isotope emits an electron and gives place to the rare ^{233}U from which the radioactive series continues until it ends at inactive ^{209}Bi.   This series gives rise incidentally to the radioactive alkali metal francium ($^{221}_{87}$Fr) which was long sought after but only discovered by Mlle M. Perey in 1939.   No wonder it was difficult to find: its half-life is only five minutes and none of its predecessors lives long enough to establish a permanent reservoir on earth (*see* Table XI, p. 411).

100. NUCLEAR REACTORS

Six months after fission was discovered, the second world war broke out. It seemed probable to all parties concerned that an atomic bomb of unprecedented strength could be manufactured from ^{235}U, and accordingly the race to achieve this end began.   But whereas the *élite* of German physicists were loath to help Hitler's war effort, the scientists of the free world, many of them Jews and/or refugees from Hitler, spent all their mental and physical energies in solving the problem.

   One of the most important steps during the experiments which led to the atomic bomb was to prove that a self-sustaining chain reaction of neutrons and fissile atoms could really take place.   Incidentally, the experimental set-up which served to prove this point has developed into the standard unit for the peaceful use of atomic energy.   It is the nuclear reactor or the "atomic pile."

   The experiment for which the reactor was constructed had to prove that given a sufficiently dense structure which contains fissile atoms, the number of neutrons produced within this structure can be made to equal the number lost during the same time by absorption and by escaping through the surface. The idea of building such a reactor almost seems absurd because one has to venture right up to the limit where a slightly greater number of neutrons would be produced than lost and this would mean an avalanche, an atomic explosion. Suppose that the time between a fission yielding two neutrons and the next fission which has already been set off by the neutrons of the first one, is 10^{-8} second. The number of fissions doubles in this time and in less than 10^{-6} second more than 10^{24} neutrons would be produced, enough to set off a gramme-atom of fissile material.

It would not have been possible to prevent such an explosion, had it not been for the fact that 0·75 per cent of all the neutrons generated during the fission of ^{235}U are not direct products of the nuclear explosion but come out of disintegrating secondary nuclei which are formed in the fission process (p. 439). These are neutrons emitted from "artificial radioactive" nuclear species and are emitted each with a specific half-life; they are "delayed," relative to the fission process. Their half-lives range up to several minutes.

Let us designate by k the multiplication factor of the neutrons within a reactor, meaning the number of neutrons generated by an average neutron which itself is going to initiate another act of fission instead of being lost through the surface or by non-fission capture in an absorbing nucleus. Clearly a k smaller than 1 means that the process works at a net loss and does not support itself. With $k = 1$, it is just self-supporting and with k greater than 1 it leads to multiplication, that is explosion.

However, because of the 0·75 per cent delayed neutrons it was possible to work with k up to 1·075 and still retain a margin of safety. In the steady state of course the explosion would occur, but this steady state was achieved with a substantial delay, with a delay so long that something could be done to diminish the number of generated neutrons in the meantime. This could be achieved by introducing into the pile a material which absorbs neutrons, cadmium for example. Thus in the practical arrangement, it is necessary to measure the neutron flux within the reactor continuously, say by the use of paraffin-coated Geiger-Müller tubes, and to have a servo mechanism which inserts or pulls out some cadmium rods from the reactor accordingly, until the desired neutron flux is stabilized and the system works self-sustainingly (with $k = 1$) at a predetermined level. There are other cadmium rods above holes in the pile, held by electromagnets which are automatically let loose to fall into the holes should anything go wrong during the operation. As a matter of fact, up to now there only occurred a single accident in an atomic reactor, in Britain, and the damage done was not excessive; there were no casualties. It was not an atomic explosion due to the reactor going super-critical but was due to material failure of parts within the pile which had been "knocked soft" by the steady flux of impinging neutrons. This effect can be controlled and is being controlled.

Uranium-235 undergoes fission even if hit by a slow, thermal-velocity neutron. In a nuclear bomb it is very important to get a large amount of fission in a short time, before the bomb volatilizes and spreads the remaining fissile material out of the sphere of action. Therefore the neutrons which are produced during this explosion are allowed to remain fast, as they are. A bomb is the extreme case of a "fast" reactor. But industrial reactors should not explode and there is plenty of time to work with thermal neutrons, in a range of velocities where capture by U-235 is highly probable. Therefore atomic piles slow down their neutrons with special "moderators." These have to be light atoms because we have seen on p. 429 that a ball, and a neutron too, lose most of their energy if they collide with particles of mass near to their own. From among the light elements in Nature only those may be used which are not very liable to react with a neutron. This is a very difficult problem, because such elements, or rather such nuclear species, are very rare indeed. The best are deuterium, carbon and oxygen; ordinary hydrogen would capture the neutron too often, yielding deuterium in the reaction. Thus in practice either heavy water or very pure graphite are used to this end; both of them are by no means easy to obtain in sufficient quantities. Many of us remember that during the war British commando troops twice destroyed the hidden German heavy-water factory in Norway and that F. Joliot-Curie risked his life to take all the French heavy-water reserves with him to Britain after the German break-through.

The purity of these moderators is so important because neutron-absorbing impurities capture neutrons which otherwise could induce fission in the reactor. Throughout the whole structure of the pile every material has to be chosen so as to decrease neutron losses to the minimum—within the limits of mechanical stability requirements. In order to decrease the number of neutrons passing through the surface of the reactor, it is surrounded with a layer of non-absorbing material which reflects part of them back into the structure; carbon and beryllium are used for this purpose.

The fissile material itself need not necessarily be pure ^{235}U or plutonium but it must contain enough of these fissile nuclei to support the chain reaction. This was why ^{235}U had to be enriched in proportion to ^{238}U by all possible means during the war. The extent to which it must be enriched depends upon the size of the reactor: less for a large one with smaller losses than for a small one. But during fission, the products of this nuclear disintegration accumulate in the "nuclear fuel" and many of them absorb neutrons so strongly that after a time there are not enough left to continue the chain reaction at the level desired. This is a contamination which gradually and spontaneously accumulates. No initial purification of the material is of any help against it.

Therefore after having worked for a time the uranium of the reactors has to be removed, dissolved and purified. It is reprocessed in remote-controlled, automatically operated chemical factories and the purified uranium is recast into metal blocks or dissolved to form uranyl nitrate, as the case may be, for return to the reactors.

In the course of this purification, however, not everything that has been separated is a net loss; on the contrary. On p. 442 we have seen that ^{238}U is capable of capturing fast neutrons of a given voltage range and combines with them to form ^{239}Np which in its turn decomposes to the relatively long-lived, fissile ^{239}Pu. If the reactor is constructed in a way which gives enough time for a fraction of the neutrons to react with ^{238}U then the reactor breeds nuclear fuel of a new kind while consuming its original fuel ^{235}U. Such breeder-reactors are designed and operated in order to produce this new, highly fissile nuclear fuel, plutonium. In the course of reprocessing the uranium, plutonium is separated by chemical operations and set apart for future use.

^{238}U is not the only material which can be used to produce new nuclear fuel; ^{232}Th captures neutrons also and goes over into beta-active ^{233}Th, which yields ^{233}Pa. This latter gives alpha-active, highly fissile ^{233}U, an isotope which occurs in very minute quantities in Nature. Thus with our thorium reserves we are able to lengthen appreciably our nuclear future.

The optimal geometrical arrangement of nuclear fuel and moderator is a very intricate problem. Its solution depends upon the amount of energy that is to be produced, upon the importance of breeding more or less plutonium, etc. One extreme case is to dissolve a salt of the nuclear fuel in heavy water at a calculated optimal concentration. Alternately, uranium rods which are enclosed in special metal coatings are fitted into holes in a complicated graphite pile. The number of metallurgical and other problems encountered in building such a structure which has to withstand the enormous impact of the neutron flux is easy to imagine.

In order to have an idea about the energies involved in the operation of a nuclear reactor let us remember that the fission of one kilogramme of ^{235}U liberates 2×10^{10} kcal, equivalent to 20,000 tons of TNT. The generation of one kilogram of plutonium per day in a pile means that the pile is operating at the rate of one thousand megawatts. This energy has to be extracted somehow from the reactor and made use of. Even this is no easy problem. The heat of fission has to be removed from the reactor and fed into a power plant by some medium, by a coolant in the pile which functions as a heating agent within the power station. This must have a very low neutron-absorbing capacity, otherwise its effect would add up with the other parasitic absorbing nuclei within the reactor and reduce its performance. It should be able to withstand high temperatures because the thermodynamic efficiency of the power station depends upon the ratio of its working temperature to that of the air around it. How difficult this problem is may best be seen if one is told that even molten alkali metals are being considered very seriously as heat-transfer agents. Surely not very agreeable materials to work with. But nuclear power stations are being built all over the world, one after another. Much is being learned from each new construction and it will only take a decade or two until nuclear energy, energy of nuclear fission, will replace a substantial part of the energy produced up to now by burning conventional fuels.

Before the reserves of fissile material on earth are nearing their end, we may hope that the opposite nuclear process, the fusion of hydrogen nuclei into nuclei of helium will become solved on a technical scale and will be able to take over. But this belongs to a later chapter.

Once a reactor is built, its controlling cadmium rods are inserted for maximum absorption and the new uranium slabs are fitted into their holes within the moderator. Concrete walls of metre thickness shield the personnel from radiation or neutrons. Everything is airtight so that no radioactive material is able to escape from the "canned" fuel elements. Then the automatic controls are set going and the cadmium rods are gradually withdrawn until, at a predetermined level of energy generation, they are moved automatically hither and thither around an equilibrium position by the servo mechanism of the controls. Heat may now be extracted and transferred to the power plant. Material which is about to be irradiated may be placed inside special compartments where it will be transformed by neutrons into valuable radioactive isotopes. Beams of neutrons may be allowed to emerge across preset tunnels into adjoining laboratories. But the reactor itself should work as an automaton as long as it is in order, and needs only supervision, no actual work, until some of its elements have to be reprocessed. Energy has become divorced from human toil and labour.

101. THE STRUCTURE OF NUCLEI

After our short excursion into the realm of nuclear technology let us return again into more abstract regions and see what knowledge, if any, we possess about the internal structure of atomic nuclei. We are eager to learn what these nuclei are composed of, what holds their components together and whether any systematic rules or even quantitative laws can be deduced from their behaviour.

In the course of the previous chapters we learned that it is possible to build protons, deuterons, alpha particles and neutrons into atomic nuclei. These same particles are also known to be ejected from the nuclei on certain occasions, but apart from these particles, negative and positive electrons are also known to be ejected. There is reason to believe too, that a minute particle without electric charge or mass, the neutrino, is ejected together with the electrons. In special cases when the positive charge of the nucleus is excessive compared with its mass it is also known to capture a negative electron from its innermost K orbital. Last but not least, we have learned that some heavy nuclei may undergo nuclear fission into two resultant nuclei plus free neutrons if attacked by a neutron, while many of them may undergo fission or spallation if bombarded with small nuclei whose energy amounts to some hundred millions of electron-volts.

Which of the aforementioned corpuscles should be regarded as "ultimate" building stones of the nucleus, as far as we are allowed to speak of ultimate results in this connexion?

First let us rule out the alpha particle as an ultimate entity, because we have seen on p. 433 that it can itself be built from smaller particles. Deuterons are ruled out too, because on one hand they explode into a proton and a neutron on impinging upon each other with sufficient energy, while on the other hand they are known to be formed from ordinary hydrogen by neutron capture.

For many years, as long as neutrons were undiscovered, it was assumed that atomic nuclei were built from protons and electrons, the latter forming a sort of electrostatic cement between the former. After all, negative electrons were known to be emitted from nuclei as beta particles—they must have been in the nuclei. Positive electron emission was not known at that time.

We have, however, come to know that neutrons do exist. And this allowed us to assume that nuclei are built from protons and neutrons. A choice had to be made between the proton–electron hypothesis and the proton–neutron hypothesis and at the present moment it seems that the neutrons have won. It is only possible to state why in broad outlines. Elementary particles have an axial property, as stated on pp. 229, 287, i.e. they spin. This gives rise to a mechanical momentum and if the particle is charged, to a magnetic moment as well. The permitted orientations of spinning particles to each other give rise to different energies which in their turn have a small but measurable effect on the energy terms of an atom, that is on the spectra. Spin momenta may thus be deduced from spectral "hyper-fine-structure" measurements. Magnetic moments may be measured directly in inhomogeneous magnetic fields. These methods give us the spin values for different particles. If particles unite, their spin momenta can only be arranged into a few "permitted" spatial combinations according to quantum mechanics, similarly to the "permitted" spatial distributions of, say, electrons around nuclei. Now it was impossible to deduce measured nuclear spins from the added spins of protons and electrons within them, but it was possible to fit theory and experiment if neutrons were assumed instead of electrons.

Thus one comes to accept that nuclei contain protons and neutrons which attract each other, while the emission of electrons and positrons from nuclei is explained by assuming that these light particles are formed during emission. In the same way light quanta are assumed not to exist within atoms but still they are emitted on certain occasions. The fact that electron or positron emission always takes some time if nuclei are made to disintegrate, while heavier particles are immediately emitted can also be taken as an argument in favour of this assumption.

However, the neutron itself is beta-active: if free, it emits an electron with a half-life of about 20 minutes, to yield a proton. So ultimately one could still say that protons and electrons are the ultimate building stones of nuclei. But this would leave no place for positrons. Summing up, it is not absolutely certain that such macroscopic ideas as one entity being "built" additively from other, smaller, entities may be retained in modern micro-physical conceptions.

Leaving open this fundamental question let us proceed to see whether something more can be said about the structure of nuclei. Necessarily, man proceeds by analogy into unknown regions and only begins to assume fundamental differences between macro and micro systems if these analogies definitely break down. So let us pose the question, what does an assembly of protons and neutrons within the nuclei resemble most: a gas, a liquid or a solid—as long as we prefer to think in the analogy of assembled molecules. Or are they arranged in a way which resembles the electronic structure of atoms?

Between gas particles there are practically no forces and there is nothing to hold gases together: they expand as long as there is room for them to do so. Nuclei hold together, so they cannot resemble gases.

Thus only the two condensed states of molecules remain as an analogy. Fluids, where the molecules are held together by attracting forces but are free to move around; or crystals where they are held together also, but where they are bound to stay in the vicinity of given positions. Let us assume now that nuclei are some sort of condensed-state assemblies of protons and neutrons. Some sort of a force must needs hold them together in this case and it is up to us now to learn as much as possible about this force.

The only force we knew of between them is the Coulombic repulsion between protons and this should decrease with the square of their mutual distances. But this, being a repulsive force, can by no means be responsible for holding them together. What other force acts between proton and proton, proton and neutron or between two neutrons? Let us remember the brilliant experiment of Rutherford in which he determined the deflexion of alpha particles going through matter (p. 255). He found that up to distances of about 10^{-13} cm between proton and nucleus the law of Coulomb explains the total deviation of the alpha particle. Below this distance the picture changes, the experimental deflexion curves can only be explained if a new force, this time an attraction (!) is postulated.

Tuve and *Hafstadt* continued this set of experiments by observing very carefully the deflexion of protons from a proton beam in hydrogen, that is by analysing the proton–proton interaction. The deduction goes exactly along the lines demonstrated in Fig. 97, but now both charges are the same positive elementary unit. Extreme deflexions are due to events when one proton by-passes the other in its immediate vicinity. Their probability has to follow a given curve if there is only Coulombic repulsion between them (p. 253). But after a certain point there are fewer deflexions than were calculated. At this distance the repulsion is weakened by another force and very soon it is more than compensated by a very quickly increasing force of attraction. This is the same result as in Rutherford's experiments, but now for the simple case of two protons and nothing else. Scattering of neutron beams in hydrogen is of course not caused by Coulombic repulsion because neutrons are uncharged. But at very "near misses" a scattering does occur and the curve connecting scattering angle with scattering direction can be explained by assuming again a force of attraction between the two particles. The magnitude and the dependence on distance of this new force is practically the same as was found between proton and proton in the preceding experiments.

Thus a fundamentally new type of force emerges in nuclear dimensions: very short-ranged and always causing attraction irrespective of whether it acts between two protons or between proton and neutron. Though scattering of neutrons by neutrons would be extremely difficult to measure, it is assumed at present for reasons of symmetry that a similar force would act to attract two neighbouring neutrons.

Quite independent observations also suggest that these new "inter-nucleonic" forces must be of very short range. If they could act appreciably over more than the diameter of one nucleon (we call protons and neutrons collectively nucleons), then the sum of forces and therefore of energy in a nucleus would be proportional to the number of possible combinations between *n* nucleons, that is to $n(n-1)/2$. (By the way, this is the formula stating how many football games must be arranged in a league between *n* football teams.) For $n = 10$ or more this expression is already very near to $n^2/2$, meaning that the energy set free when *n* nucleons combine should be roughly proportional to n^2. Now this energy is given, according to the Einstein relation between mass and energy (p. 421), by the mass defect of the nuclear species in question: the difference between the atomic mass found in the mass spectrograph and the sum of the masses of protons and neutrons within the nucleus. This should be proportional to the square of the number of nucleons involved. Actually this is not true; the difference is rather proportional to the number of nucleons, to *n* itself. This fact can be explained if we assume that each nucleon exerts its attractive force only on its neighbours. The number of neighbours being roughly proportional to the number of nucleons, this would explain that the total energy set free is proportional to this number.

It would also explain why the energy holding together a deuteron (1 proton + 1 neutron) is relatively so small, only 2 MeV (mega-electron-volts = million electron-volts), as compared with the average energy per nucleon in heavier nuclei of 8 MeV. A deuteron consists only of two nucleons that is of only one pair of neighbours. Nucleons in heavy nuclei have more neighbours. As the number of nucleons increases, the binding energy increases also and reaches a first maximum at the helium nucleus, the alpha particle. It amounts to 28·2 MeV. Fig. 160 shows the energy set free on forming different nuclei from their separate nucleons, derived of course from mass spectrographic mass deficit data. It shows that in a first approximation there is a linear relationship between *E* and *n*, though many fine irregularities remain to be explained, among others the extreme position of the helium nucleus.

There is a second quantity which is to some extent open to measurements on nuclei; it is their volume. Taking for radius the distance where pure Coulombic repulsion ceases to account for alpha-scattering in the Rutherford experiment, one is able to measure the radii of different nuclei and thus calculate their volumes. It turns out that to a first approximation the volume of a nucleus is also proportional to the number of nucleons it comprises. This agrees with what we have seen concerning energy: if the attractive force could reach farther than simply to near neighbours, one should expect that the total volume should increase more slowly than the number of nucleons because the whole system would be pulled together by forces acting all through its structure. Forces only between neighbours account for the fact that each nucleon increases the total volume by a fixed increment.

Thus the total energy as well as the total volume of a nucleus is roughly proportional to the number of nucleons within it, in the same manner as the total energy of condensation and the total volume of a condensed system of molecules is proportional to the number of molecules in the condensate. (Heats of condensation and specific volumes are practically independent of the size of the sample.)

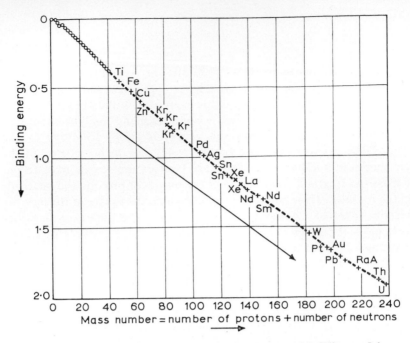

Fig. 160 (a). **The energy lost in the process of building nuclei from their component protons and neutrons**

Energy is measured as the difference between actual atomic weight and sum of the masses of the uncombined component nucleons. The energy set free is proportional to the number of nucleons in first approximation. However, when the line is less steep than the line corresponding to the energy of formation of an alpha particle (arrow), it becomes worth while to have the alpha particle outside the nucleus: the nucleus may become alpha-emitting (samarium and the heaviest elements).

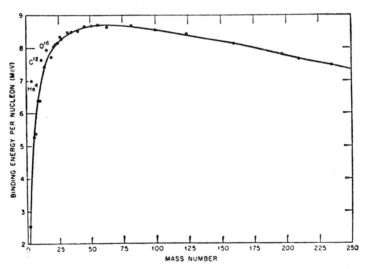

Fig. 160 (b). **Binding energy of nuclei per nucleon contained in them**

Note the extraordinarily high energies corresponding to the formation of nuclei from integral numbers of alpha particles (C^{12}; O^{16}).

(From S. Glasstone, *Sourcebook on Atomic Energy*, p. 351, Fig. 12.2 (Princeton, N.J., van Nostrand, 1952). By courtesy of the D. van Nostrand Co., Inc.)

It is worth while to have another glance at Fig. 160. An angle corresponding to the energy of formation of the helium nucleus is drawn into this graph. It is obtained by connecting the point which corresponds to helium with the zero point corresponding to hydrogen. As long as the nearly linear curve is steeper than this angle, more energy is gained by inserting the respective new nucleons into a given nucleus than would be if the last two protons and neutrons were to unite to an alpha particle outside the nucleus. However, we observe that the line becomes parallel to this critical direction in its lower third part and even becomes less steep than this direction at the upper end of the system. This means that at this upper end, among the heaviest nuclei, more energy would be set free if two protons and two neutrons were to be bound alone into an alpha particle than if these same nucleons were built into the heavy nucleus itself. And it is found that exactly these nuclei are unstable with respect to alpha particle emission: they are radioactive alpha-emitters. The critical angle of the $H \rightarrow He$ process is only once not reached in the middle of the curve, in the neighbourhood of samarium, and accordingly samarium was found to be an alpha-active element!

The question arises, why do these heavy elements, which are energetically labile, exist at all? We can only assume that they have been formed under totally different circumstances during their cosmochemical history, in the interior of stars where temperature as well as pressure surpassed all that we are able to experience on earth, and that the energy necessary for their formation was acquired by them under these circumstances. They are dying out one after the other and their dying is the process of natural radioactivity that we observe.

Are the alpha particles preformed in the nuclei or not? The current assumption is that they are not and that they only become assembled in the last instant when they are expelled from a nucleus by the act of alpha emission. The nucleus would most resemble a drop of very dense liquid according to this assumption, the zero-point motion of its nucleons acting in place of the thermal energy of ordinary liquids. *Bohr* and *Wheeler* were able to calculate, in 1939, the vibrational energy of such a "nuclear drop" which is necessary to make it divide into two smaller "drops" against its "nuclear surface tension." They arrived at the right order of magnitude for the energy of an impinging neutron which should just be able to divide, that is bring about fission in ^{235}U, ^{232}Th and ^{233}U. The meaning of "nuclear surface tension" is self-evident: just as molecules of a liquid are drawn into its interior because the more neighbours they have the more force is there to attract them, so are nucleons drawn into the nucleus where they are better encircled by neighbours. Any spreading out, elongation, of the drop or of the nucleus increases the number of molecules or nucleons, respectively, which have been separated from some of their partners to emerge into the surface. And this has cost energy.

In spite of the fact that the drop model has its great merits there are some facts which have been at one time well explained by the assumption of preformed alpha particles within the nucleus. First of all there is the remarkable stability of the nuclei composed of integral numbers of alpha particles. This stability of course manifests itself by the relatively large mass defect of these nuclear species, as if the completion of each alpha particle within the nucleus had brought to an end a process where energy has been set free. The mass defect curve of Fig. 160 shows this state of affairs. *Wefelmeier* even went so far as to postulate nuclei, which like crystals, are built from alpha particles adjoining one another. He tried to arrange alpha particles within nuclear species which are their integral multiples in the simplest manner, as spheres, and calculated the number of "nearest neighbourships" in such structures. There is one such nearest neighbourship between the two alpha particles postulated in the hypothetical $^{8}_{4}$B, three if the three alpha particles are arranged in a triangle in $^{12}_{6}$C, six if the four alpha particles in $^{16}_{8}$O are at the corners of a tetrahedron with six connecting edges between them, and so on.

TABLE XIV. "Energy of Formation" of Nuclei composed of Alpha Particles only

The energy (E) as given by the mass defect yields approximately constant increments if it is divided by the number of points (n) of contact between alpha particles—imagined as spheres—within each nucleus but yields very different increments per alpha particle (E/N).

N, number of particles	Nucleus	E, energy of formation $\times 10^{-4}$	n, number of point contacts	E/n $\times 10^{-4}$	E/N $\times 10^{-4}$
2	$^{8}_{4}$Be	$-1\cdot4$	1	$-1\cdot4$	$-0\cdot7$
3	$^{12}_{6}$C	$+77$	3	$+25\cdot6$	$+25\cdot6$
4	$^{16}_{8}$O	156	6	$25\cdot9$	39
5	$^{20}_{10}$Ne	206	9	$22\cdot9$	41
6	$^{24}_{12}$Mg	309	12	$25\cdot8$	51
7	$^{28}_{14}$Si	406	16	$25\cdot4$	58
8	$^{32}_{16}$S	488	19	$25\cdot7$	61
9	$^{36}_{18}$Ar	570	23	$24\cdot8$	64
13	$^{58}_{26}$Fe	~1000	42	$\sim24\cdot8$	77

Table XIV shows that the "energy of formation" of these nuclei as given by their mass defects is proportional to the number of these connecting edges (n) and not to the number of the alpha particles themselves (N). The value of E/n is fairly constant at 25 energy units while the value of E/N increases from 25 to 77. Only the first, the hypothetical $^{8}_{4}$B, does not fit into the series. It looks as if the single edge with its single energy increment would not be able to balance the disruptive tendency of zero-point motion.

A further very interesting consequence of this quasi-crystalline arrangement arises when 13 alpha particles have to be arranged. It is known that one sphere may be practically surrounded by 12 neighbours so as to be enclosed between them: six are around it while two sets of three form a ring each, above and below (Fig. 161). This should give an extremely symmetrical and tight arrangement; any further spheres which would accrue would have

Fig. 161. **Wefelmeier's highly symmetrical arrangements of 2, 3, 4, 5, 6 and 13 alpha particles**

Compare Table XIV.

to be built on to this sphere of 13 and would disturb its symmetry. The nuclear species corresponding to 13 alpha particles is $^{56}_{26}$Fe. Beside the alpha particles it contains four additional neutrons as is always the case whenever the Wefelmeier model postulates an *entirely enclosed* alpha particle. It is found that the maximum of nuclear stability occurs in the vicinity of iron: the total energy of formation per nucleon is the maximum in this part of the system. Thus it would set energy free if smaller nuclei were to unite into a particle of this size, while energy should be set free also if large nuclei decomposed to this extent. The fission of heavy nuclei and the eventual fusion of lighter nuclei is determined by this maximal energy of formation, or in other words, by this minimum of energy content. We shall see later on that the statistics of nuclear species in the universe show a maximum occurrence of just these medium-weight nuclei—the lightest nuclei excepted.

Be it as it may, the analogy between condensed molecules and condensed nucleons should probably not be stressed farther than this point. If we insist on a name, maybe the analogy with liquid crystals where there is mobility around a statistical order among neighbours, is the best we are able to arrive at. The best perhaps, but only as long as we do not try to construct a system of nucleons, in analogy to the electron-shell model of the atom, for the nuclei. Surely the analogy is bound to fail, because in the atom we had a concentrated mass and positive charge at the centre and only the light electrons had to be arranged around it, obeying the laws of electrostatics and of zero-point motion. In the nuclei there is no such central element; they resemble rather the atomic conception of Lenard with "dynamides" distributed across the whole volume. Thus the closed shells within a nucleus will be very different from those in an atom, but eventually some rules may emerge out of such calculations. The gamma rays emitted by radioactive nuclei can be arranged in term-schemes like the frequencies of light emitted by atoms (p. 307). This means a set of discrete energy levels in excited nuclei and will ultimately reveal their quantized structure. Nuclear species with minima of internal energy, that is with the maximum of energy set free on their condensation are found empirically from mass defects. They still have to be explained by a future, exact theory of nuclei.

102. THE INTERNUCLEONIC FORCE

After neutrons had been discovered, physicists tried to explain the attraction between a proton and a neutron by assuming that an electron oscillates between them if they are near enough. At one time the first nucleon would be a proton and the second a neutron, and at other times the reverse would be the case. The sphere of action of the electron which originally was in the neutron would have thus expanded and thereby its zero-point energy would have decreased, accounting for the bond energy of the two particles, e.g. in a deuteron. This is what may be termed "exchange energy," a much better term than resonance energy which has been so often used for a similar situation of binding energy in the homopolar bond.

However, electrons are too light; their zero-point motion allows them to move across much larger regions which are characteristic of chemical bonds between atoms and not between nucleons. In order to overcome this difficulty the Japanese physicist *H. Yukawa* postulated in 1935, in an entirely formal *ad hoc* assumption, that not electrons, but another type of charged particles "oscillate" between the two nucleons. The mass of these new particles was calculated to yield nuclear dimensions in agreement with the experiments. Yukawa found that they should be about 140 times as heavy as electrons in order to restrict their zero-point motion to the desired distance. He called them mesons, the Greek word "meso" meaning "between," because their mass was between those of protons and electrons. It was a fantastic success of such a far-fetched assumption when *C. D. Anderson, S. H. Neddermeyer, J. C. Street* and *C. C. Stevenson* discovered such particles in cosmic rays in 1936–37. We shall have to deal with these particles later on.

Beside this kind of internucleonic energy there is the energy which is gained when nucleons unite to form composite particles by spin compensation. Such a process is certainly in operation in the formation of alpha particles with zero spin. And of all the hundreds of nuclear species these species with an odd number of nucleons are always less stable than those containing an even number with spin compensation. It is *not necessary* that all spins within a nucleus should be compensated. Experimentally determined spins rise in some cases to as high as 6. But if the number of nucleons is odd, *there is no way of* compensating spin at all—at least one spin has to remain unpaired.

Let us now sum up the different forces within a nucleus. First there is the internucleonic force or energy, which as we have seen is short-ranged and acts between neighbouring nucleons. To a first approximation it is proportional to the number of nucleons, that is, to the atomic weight, A. Then comes the electrostatic repulsion between protons. This is a relatively long-range force, decreasing only with the square of the distance so that it acts between all pairs of protons present. The corresponding energy is therefore proportional to the product of charges or on the average to the square of the nuclear charge, Z^2 (Z = number of protons or atomic number within the periodic system), and inversely proportional to their average distance which again is proportional to the diameter of the nucleus. We have seen on p. 451 that this diameter is proportional to the cube root of the atomic weight (because atomic volume is proportional to atomic weight), that is the electrostatic repulsive energy is proportional to $Z^2/A^{\frac{1}{3}}$. This has to be subtracted from the first-mentioned energy of attraction. Another term to be subtracted is the surface energy because the nucleons in the surface layer have fewer neighbours than those below them. This term is proportional to the surface of the nucleus. The volume being proportional to the number of nucleons A, this gives a proportionality to $A^{\frac{2}{3}}$. If there are more neutrons than protons another energy term has to be subtracted which is proportional to the square of these surplus neutrons $(A - 2Z)^2$, divided by the volume A. This takes care of the fact that n–n and p–p attraction is smaller than n–p attraction. Finally the spin energy has to be considered giving the last, the fifth, term of the whole calculation. The proportionality constants can be made to fit by semi-empirical methods to yield the experimental function connecting the number of protons Z, and the number of neutrons, $A-Z$, with the total nuclear energy as revealed by mass-defect data.

The electrostatic repulsion proves to be the most important negative term for high atomic weights and thus it can be said that the natural radioactivity

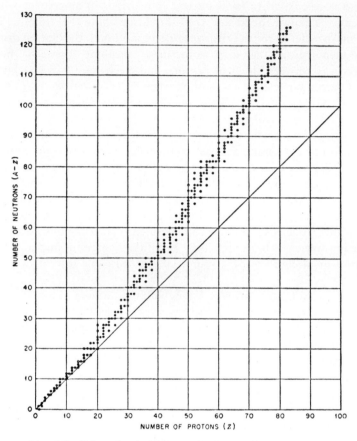

Fig. 162. **Map of existing nuclei showing number of neutrons**
v. number of protons

The heavier the nucleus, the more neutrons it needs in proportion to protons to stabilize itself against Coulomb repulsion of the protons. Nuclei along one vertical line are isotopes.

(From S. Glasstone, *Sourcebook on Atomic Energy*, p. 360, Fig. 12.3 (Princeton, N.J., van Nostrand, 1952). By courtesy of the D. van Nostrand Co., Inc.)

of heavy nuclei is due to the electrostatic repulsion between their protons which increases quadratically with their number, while the internucleonic attractive force only increases in simple proportion to the atomic weight. For a time the situation is saved by adding more neutrons than protons: At the beginning of the system their number is often the same (atomic weight = twice the atomic number) but later on the number of neutrons increases faster and in uranium we have 146 neutrons and only 92 protons. Because of the inter-nucleonic attraction these serve to bind the nucleus together in spite of the Coulombic repulsion. But at the end of the system disruptive forces gain weight as shown on p. 452 for the special case of alpha-instability. Energy is set free to take an alpha particle out of the nucleus. The last artificially produced transuranium elements are highly unstable and disintegrate with very short half-lives.

For each mass number, that is for each number of nucleons, there is an optimal ratio of protons to neutrons. As just mentioned this ratio is 1:1 for the light nuclei; one neutron is just stabilized against electron emission, by one proton, the charge oscillating between them as a meson. As the total charge of the nucleus increases, the electrostatic repulsion has to be compensated by an increasing number of neutrons so that the most stable isotopes of a given mass number come to contain more neutrons than protons. The expression "mass number" stands for the integral atomic weight irrespective of mass defects. It is equal to the sum total of nucleons, that is to the sum of protons and neutrons in the nucleus. To each mass number there belongs a set of "isobaric" nuclei which differ in the number of protons and neutrons but keep their sum constant. It is clear that isobars may change into each other if a proton within them changes into a neutron or vice versa. This may be accomplished by electron or positron emission, or by a capture of a K shell electron into the nucleus. Looking at the map of nuclear species where existing mass numbers are plotted against nuclear charge, we see that all existing species fall into a rather narrow band (Fig. 162). The stable isotopes are around the middle of this band. To their right we have the nuclei with excess neutrons which can stabilize themselves by electron emission, and to their left those with excess protons which might become stable by emitting a positron or capturing an electron. As a matter of fact this is what they do.

Plotting the internal energy of isobaric nuclei against the mass numbers yields curves like those in Fig. 163, resembling parabolae. Whenever the mass number is odd the experimental energy (mass defect) data fit on to a single curve (Fig. 163 (*a*)) while for even mass numbers there are two separate curves, one for nuclei with even-proton, even-neutron numbers, and one for odd-proton, odd-neutron numbers. The latter curve lies higher: odd-odd nuclei contain more energy and are relatively less stable because not all spins are compensated within them. It thus comes about that at odd mass numbers (where a spin is *always* uncompensated) there is a well-defined minimum to the curve, and the species lying at this minimum is *the* stable species of the respective odd mass number. The species in the middle of the curves for even mass numbers falls higher (Fig. 163 (*b*)) on the upper curve than do its two neighbours on the lower one, and thus these two come to occupy relative minima against their neighbours, both becoming stable against electron or positron emission. In accordance with practically all experience there is only one stable isobar at odd atomic numbers while there are these two for the even, differing by two mass number units. There can be no stable neighbours of isobars (Mattauch). A well-known case is the trio at mass number 40: A–K–Ca. The isotope ^{40}K is known to be capture-active;

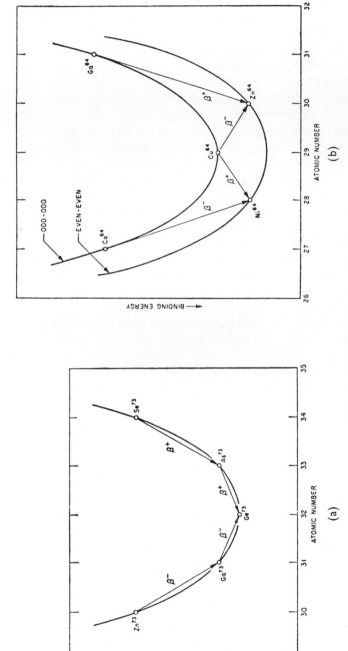

Fig. 163. **Binding energy of isobaric nuclei (different nuclear charge at identical nucleon number)**

(*a*) Binding energy values fall on a single curve for odd mass numbers, when there is an odd number of either neutrons or protons. One species, at the minimum of the curve, is the stable nucleus for a given mass number.

(*b*) Binding energy of odd-neutron odd-proton nuclei falls on a higher curve than that of even-neutron even-proton nuclei because spins are compensated in the latter case. Hence for even mass numbers neighbouring nuclei cannot be stable: one of the two neighbours necessarily has more energy than the other. Second neighbours, however, can be stable.

(From S. Glasstone, *Sourcebook on Atomic Energy*, p. 374, Fig. 12.5, p. 375, Fig. 12.6 (Princeton, N.J., van Nostrand, 1952). By courtesy of the D. van Nostrand Co., Inc.)

this is the reason why argon is so much more common than the other rare gases: it is a decay product of ^{40}K. One period below this trio we have a trio containing a beta-active, electron-emitting rubidium isotope, $^{87}_{37}$Rb, and because of this rubidium minerals always contain its decay product, $^{87}_{38}$Sr. Incidentally, two members of the trios exhibit a remarkable instance of nuclear periodicity which goes parallel to the periodicity in the electron shells of the atoms: the two alkali metals (K and Rb) have similar instability in the nuclei. Owing to the fact that elements with even mass numbers of the odd-odd type (upper curve) contain more energy, half-lives of isobars on the upper curve are shorter and their decomposition is quicker than that of isobars on the lower curve. The fact that ^{235}U is more labile than ^{238}U and undergoes fission by impact of slow neutrons is also due to its being uncompensated as far as nuclear spin is concerned. It contains an odd number of nucleons, like its even more labile isotope ^{233}U.

103. ABUNDANCE OF NUCLEAR SPECIES

Having learned how to measure the stability of nuclei by their mass defects, and having accepted some general, plausible, ideas about the reasons for nuclear stability or instability we must eventually ask ourselves whether the actual abundance of nuclear species in Nature corresponds to some extent to their being stable or otherwise. There is no question about nuclei which have been found to be radioactive: it is clear that these are being lost all the time and are transforming into their stable end-products. Most transuranium elements must have become extinct if they ever existed, for the same reason.

But what about nuclei, which seem to be stable as far as radioactivity is concerned but still differ in the amount of internal energy per nucleon? This is the quantity which matters, the internal energy or total mass defect of a nucleus divided by the number of nucleons it contains. This determines whether other nuclei could have been built from these same building stones, from these same nucleons, in such a manner that on the whole more energy would have been set free in the process.

We have mentioned already that the absolute minimum of this quantity lies around the element iron, if the lightest stable elements, hydrogen and helium, are excepted. And this agrees remarkably well with the facts that according to its density the core of the earth must consist largely of iron and nickel and that meteorites, which are probably fragments of stars, often show the same composition.

It is worth while to see once more how this minimum comes about. As with most minima, it is due to a balance of causes working in opposite directions. Because of internucleonic forces the total energy which makes a nucleus hold together should be proportional to the number of nucleons, and hence the energy per nucleon should be constant. If this were exact there would be no reason for one nucleus to be more stable and hence more abundant in Nature than any other. However, at the beginning of the periodic system all nucleons are "on the surface" of the nucleus and have fewer neighbours than they have room for around them. Gradually, as their number increases, more and more nucleons come to lie below the surface and thus are able to exert their attraction in all directions. Therefore the energy set free per nucleon increases and nuclear stability increases with it. But as the number of protons grows, a disrupting electrostatic term comes into action and thus the energy per nucleon decreases. The point where the positive action of having more nucleons surrounded, and the negative action of protons repelling each other, are balanced determines the position of minimum of the curve and it happens to coincide with the nucleus in which one alpha particle is surrounded spherically by twelve neighbours: this is the iron nucleus.

Now let us consider some actual statistics. They are difficult to obtain because the sampling has to be done throughout the Cosmos. Certainly our results will not be very accurate. But we know the composition of our Earth with a fair degree of accuracy, and the composition of the Sun can be ascertained by the absorption spectral lines (Frauenhofer lines) in its corona. The absorption of interstellar matter can also be measured to some extent, and analyses made of meteorites, which may be regarded as samples arriving from all parts of the universe. If, with all restrictions in mind, we proceed to draw a graph which represents the abundance of the elements, we arrive at Fig. 164. The difference in abundance between different elements is so great that it has to be drawn to a logarithmic scale, equal distances representing equal orders of magnitude. On the same graph, with another set of ordinate values, we draw the aforementioned energy-per-nucleon values for the same nuclei. The parallelism is striking. There are three nuclei of spectacular abundance: $^4_2$He, $^{16}_8$O and $^{56}_{26}$Fe. All three occupy special positions among the others. Helium is the first nucleus where nucleons achieve a highly condensed *spatial* arrangement with spin compensation, four points being necessary to move from a plane to a spatial configuration. Oxygen-16 is the first nucleus in which four alpha particles are arranged in the same way. And iron-56 is the nucleus in which 13 alpha particles and four neutrons are arranged in a sphere (p. 455). It is probably the nucleus which comes nearest to a sphere among those nuclei after helium and oxygen in the periodic table.

But these are only the maxima. The whole lengths of the two curves fit excellently to each other. Take for example $^{20}_{10}$Ne. It could be imagined as being built from five alpha particles, adding another one to the four in the highly symmetrical ^{16}O. If we may regard them as polyhedra, a tetrahedron has been transformed to a trigonal bipyramid. It is impossible to bring the axial pair (Fig. 161) as near as the tetrahedral partners were. Accordingly, the energy set free per nucleon is smaller. Parallel with this its abundance is smaller too, since it lies on a local minimum of the curve. We cannot understand as yet all details of the stability curves but we may rest assured that nuclear stability and abundance go hand-in-hand in the universe.

How has all this come about? This leads us of course into the vast theories of cosmogony which are certainly near to the ultimate questions which can never be answered. However, some assumptions can be made with some certainty. The abundance of hydrogen, the lightest and simplest nucleus in the universe is a fact which must be accepted. We have no idea whether or how it has been or is being created (*Hoyle*). But a glance at the

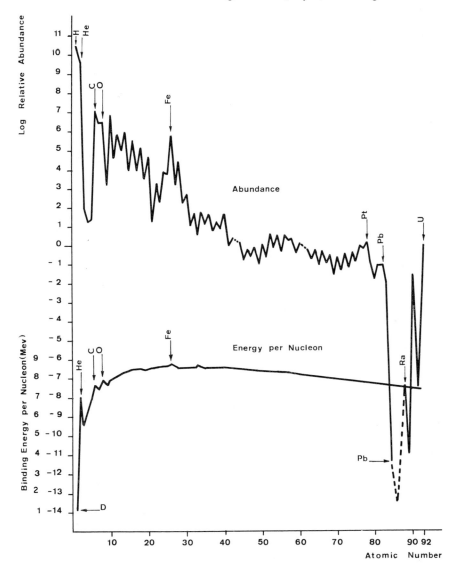

Fig. 164. **Estimated abundance of elements in the world and the energy liberated per nucleon when their nuclei are formed**

Abundance on a logarithmic scale.

chart of mass defects shows us that an immense amount of energy is set free if four protons unite under positron emission to form a helium nucleus. The relatively most simple sequence of events which could lead to this result involves the direct collision of two protons against the electrostatic repulsion between them, a process which could only occur at temperatures (kinetic energy) amounting to $10^{8}\,°K$. Such temperatures Man is now striving to reach in order to achieve nuclear fusion, the reaction just mentioned above. A possible sequence would be:

$$_1^1H \; + \; _1^1H \; = \; _1^2D \; + \; Q_1 + \; \oplus \qquad . \qquad . \qquad . \quad (103.1)$$

$$_1^2D \; + \; _1^1H \; = \; _2^3He + Q_2 \qquad . \qquad . \qquad . \quad (103.2)$$

$$_2^3He + \; _2^4He = \; _4^7Be \; + Q_3 \qquad . \qquad . \qquad . \quad (103.3)$$

$$_4^7Be \; + \; \ominus \; + \; _3^7Li \; + Q_4 \qquad . \qquad . \qquad . \quad (103.4)$$

$$_3^7Li \; + \; _1^1H \; = \; _2^4He + \; _2^4He + Q_5 \; . \qquad . \qquad . \quad (103.5)$$

The signs $\ominus$, $\oplus$ and Q denote electron, positron and liberated energy respectively.

The first postulated process of this sort which utilized throughout only known isotopes, and which would probably operate at stellar—solar— temperatures, was advanced by *H. A. Bethe* in 1939 in the U.S.A. and *C. F. v. Weizsäcker* in Germany. It begins with a carbon atom which is regenerated at the end of the process, resembling a catalytic chain of reactions in ordinary chemistry. Carbon catalyses the union of four protons to yield an alpha particle, a helium nucleus. The processes and isotopes were already known at that time—

$$_6^{12}C + \; _1^1H = \; _7^{13}N + Q_1 \qquad . \qquad . \qquad . \qquad . \quad (103.6)$$

$$_7^{13}N \qquad = \; _6^{13}C + \oplus + Q_2 \qquad . \qquad . \quad (103.7)$$

$$_6^{13}C + \; _1^1H = \; _7^{14}N + Q_3 \qquad . \qquad . \qquad . \quad (103.8)$$

$$_7^{14}N + \; _1^1H = \; _8^{15}O + Q_4 \qquad . \qquad . \qquad . \quad (103.9)$$

$$_8^{15}O \qquad = \; _7^{15}N + \oplus + Q_5 \qquad . \qquad . \quad (103.10)$$

$$_7^{15}N + \; _1^1H = \; _6^{12}C + \; _2^4He + Q_6 \qquad . \qquad . \quad (103.11)$$

One could ask why a carbon and a helium nucleus are formed in this process instead of the very stable nucleus $_8^{16}O$. It is because of the energy liberated during the process 103.11 which ruptures the $_8^{16}O$ nucleus and still allows Q_6 to escape as kinetic energy.

There is astronomical evidence for the belief that "young" stars are primarily huge agglomerations of hydrogen, brought together by the action of gravity. As the mass increases so does the gravitational field, and the mass contracts (the atoms "fall" towards each other), gravitational energy going over into kinetic energy, that is temperature. At a certain temperature, which has of course to take into account the immense energy losses by radiation also, the reaction between protons sets in and helium is produced, together with an incredible amount of energy. This brings about re-expansion and much increased emission of light, as well as other electromagnetic radiations from radio noise to gamma rays. Thus, hydrogen is the primary fuel of stars and the production of helium the primary reaction. At extremely high pressures and temperatures other nuclear fusion reactions may come about and this might be the origin of the whole array of different nuclear species that we encounter. Only this high temperature and gravitational pressure can account for the fact that nuclei beyond the stability minimum around iron (Fig. 164) come into existence, nuclei which in energy equilibrium at lower temperature and pressure should not exist at all. Just as certain chemical molecules may be synthesized which have more energy within them than others which could be built from the same atoms if only we supply the necessary energy, so are these labile nuclei formed because the immense amount of energy they consume is at their disposal. Once formed, they need a certain energy of activation to decompose and if they manage to escape from the region of high temperature in time they may go on existing, as we actually find them existing. This is the same kind of situation as occurs in chemistry. Nitroglycerine for example, should decompose to carbon dioxide, water and nitrogen, and liberate a large amount of energy, but at ordinary temperatures it does not. It has to be heated or subjected to shock before breaking up (violently), because the atoms are in a metastable state of equilibrium, but still in an equilibrium as long as energy of activation is not supplied. Then a few molecules explode and their energy of explosion starts the decomposition of their neighbours until the whole mass has exploded. In nuclear dimensions the same is true for the atomic bomb.

A mechanical analogy may serve to elucidate the idea of metastable equilibrium. A ball remains within a dish on a table until it receives sufficient kinetic energy, motion, to reach the rim of the dish, in spite of the fact that its stable equilibrium would be on the floor beside the table or rather on the earth in the street. Speaking quite strictly: at the centre of the Earth! This small energy of activation which brings it up to the rim must be invested to gain the much greater amount of energy which is set free when it eventually falls. Physicists use the term "energy trough" or "potential well" where a particle or a system remains until it receives sufficient energy to escape: from the labile molecule if we are dealing with atoms, or from a labile nucleus in the case of nucleons.

There is only one important difference between the macroscopic ball–dish, between atom–molecule systems and the nuclear systems. Nuclei are so small that the principle of indeterminacy assures some nucleons of a small probability of extending beyond their enclosing borders even if they do not possess quite the amount of energy that they would need to "climb to the rim." They "tunnel through" the wall on some relatively rare occasions, just because they have nearly been over it anyway and they are not exactly localized owing to their zero-point motion. This explanation of radioactive decay is due to *Gamow* and the word "tunnel effect" was coined by him also (1928). It was on this quantum mechanical basis that correlations between energy of disintegration and radioactive half-life (which is of course inversely proportional to the probability of decomposition) have been calculated by *Geiger* and *Nuttal* for alpha particles, and by *Sargent* for beta disintegration.

Coming back to the abundance of nuclear species we now have some idea of how they may have come into existence. Stellar evolution is an extremely interesting chapter of modern science and there is much more to be said about it—but it would exceed the competence of the author and the scope of this book. Suffice it to say that the high energy concentration in the interior of stars accounts for the existence of unstable nuclear species in the universe and it is clear that their statistical abundance must vary with the degree of their lability, that is with the surplus energy they possess above the energy content of the stable nuclei (*see* Fig. 164). Even the fact that the abundance, that is the probability with which they occur, had to be drawn on a logarithmic scale in order to become parallel with the non-logarithmic graph of energy content is in accord with what we have learned about the Maxwell distribution of molecules in different states of energy, on p. 145. It is always the *logarithm* of a probability between one state or another which is proportional to the energy difference between the two states, no matter whether we are dealing with molecules, atoms or nucleons. The deduction of probability theory in statistical mechanics remains the same.

104. NUCLEAR FUSION

The main events of nuclear fusion we have discussed already in the preceding chapter, even going into detailed hypotheses concerning the sequence of nuclear reactions which might lead to the production of helium from hydrogen. None the less the problem is so much in the current focus of interest that we must deal with it again in this separate section.

Let us state again: the absolute minimum of energy content occurs around the nuclei of iron. Heavier nuclei should disintegrate until they arrive at this magnitude and lighter nuclei should fuse until they do the same. In such a world there would be no free nuclear energy left, similarly to a world where all organic material has decomposed to carbon dioxide, water and nitrogen and which no longer possesses free chemical energy for further reactions. We have just discussed why energy of activation is needed to explode heavy nuclei (compare nitroglycerine, as a conventional case). A similar argument holds for fusion.

Just as, for instance, hydrogen or carbon and most compounds comprising the huge realm of organic chemistry are quite stable in the presence of elementary oxygen instead of reacting with it immediately, so are protons unreactive toward protons, or other light nuclei toward each other. Both have to be "heated" until the energy necessary for their activation is overcome and they may unite at last to form stable entities. The activation energy of the chemical carbon–oxygen example is consumed in liberating carbon atoms from each other and in splitting the oxygen molecules. The activation energy between nuclei is consumed in overcoming the electrostatic repulsion between them, since they are both positively charged. There seems to be no repulsion and thus no energy of activation between nuclei and neutrons, and this is the reason why neutrons react with practically any nucleus and are quickly absorbed by them. But the fact that they may collide with nuclei and transmit to them a part of their kinetic energy without always being absorbed, does indicate a kind of repulsion in some cases.

To speak of fusion today means always fusion where helium is the end product, though there is no logical reason to deny the name fusion for any other process where lighter nuclei unite. But this is the process where most energy per unit mass is set free and there are such overabundant reserves of hydrogen at our disposal that all energy problems of mankind would be solved for what amounts to eternity if we managed to tame this process which we have already set loose in hydrogen bombs.

The energy of activation has to be invested, and this means kinetic energy of the reacting particles amounting to a temperature of some $10^8\,°K$, or for that matter $°C$, because the minute difference of $274°$ between the two really vanishes in these dimensions.

The man who persisted in the belief that fusion can and should be accomplished by initiating it with the energy of an ordinary, "fission" atomic bomb was *E. Teller*. In spite of scientific doubts and political criticism he had his way and the first hydrogen bomb was exploded in 1951 by his team of American scientists. The Russian bomb followed in a few short years.

It seems that these hydrogen bombs did not start with hydrogen. According to the first set of reactions on p. 464, five steps would be necessary to arrive at helium and the task of keeping a bomb together until five successive encounters have been accomplished was certainly too difficult. The most obvious way out of the difficulty was to accomplish only the last step. $^7_3\mathrm{Li}$ is a naturally occurring isotope, much more abundant than $^{235}_{92}\mathrm{U}$ and has only to react with a single proton of appropriate energy to fall apart into two alpha particles yielding $17\cdot3$ MeV. All the energy which went into the building of the $^7\mathrm{Li}$ nucleus is left unused this way—but the last step is quite sufficient to yield a formidable bomb. $^6_3\mathrm{Li} + ^2_1\mathrm{D}$ could do the same, liberating $22\cdot4$ MeV of energy. Thus the importance of lithium isotopes is as great or even greater than that of $^{235}\mathrm{U}$ in the manufacture of atomic bombs. Such fusion reactions can only be initiated at extremely high temperatures in contradistinction to fission which may be set loose by a "cold" single neutron. Therefore the fusion type of nuclear reactions is often called "thermonuclear."

The third isotope of hydrogen, tritium, $^3_1\mathrm{H}$ (or $^3_1\mathrm{T}$) is known, and can be isolated from the reaction of two deuterons:

$$^2_1\mathrm{H} + ^2_1\mathrm{H} \rightarrow ^3_1\mathrm{H} + ^1_1\mathrm{H} + 4\cdot0\ \mathrm{MeV}\ .\qquad.\qquad.\ (104.1)$$

or by the reaction:

$$^9_4\mathrm{Be} + ^2_1\mathrm{H} \rightarrow (^8_4\mathrm{Be}) + ^3_1\mathrm{H} \rightarrow 2^4_2\mathrm{He} + ^3_1\mathrm{H}\qquad.\qquad.\ (104.2)$$

It is radioactive and emits electrons to yield $^3_2\mathrm{He}$. It has the mass $3\cdot0221$ and a half-life of 12 years. Evidently this nucleus also needs only to react with a proton of sufficient energy to fuse with it and be transformed into an alpha particle. The energy necessary to overcome the Coulombic repulsion between a tritium and a hydrogen nucleus, both having unit charge, is only one-third of that necessary to bring up a hydrogen nucleus against a lithium nucleus of three times its charge. It is "only" a question of isolating sufficient tritium for a bomb, mixing it with hydrogen and setting it loose with a fission bomb, just as TNT is exploded by the initial detonation of mercuric fulminate.

All this, however, is only of military interest and let us fervently hope that it will not be of practical interest at all but will remain academic. An atomic bomb is no nearer to an atomic motor than is a bomb filled with TNT to a heat-engine. Some other means must be devised to liberate this enormous concentration of energy in a controllable and useful manner.

In all technically advanced countries there are experiments in progress to achieve this aim. Some claims of initial success have already been made, but they have not been confirmed as yet. The process envisaged is a veritable thermonuclear reaction between hydrogen and/or deuterium nuclei at temperatures which hitherto have only been said to occur in the interior of stars. The D + D fusion will need about 4×10^8 degrees K! No wonder that the difficulties encountered in this project are so enormous. First of all an extraordinary concentration of energy has to be achieved. At temperatures beyond 4,500° all matter known to us is already molten, and not much above this point all is volatilized. There is no prospect of finding any material to contain this reaction at a million or ten million degrees which would withstand volatilization. Then there is an ever-increasing loss of electromagnetic energy by radiation: it increases with the fourth power of the absolute temperature. This means that radiation loss at a million degrees is 10^{12} times as great as at a thousand degrees. Heat loss by radiation at these temperatures would surpass losses by convection or conduction and would be the only serious loss to be considered. The gas itself that was to be reacted would acquire such an immense pressure (proportional to the absolute temperature!) that by some means or another it would have to be kept together by force. And this without any material for a container.

At first glance the difficulties seem to be unsurmountable. Nevertheless, physicists have not given up hope and there is good reason to suppose that eventually they will find their way out of the impasse. Just how this is to come about we cannot of course foretell, but the main lines of approach which are being tried in our days may be briefly mentioned.

There is good hope that sufficient energy can be concentrated to arrive at the necessary temperature. Owing to the fact that hydrogen and deuterium are gases, and that anything would be a gas at the temperature in question anyway, energy will have to be imparted to a gas. As far as we can see today this will be done in an electric discharge. Here the fundamental laws of electromagnetism come to our aid. Though charges of equal sign repel each other as long as they do not move, an attractive force appears between them as soon as they move parallel to each other. This is the electromagnetic attraction operating between parallel currents: two or more charges of equal sign in parallel motion can be regarded as currents and at the velocity of light, electro-magnetic attraction between them should just offset Coulombic repulsion. But as a matter of fact there is not much Coulombic repulsion in a totally ionized gas (generally called "plasma") because both positive nuclei and negative electrons are present at the same time. Two opposed streams, one of positive nuclei and another of electrons run mixed up one within the other, and are acted upon by the combined electromagnetic force of the whole discharge which tends to draw the whole current together. The fields generated by positive particles flying in one direction and by negative electrons flying in the opposite direction add up of course, because sign and direction are reversed at the same time. Thus there is hope, and indeed there is convincing experimental evidence, that these electrical discharges of extreme density will be drawn together into a thin thread by the electromagnetic field of the discharge itself—the so-called pinch effect. There are difficulties. Electrons are very much lighter than nuclei and they will be lost more quickly by diffusion, thus interfering with the symmetry and electroneutrality of the discharge. Inhomogeneity of the magnetic field tends to separate electrons from nuclei. Collisions between particles will tend to knock them out of the main thread even before the energy of collision has become sufficient for nuclear fusion. If fusion begins, however, its energy will become available for the process too.

In order to overcome these difficulties, very intricate systems of external magnetic fields are being devised which co-operate with the pinch effect to keep as much as possible of the discharge together at a high particle concentration. Instead of material walls which are unable to exist at such temperatures, "walls" of magnetic field are being created around the discharge. There have even been proposals that these walls of magnetic force should be made to advance towards the interior like solid pistons, thus adding energy to that of the discharge and making the plasma even hotter than it would otherwise be. Compare the gases in a Diesel engine, which are compressed adiabatically (without being able to lose their energy) until they are heated to the temperature where explosion sets in.

The radiation losses could be reduced to some extent by using reflecting walls, mirrors to reconcentrate a part of them into the region where the reaction is planned to take place.

It is impossible to foretell how much of all this will work eventually, but there is hope that the problem is nearing its first, rudimentary solution.

105. MESONS AND OTHER NEW PARTICLES

We have mentioned on p. 457 that the particles—or entities—postulated by Yukawa as causes of the exchange force between nucleons have indeed been found to occur in cosmic rays. Cosmic rays had been investigated for a decade or two by physicists who looked for the cause of the ever-present ionization of air in the atmosphere. *T. Wulf* in 1910 and *A. Glockel* at the same time made measurements on top of the Eiffel tower and in ascending balloons and found that the ionization increases with height. It should have decreased, had it been of terrestrial origin. Some years later *V. F. Hess* and *W. Kolhörster* in Austria and in Germany respectively, went up to nearly

10 km (30,000 ft) and still found that ionization went on increasing. These scientists put forward the hypothesis of radiation from without the earth and called it "altitude-radiation" (*Höhenstrahlung*), a name which was soon changed by Millikan who extended their experiments, to that of "cosmic rays," a name now generally accepted in English-speaking countries.

Artificial earth satellites of our days have been carrying instruments far outside the normal atmosphere of the earth and have proved that the earth is surrounded by at least two zones of high-energy particles carrying electric charges and evidently trapped within these regions by the magnetic field of the earth, similar to the manner in which physicists try to trap the charged particles of a plasma for thermonuclear reactions. Electromagnetic laws demand that the force acting upon a charged particle in a magnetic field should be perpendicular to both the field and its movement: this is how the particles which arrive from the sun or more distant regions of the universe are made to go round the earth, spiralling around magnetic lines of force until by collisions with other atoms they gradually lose momentum or are knocked into a direction which is sufficiently parallel with the magnetic lines of force to allow them to reach the earth. Here we have the reason why they enter the more easily the nearer they go to the magnetic poles where the lines of force are perpendicular to the earth, while they are excluded from regions near the equator (Fig. 165).

At least a great number of these radiated entities must come from the sun. This can be seen from the fact that a few hours after a solar eruption (or a sunspot) is observed, there is a marked disturbance in the magnetic field of the earth: the electromagnetic field of these incoming particles adds to that of the earth and changes its magnitude and direction. We call this a magnetic storm. At the same time as the solar eruption begins there is a general disturbance of short-wave radio communication, because the upper layers of the atmosphere have become ionized by the sun's increased electromagnetic radiation and change the absorption and reflection of our radio waves as a metal layer would. The time-lag is evidently due to the time it takes for the particles to reach the earth from the sun. Very often there is a simultaneous display of aurora borealis around the poles: the sky is lit by a fantastic array of coloured "curtains." The particles which manage to enter the lines of force ionize or excite the molecules high up in the atmosphere and these, on falling back to their normal state, emit this fanciful radiation.

Modern techniques have made it possible to identify and even photograph these cosmic-ray particles. The direction from which they come may be ascertained by using two or more Geiger-Müller counting tubes (p. 401) in a line: whenever all of them record a signal at practically the same moment it is highly probable that they have been traversed by the same particle in the determined line of sight. Such "cosmic-ray telescopes" may even be made to monitor a cloud chamber so that it gives an exposure at the right instant. (*P. M. S. Blacket* and *G. P. S. Occhialini*, 1933 in England.)

Once within the cloud chamber, a very important measurement becomes possible: a magnetic field of known intensity deflects the particles into a circular track, and from the strength of the field and the curvature of the circle it is possible to calculate their momentum (p. 402) (*D. Skobelzyn*, Russian physicist, 1927). Whenever the track of such a particle appears in a photograph taken in the cloud chamber, two other important data may be determined: the density of ionization and the length of the track. The first refers to the number of ion-pairs produced per unit length and if the total track length is known we only have to multiply by it in order to obtain the

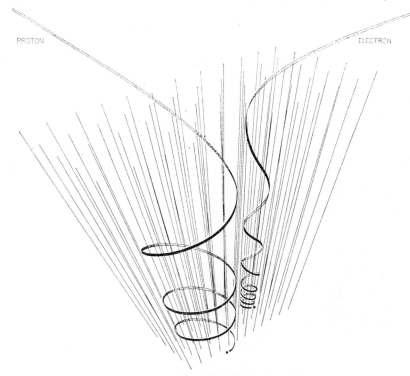

Fig. 165. **Charged particles spiralling around the magnetic field lines towards the earth, mostly towards the poles**

(From B. J. O'Brien, *Sci. Amer.*, v. 208, No. 5, p. 88. By courtesy of *Scientific American*.)

total number of ion-pairs generated by the particle of known initial momentum. Now, the energy necessary to produce a single pair of ions in the gas of the chamber is known, thus we obtain the total energy the particle has lost within the chamber before being slowed down to such a degree that it has become unable to cause any further ionization. We now possess the two relevant data which give us the mass of the particle:

$$\text{Momentum} = mv \qquad . \qquad . \qquad . \qquad . \quad (105.1)$$

$$\text{Energy} = \tfrac{1}{2}mv^2 . \qquad . \qquad . \qquad . \quad (105.2)$$

(as long as the velocity is appreciably below that of light and the relativistic correction may be disregarded). Eliminating v yields

$$m = \frac{(\text{momentum})^2}{2(\text{energy})} \qquad . \qquad . \qquad . \qquad . \quad (105.3)$$

Instead of the cloud chamber the photographic method is also very appropriate for the same measurements. A stack of films coated with photographic emulsion may be placed, say, at a high altitude in a magnetic field for some time and then developed and examined sheet by sheet to trace any tracks of interest. Again curvature of path and total ionization yield the mass of the particle.

It has long been known to physicists working with these cosmic rays that they are by no means homogeneous. They contain a "soft" component which is filtered out by a few centimetres of lead while another, "hard" (very penetrating), component can even be detected far below the surface of the ocean or within the deepest mines. The soft particles were soon identified as electrons, positrons and gamma rays, of energy as "small" as about 200 MeV. It was assumed that they are not the primary particles of the cosmic rays because they would have been absorbed or caught up by the magnetic field. They must have been generated by collision of the primary particles with terrestrial, atmospheric atoms. But what are the primary particles? A dissymmetry, very difficult to establish with cosmic-ray telescopes, between

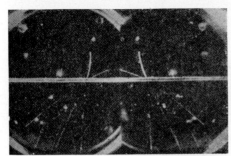

Fig. 166. **High-altitude cloud-chamber photographs of Andersen and Neddermeyer showing the tracks of mesons**

the hard components coming from east and west respectively, seems to suggest that they carry positive electricity, and they are held to be protons.

But shortly after Yukawa put forth his me on hypothesis, other, new particles were observed, *S. H. Neddermeyer* and *C. D. Anderson* obtained cloud chamber photographs at high altitude in 1935 (Fig. 166) which contained unusual tracks beginning in a lead plate within the chamber and ending after a short curved path within the gas. Curvature and total ionization of the track indicated that the mass of the particle responsible was between that of an electron and of a proton, in fact about 130 times that of an electron. Up to that time it seemed as if some obscure arrangement of Nature would only tolerate the two fundamental masses of the electron or proton, the positron having been found equal to the former and the neutron to the latter. And now suddenly a new elementary particle has been found with its mass between the two!

But this was only the beginning of the discovery of a very large family of unorthodox particles. In so far as their mass lies between the two more orthodox limits, they are called "mesons," irrespective of whether or not they really are the particles that Yukawa had postulated to account for exchange forces between nucleons. Their mass lies in the predicted order of magnitude and this is sufficient for the moment.

C. F. Powell and *G. P. S. Occhialini* found in 1947 that some photographic plates contained tracks of particles which must have been due to a particle of some 300 electron mass units. Shortly afterwards they found that after a while this track often comes to an end and a new one begins which belongs to the known, lighter meson, the mass of which has in the meantime been ascertained as lying around $200m_e$ (Fig. 167). The heavier meson is designated

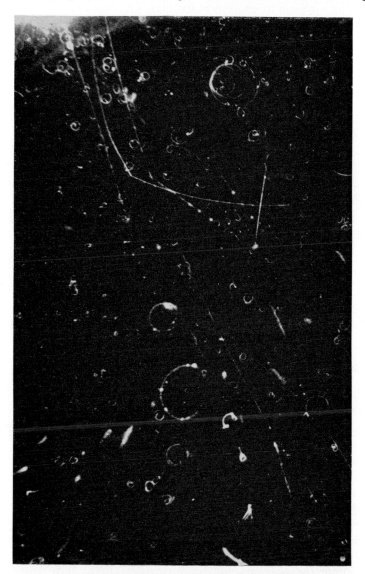

Fig. 167. **Transformation of a pi+ into a mu+ meson, the latter yielding eventually a positron**

Tracks are recorded in a photographic emulsion.

(From W. Gentner, H. Maier-Leibnitz and W. Bothe, *An Atlas of Typical Expansion Chamber Photographs*, p. 173, Fig. 160 (Oxford, Pergamon, 1954). By courtesy of the Pergamon Press Ltd.)

by the Greek letter π (pi) and the lighter one by the letter μ (mu) and exact measurements revealed that their masses are 285 and 215 m_e, respectively. From the length of the path before disintegration into a μ-meson, and from its velocity determined from the curvature of its path, it was easy to determine that the average lifetime of a π-meson was only of the order of 10^{-8} sec. Not all the energy and momentum of the π-meson was found, however, in the resulting μ-meson by analysis of its track, and hence it became necessary to postulate another particle which took up the difference. But it was not visible as an ionizing agent, hence it must have been neutral. If we remember, a similar argument in the beta-decay of radioactive elements led to the assumption of the neutrino (p. 404). It seemed possible that this imagined entity was in evidence once more. It was held more probable that it had a

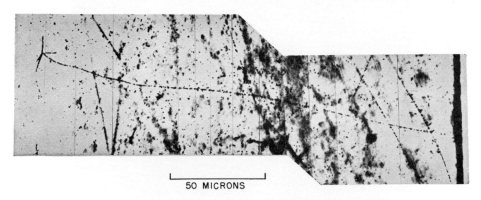

50 MICRONS

**Fig. 168. A pi meson hits a nucleus and brings it to explosion.
A "star" results in the emulsion or in the cloud chamber**

(From S. Glasstone, *Sourcebook on Atomic Energy*, p. 574, Fig. 18.6 (Princeton, N.J., van Nostrand, 1952). By courtesy of the D. van Nostrand Co., Inc.)

mass of $285 - 215 = 70m_e$ this time, because it had to take care of the mass difference as well as the difference in momentum. Owing to the fact that according to the principle of Einstein the difference of mass could be carried away in the form of additional energy by a smaller particle, even by a neutrino, the argument leading to a mass of 70 was not entirely conclusive.

It has been found, that both positive and negative π-mesons exist; the curvatures of their tracks in a magnetic field are opposite to each other. Both decay with the same period but the negative π-meson is often attracted into a nucleus of an atom and in such cases the enormous energy of the encounter brings about a nuclear explosion which manifests itself in a "star" within the photographic emulsion (Fig. 168). It is interesting to follow the track of a π-meson as it is losing speed: it spends more and more time in the neighbourhood of individual atoms and therefore is the cause of more dense ionization. The track in the emulsion is more dense towards its end. A similar phenomenon was encountered at the end of the track of an alpha particle (p. 397).

The μ-meson has also a positive and a negative variety. Their mean lifetime is about 10^{-6} sec, that is much longer than that of the π-mesons. Eventually they disintegrate into an electron and one or more neutral particles which are invoked to account for the difference in mass and momentum. Thus the life of these remarkable particles ends up in the stable negative or the labile positive electron. The latter soon finds a negative electron for itself, and they become mutually annihilated with the simultaneous emission of two gamma rays which carry away the energy of their masses

Beyond the π-mesons another set of new entities has been discovered, resulting from the interaction of a π-meson and a proton: these are the so called κ-mesons with a mass of about $970\ m_e$. Positive as well as negative "particles" of this mass have been observed and there is good reason to suppose that an uncharged variety exists also. Fig. 169 shows a bubble chamber photograph and its explanatory sketch where it is reasonable to assume that the two extravagant V-shaped tracks near the endpoint of a negative π-track are not fortuitously so near to each other but are actually caused by a neutral particle emanating near the spot where the π-meson collided with a proton (from the liquid hydrogen of the bubble chamber). The invisible particle travelling along the broken line is the product of decomposition of a heavy, transient particle yielding very quickly a positive π-meson, characterized by its track curvature and ionization. The uncharged particle travelling along the dotted line to the right gives rise to a detectable proton and a negative π-meson; the former particle is supposed to be heavier than the proton, with a mass of $2180\ m_e$ and is called a neutral lambda particle. (The proton has a mass of $1,836\ m_e$). Charged lambda particles of this mass have been identified among cosmic-ray particles and in ultra-high energy accelerators after accelerated particles have hit some target. They are found together with particles of still higher mass, up to more than double the mass of the proton. Those lighter than the proton are designated by Greek small letters (μ, π, κ, η), those heavier than the proton with Greek capital letters (Λ, Σ, Δ, Ξ, Ω). Some of them are charged positively or negatively, others are neutral. They may possess spin, that is angular momentum amounting to multiples of an elementary unit value as we have seen already in the case of the electron and of the proton. Some of their properties cannot be described by concepts of physics outside the domain of these strange particles themselves. Most of them have only very short lives, like the excited states of atoms, molecules or nuclei and decay along a series of particles until at last they end their series of decay by becoming a proton or an electron. The theory of this multitude of particles, or states, is still in its very beginning, very formal and very unlike to anything we have become used to in physics. Who knows when new kinds of such particles may be found again in the near future?

The situation is confounding indeed. Instead of the relative tranquillity of the nineteen twenties when electron and proton seemed to be the ultimate components of matter, and the still relatively serene early thirties when the picture became somewhat more complicated by the neutron and positron, we are now confronted with nearly a hundred particles. Of these only the proton and electron remain really stable.

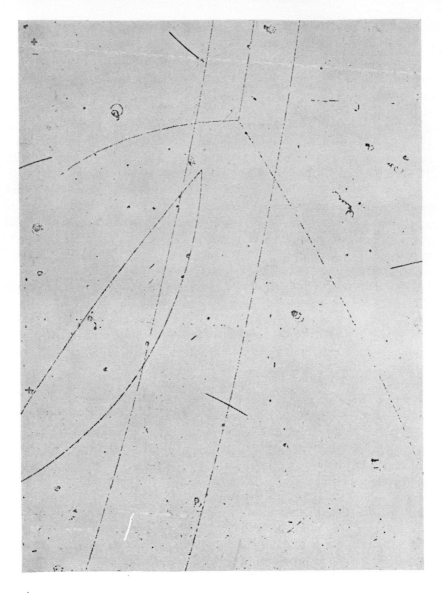

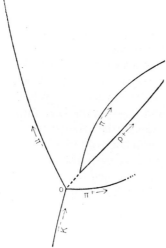

Fig. 169. **Bubble chamber photograph and its explanatory sketch of the collision between a negative κ meson and a proton at 0**

A neutral lambda particle (*broken line on sketch*) leaves no track but eventually disintegrates into a proton and a negative pi meson. Analysis of particle energies showed that the first negative pi meson recoiled from a single particle which shortly afterwards disintegrated into a positive pi meson and the lambda particle.

(From R. D. Hill, *Sci. Amer.*, v. 208, No. 1, pp. 38, 39 (1963). By courtesy of *Scientific American*.)

Very little can be said about the neutrino except that it balances the conservation laws of spin and energy wherever they seem to have been unbalanced, as for instance in beta-ray emission. Very complicated experiments in the largest accelerators showed that neutrinos traverse all matter practically unhindered and only in extremely rare cases were they found to interact with protons to yield electrons or μ-mesons. It seems very probable that two kinds of neutrino exist, one yielding electrons and the other μ-mesons in these experiments.

From among the mesons only the π-meson is capable of interacting with nuclei; this was the particle held responsible for the exchange forces between proton and neutron (p. 457).

We have reached the actual fighting area of modern physics and should not be astonished that things are topsy-turvy, as they are being discovered often in an uncoordinated manner and are being explained by a rather formal attempt to fit them into a future general theory. Many an experiment will have to be performed and probably a great genius is needed to achieve this end and enable us to see the master plan behind these strange and seemingly uncoordinated entities.

According to the Einstein relation mass and energy are commutable quantities. Mesons are particles with more energy than their stable end products so there was some possibility that mesons could be created from ordinary matter under conditions of extreme energy concentration. In 1948 *E. Gardner* and *C. M. G. Lattes* bombarded carbon with alpha particles which had been accelerated to nearly 400 MeV in the great Berkeley synchro-cyclotron, and surrounding photographic plates actually revealed the fact that π-mesons had been generated. Since then the method has been extended to produce all other "unorthodox" particles.

The relatively short lifetime of mesons makes it nearly impossible that the original cosmic ray particles entering the atmosphere were really mesons. They would have disintegrated long, long before they had reached the earth. But they must have been generated somewhere by some sort of reaction and we do not know whether this reaction has been extra-terrestrial or not. The interconversion of energy and mass allows a variety of explanations: a collision of a high-energy proton or electron from the sun or the universe with another might very well in their interaction produce matter in the form of strange particles.

We do know, however, that the energy of incident charged cosmic particles must exceed 10^9 eV (1 BeV) in order to penetrate the magnetic shield around the earth. And we have fair evidence that occasionally it may reach as much as 10^8 BeV, that is 10^{17} eV! The reason we have to postulate such tremendous energy is the occurrence of immense, wide-spread simultaneous showers of "soft" cosmic-ray particles. A cloud chamber photograph of a relatively small shower of this type is shown in Fig. 170 (*a*) where the black horizontal strips are lead plates through which the energetic particles proceed downwards. The logical explanation of such events is that originally at the top, a single high-energy particle came into collision with matter and gave rise to particles of less, but still very considerable, energy and that this process went on a number of times consecutively, in a cascade. For instance, a high-energy electron might have been decelerated in the lead plate giving rise to a "bremsstrahlung" (radiation due to "braking," slowing down, of electrons)* of such high energy that it could materialize into an electron–positron pair (p. 435), which in their turn are decelerated or annihilated with

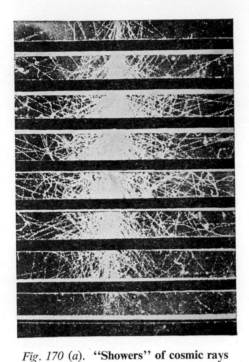

Fig. 170 (a). **"Showers" of cosmic rays**

In a cloud chamber while traversing successive plates of lead.

(From S. Glasstone, *Sourcebook on Atomic Energy*, p. 564, Fig. 18.3 (Princeton, N.J., van Nostrand, 1952). By courtesy of the D. van Nostrand Co., Inc.)

emission of light quanta, and so on until the remaining energy is insufficient to continue the process. The fact that *Auger* was able to find showers over a kilometre wide using a vast array of simultaneously registering Geiger counters, led to the postulation of the enormous primary energy of 10^{17} eV mentioned above.

---

* The German word *Bremsstrahlung* has been adopted for this phenomenon.

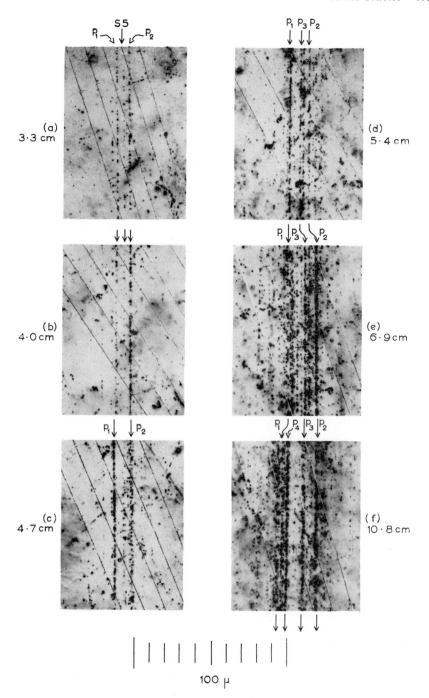

Fig. 170 (b). **"Showers" of cosmic rays.**

In a set of successive sections through a photographic emulsion. (From C. F. Powell, P. H. Fowler and D. H. Perkins, *Elementary Particles by the Photographic Method*, p. 567, Plate 15.18 (Oxford, Pergamon, 1959). By courtesy of the Pergamon Press Ltd.)

Conclusion

We have reached the end of our journey. We have seen that matter is built from molecules and that all molecules in the world are built from 92 natural and a few artificial atoms of the chemical elements. We have gained insight into the meaning of temperature by studying the movement of atoms and molecules according to the rules of statistical distribution. A whole series of independent experimental methods has led us always to the same dimension for these minute atoms of matter so that we feel sure they are really exactly as small as indicated.

But the atoms themselves have fallen victims to the dissecting scalpel of science. It became possible to extract from them an even tinier particle charged with negative electricity: the electron. The atom revealed itself as a rather empty structure of such electrons around a positively charged nucleus which carried practically the whole mass of the atom within it. The nuclear volume was found to be 10^{15} times smaller than that of the atom itself. The physical laws at these small dimensions, and those at very small and very great velocities, proved to be different from those we encountered in macroscopic physics, and these new laws play the decisive part in the building-up of atoms from electrons around the nuclei. The different chemical elements, some 100 altogether, can be fitted into a logical system with the help of these laws, one succeeding the other in an understandable sequence, as the nuclear charge, and with it the number of electrons in the atom, increases. Chemistry has been seen to depend upon the outermost electrons of these atoms and all known types of the chemical bond could be explained by applying the new physical laws of quantum mechanics together with the old laws of electricity, to these external electrons. We have seen that electrons may be localized in bonds between neighbouring atoms, may be transferred from one atom to another or may be set free to lead a life of their own within the atomic lattice of metals.

However, even atomic nuclei are not the ultimate bricks of this micro-universe. Some of them exploded by themselves, others could be caused to explode from without, and some were synthesized from smaller particles.

What, then *are* the ultimate bricks from which matter is built? Two stable entities, the positive, heavy proton and the negative, light electron, have appeared. Then a neutral particle with nearly the same mass as that of the proton: the neutron. It is not stable as long as it is alone, but seems to be stabilized within nuclei, in the neighbourhood of protons. But this is not yet the whole story. There is the positive electron which only lives as long as it has not met its negative counterpart, and the neutrino which mainly manifests itself as something that carries away energy and transmits momentum. The vast array of positive, negative and neutral mesons and strange particles follows suit; all of them labile, transient entities in a microcosm of complications. The laws which should determine just why *these* particles exist and why they are stable or unstable, have not yet even appeared on our horizon.

But in the meantime the idea of "bricks" of the material world has dematerialized. Light, our old acquaintance which so perfectly fulfilled all the expectations that we deduced from the assumption that it is a set of waves, has behaved as something which is much more similar to particles, as soon as we tried to localize its energy. And matter itself, all our beloved "particles" have flown away in waves whenever they were small and slow enough to bring their momentum times extension down to the neighbourhood of a small universal constant. Photons as well as particles have to be treated now as particles and now as waves, according to a set of rules of a very intricate theory which has made us abandon—at least for the time being—the classical concept of causality, and replace it by predictions of exactly defined probabilities. Energy and mass have become interchangeable concepts as if mass were nothing more than a special form which energy is able to assume.

Like children, we began to break up our material world into bricks. The game went on for a time and practically the whole complex of chemical formulae could be built up according to its rules. Then, as we began to descend deeper and deeper the rules began to forsake us more and more. But not entirely: some properties of nuclei could be made clear by assumptions regarding their "geometric structure." But as against these rare successes, the number of difficulties has steadily multiplied and it is certain that the idea of "bricks" from which we are able to construct the material world will never be fulfilled in its naïve form which nevertheless serves us so well in the structural formulae of chemistry.

The situation may be very disagreeable to some of us, but there it is. And, after all, who has promised us that our naïve conceptions would eventually materialize? Bertrand Russell, probably the greatest philosopher living in our time, gave a humorous summing-up of the situation. Suppose we wish to explain to somebody what a billiard ball is made of. Well, it is an intimate mixture of organic and inorganic compounds, a lot of them known to us. And the compounds? Oh, they are of course assemblies of molecules. Very well, and what are the molecules then? Molecules are built from atoms. And the atoms? They consist of small nuclei in the centre with shells of electrons around them. The nuclei are small spheres to a very good approximation. And are they built from still smaller units themselves? Oh, yes, they are built from protons and neutrons. Very well, if there is nothing known below the level of electrons, protons and neutrons, how should we imagine these seemingly last entities? Well, it is difficult to answer. In some ways they are probably small spheres with radii about as small as a millionth of a millionth of a millimetre, just imagine, as crude approximation a micro-billiard ball of this size. . . . Billiard ball! Have we not just been asking what a billiard ball is made of?

Well, we know at any rate today, that the micro-billiard balls are very different from the ordinary one and there is absolutely no *a priori* reason why they should not be different. We have not revolved in a vicious circle the whole time. What we know today is more than we knew at the time when physics was mainly concerned with objects of the size of billiard balls. Quite naturally, Man began to explore Nature at the dimensional level to which he himself belongs. He was faced with a world somewhere in its middle. In between nucleons and spiral nebulae, in between space which may or may not be finite in its three dimensions or within a world which can be described by taking time as its fourth dimension. From this starting point, which has been determined by the place of the earth within the Universe and by the time mankind has begun to have thoughts about and make observations and experiments on the world around him, the flickering torchlight of science has begun to illuminate greater and still greater portions of the Universe around us, in all its temporal, spatial and dimensional, directions. Now quickly, then more slowly and more quickly again, the light spreads around us illuminating more and more of what has been covered by darkness.

Phenomena which seemed distinct a few centuries ago such as heat and motion or electricity, magnetism and light, fuse together into a higher, harmonic unity. Millions of chemical compounds are explained as spatial and energetic arrangements of some hundred sorts of atoms. For a short time there was hope of resolving even these into two or three fundamental particles. Today we see that it is not quite as easy as that; nearly a hundred such particles still await systematic arrangement. So what next? We are going to learn more and more about them and some sort of arrangement will be forthcoming in the end.

Is the dual realm of particles as opposed to waves to be resolved some day in a higher harmony which will quiet the hearts of those who today cannot acquiesce in the fundamental difficulties of this solution? Will causality be restored or not and if not are we going to have a glimpse at the inner connection of a not strictly causal physical world, and the feeling within us that our will is free within certain limits? Nobody knows the answer.

Science goes on: the light of the small torch in our hands illuminates more and more. Our grandchildren will probably see the solution of some of the questions we have just asked as clearly as we believe we see the structure of the solar system and the sphere of the Earth. We may rest assured that we have progressed far along the infinite road which gives us more and more insight into the world of which we are small parts ourselves.

Index

Italic page numbers indicate the more important references to a subject.